Green Nanomaterials

From bioinspired synthesis to sustainable manufacturing of inorganic nanomaterials

Green Nanomaterials

From bioinspired synthesis to sustainable manufacturing of inorganic nanomaterials

Siddharth V Patwardhan
Green Nanomaterials Research Group, Department of Chemical and Biological Engineering, The University of Sheffield, UK

Sarah S Staniland
Bio-Nanomagnetic Research Group, Department of Chemistry, The University of Sheffield, UK

IOP Publishing, Bristol, UK

ISBN 978-0-7503-1221-9 (ebook)
ISBN 978-0-7503-1222-6 (print)
ISBN 978-0-7503-1836-5 (myPrint)
ISBN 978-0-7503-1223-3 (mobi)

DOI 10.1088/978-0-7503-1221-9

Version: 20191201

IOP ebooks

British Library Cataloguing-in-Publication Data: A catalogue record for this book is available from the British Library.

Published by IOP Publishing, wholly owned by The Institute of Physics, London

IOP Publishing, Temple Circus, Temple Way, Bristol, BS1 6HG, UK

US Office: IOP Publishing, Inc., 190 North Independence Mall West, Suite 601, Philadelphia, PA 19106, USA

Contents

Preface

This book aims to provide an understanding of emerging bioinspired green methods for preparing inorganic nanomaterials.

Inorganic nanomaterials are used in many applications, ranging from sun cream to catalysis, as well as the latest innovations in nanomedicine and high density data storage. In recent years, we have understandably seen a large quantity of publication activity (including books) on the safety and toxicity of nanomaterials. However, there is a distinct lack of consolidated effort in addressing the sustainability of making nanomaterials. Current methods for nanomaterial synthesis are complex, energy demanding, multistep, and/or environmentally damaging, and hence clearly not sustainable. Green chemistry has great promise for future developments, particularly in sustainable designs for materials, processes, consumer goods, etc. However, to date, green chemistry has mostly focussed on the synthesis of fine chemicals and very rarely on nanomaterials.

New bioinspired/biomimetic approaches are emerging, which harness biological principles from biomineralisation to design green nanomaterials for the future. With reference to the significant body of research on understanding biomineralisation, Ozin *et al* state in their book, *Nanochemistry: A Chemical Approach to Nanomaterials*, that 'In molecular terms, it is relatively easy to comprehend the early stages of self-organisation, molecular recognition, and nucleation that precede the morphogenesis of biomineral form. It is not obvious however, how complex shapes emerge and how, in turn, they can be copied synthetically' [1]. In this book, the aim is to address this highly sought-after aspect of how to translate the understanding of biomineral synthesis into new green manufacturing methods. We cover areas from the discovery of new green synthesis methods all the way to considering their commercial manufacturing routes.

Who is the book for? The Royal Society of Chemistry and the American Chemical Society's Green Chemistry Institute have both highlighted a 'lack of a deep bench of scientists and engineers with experience in developing green nanotechnology' [2] as a significant barrier to the development and commercialisation of green nanotechnology. This has motivated us to write this book. When any of us have been educated within a specific traditional discipline of science or engineering for our undergraduate degree, it can be very daunting to take a leap into multidisciplinary science and study within the realms of new disciplines outside our comfort zone, where the experimental approach, culture and even language can be so different, creating barriers and challenges. However, the more we work at this interface, the more we realise that these boundaries are artificial, for the purpose of our education, and do not exist in Nature. The purpose of this book is to start with basic explanations to build a foundation, so that this area of science can become accessible to students from any related discipline. We hope that this book encourages scientists and engineers to become confident in bridging the gaps between chemistry, nanotechnology, biology, engineering and manufacturing. Specifically, the book combines green chemistry and nanomaterials in a single dedicated monograph.

As such, the book is written with a wide readership in mind, including primarily academic researchers focusing on synthetic biology and nanomaterials. It is targeted towards postgraduate students (taught and research degrees) undertaking studies pertaining to advanced materials and green, sustainable and/or environmental engineering or chemistry. Final year undergraduate students specialising in nanomaterials or green processes will also find this book valuable. Indeed, various universities currently run final year electives on nanomaterials, biomaterials, green chemistry, sustainability, etc, where this book is highly suitable as a textbook. Through the authors' interactions with industry, we know that many industries wish to learn more about these green technologies. Hence, we hope to reach industrialists and raise awareness of the emerging green manufacturing routes.

What is in the book? The book starts by introducing the principles of green chemistry and engineering (chapter 1). It then highlights the special properties that nanomaterials possess, their applications and ways of characterising them (chapters 2 and 3). It describes conventional methods of synthesising and manufacturing inorganic nanomaterials (chapter 4) and highlights that these techniques cannot always deliver the specifications required for applications or be sustainable (chapter 5). This will lead to the introduction of biological and biomimetic/bioinspired synthetic methods as a solution to precisely controlled nanomaterials as well as the design of sustainable manufacturing routes (chapters 6 and 7). The book elaborates on various mechanisms and examples of green nanomaterials (e.g. the role of an organic matrix and natural self-assembly, and advantages and opportunities with green nanomaterials). It will cover two case studies of magnetic and silica materials for advanced readers (chapters 8 and 9).

How to use the book. We acknowledge this book covers many different traditional disciplines and as such we cannot go into too much depth in every area. Furthermore, this is a very current and fast-moving research area. As new methods, materials and characterisation techniques are discovered, invented and developed, fairly recent advances become old quickly. For both reasons we recommend this textbook be supplemented with more detailed, specific and contemporary science and engineering research journal papers. Indeed, in the courses we teach on this subject, the material content of this book is used to explain the background and introduce current research papers as relevant examples.

A note on ongoing discussion on the topics covered in this book: In order to allow a dialogue between the readers, the authors and the publisher, we have created a dedicated web portal in order to receive feedback from readers and to allow authors and readers to post recent updates relevant to this book. This can be accessed at https://greennanobook.com/.

References

[1] Ozin G A, Arsenault A C and Cademartiri L 2009 *Nanochemistry: A Chemical Approach to Nanomaterials* 2nd edn (Cambridge: Royal Society of Chemistry), p 23

[2] Matus *et al* 2011 *Green Nanotechnology: Challenges and Opportunities* (Washington, DC: ACS Green Chemistry Institute)

Acknowledgements

We would like to acknowledge those who helped us complete this book. SP is sincerely grateful to Professor Steve Clarson and Professor Carole Perry who have been the sources of inspiration. SS would like to acknowledge Professor Andrew Harrison, Professor Neil Robertson and Professor Steve Evans for inspiring and enabling her to begin her research in this area. We thank the scientific communities to which we belong: the networks of academics we work and collaborate with and meet at conferences, where we share and develop new ideas, converse and debate. You are a constant source of inspiration for us and for science and engineering in this field to continue to develop. Our collaborators are acknowledged for sharing their wisdom and for the many stimulating discussions over the years. In particular, SS is grateful to Dr Bruce Ward and Professor Steph Baldwin for biological training and insight. We thank many of our current and past group members who have been instrumental in providing the ammunition for this book and for their patience during the writing stages. SS thanks Dr Andrea Rawlings and SP thanks Dr Joe Manning and Dr Mauro Chiacchia for their help with conceptualising some of the complex aspects/mechanisms included in this book. We are grateful to have had support from Ms Yung Hei Tung (Jodie), Dr Andrea Rawlings, Dr Johanna Galloway and Dr Scott Bird for artwork for some of the figures, and Ms Amber Keegan for help with copyright permissions. We also thank various funding agencies for supporting projects where we developed some of the ideas underpinning this book. In particular we thank EPSRC and BBSRC for funding the SynBIM project (EP/P006892/1, http://www.synbim.co.uk and BB/H005412/1). Finally, we thank the reviewers for their insightful feedback: from the initial book proposal, to friends providing comments on early drafts (thanks to Professor Maggie Cusack, Professor Marc Knecht and Dr Fabio Nudelman) and the reviewers of the completed draft. We offer sincere thanks to the publisher for their support and patience.

Finally, we would both like to thank our families. Academia is a challenging and intense career and this is only amplified when one chooses to write a book on top of our other commitments. We are most grateful to our families for their love and support both generally and specifically over the period of writing this book. We both have young children and are especially grateful: SP to his wife Geetanjali and SS to her husband Luke and our parents, for unquestioning childcare that enabled us to achieve this body of work. We are also grateful to our children: Ninaad and Nishaad, and Owen, Alex and Joel, for their interest in our work, for making us laugh and their inquisitive nature that reminds us every day what this is all for.

Siddharth V Patwardhan and Sarah S Staniland
Sheffield, August 2019.

Author biographies

Siddharth V Patwardhan

Siddharth is currently a Professor of Sustainable Chemical and Materials Engineering at the University of Sheffield. He obtained a first degree in chemical engineering at the University of Pune (India) followed by a master's and doctoral degrees in materials science at the University of Cincinnati (USA). He gained post-doctoral experience in inorganic chemistry at the University of Delaware (USA) and Nottingham Trent University (UK). After taking up a short-term lectureship in Chemistry, he became a Lecturer in Chemical Engineering at the University of Strathclyde in 2010. He then moved to Sheffield to take up a position of Senior Lecturer, where he was promoted to a Professor in 2018.

Siddharth leads the Green Nanomaterials Research Group (www.svplab.com), with a vision to develop sustainable routes to functional nanomaterials. His group focusses on the discovery of bioinspired nanomaterials, assessing their scalability and developing manufacturing technologies for energy, environmental, biomedical and engineering applications.

Siddharth is an EPSRC Fellow in Manufacturing and a Fellow of the Royal Society of Chemistry. He has played a key role in a number of national and international networks as well as conference organisation. One such symposium relevant to this book is on 'Green Synthesis and Manufacturing of Nanomaterials', as part of the ACS Green Chemistry and Engineering Conference in 2017. Siddharth is passionate about mentoring early career researchers and has received numerous awards including Dedicated Outstanding Mentor, Teaching Excellence and recognition as a *SuperVisionary* for all-round supervision.

Sarah S Staniland

Sarah is currently a Reader of Bionanomaterials in the Department of Chemistry at the University of Sheffield. She obtained an integrated undergraduate master's degree in Chemistry followed by a doctorate in Materials Chemistry (2001, 2005) both at the University of Edinburgh (UK).

After her PhD she won a prestigious independent EPSRC Life Science Interface Fellowship (2005–8) at the University of Edinburgh, where she initiated the research in which she is currently active. This helped her transition from chemical material sciences to interdisciplinary work at the interface with biology. She took this opportunity to live and work in various places globally, from Cape Town to Tokyo, forming lasting collaborations. She then took up a Lectureship in Bionanoscience in the School of Physics and Astronomy, University of Leeds in 2008, where she was promoted to Associate Professor in 2013. She moved to Sheffield in 2013 and was promoted to Reader of Bionanoscience in 2016.

Sarah leads the Bionanomagnetic Research Group which studies the biomimetic synthesis of magnetic nanomaterials, particularly inspired from how magnetite nanoparticles are produced within magnetic bacteria. From a basis of material chemistry and PhD in magnetic materials, Sarah has moved into a multidisciplinary approach of using biology to control material synthesis. She has been invited to speak at and organised national and international conferences to promote this research area and been a board member of the Royal Society of Chemistry Materials Chemistry Division. This multidisciplinary research field requires a highly skilled, open-minded and diverse research team, which she is passionate about training, developing and mentoring and is very grateful to them all. Sarah is committed to teaching, in particular multidisciplinary science which falls at the interface between several standard degree subjects, and is always experimenting with novel methods and techniques to improve her teaching in this area. She has taught a course on bionanoscience (covering much of the material in this book) for ten years. Sarah has won three prestigious awards recently, including two for her research: the acclaimed RSC Harrison Meldola Award in 2016 and the Wain Award in 2017, as well as the Suffrage Science Award in 2017 for her work on the promotion of gender equality.

Section I

Green chemistry principles

Image courtesy of bluebay/Shutterstock.

This short section consists of a chapter on green chemistry and engineering. It introduces the 12 principles of green chemistry and various drivers for making a given process or product greener. Further, ways to improve sustainability are discussed mainly in terms of the cost of the waste produced. A brief introduction is provided on how to evaluate the sustainability or green credentials of a given process or product leading to a discussion on ways to improve sustainability. These concepts will be used in other chapters in the book in order to explore potential (un) sustainable aspects of a given method for nanomaterials synthesis. This section aims to set the scene for the book and the principles explained will be revisited in later sections of the book, in order to put them in the context of nanomaterials synthesis and manufacturing.

IOP Publishing

Green Nanomaterials
From bioinspired synthesis to sustainable manufacturing of inorganic nanomaterials
Siddharth V Patwardhan and Sarah S Staniland

Chapter 1

Green chemistry and engineering

1.1 Principles of green chemistry and engineering

1.1.1 Overview

At present, we are highly reliant on chemical processes because of the benefits they bring to us. Examples of sectors where chemical industry has impacted our lives include healthcare, transportation, communications and food, constituting total sales of around €1.5 trillion globally [1]. In recent times, society has become more aware and concerned about the negative environmental impact associated with chemical industry. As such, there have been wider legislative interventions in order to ensure greater control and monitoring of chemical processes. These changes have led to increasing pressure on the chemical industry to continue to deliver high value products in an economical fashion, while minimising or eliminating the adverse environmental burden. This new challenge has driven the development of green chemistry for sustainable chemical processing.

Sustainable development has been defined by the United Nations as '…meeting the needs of the present without compromising the ability of future generations to meet their own needs' [2]. Further, the United States Environmental Protection Agency (EPA) has extended this definition to give birth to *green chemistry* where the main goal is to 'promote innovative chemical technologies that reduce or eliminate the use or generation of hazardous substances in the design, manufacture and use of chemical products' [3]. Related to this is *green engineering*, which pertains to the design, commercialisation and the use of processes and products in an economical fashion to minimise pollution and risks to health and the environment. Utilising these concepts, 12 principles of green chemistry have been formulated [4], which are shown in figure 1.1.

Applying these green and sustainable principles has the potential to address the challenges facing chemical industry, in balancing the high value of products against the environmental burden. A common perception is that green chemistry is costly and/or can lead to costly products, thus reducing the profit for industry. However,

doi:10.1088/978-0-7503-1221-9ch1

1. Prevention
It is better to prevent waste than to treat or clean up waste after it has been created.

2. Atom economy
Synthetic methods should be designed to maximize the incorporation of all materials used in the process into the final product.

3. Less hazardous chemical syntheses
Wherever practicable, synthetic methods should be designed to use and generate substances that possess little or no toxicity to human health and the environment.

4. Designing safer chemicals
Chemical products should be designed to affect their desired function while minimizing their toxicity.

5. Safer solvents and auxiliaries
The use of auxiliary substances (e.g., solvents, separation agents, etc.) should be made unnecessary wherever possible and innocuous when used.

6. Design for energy efficiency
Energy requirements of chemical processes should be recognized for their environmental and economic impacts and should be minimized. If possible, synthetic methods should be conducted at ambient temperature and pressure.

7. Use of renewable feedstocks
A raw material or feedstock should be renewable rather than depleting whenever technically and economically practicable.

8. Reduce derivatives
Unnecessary derivatization (use of blocking groups, protection/deprotection, temporary modification of physical/chemical processes) should be minimized or avoided if possible, because such steps require additional reagents and can generate waste.

9. Catalysis
Catalytic reagents (as selective as possible) are superior to stoichiometric reagents.

10. Design for degradation
Chemical products should be designed so that at the end of their function they break down into innocuous degradation products and do not persist in the environment.

11. Real-time analysis for pollution prevention
Analytical methodologies need to be further developed to allow for real-time, in-process monitoring and control prior to the formation of hazardous substances.

12. Inherently safer chemistry for accident prevention
Substances and the form of a substance used in a chemical process should be chosen to minimize the potential for chemical accidents, including releases, explosions, and fires.

Figure 1.1. 12 principles of green chemistry, after [4]. Copyright of OUP 1998.

this is not always true. While there may be initial investment needed for the development of green technologies, which can replace existing processes, such changes can also lead to reduced costs of production and products. Besides, considering the long-term benefits of adopting green chemistry, initial costs can be offset. Some of these points and the 12 principles will be discussed further in chapters 4 and 5 with specific relevance to nanomaterials.

1.1.2 Drivers for green approaches

One of the main issues leading to adverse environmental impact is the generation of waste in a chemical process. Depending on the technological, economical and legislative frameworks, the quantities of waste produced, in particular, in relation to the amount of product produced, vary from sector to sector. For example, some industrial sectors are technologically very advanced, such that they have developed ways to minimise waste, e.g. petroleum refining. On the other hand, in some other sectors, the cost of the product is significantly higher than the loss of value from waste or cost of treating waste, and hence waste minimisation is not given due importance (e.g. pharmaceuticals). There are also examples where pro-active legislation has driven the chemical industry to find innovative ways to minimise waste. Waste relates to inefficiencies in a given process, which leads to loss of valuable resources (e.g. substances and energy) and it can cause risks to the environment and health, which ultimately increases the process costs. Figure 1.2 illustrates the financial, environmental and societal origins of the costs associated with waste.

If one considers the total costs for waste originating from all areas as shown in figure 1.2, it implies that in order to drive green chemistry, the process or product costs associated with a new green technology (N) should be lower than or at least equal to the costs of the existing (wasteful) processes (E),

$$\text{i.e. } N \leqslant E. \tag{1.1}$$

Along similar lines, the profits from the new process should be higher or at least the same when compared to the existing process. The cost of green technology includes new production costs (P') and the investment needed for developing new green technologies (G), which would reduce the waste and avoid the loss of the image of the business. On the other hand, continuing to operate using existing process

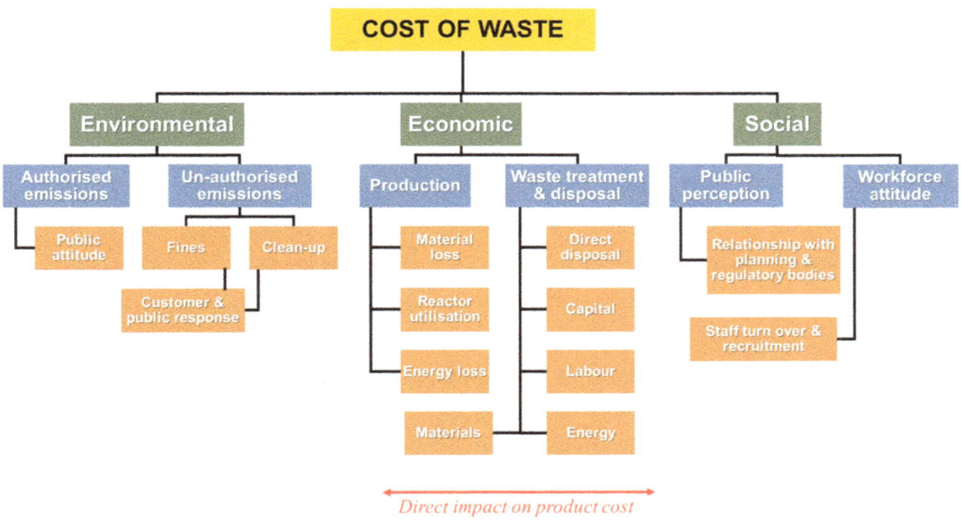

Figure 1.2. Costs associated with waste. Reproduced with permission of the Royal Society of Chemistry from [1].

includes current production costs (P) and incurs waste management costs (W) as well as costs associated with the loss of public image (I),

$$\therefore G + P' \leqslant P + W + I \tag{1.2}$$

$$\therefore G \leqslant (P - P') + (W + I). \tag{1.3}$$

This is a simplistic approach to describe various scenarios and consequences. First, if the company does not wish to invest at all ($G = 0$), then it means that the new process costs can be higher than the existing process costs, but only by the sum of the costs of waste management and those associated with the loss of image,

$$\text{i.e.} \therefore (P' - P) = (W + I). \tag{1.4}$$

Without any investment, it is not easy to obtain access to a new and sustainable process. Further, if the costs for managing the changes in the process (e.g. downtime, marketing, customer satisfaction, etc.) are considered, then in reality,

$$\therefore (P' - P) \ll (W + I). \tag{1.5}$$

This leaves very little incentive for developing a greener process and highlights the strong need for initial investment and the motivation for sustainability at all levels of the organisation. On the other hand, if a new process is developed such that there are no additional process costs ($P - P' = 0$), then the investment needed would be of the order of the cost savings from waste reduction and maintenance of the image,

$$\text{i.e. } G \leqslant W + I. \tag{1.6}$$

The production costs and the type and amount of waste generated are not just dependent on the chemical reaction. Other factors such as choice of solvent and downstream separation and purification processes play a major role. Therefore, developing greener approaches is not about simply operating at lower temperatures or using volatile solvents, for example. Sometimes it may be about balancing the priorities or perhaps radically changing the processes.

1.1.3 Estimating environmental impact

In order to make informed decisions about the need for green innovations for a given process, it is important to assess the environmental impact of that particular process. There are various methods and tools available to qualitatively and semi-quantitatively analyse the environmental impact of processes [1]. Selected methods are described below with their principles, use, advantages and limitations.

Environmental factor (E-factor) [5, 6], also known as waste-to-product ratio (equation (1.7)), is a simple measure for identifying the amounts of waste/by-products produced with respect to the mass of the product.

$$\text{E-factor} = \text{mass of waste and by-product} \div \text{mass of products}. \tag{1.7}$$

Sheldon and co-workers [5] have studied E-factors for various industries and reported great variations from one industrial sector to another (table 1.1).

Table 1.1. E-factors for various sectors. Adapted from [5].

Industry sector	Product capacity, tonnes	E-factor
Oil refining	10^6–10^8	~0.1
Bulk chemicals	10^4–10^6	<1–5
Fine chemicals	10^2–10^4	5–50
Pharmaceuticals	10–10^3	25 to >100

The main reasons for such variations appear to be related to the product value/profit margins, relevant legislations, cost of waste and market competition. The E-factor analysis of a process is simple and provides quick estimates of wastefulness, which can lead to waste minimisation campaigns. While E-factor is easy to use, it can provide misleading information in some cases. For example, consider the following reaction:

$$\textbf{Reactant 1} + \textbf{Reactant 2} \xrightarrow{Solvent\ A} \textbf{product} + \textbf{waste}. \tag{1.8}$$

In reaction (1.8), when calculating the E-factor, if water is the waste, it will be treated as any other waste, although water is not inherently toxic or hazardous. In order words, E-factor does not take into account the nature and the actual impact of the waste or by-products and can treat waste on equal grounds despite significantly different environmental impacts.

Environmental quotient (EQ) [5] has been introduced in order to address the weakness of E-factor. EQ is essentially a modified E-factor, which takes into account the environmental 'unfriendliness' quotient (Q) of the waste or by-products (equation (1.9)). **Effective mass yield** (EMY) is another similar metric (equation (1.10)), which also disregards benign substances used or produced. Therefore, both EQ and EMY help to distinguish between hazardous waste and non-hazardous waste. However, both metrics are susceptible to inconsistencies due to the vagueness around what is environmentally unfriendly or benign, leading to debate over what values to assign to individual substances.

$$\text{EQ} = \text{E-factor} \times Q \tag{1.9}$$

$$\%\text{EMY} = 100 \times (\text{mass of products} \div \text{mass of non-benign materials used}). \tag{1.10}$$

Life cycle analysis (LCA), is an extensive way of assessing the environmental impact and sustainability of a given process or product. LCA analyses the entire life cycle of a product, from the extraction of raw materials all the way to the fate of the product. In the context of nanomaterials, this has been extensively reviewed elsewhere [7]. This comprehensive analysis can overcome various pitfalls associated with the other metrics introduced above. However, performing LCA is laborious and time consuming, and it requires a large amount of process- and product-related data to be available. One important advantage of LCA is that for an alternative process or product, LCA can help differentiate between pollution/waste prevention and shifting pollution. For example, consider the reaction (1.8) shown above, and assume that the solvent A is a hazardous solvent. In order to remove the need for solvent A, an alternative reaction is available using a different precursor (equation (1.11)), where solvent B is benign:

$$\textbf{Reactant 3} \xrightarrow{\textit{Solvent B}} \textbf{product + waste}. \tag{1.11}$$

On the face of this new reaction, it appears to be 'green' because the hazardous solvent has been replaced with a non-hazardous one. However, during LCA, one needs to consider how the new reactant is derived (reaction (1.11)). Solvent C (equation (1.12)) used to produce reactant 3 may be as hazardous as solvent A. It quickly becomes clear that the alternative (reaction (1.11)) is not as green as it appeared.

$$\textbf{Reactant 1} \xrightarrow{\textit{Solvent C}} \textbf{Reactant 3 + waste}. \tag{1.12}$$

Essentially, although the need for solvent A has been removed, in order to produce the same product, solvent C is required, thus simply shifting the problem to the synthesis of the new precursor. In addition, LCA can take into account the energy required to produce the desired products.

Carbon footprint is another measure for assessing the environmental impact of a process or a product. It estimates the CO_2 equivalent emissions of greenhouse gases caused by a process or associated with a product. Carbon footprint analysis can help decide between processes where, for example, one option can produce a high quality product but require very high amounts of energy, while an alternative method could be lean on energy requirements but may produce a lower quality product.

1.2 Ways to improve sustainability

Estimation of environmental impact using the metrics discussed above can help identify the problems with an existing process or product. This can then help device targeted plans to find alternatives. Environmental impact assessment can also highlight future challenges, such as the need for technical developments in order to improve product quality for greener processes such that they can be competitive.

There are many ways to improve process sustainability, and most of them are based around finding alternatives to either the solvents used, the reactants used and/or the process conditions (e.g. temperature or pressure) [1]. Avoiding the use of solvents altogether is an excellent example of improving greenness of a process. Other alternatives include switching to solvents that are non-volatile organic compounds. The use of supercritical fluids (e.g. CO_2 and water) has also been reported as an alternative due to the ease of solvent separation and reuse. However, one should consider the energy required to operate processes under supercritical conditions (typically high temperatures and pressures; e.g. for water, the critical point is 374 °C and 220 bar). In the case of processes that require heating, alternative ways of providing energy, such as microwave or ultrasound, could be effective. These methods work on the principle of providing energy only to the desired location or chemicals without wasting energy by heating the entire medium or system.

Various ways of estimating the environmental impact, as discussed above, are powerful tools, however they require involvement and inputs from chemists and process engineers [7]. It is critically important to analyse how the modifications made to a section of a process can affect the entire process and perhaps beyond to the entire business. For example, environmental impact estimates cannot predict how changes in feedstock can affect the product value, profits and market competition due to feedstock availability. Answers to such questions can be obtained by performing techno-economic evaluation of a process [8].

1.3 Green chemistry and nanomaterials

When compared to the sectors listed in table 1.1, nanomaterials manufacturing is relatively young. Most attention has been focussed on the discovery and design of nanomaterials. Although the use of nanomaterials can be back-dated to the 9th century (see chapter 2, section 2.2), the large-scale commercial production of engineered nanomaterial developed and accelerated around the late 1990s and early 2000s [9]. This means that the manufacturing-related developments for nanomaterials are in their infancy, and for a vast majority of newly discovered nanomaterials, large-scale manufacturing does not exist. In other words, we are *happy* to have *some* process for the large-scale manufacturing of desired nanomaterials, while little or no attention is given to the sustainability of the process. However, this needs to change and the time is ripe to focus on the environmental impact of nanomaterials production, in order to apply the developments enjoyed by other sectors, listed in table 1.1, to nanomaterials production. These factors clearly stress the urgent need for developing fundamentally new production methods for nanomaterials that are green and sustainable. This change in thinking is very important, and previous experience suggests that when sustainability/green principles are included at the discovery stage, this provides the most benefits.

This book focuses on stimulating a shift towards a sustainable approach for nanomaterials—from discovery to production. Chapter 2 will serve as an introduction to nanomaterials. Their properties will be discussed and benefits offered from their use in selected applications will be presented. In chapter 3, we will provide an overview of the analytical techniques used to probe various properties of nanomaterials. Chapter 4 will present, with examples, a range of current manufacturing processes used to produce nanomaterials at large scales. We will discuss these processes for sustainability by using the theory and concept of green chemistry covered in chapter 1. Chapter 5 will summarise the benefits of using nanomaterials, while highlighting the need for greener alternatives. In chapter 6, we will detail how biology produces a range of inorganic materials from macro to nanomaterials and point out the potential for learning from biology. This learning will be consolidated in chapter 7 where strategies for developing biologically inspired green routes to produce nanomaterials will be presented. The key advantages and opportunities from these alternative routes will be identified, and further explained in chapters 8 and 9 with two case studies.

References

[1] Lancaster M 2010 *Green Chemistry: An Introductory Text* 2nd edn (Cambridge: Royal Society of Chemistry)

[2] Gro Harlem Brundtland 1987 Development and international co-operation: environment, the United Nations, A/42/427 1987, http://www.un-documents.net/wced-ocf.htm

[3] Basics of Green Chemistry [cited 2016 23rd March]; Available from: https://www.epa.gov/greenchemistry/basics-green-chemistry

[4] Anastas P T and Warner J C 1998 *Green Chemistry: Theory and Practice* (Oxford: Oxford University Press)

[5] Sheldon R A 1997 *J. Chem. Technol. Biotechnol.* **68** 381

[6] Sheldon R A 2007 *Green Chem.* **9** 1273

[7] Windsor R, Cinelli M and Coles S R 2018 *Curr. Opin. Green Sustain. Chem.* **12** 69

[8] Drummond C, McCann R and Patwardhan S V 2014 *Chem. Eng. J.* **244** 483
Lieberman M B 1989 *Strat. Manage. J.* **10** 431

[9] Vance M E, Kuiken T, Vejerano E P, McGinnis S P, Hochella M F, Rejeski D and Hull M S 2015 *Beilstein J. Nanotechnol.* **6** 1769

Section II

Nanomaterials

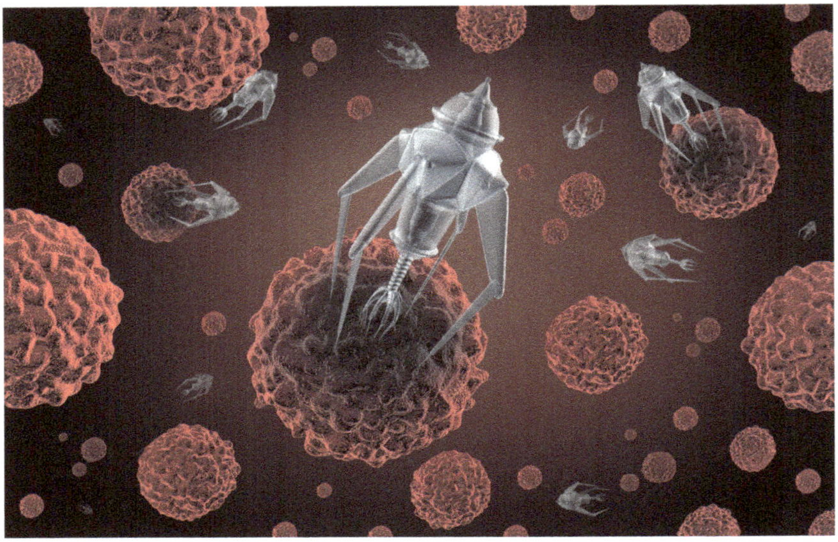

Image courtesy of Lightspring/Shutterstock.

We have learnt the benefits of green synthesis, but before we can explore sustainable bioinspired green synthesis of nanomaterials, it is important to understand what nanomaterials are, and how we can use them. Nanomaterials, nanotechnology and particularly biomedical applications of nanotechnology have captured our imagination in both science and fiction for decades, even before nanotechnology was scientifically realised: the concept of nanobots, removing diseased cells from our blood stream, is a very clear image in sci-fi literature, and excitingly it is actually something that we believe will become a reality in the coming years.

This section is devoted to highlighting special properties of nanomaterials within the scope of this book, illustrating their functions and how these properties serve several emerging and well-known applications with selected examples. Given these special features, which may not be found in bulk materials, an entire chapter focuses on giving an introduction to various key characterisation techniques used for studying and measuring properties of nanomaterials. This is followed by a description of 'top-down' and 'bottom-up' current and emerging methods for nanomaterials synthesis and manufacturing along with the physicochemical principles of nanomaterial formation. Suitable examples of the chemistry of materials are provided to help illustrate the processes.

IOP Publishing

Green Nanomaterials

From bioinspired synthesis to sustainable manufacturing of inorganic nanomaterials

Siddharth V Patwardhan and Sarah S Staniland

Chapter 2

Nanomaterials: what are they and why do we want them?

2.1 Fundamentals of the nanoscale

Although the term 'nanotechnology' is commonly used beyond science by the general public and in the media, an understanding of what 'nano' is defined in length scale, and the changes to physical properties that occur in materials in this miniaturised world, are not generally realised.

'Nano' comes from the Greek word for 'dwarf' and is the prefix of a measurement that is $\times 10^{-9}$ (or one billionth of that unit). In the context of material science and nanomaterials we are interested in the measurement unit of length: metres, and therefore the nanometre (nm), although 'nano' can prefix any unit, for example nanosecond, nanogram or nanomole. Put into context, there are one million nanometres in a millimetre and a billion in a metre. Figure 2.1 may be used to aid visualisation of these different length scales. Considering it from the opposite side, scaling up rather than down, we measure atomic bond lengths in Angstroms (Å) which are $\times 10^{-10}$ m, and so the nanometre length scale is 10× larger than the length scale at which we consider atomic reaction (chemistry) to occur. For example, the unit cell (smallest crystal unit) of sodium chloride is 0.56 nm^2, containing four of each type of (sodium and chloride) atoms. Although performing chemical reactions at this scale is an old and well-established discipline, the idea of crafting materials and components on the nanoscale and designing and creating tailored materials for smart applications is now developing as the relatively new disciplines of nanoscience and nanotechnology.

In 1959, Professor Richard Feynman delivered his seminal lecture entitled 'There's plenty of room at the bottom', in which he imagined what could be achieved if we shrunk technology down to the nanoscale. This lecture inspired material scientists and technologists to challenge themselves to venture to the smallest length scales possible for materials. The lecture has thus been defined as the birth of nanoscience and nanotechnology. As a physicist, Professor Feynman was fully aware that this shrinking

doi:10.1088/978-0-7503-1221-9ch2

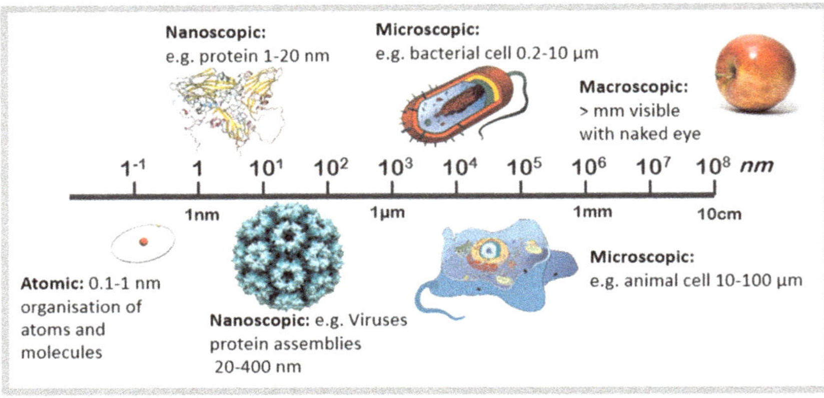

Figure 2.1. Schematic of length scale with respect to a nanometre, showing common examples. Source: Abhijit Tembhekar/LadyofHats/Alan Cann/Acharya KR, Fry E E, Logan D T, Stuart D I/; visualisation: Astrojan/ Белых Владислав Дмитриевич.

would not be a simple linear scaling issue, because smaller components will not display the same properties as the bulk material. Rather, materials at the nanoscale display rich and complex properties specifically dependent on their dimensions, proportion of surface atoms, and morphology (see the examples in the sections below).

Nanomaterial can be defined as having at least one dimension in the nanoscale (between 1–100 nm). This ranges from thin nanometre films (one dimension at the nanoscale) to nanowires (two dimensions at the nanoscale) to nanoparticles (all three dimensions at the nanoscale). Dimensions smaller than 1 nm are in the territory of clusters, made up of a few atoms, which are outside the realm of nanomaterials. We should also consider that the nanoscale can apply to voids as well as matter, and thus nanoporous materials can also be considered as nanomaterials, particularly as they have large surface areas. While 100 nm is conventionally the upper limit, this can be blurred. Materials termed nanoscale are reported over this size, if the properties displayed are still within the realms of nanoproperties (discussed in section 2.2) or, (more frequently as time goes on) it suits to classify them as a nanomaterial for political (non-scientific) reasons!

The nanoscale is a most interesting length scale for solid state materials for several reasons. First, from a philosophical viewpoint, it can be asked: how small can a particle be before it is no longer a material? Indeed, there is a valid scientific debate about whether or not a nanoparticle can or cannot dissolve, or if it is already a solute, when in a solution.

Second, nanomaterials have a large percentage of surface compared to bulk material, and surfaces have very different chemical and physical properties to the bulk. To illustrate this, let us consider two spherical particles with radii of 1 μm and 10 nm. The ratio of the external surface area to volume ($A{:}V$) of each particle is inversely proportional to its radius (figure 2.2):

$$\frac{A}{V} = \frac{4\pi r^2}{\frac{4}{3}\pi r^3} = \frac{\mathbf{3}}{\boldsymbol{r}}. \tag{2.1}$$

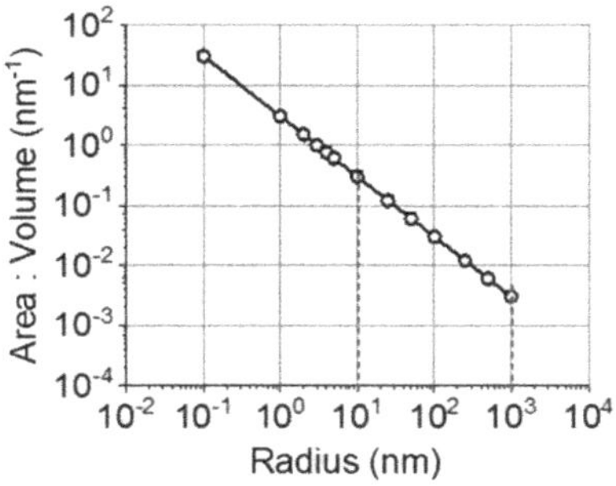

Figure 2.2. Graphical representation of how surface area increases with decreasing radius of a spherical particle (on a log 10 scale).

For these two particles, $A{:}V$ will be respectively 0.003 nm^{-1} and 0.3 nm^{-1}. This means that the 10 nm particle will have 100 times more surface for the same volume. As a result, the 10 nm particle will have 100 times more atoms on the surface available to interact with the external environment. This phenomenon leads to significantly increased catalytic activities, for example. This effect also manifests into other special effects and properties such as optical and magnetic properties, which are discussed further below.

Third, the nanoscale lies at the boundary of bulk material properties and atomic chemical properties (figure 2.3). The former is generally described by continuous transitions, while the latter must be considered as quantum mechanical. Quantum mechanics is both difficult and abstract to conceptualise but can be linked to the macroscale world by the theory of wave particle duality, where any wave can also be described as a particle with the same energy, and vice versa.

$$E = hf = \rho c. \tag{2.2}$$

These are linked by the work of both Einstein and Broglie, that shows that a particle and wave with the same energy are interchangeable (equation (2.2), where E = energy, h = Planck's constant, f = frequency of the wave, p = particle momentum = mass × velocity, (momentum of a particle) and c = the speed of light). At this stage however, it is simple to see that properties at the atomic scale are quantised because electrons (the principle particle (or wave) behind all chemistry) reside in discrete quantised electronic orbitals (defined by having an integer wavelength). This is simply described as if one considered the energy a particle can have in a defined narrow box, which is the energy that equates to any integer wavelength that will fit in the box as a standing wave (figure 2.3). These energy levels are equal distance apart as the box width remains constant (figure 2.3), however, the energy well of an atom follows a curve and as such the energy levels become closer together the higher the energy (figure 2.3). This is also directed by the shape of the energy well; the deeper the well, the further apart the energy levels, while the value of each energy level is also directed by the element (the pull of the nucleus etc). Thus, there is only one series of discrete energies that relates to that electron and its energy levels, of that particular atom. As the number of atoms in a material increases, so too does

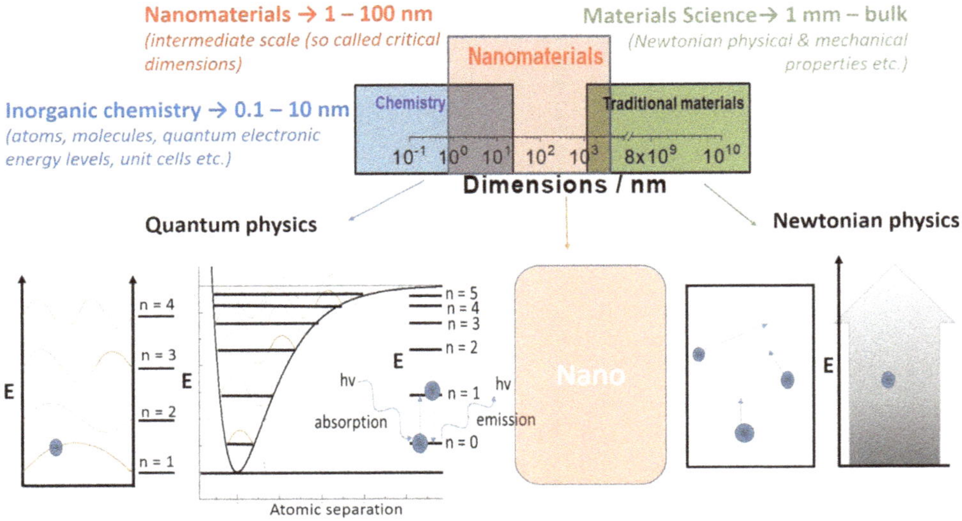

Figure 2.3. Depiction of how nanoscale materials lie at the crossover between the atomic quantum laws of physics and the macroscale Newtonian laws of physics. In quantum physics: the left shows how a particle in a very narrow box can only have the energy quantised by the size of the wavelength. The change in energy of these is the same as the box sides are parallel. For an atomic system, the change in energy between levels decreases at increasing number of *n*, due to the energy potential. Conversely, a particle in a macroscale box can move in any direction at any given energy, so there is a continuum of energy a particle can possess.

the number of interactions between atoms and electrons. The more interactions, the harder it is to define discrete energy levels as more and more energy levels occur, forming a continuum which is what we can describe in a bulk material (figure 2.3). One can similarly conceptualise the particle in a box on a macroscale. The width of the box is now much greater and the wavelength associated with a macroscale (large mass) particle is now so small, the particle can now be considered to occupy any energy, so it becomes a continuous system. In Newtonian physics a particle can have any energy, velocity and thus momentum within its box (figure 2.3). The interface where quantum states transition to continuous phases is at the nanoscale and this leads to a range of exotic behaviours, some of which are briefly described below in section 2.3.

2.2 Tangible and historical examples of nanomaterials

It is no coincidence that nanoscience has grown in parallel with the growth and development of instrumentation, imaging and spectroscopic techniques that have the resolution to allow probing on the nano- to the atomic scale (see characterisation in chapter 3). This has been essential to analyse and understand these materials as well as aid design and synthesis. However, while we may not have been able to visualise or understand them, we have been using the properties of nanomaterials for millennia, with examples stretching through history (figure 2.4). One of the earliest examples is the 4th century Roman Lycurgus cup. This beautiful glass cup, mounted

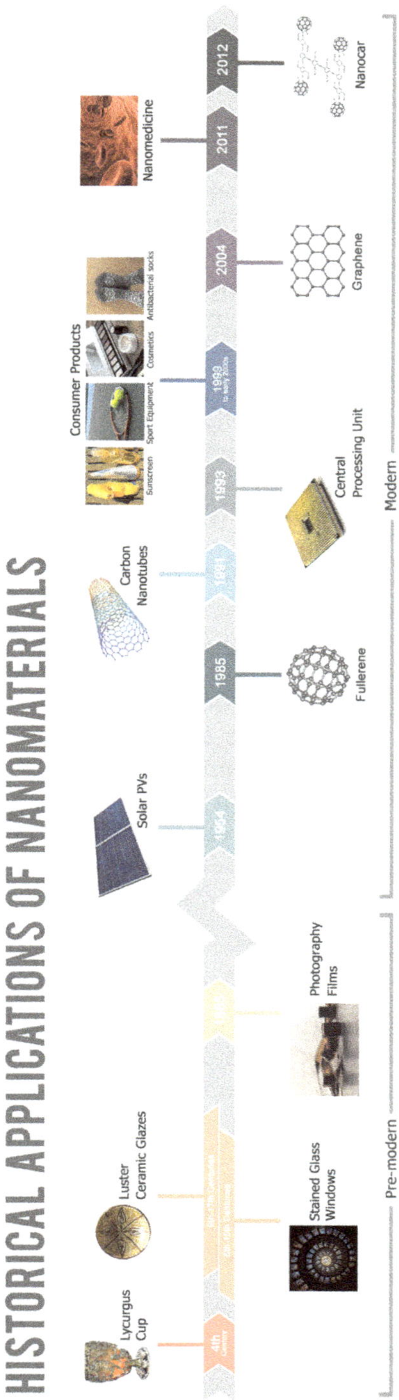

Figure 2.4. Schematic of nanoscale materials and accompanying timeline of their development.

in a carved bronze exterior, is now housed in the British Museum. It is an excellent example of nano-optics (specifically surface plasmon resonance), where, because it contains different metallic nanoparticles it looks green if illuminated from the front, because silver nanoparticles scatter the reflective light, and red if illuminated from behind, due to surface plasmon resonance from the gold nanoparticles interacting with transmitted light. It was not until 1847 that Faraday first reported nanoparticles; gold nanoparticles, as a colloidal suspension, he called a 'gold sol'. He was the first to recognise and describe the exotic optical properties gold nanoparticles possess, which is perhaps the first recognition of quantum size effects and thus quantum mechanics. Not quite so complex but equally dramatic and awe-inspiring, stained glass windows have been used for centuries in religious buildings. Again, metallic nanoparticles are responsible for the incredible colour still enjoyed today in these and also in ceramic glazes, centuries after they were created. The photo sensitivity of silver was developed during the 1800s to invent photography and film (the origin of the term 'silver screen') as we knew it up until the late 20th century. While digital imaging has now overtaken film to become the norm, many professionals still prefer traditional film for its distinctive look, and also for its better resolution. Colloid silver on film has nanoscale resolution! The later 20th century saw the explosion of nanomaterials, both in their own right (the subject of this book) and for inclusion in many consumer products, such as solar photovoltaic materials in solar panels and tyres. We even now see silver nanoparticles embedded into socks to for their antimicrobial properties, to stop them smelling! In the present day, our understanding of nanoproperties and ability to make and characterise nanomaterials is growing, and so too is our ambition for their use. The field of nanomedicine has been explored in recent years, and is the subject of section 2.4.1. While curing disease is one of the key applications, it is also the aim of nanoscientists globally to create the smallest working nanomachines, using classic engineering shapes such as axials, rings and wheels at a molecular scale, to make nanowalkers, molecular motors and nanocars [1].

2.3 Special properties offered by the nanoscale

There are many physical properties that are affected at the nanoscale, such as melting point and electronic properties. Furthermore, due to the large surface area and predominance of surface chemistry, nanomaterials are also ideal for heterogenous catalysis. The purpose of this chapter is NOT to give a complete, definitive account of all properties and applications of nanomaterials, but rather to give a flavour and thus must focus on a small selection of nanoproperties and their subsequent application.

There are two features of nanoscale materials: (1) increased surface-to-bulk volume ratio and (2) the reduction in size, which leads to the surface and quantum mechanical properties predominating over the bulk properties. Both key features are beautifully illustrated in turn by considering two optical properties of nanomaterials: the surface effect of surface plasmon resonance (SPR) and the quantum mechanical effect of quantum dot fluorescence. Furthermore, the reduction in size also affects magnetic properties. These three nanoproperties will be described in this section, with the following section showing real-world applications of these specific properties.

2.3.1 Optical: surface plasmon resonance

A surface plasmon is the interaction of the free surface electrons of a metal with electromagnetic radiation. Surface plasmons are mostly seen in inert noble metals, but all metals and even some highly doped semiconductors will exhibit a surface plasmon effect. When a metal surface is irradiated with light of a larger wavelength than the size of the particle, the conducting electrons will polarise and collectively oscillate about the surface of the particle in response to the wave, and are thus driven to form a resonance standing wave, giving a resonance peak (figure 2.5(i)). This is

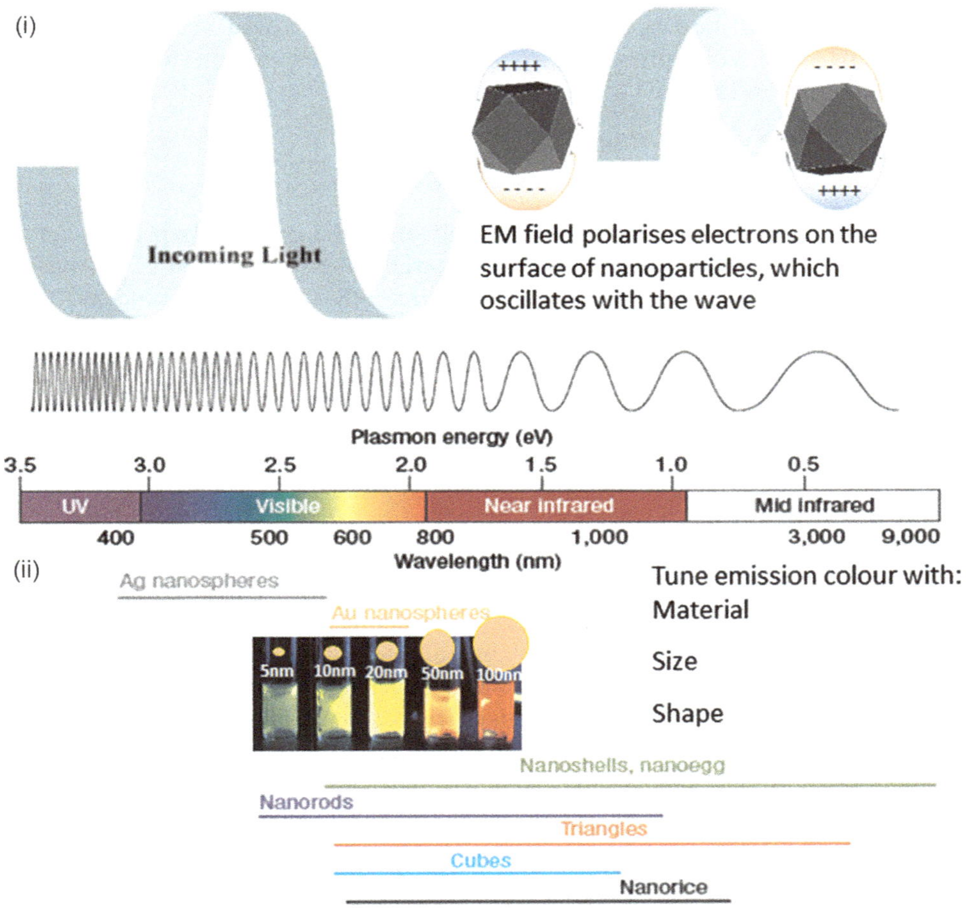

Figure 2.5. (i) Schematic of how electromagnetic wave interacts with surface electrons on a nanoparticle (adapted from figure in [2] with permission of the Royal Society of Chemistry). (ii) Regions of the electromagnetic spectrum that specific nanomaterials of different morphologies emit due to SPR (adapted from figures in [3] with permission of Springer).

known as SPR. At this peak the electromagnetic radiation is both scattered and absorbed by the metal [2]. Some of the radiation is absorbed by the nanomaterial and converted to crystal lattice vibrational energy (phonons), while the rest of the radiation is emitted in different directions (scattering).

The colour of the light emitted comes predominantly from the scattering, so we should consider what contributes to the SPR scattering. We need to think about polarisability; the ease with which charges (the conduction electrons) on the nanoparticle surface undergo charge (re)distribution and form partial dipoles. This is dependent on how far the electrons need to move, i.e. the cross-sectional area (so in turn, the radius of the particle (r)). It also depends on the nanoparticle material, and how easy electrons can move on this metal: the dielectric function of the nanoparticle ($\varepsilon = \varepsilon_1 + i\varepsilon_2$) (wavelength dependent) and the dielectric function of the medium (ε_m) in direct contact with the surface. This can be quantitatively described by the Gustav Mie theory of light scattering for any spherical particles with radius = R, for which a simplified version (when $2R \ll \lambda$) is shown in equation (2.3)

$$C_{\text{ext}} = \frac{24\pi^2 R^3 \varepsilon_{\text{m}}^{3/2}}{\lambda} \cdot \frac{\varepsilon_2}{(\varepsilon_1 + 2\varepsilon_{\text{m}})^2 + \varepsilon_2^2}. \tag{2.3}$$

Here C_{ext} is the extinction cross-section and gives the quasi-polarizability of the nanoparticle. SPR is a property of nanomaterials due to the scale of wavelength verses particle size, and the increased surface area offers a more pronounced effect. Furthermore, the wavelength of the resonance peak and thus the colour emitted by the SPR is dependent on the components highlighted above, the primary ones being the particle composition (material), and the radius. However, what equation (2.3) does not describe is the effect of shape/architecture, as the Gustav Mie equation is only for a simple spherical particle. Surface plasmons are highly dependent on the overall surface architecture, and not limited simply to size. Varying morphology affects the SPR frequency and thus emission since varying the shape alters the surface architecture and this alters the electric field distribution on the surfaces, again altering the surface plasmon resonance. This tends to result in an increase in the range of resonance frequencies; this is certainly true for cubes, rods, cages/shells and egg shapes (figure 2.5(ii)). Furthermore, all other volumes have a larger surface area to volume ratio than a sphere, so all other shapes will offer increased surface area. This is particularly true for hollow shapes like cages.

The classic historical example is gold nanoparticles. Spherical gold shows a range of colour dependence on size, from green for particles around 5 nm, up to red for particles of about 50 nm (figure 2.5(ii)). However, silver spherical nanoparticles have higher frequency SPR resonance (higher energy), emitting from green up to ultraviolet compared to gold nanoparticles of the same size and shape (gold emitting red to green), demonstrating the dependence SPR has on material as well as the size. Finally, since this is a surface effect and the surface is interacting with the surrounding media, we cannot forget the SPR frequency dependence on the dielectric constant of the media. Furthermore, SPR is affected by the locality of near materials exhibiting SPR [4]. Both are incredibly useful for applications as colorimetric indicators/sensors as very subtle changes to either the nanoparticle surface, the surrounding media, or the distance nanoparticles are from each other can be optically detected. In addition, SPR is so intense that the scattering of a single nanoparticle can be seen by the eye in an optical scattering microscope. The blue light emitted by a silver nanoparticle has a cross-sectional area a million times greater than the dye molecule fluorescein, making noble metal nanoparticles exhibiting SPR very attractive materials for ultra-sensitive sensors [4]. There have been many excellent tutorial papers and reviews on SPR that the reader is directed to for more in-depth explanation, as well as the many applications, from sensing to *in vivo* and *in vitro* biomedical applications [2–5].

2.3.2 Optical: quantum dots fluorescence

Metals conduct electricity because their mobile electrons can move through the material. Electrons are confined to their atomic orbitals in an insulator material. So what is it that allows electrons to move in metals and not in insulators? The answer is there are simply vacant orbitals (holes) to move into at the same (or very similar) energy of the mobile electrons in metals. This is described by band theory, where there is an energy band full of electrons, so there are no holes available for electron movement. This is the valance band. Above this is the conduction band. Metals have electrons in the conduction band, which is not full, so there are plenty of vacant sites to move between. Semiconducting materials have a full valance band and no electrons in the conduction band (similar to an insulator), but they differ from an insulator as they have a very small band gap between the two bands. As such they become conducting when the temperature is increased sufficiently for the electrons to be promoted into the higher energy conducting band, leaving an electron hole in the valence band (with both mobile electron and holes contributing to conduction). Examples of semiconducting materials include CdSe, InAs and GaP. Describing electronic properties with respect to bands is only possible in the bulk phase (as bands are a continuous macroscale phenomenon). As the particle size reduces in size to the nanoscale, the electrons become spacially confined (with fewer vacant orbitals to move into) and the bulk models break down. When the size of the particle becomes comparable to or smaller than the electron/hole pair (exciton) Bohr radius (usually $\leqslant$10 nm), the electrons are confined in all directions and the material is considered to have zero dimensions (hence a dot); these materials are known as quantum dots or ‘artificial atoms’. The spacial confinement results in an increase in the size of the band gap and the banded continuum (valance and conducting bands) breaking into a quantised energy level, perfectly demonstrating how the nanoscale is the intermediate between the bulk and the atomic scale. When energy is supplied to a quantum dot (in the form of electromagnetic radiation), an exciton is created as the electron is promoted into the conduction band, and this then emits fluorescence when the electron relaxes back down into the valance band, recombining with the hole. The wavelength of this fluorescence is dependent on the size of the band gap ($\Delta E(r)$) which is dependent on the band gap of the bulk material (E_{gap}), plus an increase in this size as the radius of the quantum dot (r) decreases (this relationship is shown in equation (2.4) and figure 2.6). The higher energy (shorter wavelength/higher frequency) emission is generated from relaxation across a larger band gap which occurs for smaller quantum dots. Equation (2.4) can be rearranged to equation (2.5) to show the relationship of the quantum dot size (radius r) to the change in band gap energy (h is Planck's constant, m_e is the mass of a free electron and m_h is the mass of the hole).

$$\Delta E_{\text{nano}}(r) = \Delta E_{\text{bulk}} + \left[\frac{h^2}{8r^2}\left(\frac{1}{m_e^*} + \frac{1}{m_h^*}\right)\right] \tag{2.4}$$

$$r = \sqrt{\left(\frac{h^2}{8(\Delta E_{\text{nano}} - \Delta E_{\text{bulk}})}\right)\left(\frac{1}{m_e^*} + \frac{1}{m_h^*}\right)} \tag{2.5}$$

Quantum dots have exceptional and far superior fluorescent properties in comparison to traditional organic dye molecules. (1) Due to the relationship between size and emission their wavelength can be precisely tuned. (2) They can absorb a broad range of energy but have a very narrow emission band. (3) Their fluorescence is unparalleled with respect to brightness, lifetime and resistance to photobleaching, which makes them excellent optical probes. One downside is many of the best semiconductor materials for high florescence quantum dots with the band gap at the best wavelength for visualisation are often the most toxic (such as CdSe and Cd/S). This is being addressed by coating these materials in less toxic semiconductors, such as ZnS, and also developing new quantum dot materials by doping less toxic semiconductors to tune the band gap. Again, different morphologies can offer

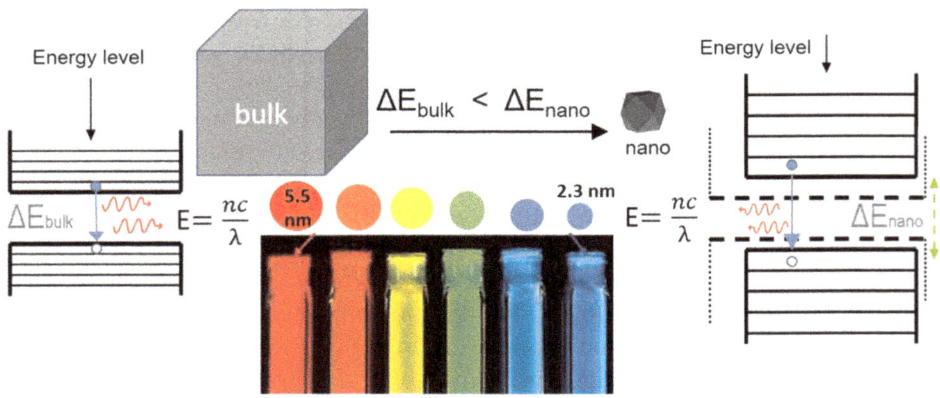

Figure 2.6. Description of how a band gap in a semiconductor increases as the size of the particle decreases. The left-hand side shows the band gap in the bulk material, while on the right an increased band gap is shown for the nanoparticle (bulk gap is shown as a dashed line on the right for comparison) as the frequency of the light emitted is proportional to this energy (and dependent on the material), the light emitted varies with size. CdSe/ZnS quantum dots of decreasing sizes from left to right are shown in the centre (image from [2] reproduced by permission of the Royal Society of Chemistry).

further tuning. Further reading on quantum dots and particularly their biomedical uses and the implications of their toxicity can be found in [6].

2.3.3 Electron spin and nanomagnetism

Nanomagnetism is another perfect illustration of a property on the boundary between bulk and atomic scale. Again we will see here how some nanoparticles behave like atoms (superparamagnets) while others behave more like the bulk (single-domain nanomagnets). Furthermore, we see how the quantum interplay between charge, electron spin and magnetic field can be harnessed in nanomaterials for spintronic applications. To understand the nanoscale properties we must first understand the interplay of electrons and magnetism at the atomic scale, appreciate how this builds to bulk scale, then interrogate the middle.

In fundamental terms, a magnetic field is generated when there is movement of charge. This is seen by the magnetic field surrounding a wire conducting electricity, which is due to the charged electrons moving in the wire (figure 2.7(Ai)). Electrons in an atom have two sorts of motion (orbital and spin (figure 2.7(Aii))) and as such have a magnetic moment: generally it is the electrons in a material that give it its magnetic properties. The moment of a single electron can be calculated from first principles, and is found to be

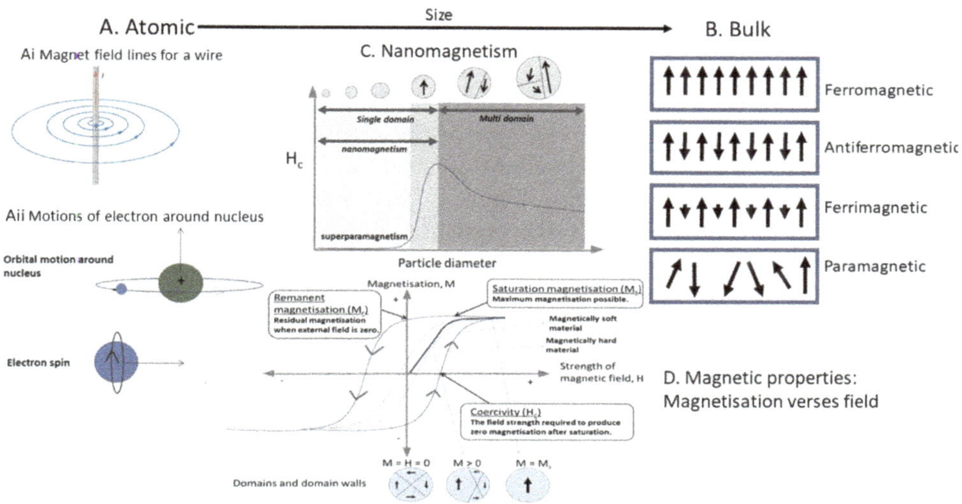

Figure 2.7. Description of magnetism from the atomic to the bulk. (A) Two demonstrations of magnetic fields generated by electrons. (Ai) Macroscale example of magnetic field lines of magnetic field generated by a wire conducting electricity due to the flow of electrons (the current *I*) in the wire. (Aii) The motion of an electron within an atom (both orbital and spin) which generates an atomic magnetic field. (B) Any unpaired electrons contribute to a bulk magnetic property depending how they are arranged in the solid. Each arrow represents the direction and magnitude of a paramagnetic atom. (C) The nanomagnetic properties, demonstrating how the largest single domain magnetism have the highest coercivity due to no loss of energy through domain wall formation, while superparamagnetic nanoparticles have near zero coercivity. (D) Description of the magnetic properties of hysteresis, plotting the magnetisation with increasing field, then reversing the field and increasing again.

$$\mu_B = \pi r^2 I = \frac{e\hbar}{2m_e} = 9.274 \times 10^{-24}\ \text{A m}^2, \tag{2.6}$$

where r is the radius of the orbital, I is the current, e is the charge of an electron, m_e is the mass of an electron and $\hbar$ is the reduced Planck constant.

Two electrons only exist in the same atomic orbital if they have opposite spin values (either + 1/2 or −1/2). The electrons are 'paired' in an orbital and these two opposite electronic spins cancel out each other's magnetic moment, so there is said to be no net overall magnetic effect. However, this is not technically true: such orbits actually have a week repulsion to the magnetic field (in the order of a million times less than the unpaired magnetic effect). If an atom has only paired electrons it is **diamagnetic**. The presence of an unpaired electron in an atom gives the atom an overall magnetic moment and such atoms are described as **paramagnetic**. The application of a magnetic field causes these dipoles to align with the direction of an external magnetic field. The fact that unpaired electrons create a magnetic field and in turn are affected by an external field is fundamental to spintronics. This new field uses the quantum physical properties of ultrathin (single atomic thick layer) 2D nanomaterials, and harnesses the effect of fields on electron spin and spin–orbit coupling to convey electronic information rather than the transport of free elections down a wire. In purely paramagnetic materials the electron dipoles are only weakly coupled, and as a result thermal energy causes these moments to randomly align. Examples of purely paramagnetic materials include many transition metals and rare-earth salts. These materials have a net magnetic moment due to incomplete electron shells.

Bulk magnetism is a cooperative effect of the alignment of many paramagnetic atoms, and as such it is a property reliant on structural order. There are three main categories of bulk magnetic ordering (figure 2.7(B)). **Ferromagnetism** is when all the paramagnetic atoms are aligned in the same direction even when a field is removed, resulting in the material having a magnetic moment. Ferromagnetism is rare, with iron, cobalt, nickel and some rare-earth metals being the only elements to be ferromagnetic. Ferromagnetism is rare as paramagnetic atoms/ions will tend to align in an anti-parallel way; this form of ordering is termed **antiferromagnetic** and will exhibit no net magnetisation as the opposing paramagnetic moments will cancel each other out. Examples include numerous transition metal compounds, such as iron manganese (FeMn). The final bulk magnetic coupling utilises the strong anti-parallel alignment, but results in a net magnetic moment simply by ensuring there is an uneven number of magnetic dipoles in each direction, so the dipoles are not fully cancelled out. A material with this magnetic ordering is called a **ferrimagnet**. Ferrimagnetic materials tend to be made up of different materials or ions, with examples including garnets such as YIG ($Y_3Fe_2(FeO_4)_3$) and ferrites such as the oldest known magnetic material: magnetite (Fe_3O_4).

Magnetic order is dependent on temperature: when the temperature is increased above a certain threshold (blocking temperature (T_B)) (also called Curie temperature (T_c) or Néel temperature (T_N) for ferr(o/i)magnetic and antiferromagnetic materials respectively), specific to that material, the thermal energy destroys the ordering and the bulk magnetism is lost. Above this temperature the material becomes paramagnetic.

It is well known that magnetic materials can become demagnetised over time. This is because a material will gradually form magnetic domains, which are small regions of local dipole alignment within the material, each separated by a domain wall. It is always worth remembering that maintaining order carries an energy cost, so long-range order will inevitably reduce to short-range order (domain), to eventual disorder at any elevated temperature if there is nothing to maintain the order (i.e. a field). Domains form to minimise the material's energy by reducing magnetic poles at the surface to reduce the magnetostatic energy. However, the creation of a domain wall requires energy to overcome this stability, so an equilibrium is reached. If a demagnetised material is placed back in a magnetic field, the domains most closely aligned to the field direction grow as the walls recede at the expense of the misaligned domains, continuing until all the dipoles are aligned with the higher field and the material is said to be magnetically saturated (figure 2.7(D)).

A material that will easily or reluctantly align with a changing magnetic field is termed magnetically soft or hard, respectively, a property that is measured by its magnetic coercivity (figure 2.7(D)). For a soft magnet, simply removing the field (field = 0) is enough to demagnetise the material. However, a hard magnet will retain its moment at zero field, and even in an opposing magnetic field (figure 2.7(D)). The electrons in such materials have a preferred (lower energy) direction of alignment, so more energy is required to realign them. Coercivity can be quantified from a hysteresis plot (figure 2.7(D)) where a material is driven to saturation by application of a cycling magnetic field through both opposing directions. Where there is a hysteresis loop, work is done to oppose the field and this energy is proportional to the area of the hysteresis curve and is dissipated in the form of heat.

If the bulk material's dimensions are reduced to the nanoscale, a critical volume is reached where it costs more energy to create a domain wall than the total magnetostatic energy of the material, and so a single domain magnetic particle is formed (figure 2.7(C)), which is also dependent on the anisotropy of the material. These are the smallest permanent magnets, and have the advantage that energy is not wasted on the formation of domain walls, so coercivity is the highest for this material (by volume). As the size is reduced further the magnetisation energy of the particles approaches the thermal energy (K_BT) and so the magnetic dipole of these particles will continually fluctuate with thermal fluctuations. These particles behave like paramagnetic atoms would (not a bulk magnet) and are thus termed **superparamagnetic**. Such particles do not display hysteresis as the particle itself flips in the field. Interestingly there is not a cut-off defined size at which a single domain magnetic nanoparticle starts to display superparamagnetic behaviour. It is entirely dependent on the temperature and the time-scale of an experiment. That is, the same single domain ferromagnetic nanoparticle could be a permanent static magnet in one experiment where the measurement was conducted over a short time-frame or the temperature is low, and superparamagnetic in another experiment when the measurement is taken over a longer time-scale and/or the temperature is raised. More detailed treatment of magnetism can be obtained from excellent textbooks [7].

2.4 Applications

There is now an ever-increasing demand for nanomaterials for numerous applications (figure 2.8) due to their enhanced/unique properties (above), increased surface area, increased reactivities and compatible size for interactions with molecular technologies. Examples range from personal care, such as TiO_2 in sunscreen cream, to audio technology with magnetic nanoparticles in dampening ferrofluids for loud speakers. More sophisticated nanomaterials are now being specifically designed for a wide breadth of challenges, from curing cancer to purifying water: nanomaterials are becoming ubiquitous in every aspect our modern life as we move towards smaller, more compact and more sophisticated devices. There are many applications in areas such as energy (generation, conversion and storage), catalysis and gas sensing. However, this chapter is designed to provide an overview and not to cover the full range of applications. As such, this section will concentrate on the emerging and

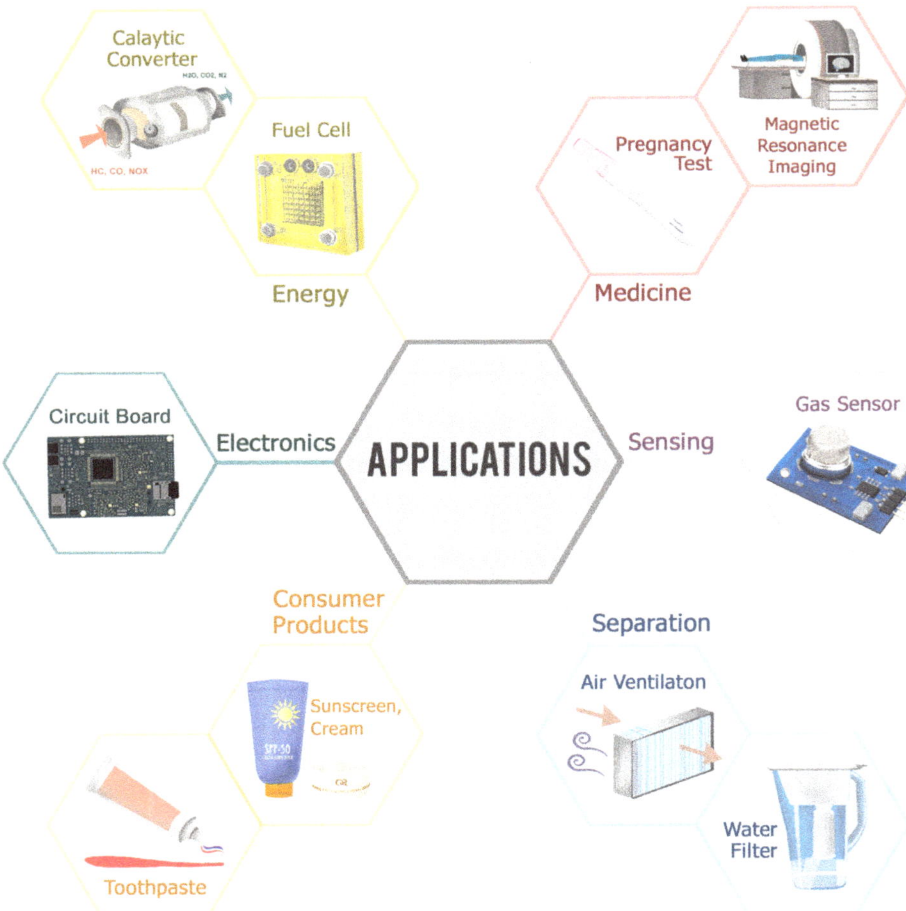

Figure 2.8. Schematic of the types of sectors and applications in which nanomaterials are used.

important field of nanomedicine, as this offers excellent examples that utilise the magnetic and optical properties described in section 2.3. While there is some cross over, we will discuss *in vivo* (in the body) magnetic materials for therapies, with some therapeutic optical materials too. Imaging enhancement material for both *in vivo* and *in vitro* (in a test tube) applications will explore materials with optical properties such as gold and CdSe, and *in vitro* bionano-sensing will be described which takes advantage of surface plasmonic properties. The final two sections will touch on different nanotechnological applications such as data storage and consumer products, to offer some breadth.

2.4.1 Nanomedicine

Healthcare utilises a range of nanomaterials, such as silver in wound-dressing (covered in later sections on consumer products), however, since the 1980s, and with a rapid increased effort in the last 20 years, a new field of biomedical research and application has developed called **nanomedicine.** Nanomedicine is the utilisation of nanotechnology in medicine. This is a really powerful concept, due once again to length scale. As can be seen from figure 2.1, medically relevant biological entities such as proteins and DNA are comparably sized to nanoparticles, making their interaction specific and optimal. Nanomedicine has utilised very specific nanoparticles for imaging disease or specific markers in areas of the body (diagnostics/imaging), or for treatment (therapeutics) or in some cases both (theragnostics) for a range of target diseases. This includes using nanoparticles to highlight specific areas of disease, for either imaging or removal (by magnetic separation), using them as vehicles to deliver drugs to specific disease sites, or using their nanoscale properties such as photothermia and magnetic hyperthermia to treat diseases, with an emphasis on cancer [8]. *In vitro* diagnostic immunoassays will be discussed as the third application, however, the overlapping field of nanomaterials in tissue engineering and regenerative medicine, while fascinating, is too large to topic to discuss here and as such is beyond the scope of this chapter.

2.4.1.1 Generic biomedical nanoparticle

When considering applications *in vivo* there are key factors to consider, the most important being efficacy, toxicity and biocompatibility (the latter two are discussed in section 2.5). These properties along with mode of action can all be controlled via either the particle inorganic core, the coating, or functional biomolecules/small drug molecules that can decorate the exterior. A generic biomedical nanoparticle is shown in figure 2.8(A).

It should be noted that not all of these three components may be strictly necessary (e.g. due to their inert properties, we will we see gold nanoparticles with no coating etc), and the properties above can be delivered by varying components (e.g. the therapeutic can be an active drug molecule on the surface, or heat treatment from the inorganic core). **The inorganic core** offers the nanomaterial's fundamental form. It could be that the particle's key property is simply to be inert and dense (to aid visualisation), but more commonly, the core offers the unique nanoproperties discussed in section 2.3 which can be utilised in the medical application. The main functions of the **coating** are: (1) to provide protection (of the particle against degradation from the environment) and (2) to protect the surroundings from toxic effects of the inorganic material. The two are related: the more stable the particle, the less likely it is to dissolve and prove toxic to the body by leakage. The coating is vital for nanomaterials with unstable or toxic cores such as CdSe quantum dots. These are regularly coated with ZnS and then again with an organic coating to prevent toxicity and increase biocompatibility. The second main function is to aid the particle's biocompatibility. Popular coatings such as dextran are cheap and increase biocompatibility. It is well known that coating with polyethylene glycol (PEG) increases blood circulation times by preventing nanoparticle removal by the immune system. The uptake of modified nanoparticles by cancer cells can be increased by coating in a pH-sensitive zwitterionic coating, that becomes charged at the cancer site (lower pH), so is more readily uptaken, increasing the accumulation at the cancer site [9]. Using biologically derived coatings provides further 'stealth' by invading the immune system and targeting cancer sites by disguising the nanoparticle as the immune system, using macrophage membranes as coatings [10]. The final function of the coating is to enable easy attachment of the third component: active biomolecules. This can be achieved through a range of chemistry, mentioned briefly in later sections. Finally, the nanoparticle can be decorated with **active biomolecules or small molecular drugs** that can be used to target the nanoparticle to the site of the disease, or can be a drug or biological therapies to treat a disease (discussed in more detail in the therapeutics section).

This field is far-reaching, so to keep the content of this section focused on the types of nanoparticles discussed in later chapters and the properties we have discussed earlier in this chapter, only imaging/diagnostics and therapeutics will be discussed here. This critical area of nanomedicine utilising nanoparticles is still in its infancy, with only a handful of nanomaterial-related nanomedicines being FDA approved so far [11].

2.4.1.2 Imaging

There is a vast range of medical imaging techniques with various purposes and advantages. Many (e.g. ultrasound scanning) are not enhanced by inorganic nanomaterials, so will not be discussed further in this section. Some (such as positron emission tomography (PET)) are starting to benefit from enhancement with nanoparticles, but expand into the science of radioisotopes, which again is beyond the scope of this section (see [20] for the development of PET with nanoparticles). Nanomaterials can act as a visualisation probe (highlighting an area of interest) in many microscopy and tomography techniques (see chapter 3 for techniques), or enhance the signal for a specific area, because of its nanoproperties such as nanomagnetism in magnetic resonance imaging (MRI) and magnetic particle imaging (MPI). The nanomaterials become diagnostic probes by being decorated with biomarkers such as antibodies that will target individual diseases, thus specifically detecting and emphasising these areas.

Probes for microscopy can be used both *in vitro* and *in vivo*. Gold nanoparticles have been used extensively on both accounts. Gold can be decorated effectively (e.g. with antibodies) by taking advantage of the strong gold–sulphur bond using a cysteine residue which contains a sulphur. Thus, gold nanoparticles coated with cancer-targeting antibodies (e.g. anti-EGFR (epidermal growth factor receptor)) can be used to visualise cancerous cells with optical microscopy *in vitro* (figure 2.9(Bi)) [12]. Similar optical microscopy can be used to visualise targeted quantum dots using confocal microscopy [6]. The density of gold nanoparticles can also be utilised as a probe for TEM imaging in what is commonly referred to as immune gold staining, where gold nanoparticles can be targeted to a specific biological site. TEM analysis will then identify structures decorated with dense gold dots, distinguishing them from the other biological tissue (figure 2.9(Bii)). Recently, a combined approach of both gold and CdSe quantum dots have been used to image a cell, highlighting different aspects (the gold to the nucleus, and the quantum dot to the insulin granules) offering multimodal enhanced imaging (figures 2.9(Biii), (Biv)) [14]. *In vitro* imaging applications are considered mainly to conduct scientific research and cross over into the area of nanosensing (section 2.4.1.5). *In vivo* imaging is where the real medical applicability lies. Again, gold nanoparticles have been used in diagnostic imaging, using their SPR properties in optical coherent tomography (OCT) [15]. This non-invasive, high resolution optical technique relies on reflective interference between emitted and backscatter light (akin to an optical ultrasound), so can penetrate biological tissue making it applicable *in vivo*. Gold nanoparticles of various morphologies (tuned so the SPR lies within the emitted light source) make ideal contrast agents for OCT as the gold nanoparticles absorb this specific light wavelength, enhancing the image at this location. Hollow nanocages have proved to be the best contrast enhancing agent, as they offer the strongest absorption in the near infrared. Figure 2.9(Bv) shows a phantom with and without enhancement with gold 30 nm nanocages [15].

Magnetic resonance imaging (MRI) is a non-invasive technique which has been used internationally since the 1970s. The technique makes use of the massive number of protons in the body. In basic terms, when a magnetic field is applied to the body (in an MRI scanner) the net magnetisation of protons in the body aligns with the direction of the field. A radio frequency is then applied giving the protons energy. When the radio frequency is removed the protons then relax back, and this relaxation can be measured. Using gradient magnets of varying strength allows the targeting of specific areas for imaging of all three dimensions. Magnetic nanoparticles have great potential as MRI contrast agents, as the interaction of the particle's local magnetic field with the surrounding protons can significantly enhance contrast *in vivo*, by increasing relaxation time, resulting in the darkening of this area of the image, without the toxicity of gadolinium-based agents. The size of the MNPs greatly affects the MRI signal, offering tunable contrast modalities [21].

Magnetic particle imaging (MPI) is a new non-invasive imaging technique that images the magnetic signal from superparamagnetic iron-oxide nanoparticles (SPIONs). Utilising the inherent magnetic properties (from the electrons) results in far greater (×22 million) intensity compared to MRI (which utilises the nuclear

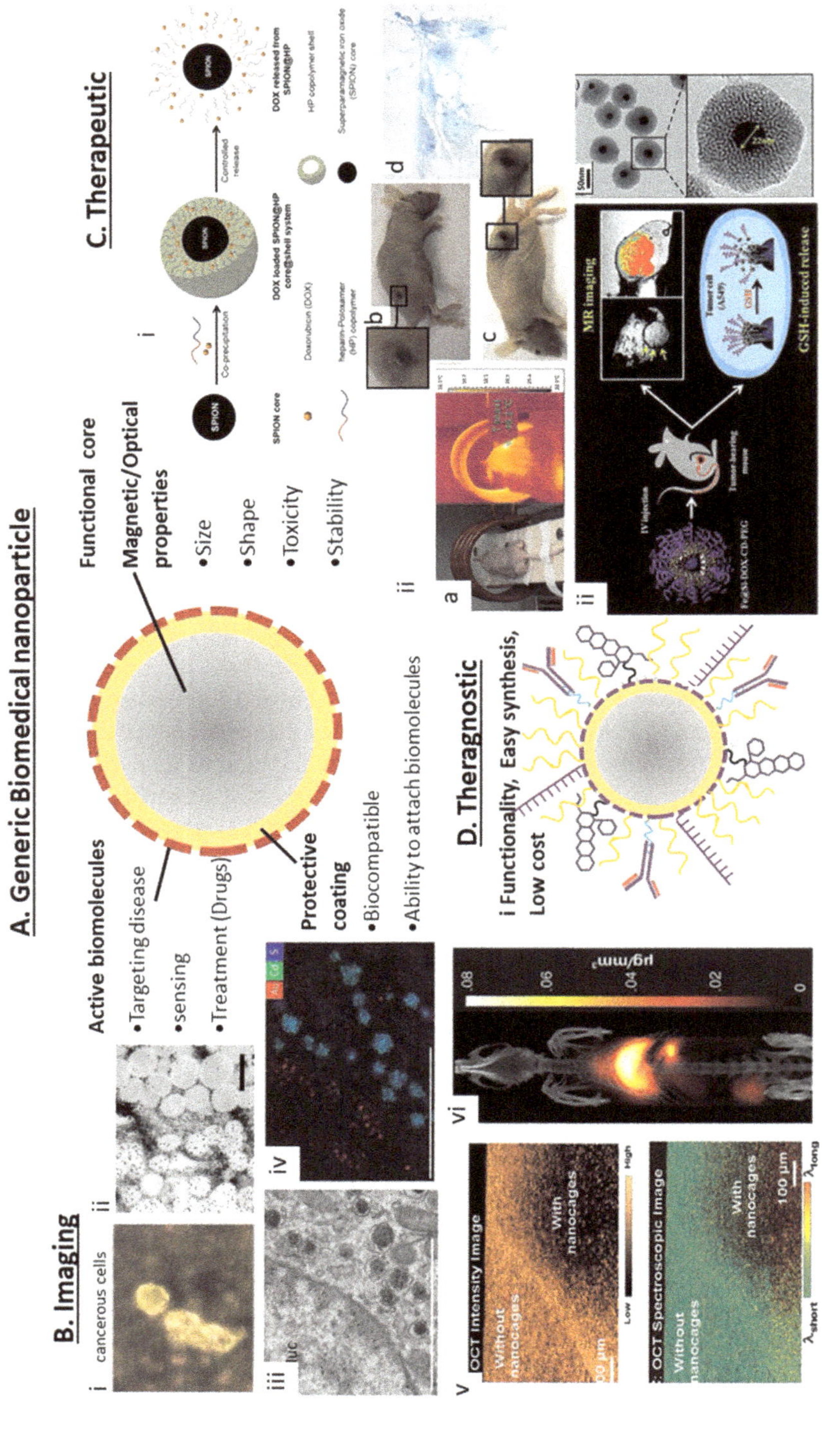
A. Generic Biomedical nanoparticle
Active biomolecules
•Targeting disease
•sensing
•Treatment (Drugs)
Protective coating
•Biocompatible
•Ability to attach biomolecules
Functional core
Magnetic/Optical properties
•Size
•Shape
•Toxicity
•Stability
B. Imaging
cancerous cells
OCT Intensity Image
Without nanocages
With nanocages
OCT Spectroscopic Image
100 μm
μg/mm²
C. Therapeutic
Co-precipitation
Controlled release
MR imaging
GSH-induced release
D. Theragnostic
i Functionality, Easy synthesis, Low cost

Figure 2.9. Summary of some inorganic nanoparticle uses in nanomedicine. (A) Schematic of and key attributes of a generic biomedical nanoparticle. (B) Some examples of how nanoparticles have been used for imaging. (i) Light scattering image of malignant epithelial cells labelled with antibody (anti-EGFR) coated gold nanoparticles (reproduced with permission from [12], copyright 2005 American Chemical Society). (ii) TEM image of human deep posterior lingual gland selectively labeled with immunogold nanoparticles. Scale bar 1 μm (reproduced from [13], copyright 2006 with permission of Elsevier). (iii) and (iv) Targeted labelling of the nucleus of a cell with gold nanoparticles (red) and insulin granules with quantum dots (green/blue) (scale bar 2 μm) (reproduced from [14]). (v) Spectroscopic optical coherence tomography (SOCT) images of a phantom (top intensity and bottom spectroscopic image) showing contrast due to nanocages (reproduced with permission from [15], copyright The Optical Society). (vi) MPI of a mouse with a lower right flank tumour 9 h after injection with 5 mg kg^{-1} of (superparamagnetic iron-oxide nanoparticle) SPION tracer. A CT scan of the skeleton is overlaid (reproduced with permission from [16], copyright 2017 American Chemical Society) and (C) shows some examples of how nanoparticles are being researched for therapeutic use. (i) Schematic showing the formation of the DOX-loaded SPION@HP core@shell system, and the sustained release at a steady rate of DOX by diffusing through the polymeric matrix as the HP shell degraded. Reproduced from [17]. (ii) An *in vivo* experiment of the treatment of a tumour with magnetosomes and a SPION control. (a) Shows the experimental setup in the alternating field coil, (b) shows the shrinkages of the tumour with treatment of magnetosome chains compared to (c) treatment with SPIONs, and (d) shows the cancer cell with magnetosome chains after 24 h (reproduced from [18] with permission from Taylor & Francis Ltd). (Di) Schematic of a multifunctional combined imaging and therapeutic nanoparticle termed a theragnostic nanoparticle. (ii) Schematic representing the dual function of Fe@Si–DOX–CD–PEG nanoparticles having both MRI contrast agents and for SMART targeted drug release. Fe@Si cores are magnetite nanoparticles coated with nanoporous silica shown in the TEM image on the right. DOX is the anticancer drug doxorubicin located in the silica pores. CD is cyclodextrin which can release the DOX when exposed to glutathione (GSH) which is present in high concentrations in cancer cells (adapted from [19]).

moment). SPIONs are targeted to a site in the body then imaged by applying a saturating magnetic field, then rostering an excitation field, which causes the SPIONs to flip, which can be detected at this location in 3D. It has been used to successfully locate and image cancer tumours (figure 2.9(Bvi)) [16] but its sensitivity means it could be used for a range of more subtle diagnostics too.

2.4.1.3 Therapeutics

Targeted drug delivery uses a nanoparticle (for this book, specifically an inorganic nanoparticle) as a vehicle to transport a drug to a specific disease cell sites. Targeted drug delivery is a game-changer for global pharmacology. Most cancer drugs' activity depends on their toxicity to cancer cells, but the obvious issue is that these are invariably toxic to healthy cells too. This means that previous therapies had terrible side effects, as the body had to cope with being poisoned overall in order to poison the cancer tumour. Targeted drug delivery means that the drugs are delivered to just the cancer cells, and thus are not administered to the whole body. This has two advantages: (1) the side effects of the treatment are dramatically reduced as the effect on healthy cells is now negligible; (2) the quantity of drug required is vastly reduced, as it is only used on the much smaller affected area. Targeting can be achieved passively or actively.

Passive targeting takes advantage of the fact that tumours have more 'leaky' vasculature and faulty drainage, so have enhanced permeability and retention (EPR). This means that tumours are susceptible to nanoparticles of between 12–500 nm accumulating preferentially within tumours, without any other 'active targeting'. However, accumulation quantities can be significantly increased by active targeting also. Active targeting can take the form of specific antibodies, as discussed above, which will target just the specific cancer cells. There are also other physiochemical properties that differentiate cancer cells from healthy cells which can be taken advantage of, such as pH. Cancer cells are more acidic, and as such a pH-responsive nanoparticle coating can be designed to release the drug only at the pH of the cancer cell, targeting the drug release.

Magnetic nanoparticles can be further exploited for targeting by utilising a magnetic field gradient to move the particles remotely *in vivo*. Magnetite or maghemite SPIONs have been most widely used due to their biocompatibility and easy availability for iron precursors. A permanent magnetic field near the tumour can enable the field gradient to concentrate and retain the drug carrying magnetic nanoparticles at the tumour site. However, depending where the tumour is, this will be more difficult when in deep tissue or internal organs. In these cases the field gradient in an MRI instrument can be used to concentrate the magnetic particles in all three dimensions at any location in the body [22].

The most studied SPION–drug conjugate is those of the intercalating chemotherapy drug doxorubicin (DOX). Doxorubicin is a well know and characterised potent chemotherapy agent, but it also has two further advantages for drug delivery design and research. The first is that it has two functional groups that can be modified to conjugate to the drug delivery vehicle, and the second is the conjugated aromatic nature of the drug makes it fluorescent, so it can be easily tracked and monitored. Researchers have designed a range of SPION–DOX nanomedicines ranging from the simple to the complex. For example, simply trapping SPIONs and unmodified DOX in polymer micelles/vesicles, or trapping unmodified DOX within the polymer coating on a SPION. Such a polymer coating can be specifically designed to change conformation or degrade to release the DOX on under the influence of a specific stimulus, or gradually over time for controlled slow release (figure 2.9(ci)) [17].

Examples of more complex systems include attachment of the DOX to the SPION through either iron coordination chemistry [23] or by conjugation to a responsive synthetic or biological polymer such as a tumour-targeting aptamer [24]. More complex systems include multiple nanoparticles such as SPION with quantum dots, gold nanoparticles and nanoporous silica for imaging and drug entrapment respectively, forming theragnostics, described in section 2.4.1.4. The details above have provided a brief sample of the extensive research being performed, to give an idea of the developments of drug delivery using nanoparticles. This field is developing all the time, and as such the information above will date rapidly. For more in-depth reviews [11, 25] of this subject, please see recent reviews and up-to-date literature.

Magnetic hyperthermia therapies are emerging non-invasive therapies, currently undergoing clinical trials, which work by causing magnetically-induced heating when magnetic nanoparticles are subjected to an alternating magnetic field (at frequencies in the range of 100–150 kHz). This heating acts to ablate and kill the diseased tissue at a higher localised temperature (above 45 °C) while lower temperatures between 40 °C–44 °C will sensitise the cell making them more susceptible to chemotherapy drugs. Lower temperatures still (39 °C–42 °C) could also be used to trigger a heat-activated drug or release a drug conjugated through a heat-activated linker [26, 27].

The heating is induced by one of two mechanisms. The alternating field can either switch the individual spins of larger single-domain MNPs, or physically flip the whole particle for smaller superparamagnetic MNPs. This switching/flipping induces heating as thermal energy which is released as the spins/particles attempt to resist the alternating field. This heat dissipates into the surrounding tissue, damaging or killing the diseased tissue or activating smart drug treatments. The first mechanism is called ferromagnetic hyperthermia, and describes the thermal energy generated by the switching dipole moments in the crystal lattice by Néel rotations. In effect the heating can be described by the area of the magnetic hysteresis loop (see equation (2.7)) so is increased with the increased magnetic saturation and coercivity of the MNPs. The second mechanism is called magnetic fluid hyperthermia, and is described by the Brownian rotation when the aligning of the magnetic moment causes the whole particle to rotate in superparamagnetic MNPs, causing shear stress as the particle flips resulting in delivery of thermal energy to the solution, which can be described by equation (2.8).

$$P_{\mathrm{FM}} = \mu_0 f \oint H \, \mathrm{d}M \tag{2.7}$$

$$P_{\mathrm{SPM}} = \mu_0 \pi f \chi'' H^2 \tag{2.8}$$

where P is the power of the heat generated, μ_0 is the permeability of free space = 2.566 3706 × 10^{-6} m kg s^{-2} A^{-2}, f is the frequency and H is the magnitude of the alternative magnetic field. The integral of $H\,\mathrm{d}M$ is the area of the hysteresis curve, while χ'' is the out-of-phase component of the magnetic susceptibility of the SPM MNP (effectively how much M lags behind H at that frequency). Within each regime the heating is largely dependent on the magnetic saturation/susceptibility and field frequency. While there is coercivity dependence in single domain ferromagnetic hyperthermia, the effect of the magnitude of the field is squared in magnetic fluid hyperthermia, so can lead to much greater heating. Furthermore, it should be noted that both mechanisms can operate in the same sample where the particles are small enough to rotate and have some coercivity, giving them both Néel and Brownian relaxation [27].

The heating power of a particle is determined by calculation of its specific absorption rate (equation (1.1)), which calculates the heating power (P) of the particles based on the mass of the MNPs in the sample (m_{mnp}). Each SARs value is for a specific field and frequency, so the intrinsic loss parameter (ILP) is the term normalised for field and frequency, both of which are shown in equations (2.9) and (2.10) respectively:

$$\mathrm{SAR} = \frac{\Delta T}{\Delta t}\frac{c}{M_{\mathrm{Fe}}} \tag{2.9}$$

$$\mathrm{ILP} = \frac{\mathrm{SAR}}{H^2 f}. \tag{2.10}$$

As with MRI and MPI the potential for MNPs in nanomedicine is far reaching, but particle size along with field and frequency must be considered, as all these factors will affect the heating power. For example, at the same frequency of 100 kHz, 8 nm iron-oxide MNPs gave the best increase in temperature of 9.3 °C in lower fields (9.6 kA m^{-1}), whereas at higher fields (23.9 kA m^{-1}) the 8 nm particles showed an improved increase in temperature of 25 °C, but MNPs of 24 nm in diameter proved more effective with increases in temperature of 55 °C. This is probably due to the combined effect of both heating mechanisms for these larger particles [28].

There are numerous examples in the literature of MNPs used to treat cancer cells from *in vitro* tissue culture to mouse model studies, both as independent therapeutics and in combination with drugs that are improved when cancer cells are heat sensitized [29]. SPM nanoparticles within polymersomes have been extensively probed for this purpose and ILP in the range of values between 1.6 and 5 nH m^2 kg^{-1} have been extensively shown to be functional for hyperthermic therapies [30]. Similarly, bacterial magnetosomes have shown excellent promise for magnetic hyperthermia with the highest recorded ILP of 23.4 nH m^2 kg^{-1} [31, 32], and with vastly increased SARs values compared to comparatively sized synthetic magnetite nanoparticles at lower fields [33]. It is not clear why these biomineralised MNP have such enhanced properties, but is has been speculated that it could be due to the monodispersity or the slightly reduced nature of the magnetite core. Magnetosomes have thus been investigated in mouse models *in vivo* and it has been found that magnetic hyperthermia using chains of magnetosomes has a much greater impact for tumour shrinkage than SPION control particles (figure 2.8(Dii)) [18].

Again, the field moves fast, so recent reviews [29] and research papers in the literature are the best sources of information.

Photothermia is another type of localised hyperthermia that takes advantage of the energy a metallic nanoparticle receives from an electromagnetic wave (see section 2.3.1), where this absorbed photon energy is transferred to phonons in the lattice and dissipated as heat. A relatively new phenomenon, the potential of photothermal therapies has only been realised over the last 15 years. As such it has not yet been subjected to the same level of theoretical treatment as magnet hyperthermia. The electromagnetic wave is provided by a near-infra-red laser (typically an 808 nm laser powered at 0.3–5 W cm^{-2}), which crucially is more transparent to body tissue and can penetrate deeper into the tissue than optical light. While the science in this area is still at the research stage, several inorganic nanomaterials are proving to be ideal materials for photothermal therapies. The front runner is gold nanoparticles (see recent reviews [8, 34]), as well as carbon nanotubes (CNTs) and MNPs such as SPION and larger magnetite MNPs. Once again, bacterial magnetosomes have proved to be excellent candidates with increases of temperature of up to 50 °C (under 1 W cm^{-2} laser power) [31]. Meanwhile, synthetically produced magnetite MNPs coated in macrophage biological coating have proved to be excellent photothermal agents. The macrophage coating renders them invisible to the immune system, offering them longer circulation time and also targeting them to the cancer cells. Furthermore, the magnetic properties of the particles mean they can be magnetically targeted to the tumour *in vivo*, with photothermal therapy resulting in tumours five times smaller by weight compared to treatment with non-macrophage coated MNPs [10].

2.4.1.4 Theragnostics

Theragnostics is quite simply the combination of both an imaging/diagnostic modality with a treatment of the disease (figure 2.8(Di)). It stands to reason that once a nanoparticle has been designed to locate a disease and visualised it, it would seem sensible and economical to provide a drug-payload or therapeutic treatment on the same particle, which has already reached its target. There is some debate as to whether or not theragnostics falls between two stools, in the sense that designing a combined 'all singing, all dancing' nanoparticle may end up doing both jobs to an okay standard, whereas actually designing a particle for one specific purpose may achieve better results. However, as we have seen throughout this section on nanomedicine, many nanomaterials, as well as coating and drugs, are multifunctional, so many theragnostic treatments are thus widely reported in the literature. For example, MNPs can use magnetic force to target the MNP to the site of a tumour, can be visualised with MRI and MPI, and can be heated with either a magnetic field or photothermia. Gold can be visualised with SPR and heated with photothermia, while both can be decorated with drugs and targeting agents such as biomarkers and antibodies and coated with biocompatible/targeting/smart activating polymers/membranes. Thus, a therapy can quite often couple as a imaging tool/diagnostic without any extra complexity and thus biological or synthetic penalty, and this is where the real advantages lies. The number of theragnostics currently being researched is too vast to describe here, so we will examine one example with multiple features. Figure 2.8(Dii) shows Fe@Si–DOX–CD–PEG [19]. Here a SPION forms a magnetic centre which can be magnetically targeted to the tumour. It can also be used as an MRI contrast enhancer offering targeting and imaging capability. The SPION is then coated in nanoporous silica, in which DOX can be loaded into the pores acting as a drug trap. This particle is then coated with PEG to provide biocompatible and cancer responsive cyclodextrin, which blocks the pores until the nanoparticle reaches the cancer site. At the site the cyclodextrin is cleaved off releasing the DOX from the silica pores [19].

2.4.1.5 Nanosensors in healthcare and medicine

Nanomaterials are ideal sensors because of their very large surface area. Furthermore, because of their small size, they are comparable in scale to many biological targets/disease markers making them particularly useful in biological sensing. Finally, the specific nanoscale properties covered in section 2.2 (particularly optical SPR) are very dependent on subtle chances at the surface of the particles, yielding very sensitive sensors.

Within medical sensoring, the central dogma for the last 50 years has been the use of antibodies raised against a specific biological target to detect that target. Initially in the 1960s this was done with radio labelling of the antibodies and detecting radioactivity. While this is a very sensitive technique, it is not very accessible and has many safety drawbacks. The 1970s saw the development of more accessible immunosorbent techniques such as the enzyme-linked immunosorbent assay (ELISA), where detection is achieved through an antibody attached to an enzyme, which when coupled to its substrate produces a detectible response (usually a definitive colour change). There are many forms of this assay but all use an antibody raised against the target to ‘capture’ the target, and an antibody linked to an enzyme detection system. Different types of ELISA include direct ELISA, where the primary antibody is already functionalised with an enzyme for detection, and sandwich ELISA, which has more steps and uses a primary antibody immobilised on the surface to catch the target and further antibodies to specifically bind to the target. Non-specific antibodies to bind to these which contain the enzyme detection system are then added. One of the best-known examples of ELISA sensing is in healthcare, and is the pregnancy test. This is a sandwich ELISA colorimetric sensor for the pregnancy hormone human chorionic gonadotropin (hCG). The test uses capillary flow to move the analyte (urine) along the test strip and over the immobilised antibodies. Most nanosensors will use antibodies in the form of an immunosorbent assay to target biomarkers and specific proteins, as they are well established, very sensitive, accessible and accurate. However, in recent years there has been a move to use nanoparticles as the detection system instead of enzyme assays, due to potential interactions of the enzyme with the sample and the higher intensity and thus sensitivity of the SPR signal. In this system antibodies can be attached to gold nanoparticles using the strong sulphur–gold bond, by simply adding a cysteine amino acid residue to the antibody. Furthermore, with SPR, there are many more sophisticated improvements that can be made on this simple assay, as this property (outlined in section 2.3.1) is ideally sensitive for nanosensing and can be tuned extensively. Furthermore, much more information, such as spacial information, can be obtained from SPR sensors, which cannot be achieved from simple colorimetric assays.

Increasingly, the SPR light emitted by gold and silver nanoparticles is serving as very sensitive colorimetric detection methods for very small quantities of target. One such target is microRNA (MiRNA). Some of these short strands of RNA are markers for the early signs of some cancers, so would be very useful to detect. However, they are present in very low concentrations. Recent work has used a DNA sandwich assay to detect the specific MiRNA [35]. Here, folded capture DNA containing the complementary sequence to the target MiRNA is immobilised on a gold surface (through a sulphur bond) and unfolds to bind the MiRNA specifically. This subtle change to the surface can be detected by a shift in the SPR signal [35]. Then a gold nanoparticle can be added with more single stranded DNA attached, designed to bind further up the capture DNA only if the MiRNA has bound, and placing the gold nanoparticle in close proximity to the gold surface, greatly enhancing the SPR refractive index signal and changing the resonance angle. This is enhanced further by adding more DNA to create a DNA supersandwich (figure 2.10(A)). This allows detection of MiRNA at concentrations as low as 8×10^{-15} Molar [35].

When two gold nanoparticles are brought close together, the effect of proximity couples their plasmons, red shifting the SPR peak to larger wavelengths as two gold nanoparticles come closer together and blue shift to smaller wavelengths as they move apart. This phenomenon can be used to sense and measure tiny distances and form a time resolved molecular ruler (figure 2.10(B)) [36]! Here the DNA is attached to one nanoparticle, thus a sulphur linker, while the other end is functionalised with biotin which allows it to bind to the streptavidin coated nanoparticle. A shift from green to orange particles is immediately seen on this coupling. The molecular ruler could be used to measure the effect of hybridising the single-stranded DNA (ssDNA) to double-stranded DNA (dsDNA) (figure 2.10(B)) [36].

2.4.2 Nanodevice technologies

2.4.2.1 Lab-on-a-chip diagnostics

There is much overlap between the last section on nanosensors, and this section on nanodevices. Indeed, much of the research in the area of nanodevices is centred around medical sensing and the future goal of completely personalised and rapid multiple diagnostics healthcare built into one small chip, which can then be easily processed in an simple hand-held electronic reader (figure 2.11(A)). Here a drop of blood could be distributed via microfluidics to multiple nanosensors to detect a

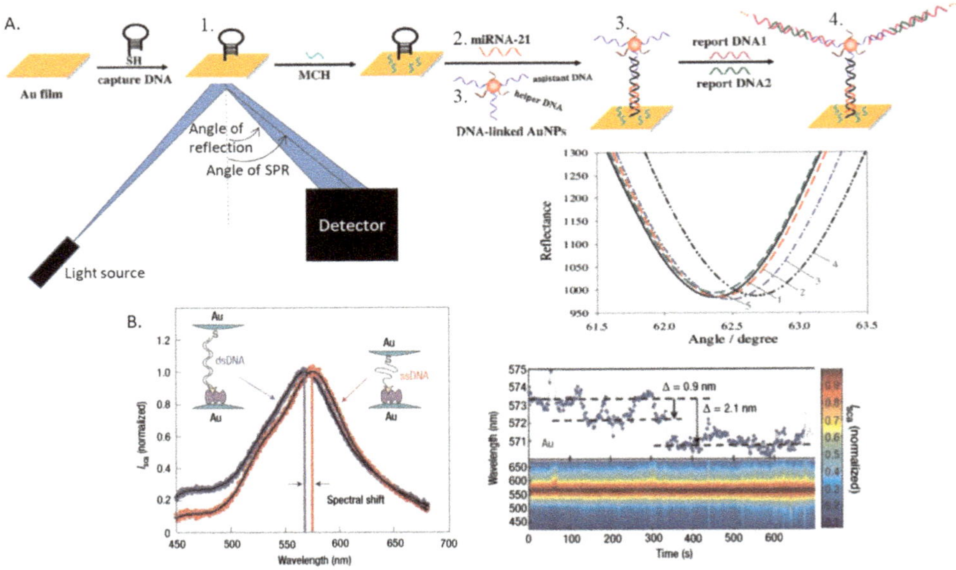

Figure 2.10. Two examples of ultra-sensitive nanosensors using SPR from gold nanoparticles. (A) A DNA supersandwich assay for the detection of very low concentrations of MiRNA. The numbering on the experimental schematic (top) corresponds to the numbering on the data (below right). A schematic showing how the data is collected from the gold surface is shown below left (adapted from images from [35], copyright 2016, with permission from Elsevier). (B) A molecular ruler measuring the extended distance of hybridising flexible ssDNA to more rigid dsDNA. (i) Example spectral shift between a gold particle pair connected with ssDNA (red) and dsDNA (blue). (ii) Spectral position as a function of time after addition of complementary DNA. Discrete states indicated by horizontal dashed lines (reproduced from [36] with permission of Springer).

complete set of diseases and wellbeing factors with unparalleled sensitivity and speed. This idea of bringing multiple nanosensors and microfluidics together on one surface ‘chip’ has been termed lab-on-a-chip and has coalesced into a dedicated research field driven by this goal, with many excellent reviews on the subject [37, 40].

2.4.2.2 Data storage technologies

In this information age, we see the collection, processing and storage of data and information becoming more important in our daily lives, and also see that the remit for electronic data storage devices goes beyond utility only in a PC. Increasingly consumer products are becoming more user-friendly, intelligent or ‘smart’, and require the ability to access and store information. The invention of the magnetic hard disk drive (HDD) in 1956 by IBM was the birth of modern information storage as we know it, and magnetic HDDs are still found within most modern PCs (figure 2.11(Bi)). They remain the most commercially viable choice for high volume data storage, something that may become more important if the current shift towards cloud computing continues.

A magnetic HDD uses the same principle of magnetic tape recording, where an electronic signal writes information in a magnetic pattern on the tape. For a

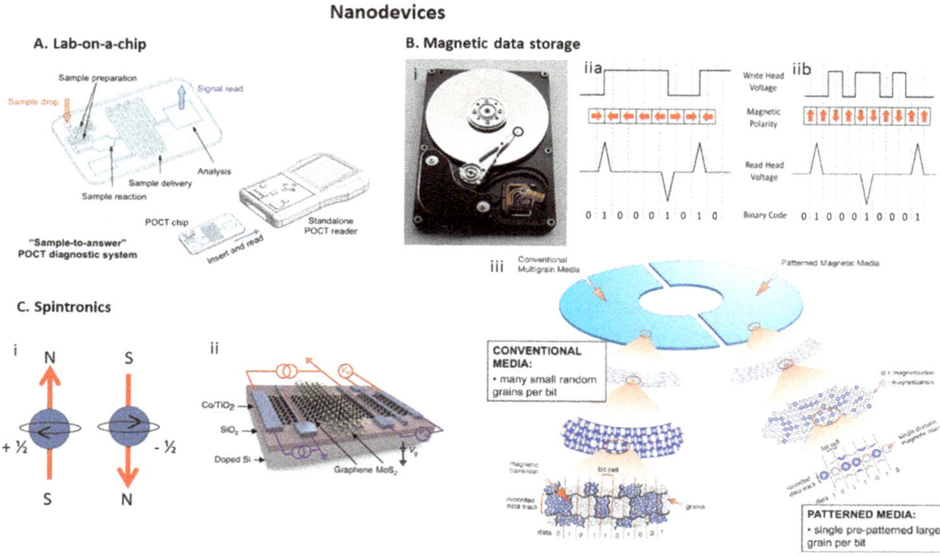

Figure 2.11. (A) Diagram of a point-of-care test chip and reader to offer immediate diagnostics (reproduced from [37], copyright 2015 with permission from Elsevier). (B(i)) Image of a hard disc drive (HDD), (ii) diagram describing how a magnetic patter can be read as binary code for (a) transverse magnetic recording and (b) perpendicular magnetic recording (reproduced with permission from Scott Bird). (iii) Schematic of conventional granular magnetic recording media (left) compared to high-density patterned media (right) (reproduced with permission from [38], copyright 2008 IEEE). (C(i)) Schematic to show the central premise of spintronics. The electronic spin can be manipulated with a magnetic field. (ii) A 2D spin field-effect switch of MoS_2 on graphene (reproduced with permission from Macmillan Publishers Ltd [39], copyright 2016).

magnetic HDD, this information is recorded in binary from 'bits'. Simply, the two binary components synonymous to either of two magnetic spins align in the same direction (0) as the neighbour, or the opposing direction (1) in longitudinal magnetic recording (figure 2.11(Biia)), or align anti-parallel (0) or parallel (1) to their neighbour in perpendicular magnetic recording (figure 2.11(Biib)). The other key difference between magic tape recording and a magnetic HDD is that an HDD can be randomly accessed rather than being read and written serially. This is possible due to the nature of the magnetic medium being a granular thin film on a spinning disc which can be accessed anywhere by a flying read–write head (figure 2.11(Biii)). The data storage capacity and economy of HDDs have vastly increased over the years: 2011 saw storage capacities of 500 Gb in^{-2} at a cost of \$0.2 per GB compared to 2 kb in^{-2} in 1956 for the first 305 Random Access Method of Accounting and Control (RAMAC), which weighed over a ton and cost \$3200 per month (equivalent to approximately \$160000 today) to lease. The revolutionary nature of computing and electronic data processing/storage was not imagined 60 years ago, when Thomas Watson (then chairman and CEO of IBM) famously said 'I think there is a world market for maybe five computers'. While the development over time has been phenomenal, there is still ever-increasing demand for miniaturisation and higher density data storage capacity. However, reducing the size of each bit is reaching its physical limit. It is likely that conventional granular recording mediums will soon reach storage capacities of approximately 1 Tbit in^{-2}, with bit sizes of approximately 625 nm^2 formed by aligning the polarities of multiple grains. At very small sizes the thermal energy is greater than the magnetic stabilisation energy with the onset of superparamagnetism, meaning the magnetic orientation and thus the data is lost at room temperature.

Two key new approaches to magnetic data storage are currently in development, with the aim of continuing the trend of increasing storage densities and capacities. These are: energy assisted recording and lithography patterned media (figure 2.11(Biii)). Energy assisted recording applies energy in the form of heat (heat-assisted magnetic recording) or microwaves (microwave-assisted magnetic recording) to the granular recording media, allowing the write head to orient the magnetic polarity of a higher anisotropy medium. Lithographically patterned media replaces the featureless granular recording medium used in current magnetic HDDs with a nanopatterned magnetic medium [38]. In nanopatterned media the size of the bit is defined by the nanopattern and not the write head, with the non-magnetic spaces reducing the risk of magnetic coupling and disorientation between neighbours, overall increasing the areal density. These patterns can take the form of discrete tracks (discreet track media) or each bit being retained on a discreet nano-island or nanoparticle (bit patterned media (BPM)) (figure 2.11(Biii)) [41]. With the use of current head materials thermally stable magnetic islands or nanoparticles suitable for BPM could be as small as 8 nm in diameter; with the addition of spaces between the islands' bit sizes could be as small as 12 nm^2. As a result it has been predicted that storage densities could be pushed beyond 50 Tbit in^{-2}. However, the development of a cost effective and industrially scalable manufacturing technique for BPM mean that currently this technology is still very much in the development stage [41].

Although common in computer systems, magnetic hard disk drives (HDDs) are not the only technologies available, with optical and solid-state semiconductor drives now found in many products. Semiconductor memories are electronic data storage devices which are compact and can operate at very high speeds due to a lack of moving parts. As such these devices have become the primary internal memories within computers. Semiconductor memories are either termed 'volatile devices' which lose stored information without a power source, or 'non-volatile devices' which will retain stored information in the absence of power. The two most common semiconductor memories, which fall into the volatile class, are dynamic random access memory (DRAM) and static random access memory (SRAM) [37]. DRAM stores bits of information inside capacitors, with each capacitor encoding a bit of information. A 1 or 0 can be formed by the capacitor being either charged or discharged, but these capacitors will discharge if the charge is not continually refreshed. Hence information stored on these devices will be lost if power is lost also. The recent trend towards portable computers, smartphones, as well as portable electronics has fuelled the development of fast and compact non-volatile semiconductor data storage [42]. One of the most successful technologies in this category is flash memory, which can store data for approximately ten years without power. Flash stores data in arrays of transistors, which make up a memory cell. As with magnetic data storage, flash memory has seen massive improvements in storage densities and capacities by scaling the size of the transistors to smaller dimensions. Again, this process cannot continue indefinitely; with the physical limits being reached, new alternative technologies that utilise the quantum physical properties of nanomaterials are being explored [42].

2.4.2.3 Nanoelectronics/spintronics

As electronically powered technology has now become vital to our lives, we need more and more efficient methods of transferring electrical energy. Furthermore, the reduction of the size of electronic technology needs consideration. In macroscale electronics electrical energy is transferred by electrons in a metal wire moving through the lattice to form a current; charge difference can build up between two points which have a potential difference between them, which is the voltage. Electronic components have been developed throughout the 20th century, such as capacitors (collecting charge, then releasing it when a critical level is reached), diodes (semiconductors that act as a valve, allowing the electrons to flow easily in one direction but not the other), and transistors (semiconductors that are able to amplify or switch electronic flow), which have allowed our lives to be transformed with ever more complex and ingenious electronic circuity in common everyday devices. However, as size reduces, so too do the physical properties of the material. As wires get thinner, the resistance in the wire vastly increases. As we miniaturise, it may be time to consider whether the movement of the charged electron, i.e. controlling electrical devices by conveying electron charge flowing in a wire, is actually the best method in smaller devices.

Spintronics is a new field of physics studying how the spin of electronics could be a better way to convey and control electronic signal in smaller electronic devices. Apart from the obvious benefit of miniaturisation, this has many advantages. In spintronics, there are additional characteristics of spin, and how the electron spin couples with the atomic orbital (spin–orbit coupling), how electrons can tunnel, and ultimately how these respond to an external magnetic field (figure 2.10(Ci)). All of these can be tuned, offering multiple electronic manipulation methods at the atomic level [43]. As recently as 2010 and 2013, graphite-like structured 2D semiconductor MoS_2 was shown to have photoluminescent and transistor behaviour, respectively. Building monolayered layered structures of such semiconductors, insulated and ferromagnetic materials are paving the way for complex spintronic effects that could be used in the nanoelectronics devices of the future. A 2D layered structure of MoS_2 on graphene combines the strong spin–orbit coupling of MoS_2 with the enhanced spin transport properties of graphene to form a spin field-effect switch capable of both transporting and controlling spin current (figure 2.10(Cii)). Such switches could be used to improve search engines and pattern recognition circuitry capability in the future [39]. However, spintronics is already in technology now: read heads on magnetic hard drives already use the effect of electron tunnelling from ferromagnetic layers through an insulating layer in the material. Combining the set-up of semiconductor RAM devices with magnetic domains for the data storage (instead of electronic charge) and utilising the electronic tunnelling of spintronics has led to the development of magnetoresistive random-access memory (MRAM), which has been heralded to out-compete all other forms of data storage in the future (currently not for high density, but for speed and longevity of retention) [44]. A MRAM is made up of an array of magnetic tunnel junction that holds a bit of information. Each magnetic tunnel junction is composed of a complex layered structure of two ferromagnetically orientated thin films separated by an insulating layer, with the bottom ferromagnetic film 'pinned' in orientation by a further antiferromagnetic thin film below.

The key issue with the development of all the complex nanodevice technologies described above is fabrication. Producing the correct material, of the correct phase and orientation in the correct layered or nanoscale morphological structures at the nanoscale is very problematic. Adding to that the ever-increasing need to fabricate such intricate complex materials and systems in a reproducible and sustainable way, for them to be industrially viable for applications, presents a formidable challenge.

2.4.3 Consumer products

A consumer product inventory has listed in total 1829 products containing nanomaterials, using in total 47 types of nanomaterials (www.nanotechproject.org/cpi) [45]. These products are produced by over 600 companies, which are spread over more than 30 countries globally. Of the nanomaterials used, titania, silica, zinc oxide and silver are the most common both in terms of total mass, as well as the number of products they are found in. When considered on mass basis, titania, zinc oxide and silica are the top three most manufactured nanomaterials.

2.4.3.1 Titania and zinc oxide

Titania and zinc oxide form the active components of sunscreens due to their ability to specifically interact with UV light (absorb or reflect) but not visible light. As discussed in the optical properties section above, it is their size and specific chemistry that enables them to block UV rays (both UV-A and UV-B in the range of 320–400 nm and 290–320 nm, respectively). The interactions of these two materials arise from their band gaps, which allow them to absorb UV rays; the specific wavelengths of absorption depend on the actual band gap (which in turn depends on the chemistry and sizes) of the materials. Generally, titania is most effective in blocking the UV-B range, while ZnO blocks the UV-A range. Using nano-sized (<100 nm) titania or ZnO makes them transparent in visible light, as well as enabling control over their band gaps (which become dependent on the particle sizes, as seen in section 2.2).

2.4.3.2 Silica

Silica is another widely used nanomaterial in applications such as toothpastes, tyres and cosmetics. One of the key properties in its success is its inertness under most conditions of use. Silica is resistant under acidic conditions and most organic reagents—it can only be degraded by strong caustic solutions ($pH > 10$) or hydrofluoric acid. As silica can withstand high temperatures similar to glass, it provides excellent improvement in thermal properties when blended with polymers. This, coupled with its added mechanical strength, means silica is an ideal component in tyre and composite applications. Further, silica (in most forms) has been 'generally regarded as safe (GRAS)' by regulating bodies in the USA and Europe, and hence is widely used in food and cosmetics. In those applications, silica improves the rheological (flow) properties of food and cosmetic products. Finally, its specific porous structure and the ability to absorb water/moisture makes it highly useful in desiccant applications where even a small amount of moisture needs to be avoided (e.g. electronics packaging and table salt).

2.4.3.3 Silver nanoparticles

Silver nanoparticles are another major nanomaterial used in consumer products, mainly for antimicrobial protection. The use of particulate silver in healthcare for the treatment of wounds has been recorded as far back in time as Ancient Greece. Silver is a well known antimicrobial agent, being used in antibacterial, anti-fungal and antiseptic creams, dressings and medical equipment coatings. While silver is an inert metal to us, it is highly toxic to microbes by damaging enzymes in pathogen membranes. The advantage of nanoparticulate silver is that the increased surface area increases the relative activity. It is now common to find silver nanoparticles in simple cheap wound dressings such as sticking plasters. The antimicrobial effect of silver nanoparticles is used in a wide range of consumer products, such as coatings and clothing.

The nanomaterials outlined above can also be used in air and water filtration/ purification media for domestic use. They can treat air or water by removing pollutants via adsorption (e.g. silica), disinfecting (e.g. silver) or photocatalytic destruction (e.g. titania). They are also used in coatings for optical or mechanical property enhancement (e.g. controlling gloss or providing scratch resistance).

2.5 Nanomaterial biocompatibility and toxicity

Just as the special features of nanomaterials enable new functions and applications as discussed above, the same features can lead to undesirable toxic effects on biological systems and to the environment. Although nanomaterials can cause toxicities of varying nature and degree, results from studies indicate that these effects can be mitigated or controlled by developing a mechanistic understanding of the effects, and hence nanomaterial toxicity is unlikely to become a major problem [46].

Before diving into learning about how nanomaterials can cause toxicity, it is important to point out a key difference between large (not-nano) objects containing nanoscale active areas, and nano-objects where the active area is a discrete nanoparticle [47]. An example of the former is a mm-wide electronic chip with nanoscale junctions, where the nano-features do not lead to any added toxicity. An example of the latter case is nanoparticles used in cosmetics or as drug delivery vehicles where individual particles are in contact with the external environment and hence capable of invoking responses due to nanoscale effects. This section only covers such cases.

Biocompatibility refers to efficacy without any adverse effects, such as toxicity. Toxicity of nanomaterials to a life form is typically caused by the ability of nanomaterials to release free radicals or cause oxidative stress when in contact with tissues or cells. These effects are detailed in table 2.1 and they arise from a range of physicochemical properties of nanomaterials. These properties can be categorised in three forms as discussed below.

Chemical composition and functionality: The chemical composition of the core and the surface are important. Let us consider each of these in turn. Comparing only the chemical composition, materials made up of cadmium are highly toxic because it affects cell functions such as proliferation and death by damaging lipids, DNA, proteins and enzymes and potentially causing gene mutations [48]. On the other hand, materials made from silica are generally biocompatible, making it widely applicable in food and cosmetics (see exceptions below) because silicon has been viewed as an element essential to life, including bone and tissue formation in animals and a defence against biotic and abiotic stresses in plants and animals (e.g. toxicity from metals such as aluminium) [49].

Surface chemistry and functionality (in combination with size effects) are very important factors determining the toxicity/biocompatibility of nanomaterials

Table 2.1. Toxic effects of nanomaterials. Note that most of these are supported by limited experimental or clinical evidence. Table adapted from [46].

Effects	Outcomes
Reactive oxygen species generation	Protein, DNA and membrane injury, oxidative stress
Oxidative stress	Phase II enzyme induction, inflammation, mitochondrial perturbation
Mitochondrial perturbation	Inner membrane damage, permeability transition (PT) pore opening, energy failure, apoptosis, apo-necrosis, cytotoxicity
Inflammation	Tissue infiltration with inflammatory cells, fibrosis, granulomas, atherogenesis, acute phase protein expression (e.g. C-reactive protein)
Uptake by reticulo-endothelial system	Asymptomatic sequestration and storage in liver, spleen, lymph nodes, possible organ enlargement and dysfunction
Protein denaturation, degradation	Loss of enzyme activity, auto-antigenicity
Nuclear uptake	DNA damage, nucleoprotein clumping, autoantigens
Uptake in neuronal tissue	Brain and peripheral nervous system injury
Perturbation of phagocytic function, 'particle overload,' mediator release	Chronic inflammation, fibrosis, granulomas, interference in clearance of infectious agents
Endothelial dysfunction, effects on blood clotting	Atherogenesis, thrombosis, stroke, myocardial infarction
Generation of neoantigens, breakdown in immune tolerance	Autoimmunity, adjuvant effects
Altered cell cycle regulation	Proliferation, cell cycle arrest, senescence
DNA damage	Mutagenesis, metaplasia, carcinogenesis

because they determine the reactivity and the energy of the surface. It is the surface of nanomaterials that will interact with the external environment. Certain features, such as transition metals on the surface, can lead to excessive generation of reactive oxygen species (ROS), e.g. free radicals, which can overwhelm the natural antioxidant defences of a given biological system and then lead to severe responses and injuries. Some metals, due to their specific electronic structure and bonding properties, are also able to catalyse unwanted reactions as detailed in figure 2.12. It is therefore essential that the core inorganic nanomaterial is either not toxic, or is covered with a protective coating that will prevent it from degrading, or toxic ions leaching out into the body. For example, quantum dots with polymer coating have been shown to have improved chemical stability and are less toxic [50]. Similarly, the surface charge (measured as zeta potential) and hydrophobicity affect nanomaterial toxicity significantly. Figure 2.13 shows results from the screening of 130 nanoparticles and evaluates the toxic effects of the particle sizes (further discussed below), their zeta potential and the surface hydrophobicity. The study showed that positively charged small nanoparticles with low hydrophobicity can cause severe cytotoxicity, due to a combination of their strong interactions with various biological components as well as their inability to 'clear-out' of the system. Table 2.2 gives an extensive list for advanced readers of nanomaterials of varying chemical compositions and the toxicity caused by them. In most cases, potential strategies to mitigate the toxicity include some form of surface passivation, by either tethering biocompatible molecules on the surface, reducing the surface charge and/or modifying the surface chemistry/structure [51].

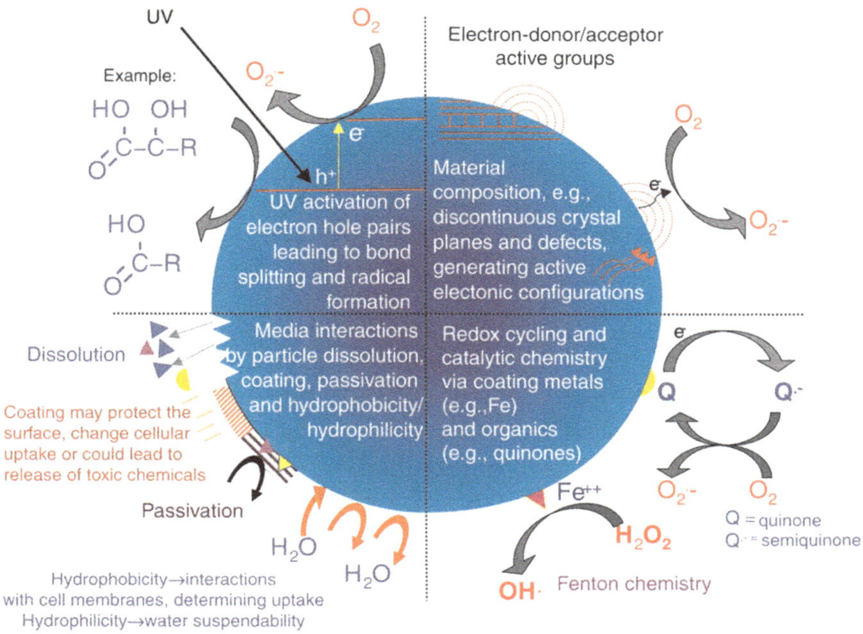

Figure 2.12. Pathways and mechanisms of nanomaterial–biological tissue interactions. Image from [46], reprinted with permission from AAAS.

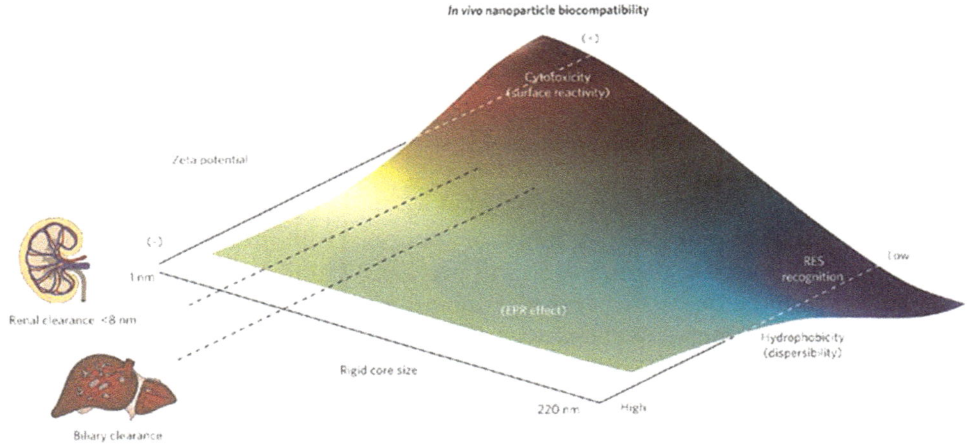

Figure 2.13. A 3D qualitative phase diagram showing biocompatibility trends as a function of the properties of nanoparticles using data from *in vivo* screening of around 130 nanoparticles. Red = likely toxic, blue = likely safe, and blue–green–yellow = intermediate levels. RES = reticuloendothelial system and EPR = enhanced permeation and retention. Image taken from [51], copyright 2009 with permission of Springer.

Physical features: Particles and materials in the nanoscale influence and interact at the cellular, subcellular and protein levels given the overlap of the length scales (figure 2.1). Examples include wrapping proteins or DNA on the nanoparticle surface, disrupting protein–protein assemblies by the insertion of nanoparticles, and disrupting lipid–bilayer structures in cell walls. Further, given the size, nanoparticles can travel through the body and penetrate various membranes, causing deposition or accumulation in certain areas, thereby triggering biological (e.g. immune) responses. A key physical feature of nanomaterials that strongly influence their toxicity is the total available surface for interactions with biological systems. As we have seen in figure 2.2, with the size decreasing, the surface area increases for a constant volume of particles. This allows significantly more opportunities for interactions with cellular and subcellular components, thus enhancing the toxic effects.

Some of the physical effects can be illustrated by discussing a classic case of silicates/silica. As discussed in the section above, silica in its amorphous phase is typically not toxic, even at nanoscales, and hence it has been 'generally regarded as safe (GRAS)' by regulating bodies in the USA and Europe for use in food and cosmetic applications. However, quartz, a crystalline form of silica, of any shape, causes severe lung inflammation, even at low exposures and at μm size [47]. The crystalline nature of the surface induces the formation of ROS and oxidative stress in biological systems. Asbestos, a naturally occurring fibrous silicate mineral, is well known for its toxicity, leading to fatality. It is the shape of the asbestos that causes its adverse effects and not the chemical composition. The nanoscale diameter of the fibres and microscale lengths mean that asbestos can find its way into lungs, but can get easily trapped. This peculiar shape makes some of the immune responses ineffective, leading to scarring (asbestosis).

Table 2.2. A list of chemically distinct nanomaterials with their cytotoxicity. Table adapted from [51].

Nanomaterial	Cytotoxicity mechanism
TiO_2	ROS production mediated by electron–hole pairs. Glutathione depletion and toxic oxidative stress as a result of photoactivity and redox properties. Nanoparticle-mediated cell membrane disruption lead to cell death and protein fibrillation
ZnO	ROS production. Dissolution and release of toxic cations. Lysosomal damage. Inflammation
Ag	Dissolution and Ag^+ release inhibits respiratory enzymes and adenosine triphosphate (ATP) production. ROS production. Disruption of membrane integrity and transport processes
Au	Disruption of protein conformation
CdSe	Dissolution and release of toxic Cd and Se ions
SiO_2	ROS production by surface defects in crystalline silica and impurities in silicates. Protein unfolding and membrane disruption
Fe_3O_4	ROS production and oxidative stress. Liberation of toxic Fe^{2+}. Disturbance of the electronic and/or ion transport activity in the cell membrane
CeO_2	Protein aggregation and fibrillation
SWCNT, MWCNT	Frustrated phagocytosis causes chronic tissue inflammation and DNA oxidative injury. Generation of ROS due to the metal impurities trapped inside carbon nanotubes (CNTs). Pro-inflammatory effects due to oxidant injury. Granulomatous inflammation due to hydrophobic CNT aggregation. Interstitial pulmonary fibrosis due to fibroblast-mediated collagen production
Fullerenes	ROS production (spontaneous or photoactivated). Hydrophobic surface increases aggregation but promotes intramembranous localization
Dendrimers	Membrane damage, thinning and leakage. Damage to the acidifying endosomal compartment by the proton sponge effect that allows entry into the cytosol
Al_2O_3	ROS production. Pro-inflammatory response
Cu/CuO	DNA damage and oxidative stress
MoO_3	Membrane disruption

Bioavailability: This includes the fate of nanomaterials in the body which depends on aspects such as the solubility/biodegradation of nanomaterials and the body's ability to clear them via various mechanisms (e.g. see figure 2.13). The exposure levels and the routes of exposure are also important factors affecting bioavailability. If certain particles, which can be toxic, are readily removed/excreted by the body then they offer little threat. Similarly, if nanoparticles are able to break down/dissolve *in vivo*, they can be safe. For example, in addition to the shape effect, the inability of asbestos to dissolve or break down poses a sustained stress/injury, while amorphous silica can dissolve in the body and can be easily removed [52]. On the other hand, dissolution of particles where the toxic core is protected by biocompatible coatings can induce toxic responses. In such cases, the anticipated fate is removal by the body before any leaching of the core happens.

2.6 Summary: key lessons from nanomaterials, nanoproperties and applications

Summary of content

- Nanomaterials are defined as having at least one dimension in the nanoscale range, be this nanosheets (one dimension), nanowires (two dimensions) or nanoparticles (three dimensions). It should be noted that nanoporous materials are also considered nanomaterials due to many shared properties.
- The breadth of types of nanomaterials, nanoproperties, and application for nanomaterials are too great to cover even superficially in one chapter. Thus this book only considers inorganic nanomaterials, and mainly focuses on nanoparticles.
- In this chapter we have focused on the optical and magnetic properties of mainly gold, silver, CdSe, and magnetic iron oxide and considered mainly nanomedical applications of these materials. Nanodevices and consumer products of a range of other nanomaterials are also considered briefly.

Key lessons

- Nanomaterials have a very high surface area-to-volume ratio. This means surface chemistry and physics dominates their properties.
- The large surface area is also useful for increasing the activity of surface reactions (i.e. heterogenous catalysis) and ideal for delivery of an active species attached to the surface (i.e. high loading of drugs, etc).
- Small size also lends itself to nanomedical applications as it warrants access to everywhere in the body and it also makes nanomaterials a comparable size to biological targets such as proteins.
- Nanomaterials lie at the boundary between atomic level quantum physics and bulk material Newtonian physics, and thus exhibit properties between the two.
- Surface plasmon resonance (SPR) is a surface effect of how an electromagnetic wave polarises and thus oscillates free electrons on metal surfaces. This absorbs and scatters light dependent on the polarizability of the electrons in that material, the size and shape of the particle and the surrounding media.
- SPR can be used to heat a particle where the energy absorbed is converted to phonons (lattice heat) and this can provide photothermal therapies when the particles are irradiated with a laser.
- SPR has proved very useful for increasing the sensitivity of nanomedical sensing and has also been used to create tools such as molecular rulers.
- Quantum confinement is a property of semiconductors which occurs when the particle becomes comparable to or smaller than its Bohr radius. At this size the electrons are confined in all directions which results in a bigger band gap (proportional to the radius of the particle) and strong fluorescence at this energy. These particles are quantum dots.

- Quantum dots have proved very useful for *in vitro* imaging. They are ideal as they have very intense fluorescence which does not fade, and the colour can be tuned with materials and particle size. Concerns over toxicity of several semiconductor materials such as CdSe quantum dots means they are slower to be used *in vivo*.
- At the nanoscale, magnetic particles are single domain and just before they become multidomain they have the highest magnetic coercivity for that material. This is because there is no energy wasted in creating the domain wall. As the size decreases they become superparamagnetic, which means the particles exhibit no coercivity.
- Iron oxide magnetic nanoparticles are ideal for nanomedical therapies. The magnetic properties can be used to magnetically target the particles and their payload to a site in the body using a magnetic field gradient. The magnetic nanoparticles can be heated by application of an alternating magnetic field. The heat is provided from the internal magnetic resistance to the flipping field for magnetic nanoparticles with coercivity, or the external friction of a superparamagnetic particle against their media as they flip. The heating can provide hyperthermic therapies. Magnetic nanoparticles can be used to enhance MRI imaging by providing a local magnetic field to increase relaxation times, and superparamagnetic iron oxide nanoparticles can be used in magnetic particle images which has increased intensity by millions of times compared to MRI. Magnetic nanoparticles can also be used *in vitro* for cell separation and sample concentration for biosensing.
- Nanotoxicity and biocompatibility must be considered when designing all types of nanomaterials, not just those for biomedical use.
- Nanotoxicity must be considered throughout the whole life cycle of the production of the material, the use of the material and the disposal of the material.
- The section on nanodevice technology demonstrates the real power sophisticated nanomaterials platforms have to revolutionise technology, from personalised point-of-care diagnostic healthcare chips, to ultra-high-density data storage and spintronic electronics of the future.
- A broader range of materials are discussed in consumer products to give a flavour of the breadth of uses for various nanomaterials.
- As one of the largest and expanding fields in nanoscience, nanomedicine was the focus of the applications section. In medicine and healthcare we are seeing incredible developments of smart and sophisticated multifunctional nanomaterials.
- Across the applications it is clear that as the design of our nanomaterials becomes more sophisticated, we are challenged to see if the synthesis of more intricate nanomaterials with more demanding specifications can keep up and deliver these materials precisely and consistently. This is ever more important for nanomaterials for *in vivo* medicines. These need to be precise and consistent, but also produced in a non-toxic way that ensures the nanomaterials are biocompatible and safe for medical use.

References

[1] Erbas-Cakmak S, Leigh D A, McTernan C T and Nussbaumer A L 2015 *Chem. Rev.* **115** 10081

[2] Eustis S and El-Sayed M A 2006 *Chem. Soc. Rev.* **35** 209

[3] Lal S, Link S and Halas N J 2007 *Nat. Photonics* **1** 641

[4] Anker J N, Paige Hall W, Lyandres O, Shah N C, Zhao J and Duyne R P V 2008 *Nat. Mater.* **7** 442

[5] Hu M, Chen J, Li Z-Y, Au L, Hartland G V, Li X, Marquez M and Xia Y 2006 *Chem. Soc. Rev.* **35** 1084

[6] Wang Y, Hu R, Lin G, Roy I and Yong K-T 2013 *ACS Appl. Mater. Interfaces* **5** 2786

[7] Blundell S 2001 *Magnetism in Condensed Matter* (Oxford: Oxford University Press)

[8] El-Sayed I H 2010 *Curr. Oncol. Rep.* **12** 121

[9] Piao J-G, Gao F, Li Y, Yu L, Liu D, Tan Z-B, Xiong Y, Yang L and You Y-Z 2018 *Nano Res.* **11** 3193

[10] Meng Q-F *et al* 2018 *Nanotechnology* **29** 134004

[11] Bobo D, Robinson K J, Islam J, Thurecht K J and Corrie S R 2016 *Pharm. Res.* **33** 2373

[12] El-Sayed I H, Huang X and El-Sayed M A 2005 *Nano Lett.* **5** 829

[13] Piludu M, Lantini M S, Cossu M, Piras M, Oppenheim F G, Helmerhorst E J, Siqueira W and Hand A R 2006 *Arch. Oral Biol.* **51** 967

[14] Scotuzzi M, Kuipers J, Wensveen D I, de Boer P, Hagen K W, Hoogenboom J P and Giepmans B N G 2017 *Sci. Rep.* **7** 45970

[15] Cang H, Sun T, Li Z-Y, Chen J, Wiley B J, Xia Y and Li X 2005 *Opt. Lett.* **30** 3048

[16] Yu E Y, Bishop M, Zheng B, Matthew Ferguson R, Khandhar A P, Kemp S J, Krishnan K M, Goodwill P W and Conolly S M 2017 *Nano Lett.* **17** 1648

[17] Thi T T H, Tran D-H N, Bach L G, Vu-Quang H, Nguyen D C, Park K D and Nguyen D H 2019 *Pharmaceutics* **11** 120

[18] Alphandéry E, Chebbi I, Guyot F and Durand-Dubief M 2013 *Int. J. Hyperth.* **29** 801

[19] Lee J, Kim H, Kim S, Lee H, Kim J, Kim N, Park H J, Choi E K, Lee J S and Kim C 2012 *J. Mater. Chem.* **22** 14061

[20] Goel S, England C G, Chen F and Cai W 2017 *Adv. Drug Deliv. Rev.* **113** 157

[21] Jun Y-W *et al* 2005 *J. Am. Chem. Soc.* **127** 5732

[22] Muthana M *et al* 2015 *Nat. Commun.* **6** 8009

[23] Gautier J, Allard-Vannier E, Burlaud-Gaillard J, Domenech J and Chourpa I 2015 *J. Biomed. Nanotechnol.* **11** 177

[24] Yu M K, Kim D, Lee I-H, So J-S, Jeong Y Y and Jon S 2011 *Small* **7** 2241

[25] Ulbrich K, Holá K, Šubr V, Bakandritsos A, Tuček J and Zbořil R 2016 *Chem. Rev.* **116** 5338

Mou X, Ali Z, Li S and He N 2015 Applications of magnetic nanoparticles in targeted drug delivery system *J. Nanosci. Nanotechnol.* **15** 54

[26] Dobson J 2006 *Nanomedicine* **1** 31

[27] Pankhurst Q A, Connolly J, Jones S K and Dobson J 2003 *J. Phys. D - Appl. Phys.* **36** R167

[28] Li Z, Kawashita M, Araki N, Mistumori M and Hiraoka M 2011 *Bioceram. Dev. Appl.* **1** 1

[29] Das P, Colombo M and Prosperi D 2019 *Colloids Surf.* B **174** 42

[30] Kallumadil M, Tada M, Nakagawa T, Abe M, Southern P and Pankhurst Q A 2009 *J. Magn. Magn. Mater.* **321** 1509

[31] Sangnier A P, Preveral S, Curcio A, Silva A K A, Lefèvre C T, Pignol D, Lalatonne Y and Wilhelm C 2018 *J. Control. Release* **279** 271

[32] Hergt R, Dutz S, Müller R and Zeisberger M 2006 *J. Phys. Condens. Matter.* **18** S2919

[33] Hergt R, Hiergeist R, Zeisberger M, Schuler D, Heyen U, Hilger I and Kaiser W A 2005 *J. Magn. Magn. Mater.* **293** 80

[34] Vines J B, Yoon J-H, Ryu N-E, Lim D-J and Park H 2019 *Front. Chem.* **7** 167

[35] Wang Q, Liu R, Yang X, Wang K, Zhu J, He L and Li Q 2016 *Sens. Actuators* B **223** 613

[36] Sönnichsen C, Reinhard B M, Liphardt J and Alivisatos A P 2005 *Nat. Biotechnol.* **23** 741

[37] Jung W, Han J, Choi J-W and Ahn C H 2015 *Microelectron. Eng.* **132** 46

[38] Dobisz E A, Bandic Z Z, Wu T and Albrecht T 2008 *Proc. IEEE* **96** 1836

[39] Yan W, Txoperena O, Llopis R, Dery H, Hueso L E and Casanova F 2016 *Nat. Commun.* **7** 13372

[40] Lafleur J P, Jönsson A, Senkbeil S and Kutter J P 2016 *Biosens. Bioelectron.* **76** 213
Samiei E, Tabrizian M and Hoorfar M 2016 *Lab. Chip* **16** 2376
Syedmoradi L, Daneshpour M, Alvandipour M, Gomez F A, Hajghassem H and Omidfar K 2017 *Biosens. Bioelectron.* **87** 373

[41] Thomson T and Terris B D 2011 *Developments in Data Storage* (New York: Wiley), p 256

[42] Chong T C and Piramanayagam S N (ed) 2011 *Developments in Data Storage: Materials Perspective* (Piscataway, NJ: Wiley-IEEE Press), p 352

[43] Premasiri K and Gao X P A 2019 *J. Phys. Condens. Matter.* **31** 193001

[44] Bhatti S, Sbiaa R, Hirohata A, Ohno H, Fukami S and Piramanayagam S N 2017 *Mater. Today* **20** 530

[45] Vance M E, Kuiken T, Vejerano E P, McGinnis S P, Hochella M F, Rejeski D and Hull M S 2015 *Belstein J. Nanotechnol.* **6** 1769

[46] Nel A, Xia T, Mädler L and Li N 2006 *Science* **311** 622

[47] The Royal Society and The Royal Academy of Engineering 2004 Nanoscience and nanotechnologies: opportunities and uncertainties, Joint report https://royalsociety.org/~/media/Royal_Society_Content/policy/publications/2004/9693.pdf

[48] Stohs S J and Bagchi D 1995 *Free Radic. Biol. Med.* **18** 321

[49] Farooq M A and Dietz K J 2015 *Front. Plant Sci.* **6** 14

[50] Hoshino A, Fujioka K, Oku T, Suga M, Sasaki Y F, Ohta T, Yasuhara M, Suzuki K and Yamamoto K 2004 *Nano Lett.* **4** 2163

[51] Nel A E, Mädler L, Velegol D, Xia T, Hoek E M V, Somasundaran P, Klaessig F, Castranova V and Thompson M 2009 *Nat. Mater.* **8** 543

[52] He Q, Shi J, Zhu M, Chen Y and Chen F 2010 *Micropor. Mesopor. Mat.* **131** 314

Chapter 3

Characterisation of nanomaterials

3.1 Introduction

There are numerous methods for characterising materials, based on their chemical composition and structure as well as their chemical and physical properties and functions. Chemical composition can be interrogated by mass spectrometry and inductively coupled plasma spectroscopy (ICP), while structure can be determined by powder x-ray diffraction (XRD), nuclear magnetic resonance (NMR), and crystallography. Chemical properties can be found by infra-red (IR) spectroscopy, UV–vis, and physical properties assessed by porosimetry and stress–strain measurements. The key to characterisation and choosing the right technique is to ensure the characterisation technique is answering the characterisation question that you are asking. Critically for nanomaterials, all the methods above give information about the whole material, and no information specifically at the nanoscale. Therefore, the birth of nanoscience and engineering has developed hand in hand with nanoscale characterisation techniques (particularly microscopy), as one cannot progress without the development of the other. Namely, until one is able to 'see' at the nanoscale we cannot begin to develop experiments to utilise and understand nanomaterials. Figure 3.1 shows schematically some commonly used characterisation techniques and the features of materials they can probe. Typical examples of materials/systems in each length scale are also shown.

Thus, the development of electron microscopy has been essential to initiating nanoscience. In this section we will briefly describe characterisation specifically for nanoscale features, such as electron and force microscopy, nanoscale specific spectroscopy and scattering techniques. It should however be noted that these techniques can and should be complemented by other characterisation techniques (such as those listed at the beginning of this section) to give a full understanding of the material and its properties.

doi:10.1088/978-0-7503-1221-9ch3

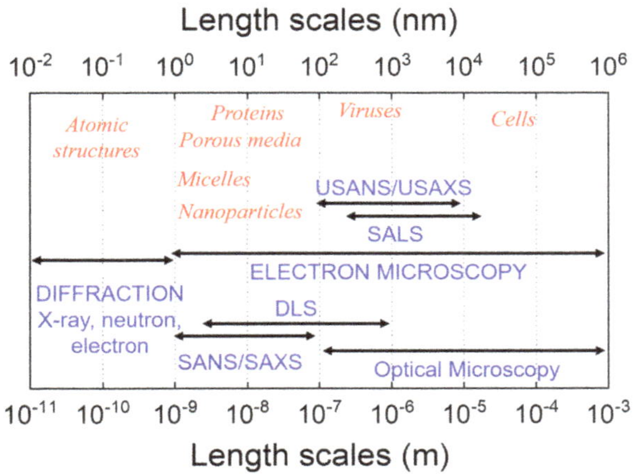

Figure 3.1. Schematic showing the different characterisation techniques and the length scales/materials they probe.

3.2 Microscopy

3.2.1 Optical microscopy

Most people equate microscopy with optical microscopy, where a sample is viewed on an illuminated stage (with either natural or artificial light) using a series of lenses to bend the light and magnify the sample's features. While light microscopy is an excellent characterisation technique, it has a physical limit to its resolution, which is the wavelength of light. The diffraction limit ($d = \lambda/2$) means that when measured at wide angles, light cannot be used to observe something smaller than half its own wavelength ($\approx$250 nm for green light) [1]. In this respect optical microscopy cannot be considered a characterisation method for nanoscale materials. However, over the last 30 years there has been a concerted effort to surmount this physical limit, and to develop super-resolution light microscopy. There have been several approaches, and some of these have been combined. One approach exploits the evanescent wave produced by confining excitation laser light through a hole smaller than its wavelength. This can be used to visualise a sample with the resolution dependent on the size of the aperture (hole), not the wavelength, if held very close to (a few nm away) from the sample surface. This technique, near-field scanning optical microscopy (NSOM or SNOM), can give resolutions down to 20 nm [2]. However, the close distance to the sample surface and relatively large probe means there are limitations to the sort of samples that can be imaged, furthermore there is a very shallow depth of field, and the method can have very long scan times.

Other methods of super resolution light microscopy use fluorescence microscopy. By performing fluorescence microscopy, but only on optically distant fluorescent probes (so they can be resolved), a resolution of 200 nm can be achieved in spectral precision distance microscopy (SPDM) [3]. Similarly, the signal generated from a dye bound to DNA can be used in binding activated localization microscopy (BALM) to image DNA in the nucleus, while a fluorescent probe in cold conditions to slow down the fluorophores chemistry can be used to obtain molecular level resolution (if compared/combined with structural information from election microscopy or crystal structure data) in cryogenic optical localisation in 3D (COLD) [4]. Further to this, several techniques have been developed that visualise fluorescent probes at different times using photoswitchable fluorophores. Thus, only distant fluorophores are resolved at each time point, allowing a high-resolution image to be built up over time. Techniques include stochastic optical reconstruction microscopy (STORM), and (fluorescence) photo active localization microscopy (PALM and FPALM), which along with sophisticated computational deconvolution now have resolution limits down to around 30 nm.

Advantages of optical microscopy are that it can use liquid ambient environments, can be dynamic and has a huge size range. However, samples usually need to be tagged with specific probes for this method and the resolution, while improving, is still not adequate for imaging many native nanoscale features.

3.2.2 Electron microscopy

Another way to overcome the diffraction limitation of light is to not use light as the vehicle for imaging. Election microscopy uses accelerated electrons to visualise the sample, similar to the way photons (light) are used in optical microscopy. Accelerated electrons have a wavelength much smaller than that of visible light (2.5 pm (at 200 keV)), resulting in electron imaging resolution being practically unrestricted by the wavelength. Additionally, electrons interact more strongly with matter than photons, so a stronger signal can be obtained from samples with reduced dimensions (figure 3.2). The source of electrons for imaging is from an elemental solid target such as a a tungsten filament, which releases electrons on heating (thermionic emission) or by quantum tunnelling when exposed to a strong electric field (field emission). However, it is becoming more common to use a Schottky-style field emission gun as the source of electrons in modern transmission electron microscopes. The electron beam is then focused down the column onto the sample using a series of lenses. These are analogues to optical glass lenses, but are in this case electromagnetic lenses, using the electromagnetic field to direct and focus the charged electrons (see figure 3.3 in section 3.3.2.4).

Electron microscopy has the added advantage that the electrons can be used to provide structural and elemental information about the sample *in situ* at the same resolution of the electron beam. The electron beam can be diffracted by the sample to give a crystal diffraction pattern that can be assigned to a crystal structure. Furthermore, the electron beam can ionise the sample by 'knocking out' inner shell electrons of atoms in the sample. Electrons in higher energy levels then drop into the vacant lower energy states, potentially emitting x-rays as part of the de-excitation process, critically at the energy (wavelength) of that atomic energy level transition which is element specific. Energy dispersive x-ray analysis (EDAX) measures the energy of x-rays emitted from a sample impacted by an electron beam, to determine the atomic species that are present. As a result, the x-rays that are produced have characteristic energies and wavelengths depending on the atomic species that are

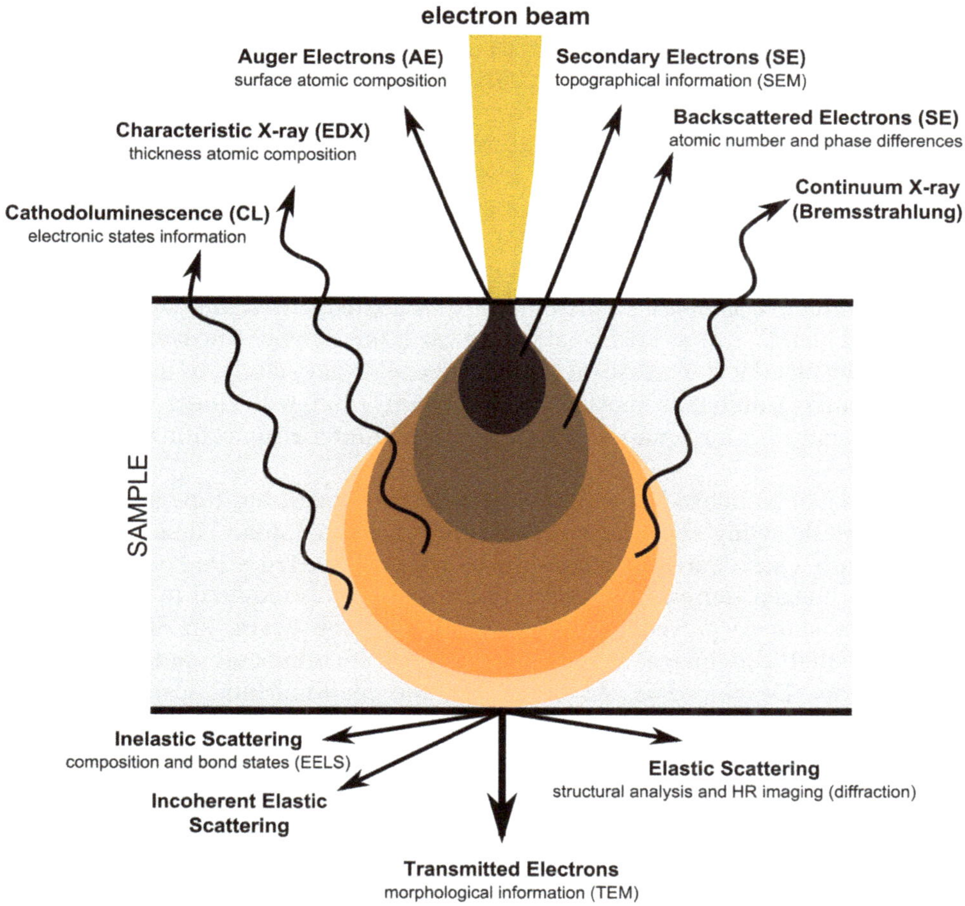

Figure 3.2. Schematic of high energy electron interactions with a sample during electron microscopy, showing the depth from which such waves can penetrate. Wikimedia Commons images © Claudionico under Creative Commons Licence.

present (figure 3.2). Electron microscopy is thus a very powerful tool with the ability to give topological, structural and elemental information at close to atomic level resolution. However, one disadvantage is that all optical paths in an electron microscope have to be held under high vacuum.

3.2.3 Scanning electron microscopy

Scanning electron microscopy (SEM) has become one of the most important and widely used techniques for the characterisation of surfaces containing nanoscale features or nanomaterials on a surface. SEM gives an image of the topology of the surface (figure 3.2) (i.e. it does not penetrate the sample) as the beam is scanned across it in a raster scan. Primary electrons are accelerated (approx. 0.1–40 keV) and focused onto a sample to interact with a volume that is dependent on the energy of the primary beam, and the atomic number of the sample. After electron–surface interactions, secondary electrons escape from the sample with a kinetic energy <50 eV. Secondary electrons are most likely ionised electrons from atoms close to the sample surface, or primary electrons that have lost almost all their energy through scattering. Backscattered electrons are also produced as a result of primary electrons undergoing large deflections, leaving the surface with only small changes to their initial kinetic energy. Furthermore, atoms within the sample can undergo inner shell ionisation which can be utilised to collect (by EDAX) chemical information about the sample, as discussed above (figure 3.2). Electric charge will thus build up on non-conductive samples, so these need to be coated with a thin layer of an inert conducting metal, such as gold or silver, to earth the sample. Furthermore, coating with a heavy metal will increase the signal of the secondary electrons (as an increased atomic number) and thus increase image quality, but will hinder EDAX, so if elemental analysis is necessary no coating or a non-interfering coating (e.g. carbon) is required.

Samples can be imaged in a variety of different modes, but topographic images obtained by detecting secondary electrons are most common. These are normally detected with a scintillator–photomultiplier system placed at a shallow angle to the side of the imaged sample. While electron microscopy is required to be performed under a vacuum to prevent scattering of the electron beam, an SEM has been developed that can image in low pressure gas environments and humid/liquid environments. Developed in the late 1980s, the environmental scanning electron microscope (ESEM) can image wet samples in atmospheres by separating the beam column (kept under vacuum) and the sample chamber (in the environmental condition), and using a secondary electron detector modified to collect secondary electrons in humid environments. This is a very important advancement for SEM, particularly for the analysis of biological samples. First, a conductive coating is not needed, allowing better image detail and easier chemical analysis, and second, samples can be observed in a range of environments, such as a biological sample in its native environment.

3.2.4 Transmission electron microscopy

Transmission electron microscopy (TEM) is similar in principle to SEM, and has also become one of the key techniques available for imaging very fine features and ultrathin structures on the nanoscale. TEM operates at much higher accelerating voltages than SEM (typically 100–400 keV, but can be up to 1 MeV), ultimately leading to greater resolution. It is now routine to resolve crystal lattices and even atomic level features. As with SEM, a TEM focuses electrons through the application of electric or magnetic fields onto a sample in high vacuum conditions (figure 3.3). However, this sample must be thin (no more than a few nanometres for hard materials, <1 μM for softer samples), as the technique relies on the electrons being transmitted *through* the sample (figure 3.2) and are detected underneath the sample with a phosphor screen or a charged coupled device to form a two-dimensional projection of the sample. A TEM can image in a number of different

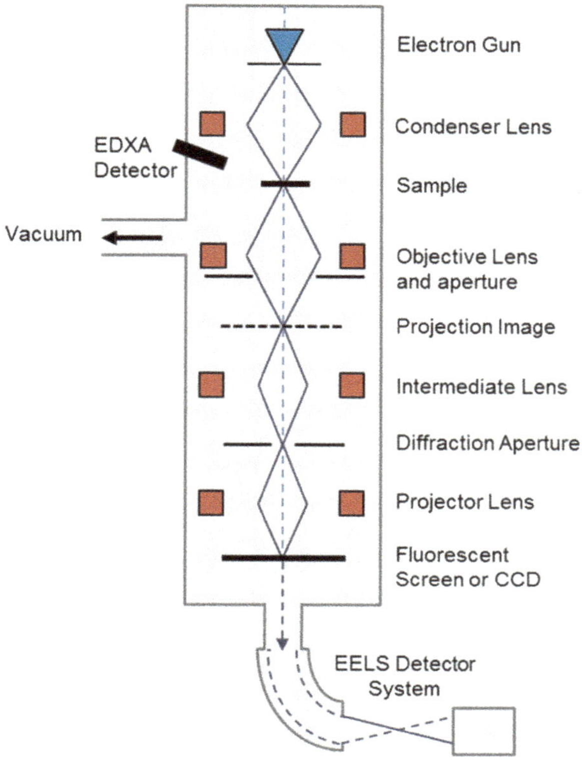

Figure 3.3. Schematic showing the typical layout of a TEM. Electrons are focused onto a sample through the use of electromagnetic lens systems (red boxes) under high vacuum conditions. After passing through the sample, the electron beam (blue line) is then focused to form an image. The typical locations for EDXA and EELS detection systems are also shown. Adapted from [5], copyright of Oxford University Press 2000.

modes; a bright field image can be generated as electrons undergo a number of different scattering processes when interacting with the sample. Regions that are more dense or thicker will lead to greater scattering, which will not be detected and lead to these regions appearing darker on the final image. Crystalline samples also diffract the electron beam, which can be used to form a dark field image if the transmitted beam is excluded. Larger soft samples can be imaged by thin sectioning, which is where the sample can be set in a resin then sliced into thin sections (50–200 nm thick depending on requirements). Similar to SEM, samples with low atomic number (particularly biological samples) can be stained with a heavy metal solution such as uranyl acetate to increase the contrast. Also similar to SEM, there have been recent advances in the development of methods to image samples within a liquid/gas environment. Fluid cell TEM (or environmental TEM/*in situ* TEM) again utilises a separate sample chamber so the electron beam column can be retained under high vacuum, while the sample chamber itself comprises a very thin chip with a fluid chamber, with an electron invisible nitride window above and below the sample for the electrons to be transmitted through. Again, using such systems, living biological cells and chemical reactions such as nanoscale crystallisation events can be observed in nanoscale detail in real time.

3.2.5 Atomic force microscopy

Like electron microscopy, scanning probe techniques have become one of the most well used methods for characterising surfaces and nanomaterials with nanoscale resolution. Atomic force microscopy (AFM) is a topological method that is able to map a surface or nanoscale feature on a surface by reading the force of attraction/ repulsion between a probe and the surface (figure 3.4). This simple method means that AFM has the ability to image materials without the need for high vacuum conditions, and has become one of the most versatile characterisation tools in nanoscience. AFM images the topography of a sample by scanning an extremely sharp tip (the sharper the tip, the better the resolution) mounted on the free end of a cantilever over the surface in sequential line scans. Several forces act between the tip and the surface, causing the cantilever to deflect. The most common are attractive van der Waals forces, when the tip is between 10–100 Å from the surface, and repulsive electrostatic forces, when the tip is <2 Å from the surface. A topographical image of the scanned sample is built from measurements of the cantilever positions, usually with a laser beam that is reflected off the back of the cantilever and detected with a position-sensitive photodetector. AFM can be operated in a variety of different modes, including constant-height, where the scanner height is fixed and an image is built from measurements of the cantilever deflection, and constant-force, where measurements of the cantilever deflection are used in a feedback loop to maintain it at a constant force by moving the scanner over the surface topology. AFM can also be operated in non-contact modes, such as the tapping mode, where a

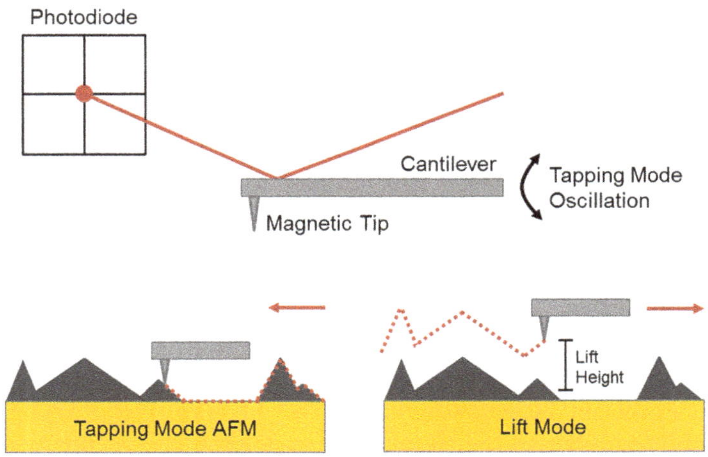

Figure 3.4. Schematic of AFM and MFM. The defection of an AFM cantilever is detected by the reflection of a laser off the cantilever and onto a photodiode. The surface is first imaged with tapping mode AFM, before the cantilever is raised a certain lift height and the surface is retraced to record magnetic information. Reproduced with permission from [6, 7].

stiff cantilever is vibrated at close to resonant frequency so that the tip just touches the sample surface at the bottom of its movement. When the tip approaches the surface it causes a shift in the frequency of vibration, but with the aid of a feedback system the AFM vibrates the cantilever at a constant frequency by moving the scanner up and down. This provides a method of recording surface topography, but due to the fact the tip is not dragged laterally across the surface, it causes less damage than contact mode AFM.

Furthermore, the tip can be functionalised to obtain more information. For example, it can be magnetised, to not only assess the topology of a sample but also to characterise the magnetism at high locality resolution. Magnetic force microscopy (MFM) first uses tapping mode to obtain a topological image of the sample, before being retraced in lift mode at a constant lift height of 50–200 nm above the surface. The magnetic interactions between the magnetised tip and the sample cause vertical frequency to be felt by the cantilever's oscillation, which can be mapped as repulsive and attractive magnetic interactions over the sample surface. Additionally, the tip can be chemically functionalised with a molecule that can interact specifically with a certain chemistry on the surface, again giving highly localised specific chemical information. The tip can be used to measure forces: if the tip is functionalised with a molecule that binds to a different molecule on a surface, the force required to pull them apart can be measured (single molecule force microscopy); rheology and elasticity can be probed observing deformation under different applied forces, and so on.

3.3 Spectroscopy applied to nanomaterials

Spectroscopic and spectrometric techniques are generally applied to bulk materials. However, with modifications, they can be used specifically to probe surfaces within μm to nm depths. As nanomaterials have a much greater proportion of the material on their surface, these surface-specific spectroscopic methods become powerful in characterising nanomaterials. In this section, selected methods are considered. The aim is to provide a brief overview of the technique, an explanation of how it is valuable in characterising nanomaterials, and a suitable example to illustrate the discussion.

3.3.1 Mass spectrometry

A specific type of mass spectrometry—time-of-flight secondary ion mass spectrometry (ToF-SIMS)—has been developed to investigate the chemical composition and functionality of topmost layers (~1 nm). In the case of nanomaterials, ToF-SIMS provides key information on functional performance of a given nanomaterial as well as its potential toxicity. Examples include the assessment of surface chemistry to predict cell-binding and proliferation for a bioglass implant, stability of zinc oxide nanoparticles in a cosmetic formulation, and selective capture of carbon dioxide on amine functionalised silica.

In mass spectrometry, a sample surface is subjected to an ion beam, which removes molecules or more typically charged fragments of molecules (called

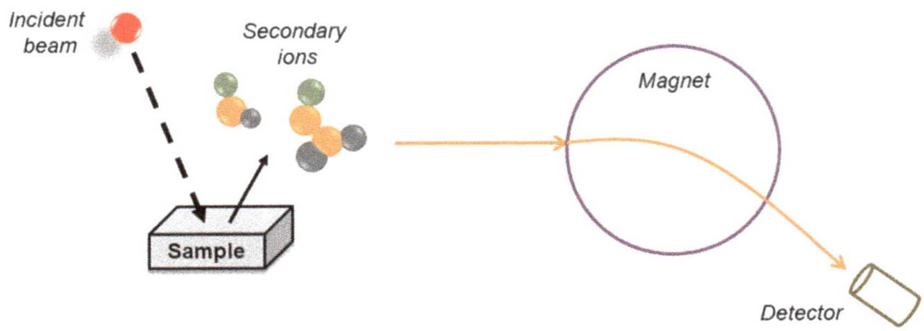

Figure 3.5. Schematic showing an incident ion beam generating secondary ions from a sample, which travel through a magnet before being collected by the recorder. Image adapted from [5], copyright of Oxford University Press 2000.

secondary ions). These secondary ions are made to travel through a varying magnetic field before being collected by a detector (see figure 3.5). Typically the time required for the flight of secondary ions to the detector is also measured, and the technique is therefore called time-of-flight secondary ion mass spectrometry, or ToF-SIMS. The output of such measurements provides a relative abundance of ions collected using the following relationship:

$$\frac{m}{z} = \frac{B^2R^2}{2V} \tag{3.1}$$

where m and z are the mass and charge of the secondary ion, V is the acceleration potential of ions, B is the magnetic flux density of the magnet and R is the radius of the arc in which the secondary ions move. The varying magnetic field enables the separation of the secondary ions based on their charge and size (m/z). This information, when analysed for the origin of the secondary ions, can provide information on the surface chemistry of nanomaterials. One point to note is that the secondary ions can undergo breakage during their flight and result in fragments that are detected. For a multifunctional sample, the analysis can become complex due to overlapping signals from fragments of various secondary ions. This technique is less widely used for characterising nanomaterials on their own but is powerful in probing the interface between nanomaterials and a foreign molecule such as a reactant in catalysis, a pollutant or a protein.

Let us illustrate this with an example of a biomaterial [8], where researchers designed experiments to correlate the chemical functionality of biomaterials surfaces with cell proliferation. They created an array of materials by systematically varying the surface chemistry and then measured the molecular details using ToF-SIMS. The materials were then subjected to cell cultures. They observed that even for the same chemical precursors used for surface functionalisation, the processing affected the final surface functionalities. Further, they correlated the chemical information of the surface measured by ToF-SIMS with the proliferation of cells on those surfaces. Most importantly, it was found that the ions with tertiary-amine moiety (e.g.

$C_2H_6N^+$) were inhibiting cell growth, while oxygen-containing ions (e.g. CHO_2^-) amongst other ions were able to promote cell growth.

3.3.2 Infra-red spectroscopy

IR spectroscopy is well-known for qualitative and quantitative investigations of chemical bonds. The reason for this is that chemical bonds exhibit characteristic vibrations due to specific types of deformations. The frequencies of these vibrations match with the IR frequencies and the absorption of incident IR by a given sample occurs only in specific frequencies. The frequency of IR depends on the geometry of the molecule, the mass of the atoms involved the chemical bond (and hence the bond strength). The vibrations occur in stretching and bending modes. Examples of some of these vibrations are shown in figure 3.6.

IR spectroscopy has therefore been used for qualitative analysis such as identification of molecules or chemical functionalities present in a given sample, including a nanomaterial. This can provide information on the chemical composition of a nanomaterial and detect any impurities present. Further, as the absorbance of IR energy is proportional to the molar concentration of the bonds present, a quantitative analysis is also possible, where changes or differences in bonds of interest can be quantified. These methods are again powerful in studying interactions of nanomaterials with foreign molecules including solvents.

In order to specifically probe the surface of nanomaterials, two modifications of IR spectroscopy have been developed: attenuated total reflection Fourier transform IR spectroscopy (ATR-FTIR), and diffuse reflectance IR Fourier transform spectroscopy (DRIFTS). Both techniques depend on the IR beam penetrating a given sample (to ~1 μm), thus causing interaction with the chemical bonds, predominantly on the surface. In DRIFTS, the IR beam is scattered upon reaching the sample surface and results in diffused reflection. In ATR-FTIR, the instrumentation is slightly different. The incident IR beam is passed through a crystal of high refractive index. This results in total internal reflection and creates an evanescent wave that has the ability to extend beyond this crystal. The sample to be measured is placed on this crystal, and the extending evanescent wave is absorbed at certain frequencies by the sample. This causes a reduction (attenuation) of the wave. The

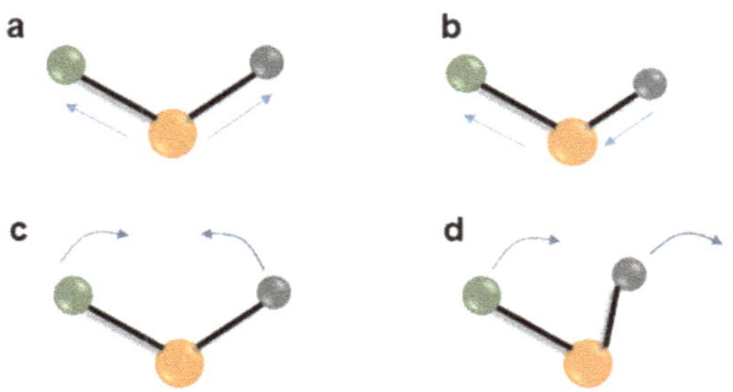

Figure 3.6. A schematic representation of stretching and bending modes in a chemical bond. (a) Symmetric stretching, (b) asymmetric stretching, (c) scissoring and (d) rocking. Reproduced from [5], copyright of Oxford University Press 2000

detector measures this attenuation of the total reflectance of the incident IR beam, which provides information on chemical bonds present in the sample. Some of the key advantages of both these techniques include the ability to probe surface and interface effects in nanomaterials and thin films. Furthermore, both require little or no sample preparation when compared to *routine* IR spectroscopy; samples can be simply placed on the holder and analysed, making this analysis rapid. Finally, ATR-FTIR can also sample liquids, which enables *in situ* monitoring of dynamic processes as well as measuring samples in their *real* environments.

Let us consider these techniques with examples, starting with ATR-FTIR. One of the key features controlling the effectiveness of a biomaterial implant is its biocompatibility. In particular, whether the host *body* accepts the implant depends on the interactions of cells with the surface of the biomaterial. However, before cells can interact, it is the proteins from the surrounding environment that adsorb on the biomaterial—depending on these proteins and how happy they are on the surface, the cells may or may not attach, which in turn decide the fate of the implant. ATR-FTIR can be used to study the protein adsorption on nanomaterials which can lead to understanding of the performance of biomaterials when implanted. One study used ATR-FTIR to investigate the interactions of bovine serum albumin (BSA) and bovine fibrinogen (Fg) on silica nanoparticles with sizes of 15–165 nm [9]. In particular, the IR spectra arising from the amide bond vibrations (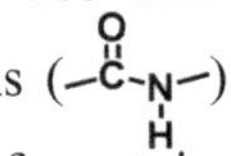) of the proteins can be used to infer the conformation and stability of proteins when adsorbed on nanoparticles. One of the amide bands, centred at 1700–1600 cm^{-1} as shown in figure 3.7, can be used to measure changes in the secondary structure of the protein. The components of this amide-1 band can be deconvoluted to determine the composition of α-helices, β-sheet and disordered structures within the protein structure, for example. This analysis revealed that BSA, which is a globule-like protein, was stabilised on smaller particles, while it was highly denatured when adsorbed on larger particles. On the other hand, Fg, which is rod-like in shape, retained its structure on larger particles but was denatured when adsorbed on small particles (i.e. with higher radius of curvature). This study shows how ATR-FTIR can be used to probe the interactions between nanomaterials and proteins for developing biomaterials. In particular, this work demonstrated the nanoscale effects that control protein (and in turn cell) adsorption on surfaces.

DRIFTS has found interesting applications in studying surface chemistry of a range of materials, particularly nanomaterials. Consider nanoporous carbons as an example, which find applications in catalysis and adsorption. Success depends on the porosity as well as the surface chemistry of the pores. DRIFTS was applied to probe the surface chemistry and its dynamics under heating of this technologically important porous material [10]. When used in combination with other techniques, DRIFTS was able to probe key chemical functionalities on the surface of carbons, such as carboxylic, carbonyl and phenolic.

Figure 3.8 clearly shows three peaks centred around 1750 cm^{-1}, 1600 cm^{-1} and 1250 cm^{-1}. These peaks correspond to C = O stretching (found in carboxylic anhydrides), C = O vibrations of quinone groups and C–O stretching associated

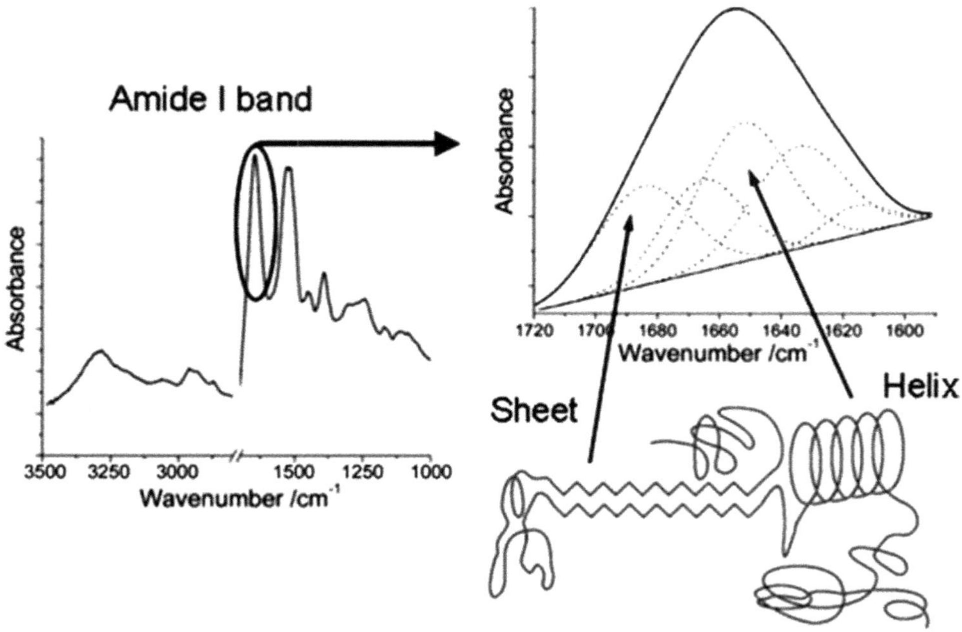

Figure 3.7. ATR-FTIR spectrum obtained for a protein (left) showing the amide-1 band of interest. The images on the right show how this band can be further analysed for underlying constituent peaks (shown with dotted lines) in order to infer secondary structures of a given protein (e.g. beta sheets and helix). Image taken with permission from [9], copyright 2006 American Chemical Society.

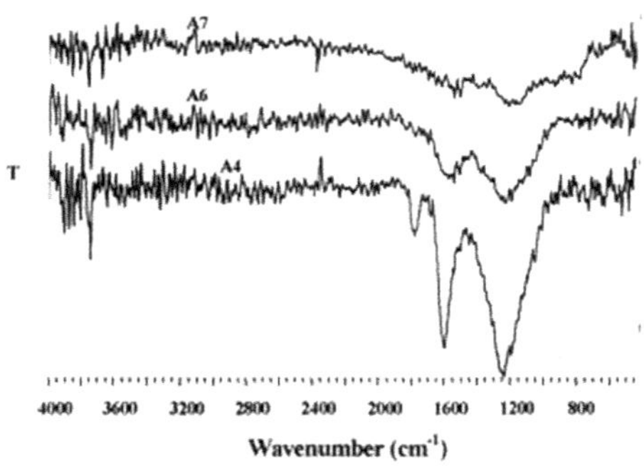

Figure 3.8. DRIFTS spectra obtained for porous carbons (sample A4 is without any treatment, A6 and A7 are heat treated samples). Image taken from [10], copyright 1999 with permission from Elsevier.

with various functionalities, respectively. It can be seen that upon heat treatment, the C = O peaks progressively decrease, finally disappearing completely (sample A7). The C–O peaks, although reducing upon heat treatment, remain to some extent. These findings are not only useful in deciding on the correct samples and heat treatments needed for a given application, but they also help understand the molecular interactions controlling catalysis and adsorption.

3.3.3 X-ray photoelectron spectroscopy

Incident x-rays on a given material can cause multiple interactions and outcomes. One interaction is that the energy from the x-rays causes electrons from the top layer (a few nm deep) to escape; these escaped photoelectrons are detected and their kinetic energy is measured. The binding energy for these photoelectrons is the difference between the incident energy from the x-rays and the measured kinetic energy. The binding energy depends on the electronic configuration and the element itself, thus binding energy can be used to identify the chemical composition of the surface of a given sample. X-ray photoelectron spectroscopy (XPS) provides quantitative data and hence the chemical composition of surfaces can be fully quantified using XPS. Further, as the binding energy changes upon chemical bonding with another element, subtle shifts in binding energy can be used to study the oxidation states of elements as well and their bonding/co-ordination with other elements.

XPS can be used to understand surface chemistry in a range of applications, such as studying coatings. In the development of inorganic coatings [11], the surface was needed to be pre-coated with polyamines. A method was developed for coating surfaces with polyamines and figure 3.9(a) shows the XPS data before and after polyamine pre-treatment. The appearance of the peaks associated with binding

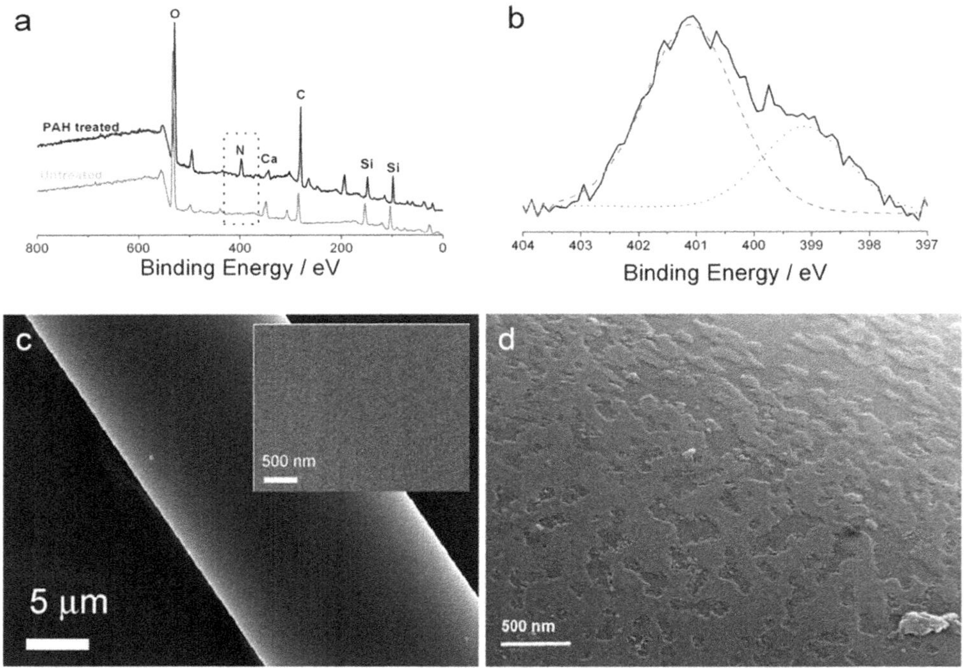

Figure 3.9. XPS (a, b) and SEM (c, d) analysis of glass-fibre surface before and after polyamine coatings. The polyamine used in this case was polyallylamine hydrochloride. Image taken with permission from [11], copyright 2007 American Chemical Society.

energy of electrons from nitrogen provided the evidence for the formation of amine coatings only in the treated sample. This was also confirmed by the SEM results, which showed smooth and coated surface before and after the polyamine treatment, respectively. The significant increase in carbon peaks also confirmed the presence of polyamine coatings. Further, as the peaks from the surface (Si, Ca and O) were also detectable, it was clear that the polyamine coating was very thin (within a few nm). Taking a closer look at the nitrogen peak (figure 3.9(b)) and performing peak deconvolution indicated that there were two sub-peaks centred at ~399 and ~401 eV. These peaks represent the amines in uncharged and charged/protonated ($\equiv N^+$) states, respectively. The quantification of the area under these peaks suggested that there were two charged sites per an uncharged site. Incidentally the charged sites were desirable for the given application. This example illustrated how XPS is a valuable tool in studying the chemical compositions of nanomaterials, all the way down to oxidation states of constituent elements. This feature makes XPS powerful in, for example, monitoring catalytic reactions where oxidation states of metal surfaces control the reaction pathways and outcomes.

3.4 Diffraction and scattering techniques

A selection of characterisation techniques that utilise how a probing electromagnetic wave (be it light or x-ray) is deflected/reflected/scattered by a sample are discussed in this section. While the light and x-ray scattering techniques give detailed information about size and shape at the nanoscale, XRD is a bulk method that gives details about the crystalline structure of any material. However, it can have specific implications for particle sizing at the nanoscale.

3.4.1 X-ray diffraction (XRD)

A crystalline material by definition is composed of a repeating ordered structure of atoms, the smallest repeat of which is called the unit cell. This regular 3D pattern of atoms can be probed, measured, and reconstructed using the constructive interference of x-rays with this crystal lattice. The Bragg father/son scientist duo won the 1915 Nobel prize for their work understanding how x-rays interact with crystalline materials and how they developed this to XRD as an analytical method.

XRD works by an incident x-ray beam striking a crystalline solid surface at an angle (θ), resulting in the production of multiple secondary reflected x-rays. In most directions these waves will cancel out due to destructive interference. However, they will add via constructive interference in certain directions based on where the atoms lay in the material and the in phase reflections these cause resulting in a diffraction pattern. This constructive interference is given by Bragg's law:

$$2d \sin \theta = n\lambda, \quad (3.2)$$

where d is the crystal lattice spacing, λ is the wavelength of the incident x-ray and n is the order (as shown in figure 3.10).

On a simple level, the x-ray hits the sample at varying angles, which comes to no consequence, but at the angle at which it hits each of the atoms that are regularly

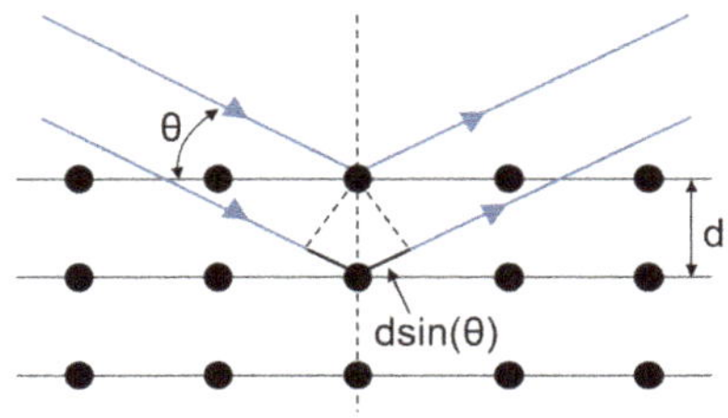

Figure 3.10. An x-ray beam incident on a symmetrical plane of atoms results in constructive interference when their path-length difference ($2d \sin \theta$) is equal to an integer number of the incident wavelength ($n\lambda$).

spaced (d-spacing) and is diffracted with their wave in phase with the other diffracted x-rays, an intense diffracted x-ray signal will be recorded. Thus an x-ray diffractometer records the intensity of the secondary reflected waves over a range of angles, either by rotating the specimen and detector while keeping the incident beam fixed, or by keeping the specimen fixed while rotating the detector and the incident beam [1]. Crystallographic planes produce characteristic peaks in the recorded spectrum at precise angles, which can be converted back to d-spacing with the use of equation (3.7); as such crystal structure can be determined from the range of d-spacings calculated, or if the material is already known, the full spectrum can be compared to a database to identify the sample. XRD can be performed on a single crystal, or powder XRD can be performed on fine powder crystalline samples. In the latter, the powder is laid on a sample plate, with an increased sample size increasing the intensity of the peaks. With XRD, the fine powder is randomly orientated giving a signature XRD pattern for any one material. As the material is orientated more in any one direction, the peak intensity of the dominant reflection will become more intense.

XRD is also a useful sizing method for crystalline nanomaterials. The average size of particles can be obtained from the peak broadening. The larger the single crystal, the more intense and narrow the peak. As the size decreases and the number of particles increases, the peak height reduces and the width broadens.

$$D = \frac{K\lambda}{\beta \cos \theta}. \tag{3.3}$$

The Scherrer equation (3.3) can be used to calculate the minimum crystallite size, where D is the size of the smallest crystallites, K is a shape factor which is normally between 1 and 0.9 for spherical particles, but does vary with different shapes, λ is the wavelength of the x-ray beam, β is the full width at half maximum of a selected peak and θ is the angle of this peak. With good quality nanoparticles this can normally be equated to the nanoparticle size. However, peak broadening can occur due to the instrument, and imperfections in the nanocrystals can give a smaller particle size, so this technique is best used in conjunction with other techniques.

3.4.2 Dynamic light scattering

Dynamic light scattering is the measurement of fluctuations observed in light scattered from a particulate sample. These fluctuations can then be related to diffusion within the sample, and then this information can be used to calculate particle size. Due to Brownian motion occurring within the sample, there is constructive and destructive interference which has an effect on the light intensity and scattering over time, this is called the translational diffusion coefficient (D) (equation (3.4)). This takes into account solution viscosity (η) Boltzmann's constant at absolute temperature (kT), and the hydrodynamic diameter (D_{H}) of the particle; this assumes that all particles are solid spheres.

$$D = \frac{kT}{3\pi\eta D_{\mathrm{H}}}. \tag{3.4}$$

This is due to the fact that smaller particles diffuse faster through the solution, causing more fluctuations than bigger, slower diffusing particles. The technique works by way of a polarised beam, which is then scattered by density or concentration fluctuations in the sample. A detector picks up speckle caused by scattered light interfering with the sample. The time dependence of this scattering at a specified angle is used to form an autocorrelation function. For monodisperse samples this comes from the exponential decay arising from the diffusion within the sample, which is then related to the hydrodynamic diameter of the particle. If a sample is polydisperse the correlation function is based upon a distribution of decay rates. For vesicle samples a concentration of 400 μl ml^{-1} was used, with a scattering angle of 173°. Samples were generally scanned three times, with each scan having 10–14 runs. Data was analysed using Malvern Zetasizer software. It is also possible to estimate the polydispersity index (PdI) (equation (3.5)) of the sample, which relates to the standard deviation (σ) of a Gaussian distribution applied to the data (Z_{D}^2):

$$PdI = \frac{\sigma^2}{Z_{\mathrm{D}}^2}. \tag{3.5}$$

3.4.3 Small angle scattering

Small angle scattering (SAS) is an advanced technique for studying nanoscale features of a given sample at multiple levels (e.g. primary particles, secondary particles, aggregates, etc). Depending on the energy source used (light, neutrons or x-rays), specific length scales can be mapped. As small angle x-ray scattering (SAXS) can typically probe sample features in the range of 1–100 nm, this section will focus on SAXS, although the fundamental principles apply to other SAS techniques. The scattering of incident x-rays occurs due to the difference in the electron density of a given sample. The advantages of using SAXS for characterising nanomaterials include:

- It provides a global average of properties (e.g. particle size), as opposed to local information typically obtained by microscopic techniques.
- It can probe features that are buried inside, which typically cannot be 'viewed' by microscopy and surface techniques.
- It can quantify a range of properties for nanomaterials, such as sizes of features at multiple length scales, shapes, relative concentration, polydispersity, morphology and structure.

Having said that, the analysis of SAXS data is extremely complicated, dependent on developing relevant models and requiring complementary characterisation. Therefore, this section is limited to an overview of what SAXS can offer in the context of nanomaterials, avoiding the details of the underpinning physics and the model-based analysis of the data. The length scales (or the scattering vector q) probed directly depend on the wavelength of the source and the measurement angles, as given by the following equation:

$$q = 2\pi/d = 4\pi \sin \theta/\lambda \tag{3.6}$$

where d is the feature size or characteristic length, θ is the measurement angle ($\ll 1°$ in SAS) and λ is the wavelength of the incident x-rays. This means that x-rays generated from the Cu Kα line, which has a wavelength of 0.154 nm, have the limit of detection of ~45nm if measured at an angle of 0.2°. In order to measure wider length scales, the source can be changed to lower wavelengths (or higher energy, e.g. a synchrotron radiation source) and/or the detection angles can be reduced (which requires advanced instrumentation). A typical set of data obtained from SAXS measurements shows the intensity of scattering measured at different angles (or q), as shown in figure 3.11.

Using a pre-defined model, the data is fitted to obtain parameters in the following relationship:

$$I(q) \propto q^{-\alpha} \tag{3.7}$$

where α is the fractal dimension and provides information on the shape/morphology of the sample such that the values of α of 3, 2 and 1 represent spheres, disks and rods respectively. Note that high q means smaller sizes, since q is inversely proportional to characteristic length. Figure 3.11 shows that SAXS is able to probe a wide range of length scales. Further analysis of the SAXS data shown in figure 3.11 provided sizes of primary and secondary particles, which were found to be in the range of 5–10 nm and 100s nm respectively. The analysis also identified fractal dimensions and connectivity, which provide an understanding of their formation and porous properties.

3.5 Porosimetry

The porosity of nanomaterials is of vital importance in understanding their properties as well as performance. This becomes particularly critical in applications such as catalysis and pollution control, where the effectiveness of these materials is

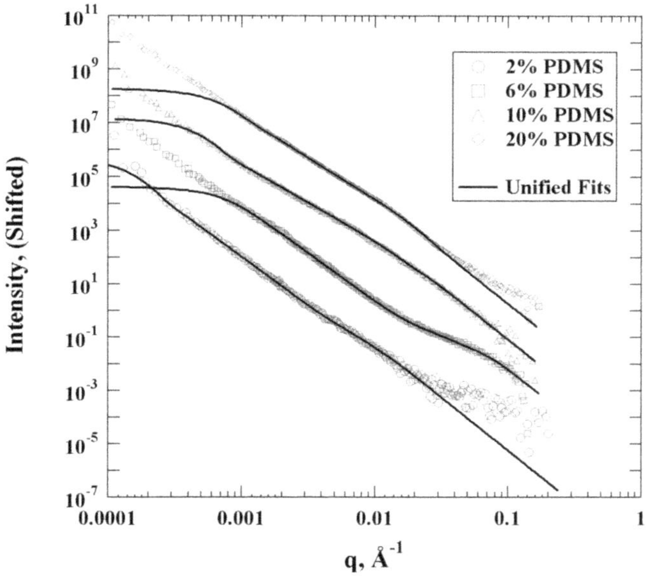

Figure 3.11. Ultra small angle x-ray scattering data obtained for a polymer filled with nanoparticles at varying amounts. Image taken from [12], copyright 2006 with permission from Elsevier.

controlled by the surface area, pore volume, pore sizes and size distribution, pore shapes and network of pores. Although some of these properties can be estimated by SAXS, it does not provide information on the accessibility of pores (i.e. it does not differentiate between open and closed/buried pores), as well as their adsorption characteristics for a specific adsorbent. Therefore, studying adsorption of a probe molecule on to porous nanomaterials, called porosimetry, has become a powerful technique. There are generally two types of porosimetry techniques: mercury intrusion porosimetry and gas adsorption. The former is more suitable for measuring macroscopic length scales (≫100 nm) and it operates under high pressures, which can at times crush or break down nanomaterials. As such, gas adsorption is commonly used for probing the porosity of nanomaterials which contain (mainly) internal and some external porosity, in the range of a few nm or less.

Porosity at such length scales is classified into three categories [13]:

- *Macro*pores with pore widths >50 nm.
- *Meso*pores with pore widths between 2 nm and 50 nm.
- *Micro*pores with pore widths <2 nm.

It is generally the micro- and meso-porosity that are of interest in many cases that involve the adsorption of small molecules (e.g. reactants or pollutants) as a key step in the given application. Gas adsorption can be used in manometric (or volumetric) or gravimetric modes, with the former being the most common. Typically, the probe molecules used include N_2, Ar, and Kr, with nitrogen being the most common. In a typical measurement, the probe gas is dosed in a chamber containing the sample and

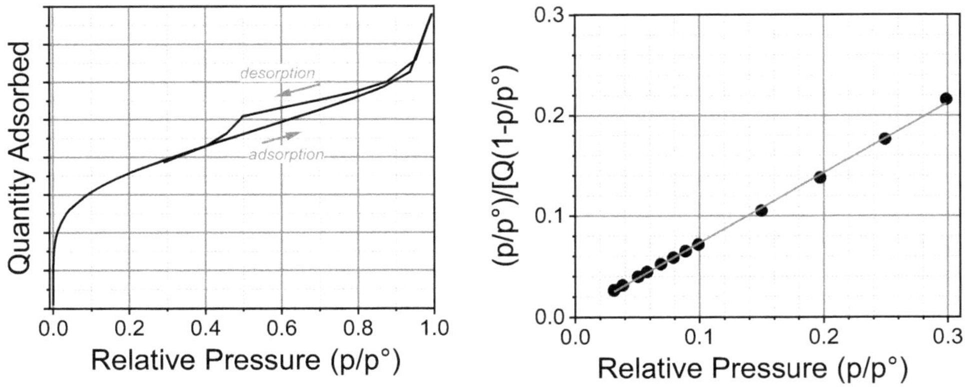

Figure 3.12. Typical gas adsorption data collected (left) and analysed using a BET method (right) to determine the BET surface area.

allowed to equilibrate at a constant temperature and partial pressure (p/p^o, where p is the equilibrium pressure and p^o is the saturation vapour pressure). The amount of gas adsorbed is measured before increasing the partial pressure in stages until saturation vapour pressure is reached. A similar procedure is used in desorption mode, where the amount of gas desorbed is measured as a result of sequentially reducing the pressure. These data are used to produce an adsorption isotherm, as shown in figure 3.12. This isotherm has various features, including initial sharp adsorption due to micropores, and the presence of hysteresis due to complex pore structures (e.g. pore networks and pore blocking effects). The details of the isotherm types, their physical significance and analysis are outside the scope of this section (they are also widely described in depth elsewhere, see [13]). However, it is worth describing the method used to obtain a specific surface area (a_s) for a given sample once the amount adsorbed (Q) at a range of partial pressures is measured (as shown in the left graph in figure 3.12). The procedure developed by Brunauer–Emmett–Teller (the 'BET method'), is the most widely used method to estimate a_s, and is based on monolayer adsorption capacity (Q_m). The data from the isotherm is converted in to a BET equation as follows:

$$\frac{p/p^o}{Q\left(1-\frac{p}{p^o}\right)} = \frac{1}{Q_m C} + \frac{C-1}{Q_m C}\left(\frac{p}{p^o}\right) \tag{3.8}$$

Using this equation in the monolayer adsorption region (typically between the p/p^o range of 0.05–0.3), a linear BET plot is created by plotting the left hand side of this equation against the partial pressure as shown in figure 3.12 (right). The straight line in this plot provides the value of the monolayer capacity (Q_m) and an indication of the monolayer adsorption energy (via C).

The BET specific surface area is then calculated using the following equation:

$$a_s = Q_m L \sigma_m / m \tag{3.9}$$

where Q_m is obtained from the BET plot, L is the Avogadro constant, m is the mass of the adsorbent sample used in the measurement, and σ_m is the molecular cross-sectional area of the probe molecules (taken as 0.162 nm^2 for N_2). A number of methods exist for determining the pore sizes and their distribution and due to the complex and material-specific nature of the analysis, readers are directed to specialised texts [13, 14].

3.6 Summary: key lessons for characterisation of nanomaterials

This chapter has given a brief overview of several different characterisation techniques used by researchers to understand and assess materials produced. It should be noted these are brief introductions, and more in-depth literature can and should be sought from the references in this chapter, and from more detailed, focused papers for those intending to specialise in any specific technique. This chapter gives an assessment of such characterisation with nanoscale materials in mind, and a selection of the most suitable methods. However, after reading this chapter and before attempting to characterise a nanomaterial, or assess/review if methods used by others to characterise a new material are adequate/suitable and valid, one should first consider what the purpose of characterisation is. On the most basic level, it is to 'see' the material: the structure, the shape and size, the homogeneity, the chemical elements present. Characterisation is also undertaken to obtain the 'feel' of the material: to understand it in relation to the chemical and physical properties. This may require more experimental thought and design on the nanoscale than on the macroscale, as the materials themselves are not visible to the naked eye. A few key lessons to think about beyond this chapter are as follows.

- Design the experiment and analysis well. Think: what do you want to find out? What question is being asked of the material produced? Then ask: what is the right characterisation technique to find that out? Sounds obvious, but so often researchers will simply go through the motions of repeating the same analysis for everything they produce. The mistake here is that all the design and imagination goes into the synthesis of the new material, with little consideration for the design of analysis. However, the analysis is equally, if not more important, if you want to see and understand what has been made with any detail and robust rigour.
- As has already been inferred, it is essential to completely understand the technique and methodology of characterisation that is being used. Make sure to read around the techniques chosen. This is key on so many levels. It is key to the design of an experiment, to ensure that the strengths and limitations of the method are understood. Will this technique really answer the question being asked? It is key to interpreting the data: an anomalous result may be a false negative, and to an untrained/uneducated eye lead to the whole experiment being abandoned, but a basic understanding of the technique could easily show this to be an artefact of the preparation method or analysis conditions, and correction could be made, to optimise (otherwise discarded) results.

- Use complementary techniques. Never rely on just one technique and never trust others' research that does. All techniques have their strengths and limitations and as such a full interpretation of a material cannot be presented with data from only one technique. Again, this comes back to designing the analysis. For example, if the aim is to assess the size and shape of a population of nanoparticles, TEM is great for seeing the shapes, but only in a 2D projection. In addition, artefacts of drying can lead to shrinkage of some materials and incorrect sizing. 3D information can be obtained from cryo tilting, holography, and reconstructions of thin sections, and these methods are more accurate for true size. However, the limitation here is selectivity. The whole population is not considered. Scattering techniques are better in this respect, but one must understand the method to understand if the data is giving the particle size or the solvated size. What is assumed? That the particle is spherical? Can and should mathematical constants used to calculate the size be adjusted for the sample? Are the answers obtained realistic in relation to other results? And so on; e.g., can the results from both microscopy and scattering agree (considering everything that is now understood)?
- Do not rely on us (this chapter!): we all get old and time moves on. Characterisation techniques are improving at an astonishing rate. The 'go to' techniques that most people teaching used for their undergraduate and postgraduate research projects will be very old now. Unless supervisors/mentors specialise in the latest cutting edge characterisation, you should look further afield too. Look up the latest instrument for the question you want to ask, for the analysis you have designed (see above). Then look up who has that instrument and how you can access it. Do not just use what is in your lab because it is there. The chances are it will be a good characterisation method, because it is in a nanomaterials research lab, but only use it because it is the correct instrument to use, not just because everyone else in the group uses it! Use the characterisation methods that are right for the material and answer the questions you want to ask, with more than one method per question, to gain a full and robust analysis.

References

[1] Kelsall R, Hamley I W and Geoghegan M 2005 *Nanoscale Science and Technology* (New York: Wiley)

[2] Oshikane Y, Kataoka T, Okuda M, Hara S, Inoue H and Nakano M 2007 *Sci. Technol. Adv. Mater.* **8** 181

[3] Lemmer P, Gunkel M, Baddeley D, Kaufmann R, Urich A, Weiland Y, Reymann J, Müller P, Hausmann M and Cremer C 2008 *Appl. Phys.* B **93** 1

[4] Weisenburger S, Boening D, Schomburg B, Giller K, Becker S, Griesinger C and Sandoghdar V 2017 *Nat. Methods* **14** 141

[5] Duckett S and Gilbert B C 2000 *Foundations of Spectroscopy* (Oxford: Oxford University Press), 90

[6] Bird S 2016 A bioinspired approach to Data storage *PhD Thesis* University of Sheffield

[7] Galloway J 2012 Biotemplated arrays of nanomagnets using the biomineralisation protein Mms6 *PhD Thesis* University of Leeds

[8] Mei Y *et al* 2010 *Nat. Mater.* **9** 768

[9] Roach P, Farrar D and Perry C C 2006 *J. Am. Chem. Soc.* **128** 3939

[10] Figueiredo J L, Pereira M F R, Freitas M M A and Órfão J J M 1999 *Carbon* **37** 1379

[11] Pogula S D, Patwardhan S V, Perry C C, Gillespie J W, Yarlagadda S and Kiick K L 2007 *Langmuir* **23** 6677

[12] Patwardhan S V, Taori V P, Hassan M, Agashe N R, Franklin J E, Beaucage G, Mark J E and Clarson S J 2006 *Eur. Polym. J.* **42** 167

[13] Thommes M, Kaneko K, Neimark Alexander V, Olivier James P, Rodriguez-Reinoso F, Rouquerol J and Sing Kenneth S W 2015 *Pure Appl. Chem.* **87** 1051

[14] Cychosz K A, Guillet-Nicolas R, Garcia-Martinez J and Thommes M 2017 *Chem. Soc. Rev.* **46** 389

Chapter 4

Conventional methods to prepare nanomaterials

4.1 Top-down and bottom-up methods

Existing methods for producing nanostructured materials are generally categorised as 'top-down' or 'bottom-up'. Top-down methods involve starting with a large 'piece', which is subsequently reduced and shaped to a desired size and form (figure 4.1). Conceptually, this is analogous to wood carving, where a final sculpture or artwork is made by carefully removing unwanted parts using specialised tools (e.g. a chisel). In the context of nanomaterials, the starting point for top-down methods is a bulk material, which can be fragmented to smaller sizes using etching or milling, until a desired size, or the limit of the method, is reached. Sometimes, these are also called destructive methods, because they rely on material removal. On the other hand, bottom-up techniques start with smaller objects, which are assembled together into a desired form using specialised mechanisms (figure 4.1). This is similar to using bricks to build a building or creating Lego structures. Hence they can be viewed as constructive methods, as they rely on material addition. In this case, the particle formation pathways all the way from atoms/molecules to cluster and nanoscale structures are controlled by modulating the reaction chemistry, nucleation and growth.

Table 4.1 lists selected examples of the most commonly used top-down and bottom-up methods for producing nanomaterials—some of these are currently commercially used while others are in the development stages. Figure 4.2 shows various processing techniques as a function of the sizes of objects/features they can produce, and the progress made with time. In the early days, *small* objects/features were made using precision machining techniques. Soon after, new methods such as lithography started to emerge, which pushed the boundaries towards sub-micron features. Simultaneously, bottom-up chemical approaches were developed. As both approaches moved into the nanoscale (1–100 nm), techniques using a combination of the two approaches were discovered. The sections below provide an overview of selected methods from both top-down and bottom-up approaches, with

doi:10.1088/978-0-7503-1221-9ch4

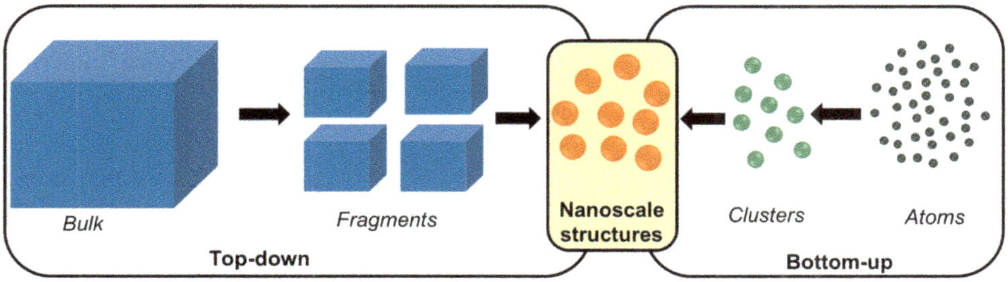

Figure 4.1. Schematic representation of top-down and bottom-up approaches to producing nanomaterials.

Table 4.1. Selected examples of the most common methods for manufacturing and synthesising nanomaterials. Adapted from [1].

Top-down methods	Bottom-up methods
Lithography	*Vapour-phase methods*
Photolithography	Molecular beam epitaxy
E-beam lithography	Atomic layer deposition
Immersion lithography	Pulsed laser deposition
Lithography with particles	Sputtering
Nano-imprint lithography	Evaporation
Soft lithography	Laser ablation
	Flame synthesis
Etching	*Liquid-phase methods*
Wet etching	Precipitation
Dry etching	Sol-gel
	Solvothermal synthesis
	Sonochemical synthesis
	Microwave-assisted synthesis
	Reverse micelle
	Electrodeposition
Electrospinning	*Self-assembly methods*
Mechanical attrition/milling	Electrostatic self-assembly
	Self-assembled monolayers
	Langmuir–Blodgett films

suitable examples. The purpose is to give an idea of current and future industrial methods, rather than providing an exhaustive list and in-depth description of all available methods. In particular, we focus on methods that are most widely used for commercial production, as shown in figure 4.3.

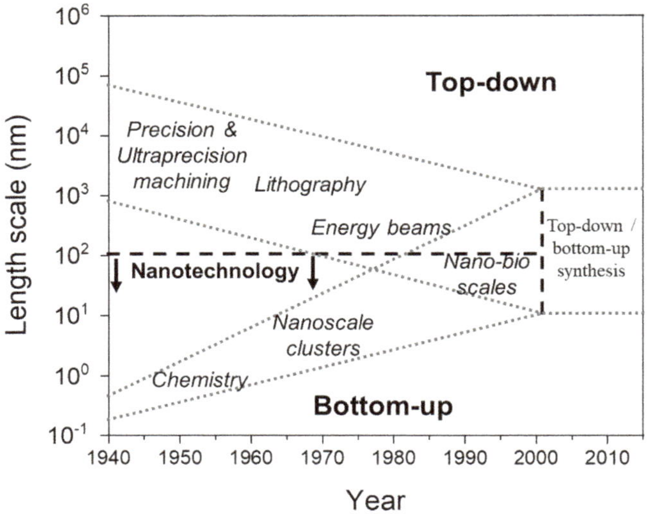

Figure 4.2. Various manufacturing techniques shown as a function of the length scales they can achieve and the progress made over time. Image adapted from [2].

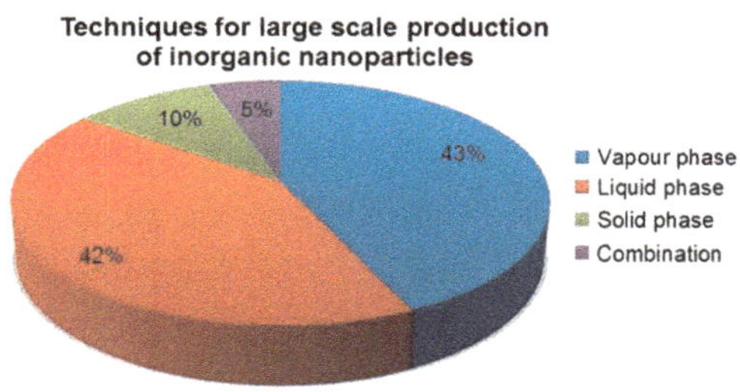

Figure 4.3. Distribution of techniques used for inorganic nanomaterials at large scales. Image adapted from [3].

4.2 Top-down methods

Top-down approaches typically start with a bulk 3D material in the form of ceramic/metal powders (≫μm sizes) or a bulk 2D surface in the form of a pre-deposited thin film (≈μm thick). Using various techniques listed in table 4.1, nanostructured patterns are created on the films, or the powders are reduced to smaller sizes, with varying levels of control over the process. Most commonly, in the case of films and surfaces, some form of lithography is used.

Lithography involves patterning a surface using light, ions or electrons [1]. Lithography either increases or decreases the stability/solubility of selected areas. Upon subsequent etching, the less stable/more soluble areas are removed. A mask is used to selectively remove parts of the surface, thus leaving behind a desired pattern. Although electron- and ion-lithography can create features in the nanoscale (even <10 nm), their speed is too slow for production timescales. Instead, optical (or photo-) lithography is widely used due to its fast production rates and economic operations. Research on photolithography in recent years has enabled patterning features at 10s of nm length scales [4]. Lithography is mostly limited to high value or high performance applications, such as electronic chips. It also requires specialised environments such as a vacuum and high energy source.

Mechanical attrition constitutes size reduction using physical techniques such as crushing and milling, which has been used for production of 'fine' powders since the 1970s. Traditionally, these methods are used for reducing the size of particulate solids (e.g. minerals and ores) from ≫mm to the mm–μm range, while specialised techniques can produce particles in the sub-μm range (e.g. a colloid mill) [5]. These methods have the advantage of operating under low temperatures and can be solvent-free. There is the possibility of operating in reactive or unreactive mode. In the latter, the precursors are milled to desired sizes typically using ball mills. When operated in reactive conditions, metal precursors are milled in the presence of a reactive gas (e.g. O_2) and allowed to react, thus forming nanocrystallites of metal oxides. These methods are also called mechanochemical processing, and in recent times they have been used for producing sophisticated materials such as metal–organic frameworks.

Mechanical attrition has been also used to produce new alloys and mixtures at large scales. These techniques operate under the principle of mechanical breakage of solids by imparting energy (via impact). The energy required for a given size reduction operation will be directly proportional to the increase in surface and the reduction ratio. It is clear that mechanical or physical size reduction operations consume large amounts of energy, and as the desired product sizes push beyond the submicron range, the energy required will be substantially higher. Another drawback of these methods is the limited control over particle sizes, shapes and size distribution. Monodisperse particles in the sub-micron range are seldom possible with mechanical attrition.

4.3 Bottom-up methods

In order to synthesise (nano)materials from a bottom-up approach, there are fundamental chemical principles and methods involved in building materials from atoms and molecules. These principles are described in depth elsewhere [6], and here we provide a summary of these principle reactions by focusing on the physical state of the precursors. A number of methods exist that involve the reaction in a single phase—either in solid state, solution or gas phase. One reaction proceeding in solid state is ***metathesis exchange***. A metathesis exchange reaction starts with precursor salts (e.g. a metal chloride and a sulphide) producing a metal sulphide. A ***single source reaction*** also occurs in solid state where a precursor, typically a metal complex, is thermally converted to a desired material. Both types of method produce by-products that are solids or gases (e.g. NaCl or CO_2). Solid state methods rely on heating to provide the activation energy for the reaction and hence typically occur at high temperatures (sometimes in excess of 1000 °C).

There are a number of reactions that proceed in solution state, and these occur at lower temperatures compared to solid state reactions. The increased mobility of molecules in the liquid state helps overcome the activation energy at lower temperatures. Reactions include a thermally driven ***direct reaction*** between stoichiometric amounts of precursors (e.g. chemical precipitation), which is very common given its simplicity. Another commonly used solution method involves ***sol–gel chemistry***, where metal alkoxide solutions in a water–alcohol mixed solvent are hydrolysed in the presence of a suitable catalyst, with further polycondensation leading to the formation of sol first and then a gel. Further, this gel is typically thermally processed to remove the solvent and complete the condensation reaction to produce an oxide material (see section 4.5). ***Hydrothermal and solvothermal*** syntheses are also solution phase methods, which occur at high temperatures and pressures. While hydrothermal synthesis occurs in aqueous medium, solvothermal methods use non-aqueous solvents. In the majority of the solution phase methods, given that the products coexist in the solvent with the unreacted precursors and by-products, the downstream separations become important to obtain pure materials.

Mixed-phase syntheses are also common where the reactants are in different phases. These include reactions between solids and liquids (e.g. ***ion exchange*** reactions between an oxide in a melt of a salt or ***intercalation*** of dissolved ions into a dispersed solid) and between solids and gas (e.g. ***chemical vapour deposition*** onto a hot solid substrate or ***electrochemical*** synthesis/deposition from a solution onto a substrate). Improved reactivities in a gas/liquid phase in combination with heating provides the activation energy for these types of reactions.

4.4 Nucleation and growth theory

How do atoms aggregate to form nanoparticles, is the ultimate bottom-up synthesis question. Specific for the liquid phase, but also relevant to some forms of vapour phase synthesis is the question of how particles nucleate and grow (described in section 4.5). It is important to understand how this process occurs so one can understand the size, shapes and distribution (homogeneous or polydispersed) profiles that these particle populations establish.

Nucleation is the starting point in the formation of either a new material or thermodynamic phase via the aggregation/assembly of the smallest building blocks of that material, be they atoms/ions/monomers or molecules. The nuclei are the smallest[1] clustering of these building blocks and can be considered the first step of the particle formation process. Nucleation occurs when the small units such as atoms or molecules come into close proximity and aggregate. This occurs primarily by increasing the concentration, increasing the likelihood of atoms coming into contact with one another. Once nucleation occurs and particles begin to grow, the soluble atoms are 'used up' in forming the nanoparticle, so the concentration begins to drop off until growth ceases once the atoms have reached too low a concertation to continue to grow the particles. This theory of nucleation and particle growth is over half a century old and is still a good model (figure 4.4). There are two forms of nucleation: homogeneous and heterogeneous, detailed below.

[1] It should be pointed out that research in this field has now identified a smaller entity still in the liquid phase, called a 'pre-nucleation cluster'.

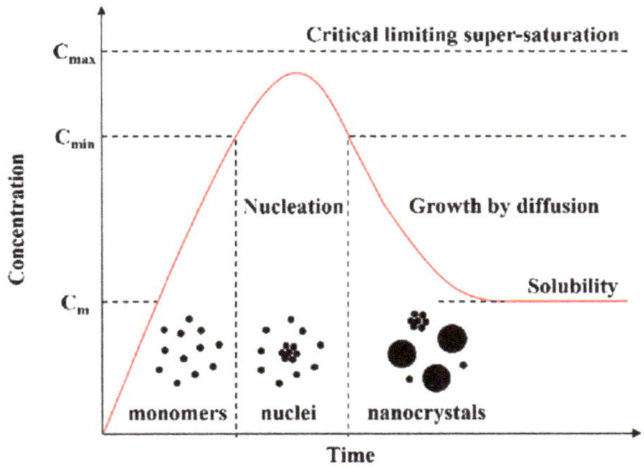

Figure 4.4. Concentration of a monomer over time, as the concentration is increased, C_m is the saturation concentration, C_{min} is the minimum concentration for particle nucleation, C_{max} = critical limiting super-saturation. Reprinted with permission from [7], copyright 1950 American Chemical Society.

4.4.1 Homogeneous nucleation

Homogeneous nucleation describes the formation of all nuclei at the same time in a bulk solution. This occurs spontaneously from a highly saturated solution or vapour. The key to homogeneous nucleation is high concentration. The nucleation is driven solely by monomer species interacting enough to aggregate. So what are the driving forces in homogeneous nucleation? We can consider this thermodynamically with the Gibbs free energy of the forming nuclei. There are two competing energies:

The volume free energy—this is negative, thus favourable and describes the energy of the bulk within the forming particle.

The solid/liquid interfacial energy—this is positive, thus unfavourable and describes the energy associated with the interaction of the surface of the forming particle with the surrounding media.

Thus the volume free energy favours the growth of the particle, while the interfacial energy favours the dissolution of the particle. The more surface area there is compared to the bulk particle, the more likely the particle will not form, as dissolution is favoured. This is why a sugar cube is slower to dissolve than smaller granular sugar. This is summarised in the following equation (4.1):

$$\Delta G = \underbrace{\boxed{A_{SL}\gamma_{SL}}}_{\text{Interfacial term}} + \underbrace{\boxed{V_s \Delta G_V}}_{\text{Volume term}} \tag{4.1}$$

where A_{SL} is the surface area term ($= 4\pi r^2$), and V_s is the volume term $\left(= \frac{4\pi r^3}{3}\right)$. Therefore, at small particle size (with small r) the interfacial term will dominate and the nuclei formed will redissolve back into the solution, but at a critical nucleus size

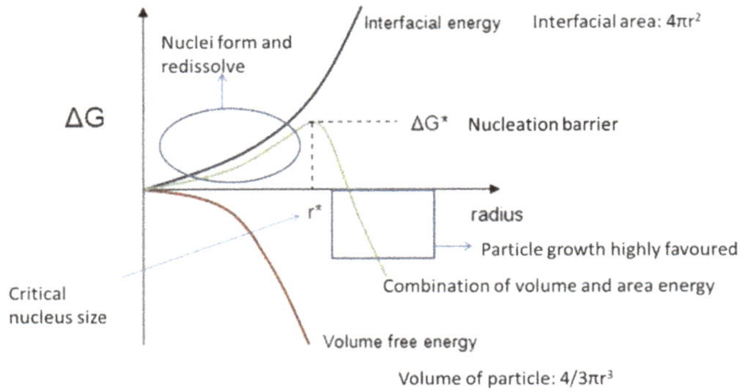

Figure 4.5. The free energy versus radius of homogeneous nucleation.

(r^*) the free volume energy term begins to dominate and the total free energy of nucleation begins to fall, quickly resulting in it being more favourable to form and grow particles (figure 4.5).

The theory shows that once the radius of the nuclei has surpassed the critical radius ($r > r^*$), the particle will grow. This also shows that the critical radius is the lower limit of the particle size. Thus if nanoparticles smaller than the critical radius are required, the surface energy must be reduced, or they will be unstable and redissolve. This can be done in a number of ways, such as using a different solvent, or by changing the surface in other ways, e.g. adding an additive to bind to the surface.

The rate of nucleation (J) is thus dependent on the thermodynamic nucleation barrier (ΔG^*) and the concentration of monomer species. This concentration is defined within the kinetic constant K, which represents the number of monomer species able to interact (per unit volume). This is the collision frequency (equation (4.2)):

$$J = K\exp(-\Delta G^*/kT). \tag{4.2}$$

The equation shows that at a given concentration and temperature, the nucleation barrier is overcome, and then all the nuclei start to grow into particles spontaneously. The higher the concentration and the lower the energy barrier, the higher the rate of spontaneous nucleation.

It should become apparent on reflection that these are difficult and artificial circumstances to create (although it can be done with a supersaturated NaCl solution) and thus homogeneous nucleation is rarer than heterogeneous nucleation based solely on increased concentration of monomers. In reality, other factors to promote aggregation apart from just concentration such as chemical reactions resulting in decreased solubility, will drive nulceation. However, the theory is very useful to understand the process of both types of nucleation.

4.4.2 Heterogeneous nucleation

Heterogeneous nucleation describes nucleation on a substrate, such as a container wall or the surface of impurity particles. It is much more common, as it does not require a saturated concentration of the crystallising species. Nucleation can occur at much lower concentration because the nucleating surface acts to lower the interfacial energy, effectively reducing the nucleation barrier. Basically, a particle forming on a surface is less exposed to the surrounding medium, has less surface atoms compared to bulk and thus has a lower interfacial energy. By this rationale, while a flat surface is good, a step offers even less exposure to the solution (two sides are contacting the substrate so in the 'bulk'), and a crack even less (figure 4.6).

This explains why you can encourage crystallization of a solution by scoring a glass flask that contains it with a spatula. This makes a rough groove in the glass to promote crystallisation in this groove.

As a simplistic model, the reduction in energy barrier can be calculated for a flat surface by multiplying the homogeneous free energy barrier by a (fractional) shape factor, as demonstrated in figure 4.7 and shown in equation (4.3).

$$\begin{aligned} \Delta G^*_{\mathrm{het}} &= S(\theta) \times \Delta G^*_{\mathrm{hom}} \\ \text{Shape factor } S(\theta) &= \frac{(2 - 2\cos\theta + \cos^3\theta)}{4} \\ &= \frac{(2 + \cos\theta)(1 - \cos\theta)^2}{4} < 1 \\ (\theta = 0 &\text{ or } 180^\circ \text{ is homogeneous nucleation}). \end{aligned} \tag{4.3}$$

4.4.3 Growth

How particles grow once nucleated is highly dependent on the species concentration, the type and rate of nucleation that has occurred, and any other 'additives' that are present in the media/solution.

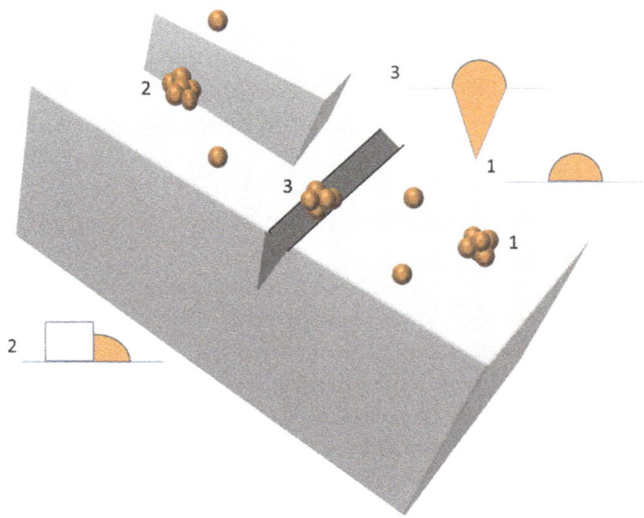

Figure 4.6. Showing heterogeneous nucleation in different surface types. (1) A flat surface, (2) a step and (3) a crack.

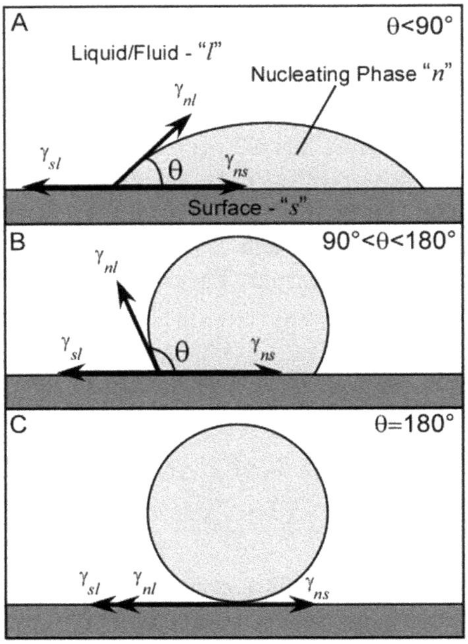

Figure 4.7. Schematic of the possible interactions of a nucleus with a surface at various contact angles according to the Young equation. (A) shows a favourable interaction at $\theta < 90°$, (B) shows an unfavourable interaction at $90° < \theta < 180°$, and (C) shows no interaction ($\theta = 180°$), which is the same as homogeneous nucleation (in the absence of a surface). Figure reproduced from [8], copyright 2012 American Chemical Society.

Nucleation defining growth: At one extreme, if a rapid, spontaneous, **homogeneous nucleation** occurs, all the particles will grow quickly, at the same time and the same rate until the concentration of the species in solution drops below a value where particle formation ceases. This will **result in homogeneous particle population of the same size**. The size will be dependent on the growth rate and concentration. If many particles are nucleated very rapidly, then the concentration will reduce rapidly, resulting in a population of many small particles, whereas a growth of less particles will result in a population with a larger particle size. The key, though, is that a homogeneous population is achieved if all the nuclei are spontaneously formed at the same time.

Heterogeneous nucleation can occur at various concentrations depending on the surface and the shape/roughness of the surface. This means that particles can nucleate at different concentrations and different times in the same solution. Thus some particles will be growing while others will still be nucleating in the same solution, leading to a **more heterogeneous size population.** A phenomenon called **Oswald-ripening** can occur which reduces the polydispersity of particle sizes somewhat. This is where the more unstable, smaller nuclei will dissolve back into the solution enabling the larger particles to grow further.

4.4.3.1 Additives restricting growth

The addition of either a soluble or insoluble additive can restrict particle growth and thus change a particle's shape/morphology. An **insoluble additive** will take the form of a **template or mould** for the particle to grow to fill or coat, depending on the template. Furthermore, crystallisation in confined spaces has an impact on the resultant material as well as its shape [9]. Several biological templates will be discussed in this context in chapter 7. Soluble additives such as small molecules can also define a particle's morphology by restricting growth of some facets over others.

A particle will nominally grow spherically at first when it is simply a cluster of atoms, as this reduces the surface energy. However, after this nucleation stage most crystals will inherit the shape defined by the crystallographic unit cell for that crystal structure. These will have defined crystal faces as facets of the particles. If an additive has a particular propensity to bind strongly to one particle face over another, the morphology of the growing crystal can be changed. A crystal face with a bound additive will have a lower surface energy, it will be more stable and thus grow more slowly than higher energy surfaces. In fact, molecules designed to stop ice forming do so by binding to the ice nuclei crystals and suppressing ice crystals from forming completely. However, if the additive binds to only one type of face only, the growth of this face is suppressed, meaning the other faces grow more rapidly and ultimately grow into extinction, leaving the additive-bound stable, slow growing face to dominate. This can be conceptually difficult to grasp, and is aided by a diagram (figure 4.8). Again this idea is expanded on with biological face-selective additives in chapter 7.

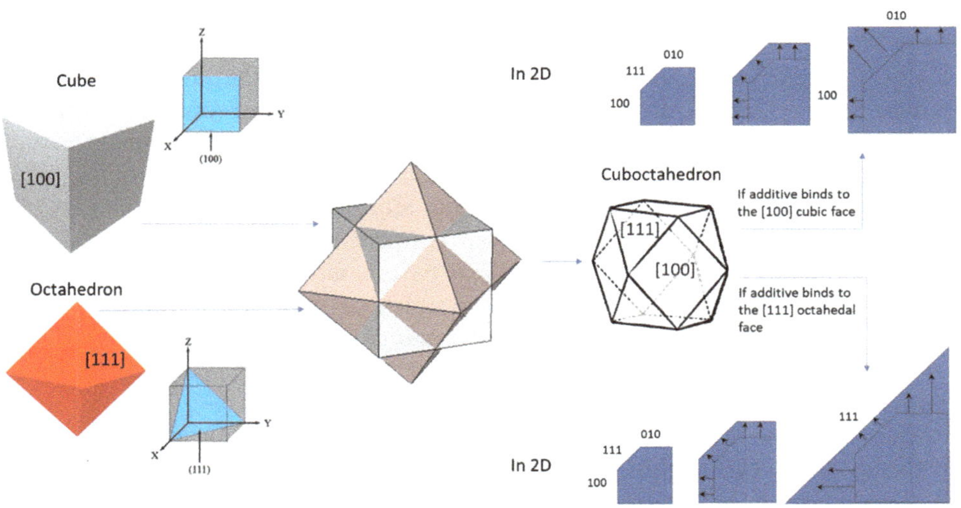

Figure 4.8. Illustration of how a cubooctahedral is formed by combining a cube and an octahedral (by cutting of the corners), and how an additive will direct the formation back to either a cubic or octahedral morphology.

4.5 Conventional bottom-up methods

This section focuses on the methods that are currently in use for large scale production of nanomaterials, while the next section is devoted to emerging methods that are showing promise for future commercial implementation. Bottom-up techniques can be broadly divided into two categories: (a) vapour phase condensation, and (b) chemical precipitation from salt solutions (table 4.1) [10]. Depending on how the energy is provided to the system or the precursor (thermally or mechanically), various sophisticated methods can branch out of these two broad categories. Examples of chemical precipitation include precipitation from an organometallic precursor or spray conversion, while examples of vapour phase methods include laser ablation, microwave plasma and spray pyrolysis. A selection of the most common methods are described below.

4.5.1 Vapour-phase method

Typically, vapour-based methods include the evaporation of metal or other precursors, followed by their deposition under vacuum on a cold surface [11]. In the case of oxides and ceramics, the chemical conversion is induced thermally, e.g. using flame process. Suitable gases can be introduced to achieve the desired composition or chemical conversion. E.g. O_2 for oxides, NH_3 for nitrides and CH_4 for carbides. This is a versatile method, which can use vapour, aerosol or fog (liquid in gas) as a feed. Flame synthesis, in particular, is widely used industrially to manufacture materials such as titania and silica despite the high temperature operations. A key advantage of this method is convenient separation of the product from the reactants and by-products, because only the product is in solid state, while other compounds are in a gas phase.

4.5.2 Solution processing

This is a simple chemical synthesis/precipitation method that is widely used to produce large quantities of nanomaterials. Aqueous solution processing starts with metal salts (e.g. chloride or nitrate) dissolved in water, followed by neutralisation, displacement and/or thermal conversion reactions to form nanoparticles, which agglomerate and precipitate. The conditions required for this method vary, but typically require low temperatures (room temperature or with slight heating to ~80 °C). The colloidal stability is a key parameter that controls the particle formation and precipitation. Typical examples of materials produced include oxides such as silica, titania, zinc oxide and manganese oxide. Although practiced widely, this method lacks the versatility to produce high quality and/or multi-component nanomaterials of desired composition.

Aqueous solution processing is also used to produce metallic nanoparticles by chemically reducing aqueous metal salt solutions. The particle sizes are controlled via the choice of the precursor, the reducing agent used, the solvent and the temperature. Addition of surfactants or capping agents is commonly used to arrest nanoparticle growth and avoid aggregation (detailed in section 4.6.5). From a commercial viewpoint, the downstream processing is challenging and costly (e.g. the separation of nanoparticles, or the removal of unreacted precursors).

4.5.3 Spray conversion

This method is used to produce multicomponent nanomaterials. This can be achieved by producing each constituent particle first and then combining them, typically via sintering. Alternatively, in spray conversion, the aqueous solutions of the precursor salts of each constituent are mixed and then spray dried in order to produce a 'composite' precursor powder. This precursor is then thermochemically converted to produce the desired multicomponent nanomaterials. Important examples include tungsten carbide and cobalt composites [10].

4.5.4 Sol–gel method

This is a highly versatile solution-based, low-temperature (~60 °C–120 °C) method useful for producing a range of multicomponent oxides and composites [11]. Sol–gel processing works on the principle that metals can react with water (hydrolysis) forming metal hydroxide colloids ('sol'), which can further react or condense to form a gel under controlled conditions. The use of catalysts for both the hydrolysis and condensation steps, the co-solvents and the precursor chemistry play crucial roles in controlling the sol–gel process. This route has been widely adopted at commercial scale in the production of high value nanomaterials such as oxides and ceramics (e.g. silica, titania, ZnO), particularly where multicomponent materials with a high level of control over their properties is required.

4.6 Emerging bottom-up methods

In recent decades, a range of new methods have been invented and developed for the bottom-up synthesis of 2D and 3D nanomaterials. A predominant common feature of these methods is the use of self-assembly of atoms, (bio)molecules or clusters composed of organic, inorganic or hybrid species. Before discussing individual methods, let us first grasp the principles behind self-assembly.

4.6.1 Principles and overview

Self-assembly can be defined as (after [12]) a spontaneous process (under the right conditions/trigger) which results in 'aggregates' of length scales larger than its components, such that there is a higher order, stable structure and/or specific properties, typically not present in its components. These special features are rendered due to the ability of the components to recognise specific binding sites, resulting in selectively positioning themselves into the final non-covalently connected structures. The self-assembled structures can be static or oscillating between different states. They find vast applications in templating nanomaterials; we will describe this further with an example later in this section.

Self-assembly is extremely important in bottom-up fabrication used for both 'soft' and 'hard' materials [12]. Components that can self-assemble, fall between the sizes that can be controlled chemically and those that can be manipulated by conventional manufacturing. Hence self-assembly offers a crucial technique to bridge top-down and bottom-up methods (described below with the example of soft lithography). As self-assembly requires components to be mobile, it typically occurs in the fluid phase.

As such, significant progress has been made in the synthesis of nanomaterials using bottom-up approaches which employ sophisticated techniques such as self-assembly, nano–bio interactions, and templated synthesis [6, 12, 13]. Self-assembly is an integral part of bottom-up synthesis approaches. It allows materials and structures to organize themselves at different length scales, using a concerted combination of weak and noncovalent chemical interactions—hydrogen bonds (water-mediated), ionic bonds (electrostatic interactions), hydrophobic interactions and van der Waals interactions—between individual molecular building blocks. Some great examples of self-assembly are biological building blocks, ranging from nucleotides and amino acids to sugars and lipids, that self-assemble to form nucleic acids, proteins, peptides, and supramolecular structures such as membranes and vesicles. From the industrial manufacturing point of view, cooperativity in the assembly of the building blocks at different length scales is an important principle. It offers the opportunity to control the material architectures at various length scales. Manipulation of molecules into functional nanomaterials using self-assembly requires a deep understanding of the following:

- The complex nature of the weak and non-covalent chemical interactions that lead to thermodynamically stable nanostructures.
- Pathways for assembling building blocks into larger and more complex nanomaterials starting from a limited number/type of precursors.
- The chemically-coded information in the molecular building blocks to obtain self-assembly (conformation and charges).

Self-assembling nanomaterials range from natural biopolymers (DNA, proteins) to small-molecular hydrogels and organic–inorganic composite materials such as metal organic frameworks (MOFs) [14]. Nanomaterials from the assembly of polypeptides and DNA with inorganic clusters/particles have been produced by several research groups for an array of applications, ranging from gene therapy to electronics [15].

Accordingly, small molecules exhibiting molecular recognition such as oligopeptides or PEG can be used to take advantage of their similar gelation behaviour to biomolecules. These materials are particularly useful in biomedical applications due to their often biocompatible nature, leading to exploration for drug delivery or as medical adhesives. In addition to their greater abundance than biomolecules, these gels are made at ambient conditions with only mild initiation required (e.g. oxidative conditions [16] or enzymatic decomposition [17]), therefore representing a low environmental and energetic cost of manufacture.

With this background information on principles involved in bottom-up methods, in particular the role of self-assembly, let us consider selected examples that have been of enormous interest for fabricating/producing nanomaterials. As these methods are of emerging nature, there are limited commercial products from these methods, however, they have great potential in future applications.

4.6.2 Soft lithography

Soft lithography uses a combination of traditional lithography, self-assembly and chemical synthesis to produce patterned features, mainly on surfaces, creating two-dimensional materials. Typically, these features can span the sub-micron to nano-scales and therefore they are of considerable interest, as traditional lithography cannot reach these length scales without difficulties. Further, soft lithography enables the patterning of features with chemical (charged, acidic or basic), physical (wetting, hydrophobic or hydrophilic), biological (proteins or cells), electrical (conductivity), etc functionalities [6].

Briefly, soft lithography works by using traditional photolithography to create a master stamp (or a mask or mould) made up of a polymer (polydimethylsiloxane (PDMS)) [18]. This stamp is then used to pattern the surface of interest via patterning with chemical 'ink' (e.g. micro-contact printing). Subsequent sequential chemical treatments can create surfaces patterned with the desired functionality. The following procedure describes the entire process step by step, and this is also summarised in figure 4.9.

1. The first step in soft lithography is to create a ***mask*** with the designed pattern. This can be achieved by a suitable computer aided design (CAD) package and traditional lithography.
2. Next, a photoresist layer is created on a substrate such as silicon wafer.
3. The photoresist layer is then selectively exposed to UV light via the ***mask*** to cure the exposed areas.
4. Unexposed (or exposed) areas are then washed away by chemical treatments to create a '***master***' for the desired stamp.
5. This master is surface treated with fluorosilane to reduce its stickiness and allow easy removal of the stamp once casted.
6. The master is then filled with PDMS pre-polymer, cured and then released to obtain a ***stamp*** with the pre-defined pattern.
7. The patterns on the PDMS stamp are then coated with a chemical ink (typically alkane thiols) via simply dipping the stamp in the ink.
8. The substrate desired to be patterned is pre-coated with gold (and titanium) to allow the ink to be transferred via thiol–gold bonds.
9. The ink is then ***printed*** on the substrate to create a chemically patterned surface. One common way of achieving this is micro-contact printing, which means simply stamping the chemical ink carefully on the substrate.
10. The parts of the substrate that remain uncoated with the ink are ***etched*** away by washing with a cyanide and hydrofluoric acid.
11. Finally, subsequent ***finishing*** treatments can be performed to produce a patterned surface with the desired chemical and physical properties. These treatments can include depositing additional chemicals or activating the surface for its use.

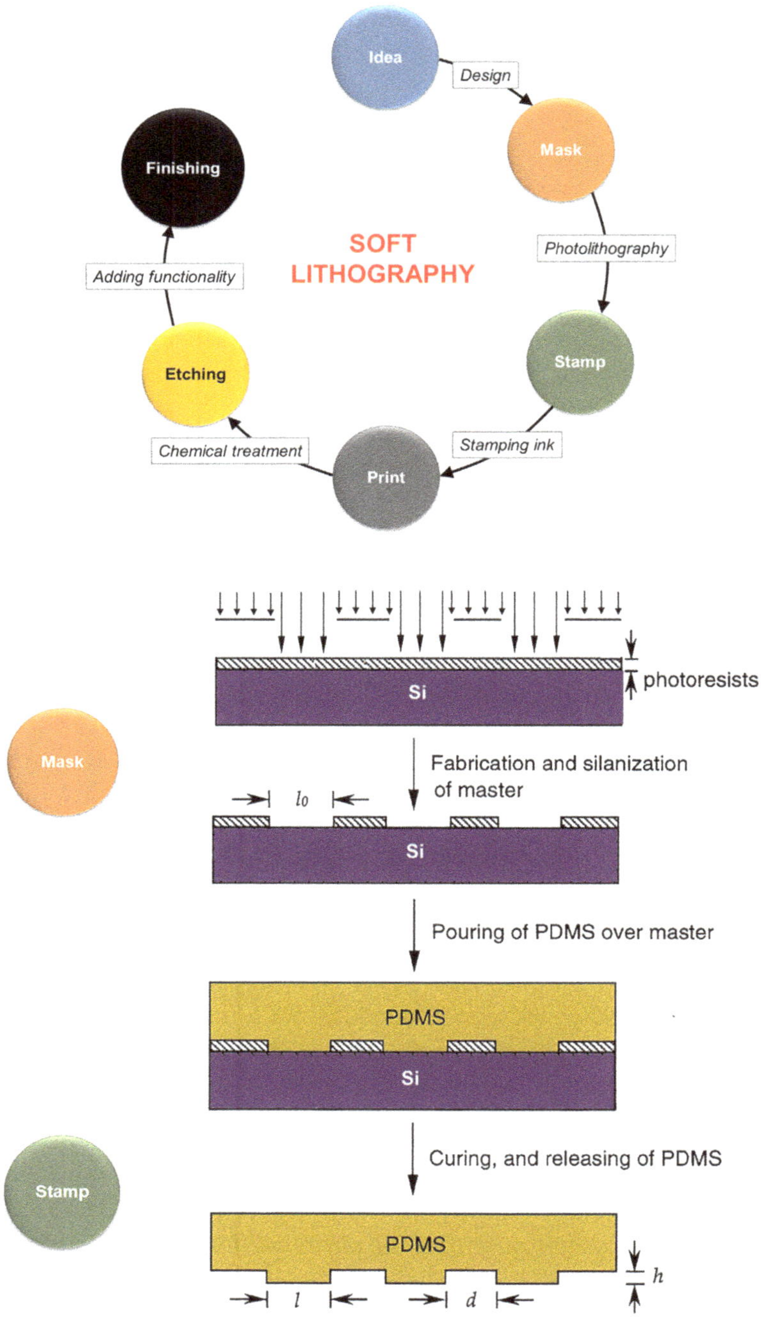

Figure 4.9. Overview of the steps involved in soft lithography (top); schematic showing how a PDMS master stamp is made (middle) and used for patterning surfaces (bottom). Adapted with permission from [6, 18, 19], copyright Royal Society of Chemistry, John Wiley & Sons and Annual Reviews. Typical values for h, d, l_0 and l are 0.2 ± 20, 0.5 ± 200, 0.99l and 0.5 ± 200 μm, respectively. Stages shown in the top image are clearly identified in the rest of the figure. (The figure is continued on the next page.)

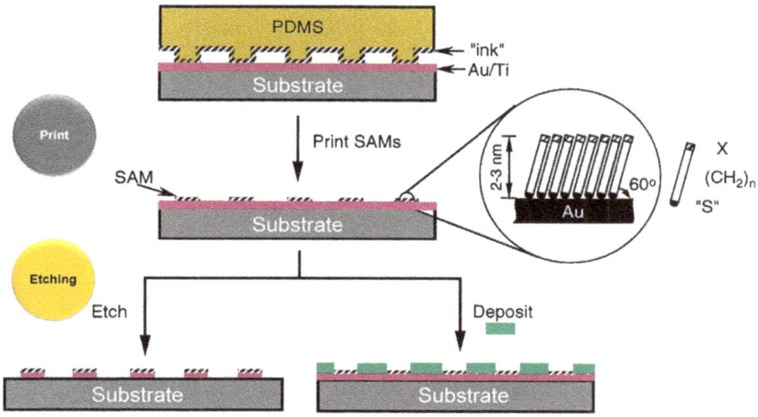

Figure 4.9. (Continued.)

The selection of the ink (from alkane thiols of a variety of functional groups to even cells) and chemical treatments in steps 10 and 11 (including the growth of nanoparticles/crystals) enable the generation of a range of surface functionalities. A clear novelty of soft lithography is how elegantly it combines top-down methods (steps 1–6 and 10) with bottom-up methods (steps 7–9 and 11).

4.6.3 Dip-pen nanolithography

An obvious limitation of soft lithography is its inability to reach nanoscales readily when patterning surfaces (see typical dimensions in figure 4.9). In order to address this challenge, nanoscale quills can be used to directly 'write', instead of PDMS stamps, to transfer/create patterns onto substrates. A tip of a scanning probe microscope, which is ~10 nm or smaller in size, is an ideal nano-quill. This has resulted in a new method called dip-pen lithography, which is a further extension of soft lithography. Dip-pen nanolithography (DPN) works by depositing molecules from their solution droplet adhering to the tip or the pen directly on the substrate, as shown in figure 4.10. DPN has opened up the possibility of patterning down to 10 nm. A diverse range of inks have been used for DPN, such as molecules, enzymes, nanoparticles, reactive precursors and viruses, for patterning various substrates.

4.6.4 Layer-by-layer self-assembly

This method is driven by electrostatic self-assembly between oppositely charged species containing multiple ions. These polyions can be deposited one layer at a time on charged surfaces. Although a single electrostatic 'bond' is much weaker than a covalent bond, the formation of multiple electrostatic attractions provides robustness to these films/layers. A typical procedure for performing layer-by-layer (LbL) assembly includes starting with priming the surface of a substrate to render electrostatic charge. This is typically achieved by chemical treatment with reactive

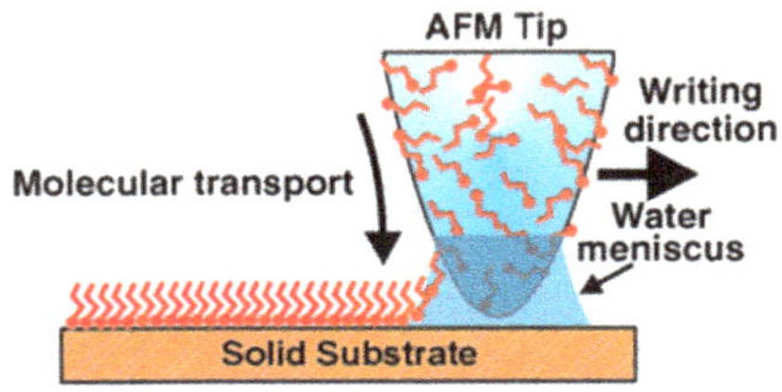

Figure 4.10. An overview of DPN. Image reproduced with permission from [20], copyright 2004 John Wiley & Sons. AFM stands for atomic force microscope, a form of scanning probe microscope.

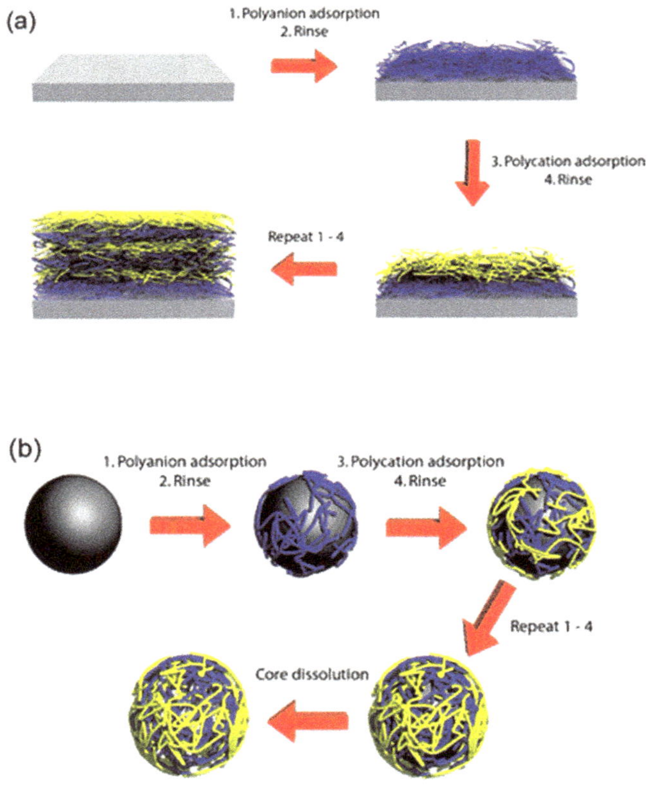

Figure 4.11. Overview of LbL assembly. Image taken from [21], reproduced by permission of The Royal Society of Chemistry.

molecules such as silanes and thiols. Next, the surface is treated with a polyion, rinsed and then coated with the next polyion of opposite charge, and rinsed (see figure 4.11(a)). This process can be repeated multiple times to obtain multilayers. An interesting feature is that the substrate does not need to be a flat surface; LbL works even on particles, as shown in figure 4.11(b).

The versatility of this method can be appreciated further once we consider the possibility that the polyion species could be polymers, proteins, peptides, polynucleic acids, viruses, colloids or functional nanoparticles. By manipulating the conditions (pH, ionic strength and solvents) and selecting the desired chemistry of the polyions (composition and structure), the LbL growth and properties of the final structures can be controlled. Some examples of this method include the fabrication of:

- conducting layers, by using electroactive nanoparticles or polymers,
- stimuli responsive surfaces, by exploiting the swelling behaviour of polymers under certain conditions to adsorb or release active components (e.g. pollutant removal or drug delivery),
- porous multilayers, by selectively dissolving colloids from a colloid–polyelectrolyte assembly and then using the pores to host enzymes.

4.6.5 Solution synthesis of nanoparticles

A number of solution-based synthesis methods exist for nanoparticles. They use the principles of direct reaction in combination with solvothermal or hydrothermal synthesis. When performing a direct reaction, for example in chemical precipitation, a suitable precursor (generally a salt) is dissolved in water and 'neutralised' to induce the precipitation of particles. Although widely used in industry, this method offers limited control over particle sizes, their polydispersity and structure (which is important for crystalline materials). Given the importance of such methods for producing nanomaterials in bulk, in order to address these issues, an 'arrested nucleation and growth' technique has been developed for the controlled synthesis of nanoparticles. This method uses an organic capping agent to rapidly trap nuclei and allow controlled slow growth of these nuclei into nanoparticles. The capping agents also help improve the colloidal stability of the nanoparticles thus formed. This method has been extensively used in the past two to three decades for the synthesis of a range of metallic nanoparticles, which is reviewed elsewhere [6, 22, 23].

With the examples of two semiconductor nanoparticles (CdSe and CdS), we describe this synthesis method. Typically, a capping agent is selected such that it can also act as a solvent. Trioctylphosphine oxide (TOPO) is a commonly used capping agent cum solvent in the synthesis of CdSe nanoparticles. The synthesis occurs by heating dimethyl cadmium in TOPO to 300 °C, followed by rapid addition of tributylphosphine selenide. Over several hours of growth period, TOPO capped nanoparticles are obtained by washing steps and antisolvent precipitation using acetone. In such synthesis, high temperature is required to generate high quality crystallinity, while the solvent/capping agents control the particle sizes and polydispersity. CdS nanoparticles are also synthesised in a similar fashion where oleylamine is used instead as a capping agent solvent. Sulphur is directly added to a pre-heated solution of $CdCl_2$ in oleylamine [22]. The growth occurs slowly over up to 6 h at temperatures of 140 °C–160 °C, to form nanoparticles, which can be again isolated with antisolvent precipitation using ethanol. These are some of the early and pioneering examples of controlled nanoparticle syntheses in solution. There have been a very wide range of investigations reported where modifications are made to these methods. Modifications include the substitution of the capping agents with more intricate chemistries and/or compatibilities with solvents of choice. A range of surfactants and polymers have been used as capping agents. Other modifications include a two-step synthesis where in the first step a metal-complex is prepared, which is then reacted in the second stage, allowing further control over the particle formation.

4.6.6 Templated synthesis

Aside from using self-assembly to produce a material itself, self-assembling templates are often used to impart long-range order in higher-volume materials (mostly inorganic). Templated synthesis enables the formation of nanoporous materials. These can be viewed as nano-sponges where the pores can be of the sizes of small molecules (<2 nm, called micropores) or larger (between 2–50 nm, called mesopores, and >50 nm, called macropores). Zeolites (microporous aluminosilicate networks), also called molecular sieves due to their ability to 'filter' molecules based on their sizes, are often templated around single molecules such as small alkylammonium ions. Materials templated around self-assembling surfactant molecules have begun to introduce mesopores and more complex structures into materials. The templates self-assemble into micelles or liquid-crystals, leading to porous materials with either larger porosities than are available through simple molecular templating alone, or lamellar/interconnected structures [12].

These features are best illustrated with an example of one of the most studied materials—mesoporous silicas (see figure 4.12), although we note that similar strategies have been applied to many other nanoporous materials. In a typical

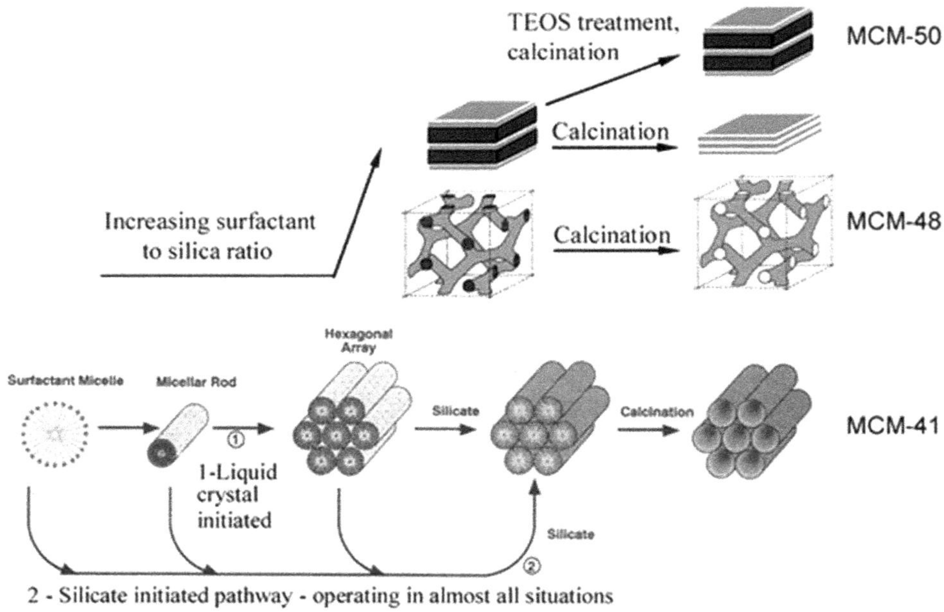

Figure 4.12. Synthesis of porous materials templated around self-assembling surfactants. Image reproduced by permission of The Royal Society of Chemistry from [24].

synthesis, cetyltrimethylammonium bromide $[(C_{16}H_{33})(CH_3)_3N^+Br^-]$ is used as the template. Depending on the template concentration, ionic strength of the system, the solvents used and the temperature, the template can self-assemble into spherical micelles, micellar rods or lamellae; these rods can further pack in to hexagonal or cubic arrangements [24]. When treated with the inorganic precursor, these assemblies can template a framework around them, leading to porous materials upon template removal (typically by calcination or solvent reflux) [25]. An interesting point to note is that contrary to earlier beliefs, the surfactant assembly is affected by the presence of inorganic precursors and hence the templated synthesis includes complex cooperative assembly between the template and the inorganic species, making it more interesting [26].

4.7 Summary: key lessons about conventional routes to nanomaterials

In this chapter, we have learnt that there are a number of methods for producing nanomaterials, and that they are broadly classed into top-down (or destructive) and bottom-up (constructive) methods. Bottom-up methods are becoming more popular due to the various advantages they offer, including the ability to tailor the composition, size and structure of nanomaterials. Generally speaking, they intrinsically produce less waste. The key features in bottom-up methods are understanding and applying nucleation and growth principles. The utilisation of self-assembly (e.g. templated synthesis) and the combined use of top-down and bottom-up strategies (e.g. soft lithography) is creating valuable emerging methods for nanomaterials synthesis, which have great potential for less wasteful commercial implementation.

References

[1] Sengul H, Theis T L and Ghosh S 2008 *J. Ind. Ecol.* **12** 329
[2] The Royal Society and The Royal Academy of Engineering 2004 Nanoscience and nanotechnologies: opportunities and uncertainties, Joint report https://royalsociety.org/~/media/Royal_Society_Content/policy/publications/2004/9693.pdf
[3] Charitidis C A, Georgiou P, Koklioti M A, Trompeta A-F and Markakis V 2014 *Manuf. Rev.* **1** 11
[4] Neisser M and Wurm S 2015 *Adv. Opt. Technol.* **4** 235
[5] Backhurst J R, Harker J H, Richardson J F and Coulson J M 2002 *Coulson and Richardson's Chemical Engineering* 5th edn (Oxford: Butterworth-Heinemann)
[6] Arsenault A C, Ozin G A and Cademartiri L 2009 *Nanochemistry: A Chemical Approach to Nanomaterials* 2nd edn (Cambridge: Royal Society of Chemistry)
[7] LaMer V K and Dinegar R H 1950 Theory, production and mechanism of formation of monodispersed hydrosols *J. Am. Chem. Soc.* **72** 4847–54
[8] Hamilton B D, Ha J-M, Hillmyer M A and Ward M D 2012 *Acc. Chem. Res.* **45** 414
[9] Wang Y-W, Meldrum F C and Christenson H K 2013 Confinement leads to control over calcium sulfate polymorph *Adv. Funct. Mater.* **10** 6
[10] Kear B H and Skandan G 2000 *Ullmann's Encyclopedia of Industrial Chemistry* (Weinheim: Wiley)

[11] Scherer G W and Brinker C J 1990 *Sol-gel Science* (New York: Academic)
[12] Overton T, Weller M T, Rourke J and Armstrong F A 2014 *Inorganic Chemistry* (Oxford: Oxford University Press)
[13] Service R F, Szuromi P and Uppenbrink J 2002 *Science* **295** 2395
[14] Loweth C J, Caldwell W B, Peng X G, Alivisatos A P and Schultz P G 1999 *Angew Chem. Int. Ed.* **38** 1808
Mirkin C A, Letsinger R L, Mucic R C and Storhoff J J 1996 *Nature* **382** 607
Paul Alivisatos A, Johnsson K P, Peng X, Wilson T E, Loweth C J, Bruchez M P and Schultz P G 1996 *Nature* **382** 609
[15] Niemeyer C M 2001 *Angew Chem. Int. Ed.* **40** 4128
Aida T, Meijer E W and Stupp S I 2012 *Science* **335** 813
Zhang S G 2003 *Nat. Biotechnol.* **21** 1171
Zhang Z L, Horsch M A, Lamm M H and Glotzer S C 2003 *Nano Lett.* **3** 1341
[16] Lee B P, Dalsin J L and Messersmith P B 2002 *Biomacromolecules* **3** 1038
[17] Sadownik J W, Leckie J and Ulijn R V 2011 *Chem. Commun.* **47** 728
[18] Xia Y N and Whitesides G M 1998 *Angew Chem. Int. Ed.* **37** 550
[19] Xia Y N and Whitesides G M 1998 *Annu. Rev. Mater. Sci.* **28** 153
[20] Ginger D S, Zhang H and Mirkin C A 2004 *Angew Chem. Int. Ed.* **43** 30
[21] Quinn J F, Johnston A P R, Such G K, Zelikin A N and Caruso F 2007 *Chem. Soc. Rev.* **36** 707
[22] Joo J, Na H B, Yu T, Yu J H, Kim Y W, Wu F, Zhang J Z and Hyeon T 2003 *J. Am. Chem. Soc.* **125** 11100
[23] Daniel M C and Astruc D 2004 *Chem. Rev.* **104** 293
[24] Kresge C T and Roth W J 2013 *Chem. Soc. Rev.* **42** 3663
[25] Patarin J 2004 *Angew Chem. Int. Ed.* **43** 3878
[26] Huo Q, Margolese D I, Ciesla U, Demuth D G, Feng P, Gier T E, Sieger P, Firouzi A and Chmelka B F 1994 *Chem. Mater.* **6** 1176

Section III

From biominerals to green nanomaterials

Image courtesy of Kateryna Kon/Shutterstock.

The previous section has highlighted the need for and advantages of nanomaterials, as well as demonstrating current production methods for such materials. The first section described the principles of green chemistry. This section begins by bringing together these two concepts and evaluating existing methods using green principles. Key challenges and areas of improvement are identified, and the use of biology as a source for inspiration is proposed. From here the production of inorganic materials within biology at various scales with incredible precision is evaluated in the next chapter, to gain insight into and a better understanding of the concepts and tools that Nature utilises. The final chapter of this section highlights the importance of learning from biomineralisation in order to discover new and green routes to sophisticated nanomaterials, by providing a platform for designing synthetic methods that offer the control and sophistication seen in biomineralisation yet without the need for native biological components or machineries. The key rules and insights obtained from biomineralisation are systematically identified, with examples and lessons learnt explored to develop green bioinspired routes to make nanomaterials. Moreover, the key challenges faced in scaling up and making these routes commercially viable are discussed, as these are better solved at the discovery phase.

IOP Publishing

Siddharth V Patwardhan and Sarah S Staniland

Chapter 5

Green chemistry for nanomaterials

In this chapter, we aim to combine our knowledge from previous chapters on green chemistry principles and nanomaterials synthesis. We will evaluate how green existing nanomaterials synthesis/manufacturing methods are, and understand the reasons behind them. With selected examples of nanomaterials, the analysis of sustainability is illustrated in this chapter. It forms a platform for the remaining book by identifying ways to discover green routes to nanomaterials.

5.1 Sustainability of nanomaterials production

There are two major challenges with nanomaterials—safety and sustainability. The toxicity of nanomaterials has been widely investigated as discussed in chapter 2, however, the environmental burden from their production is largely ignored [1]. As described in chapter 4, at present, nanomaterials are manufactured using top-down (lithography, milling and etching) or bottom-up (vapour deposition, sol–gel, precipitation, pyrolysis, solvothermal) approaches [2, 3].

The success of a process relies on controlling the key specifications for a desired nanomaterial. These include nanoscale features such as particle size, particle size distribution, the degree of agglomeration and purity. At the commercial stage, the cost of the nanomaterial production also plays a key role. If the cost of the nano-product is more than traditional or 'current-best' materials, then the nanomaterials need to perform significantly better, thus providing an added value for the additional cost.

Bottom-up approaches such as hydrothermal or sol–gel synthesis are promising, but they suffer from many problems leading to high costs, an extremely adverse environmental impact (hazardous waste and toxicity problems) and unsustainable production. Indeed, the environmental impact and sustainability analysis performed using E-factor (waste-to-product ratio) revealed that the current nanomaterials production methods are up to 1000 times more wasteful when compared to the production of pharmaceuticals and fine chemicals (see table 5.1) [4]. Despite being

doi:10.1088/978-0-7503-1221-9ch5

Table 5.1. A comparison of wastefulness of various manufacturing sectors. Adapted from [4].

Sector	Waste: product ratio (E-factor)
Bulk chemicals	<1 to 5
Fine chemicals	5–50
Pharmaceuticals	25 to >100
Nanomaterials	100–100 000

highly wasteful, and hence expensive (as described below), these approaches predominate the current manufacturing processes for nanomaterials. Significant progress has been made in the synthesis of nanomaterials using bottom-up approaches which employ sophisticated techniques such as self-assembly and nano–bio interactions. However, these recent developments still suffer from multi-step, laborious methods leading to uneconomical scale-up and/or high wastefulness.

In addition to existing nanomaterials, new nanomaterials are being invented (e.g. quantum dots, metal organic frameworks, carbon nanotubes), but due to a lack of sustainable manufacturing methods, these materials are difficult to scale-up, and uneconomical to commercialise [4, 5]. Indeed, a recent article in *Chemistry World* clearly highlighted these scale-up issues with new materials or processes [6].

5.2 Reasons behind unsustainability

Specific issues with existing methods for manufacturing nanomaterials are shown in figure 5.1, and are also listed below and described with suitable examples [3].

1. **High purity requirements**: Ultrapure precursors, reagents and solvents are typically needed, which demand high energy for purifying reagents and stake into scarce resources (e.g. potable water). Further, due to extremely low tolerance for contamination, the recycling of solvents and unreacted reagents becomes difficult, if not impossible.
2. **Low yields**: Although some progress has been made recently to improve reaction yields, the material utilisation efficiencies over the entire process (reaction, separation, finishing) are poor for nanomaterials, ranging between 1%–20% (typically 3%–10%), causing wastefulness; such methods are not suitable for scale-up.
3. **Repeated/sequential processing**: Many processes for nanomaterials synthesis/manufacturing require sequential steps (e.g. layer-by-layer deposition) including repeated washing for purification. Due to low efficiencies in each step, the waste produced multiplies.
4. **Toxic or hazardous chemicals**: Each of the current methods for nanomaterials either use or produce toxic or hazardous chemicals including greenhouse gases (e.g. the use of metal halides or alkoxides as precursors in bottom-up methods, and emission of PH_3 gas from thin film deposition).

Figure 5.1. Summary of the issues associated with current manufacturing methods for nanomaterials. Courtesy of Yung Hei Tung.

5. **Need for specialised environments**: Specialised environments such as ultrahigh vacuum (<5 torr), or high temperatures to melt metal salts (>1000 °C), are typically required for nanomaterials production, which create an additional cost and energy penalty.
6. **High consumption of water and energy**: Most of the issues listed above dictate, directly or indirectly, high consumption of water and energy. It has been estimated that to create a 1 cm^2 nano-device, 10–60 L of water and 1 kWh energy are required, which create an enormous burden on the environment.

5.3 Evaluation of sustainability for selected methods

5.3.1 E-factors for solution methods

In order to illustrate these issues, we provide examples of nanomaterials synthesis. The E-factor results shown in table 5.1 consider three types of nanomaterial—carbon nanotubes, titania and gold nanoparticles—and selected methods for each [4]. It was found that for titania produced using an industrial hydrolysis–calcination method (a hybrid of sol–gel and vapour-phase synthesis) resulted in an E-factor of 17800. This means that to produce 1 kg of titania using this method, nearly 18 tonnes (yes, tonnes!) of waste is generated. For solution synthesis of gold nanoparticles, the E-factor was between 3320 and 99400 for chemical processing, depending on the capping agent and solvent used. A key finding was that downstream processing such as purification increased the waste produced significantly. This means that when reporting and evaluating the sustainability of a method for nanomaterials synthesis, we should quantify the washing steps and avoid phrases such as 'washed with copious amounts of solvent', 'rinsed until no further change in conductivity was recorded', or 'the antisolvent was added until precipitation occurred'. In the case of semiconductor nanoparticle synthesis (section 4.6.5), simply noting the temperatures required (>140 °C), the duration of the synthesis (>6 h) and the chemicals involved, it is clear that these methods are unlikely to be regarded as sustainable. Attempts to improve resource efficiency sometimes result in two-step methods, where in the first step a metal complex is prepared, which lowers the temperature and/or time to synthesise nanomaterials in the second step. As explained in chapter 1 (equations (1.8), (1.11) and (1.12)), such modified methods may indeed be more unsustainable as they overlook the environmental burden of the first step and only focus on the improvements in the second step. An example of this is the synthesis of CdSe quantum dots using a 'single-source precursor' Cd $(S_2CNMeHex)_2$ containing a complex of Cd and Se [7]. Heating this precursor in tri-*n*-octylphosphine (TOP) and tri-*n*-octylphosphine (TOPO) produced high quality CdSe nanoparticles, perhaps with properties better than a common solution synthesis explained in section 4.6.5. However, when one looks at the synthesis of the single-source precursor, which uses carbon diselenide, the authors noted 'carbon diselenide is a very toxic, evil-smelling liquid and extreme caution must be observed during preparation and manipulation.' This clearly means that the first step of this two-step synthesis is highly unsustainable, thereby perhaps not improving the green credentials of the overall method.

5.3.2 How green is soft lithography?

Soft lithography, described in chapter 4, has been very popular in recent decades and is considered to be a revolutionary method for producing nanomaterials. It is important to reiterate that soft lithography has truly extended surface patterning to length scales that were not readily accessible using traditional lithography or other methods. Therefore, let us consider its sustainability in order to illustrate the points made above. At this point, we note that we could not find a single published article related to the life cycle or environmental analysis of soft lithography [1]. We will therefore have to use qualitative information in order to assess this method. The first thing that strikes one when looking at the soft lithography process (see section 4.6.2 and figure 4.9), is how rather lengthy the procedure is. It requires more than 11 steps, which means that the method is bound to suffer from the issues faced by sequential and repeated processing. Next, consider steps 4, 10 and 11, which pertain to some sort of washing or etching. As soft lithography is based on sequential addition and selective removal of layers, these steps are bound to produce relatively large amounts of waste (compared to the actual material un-washed/retained as a pattern), resulting in high E-factor. Further, it is remarkable to note the variety of chemicals required for this method; most of these are extremely hazardous and require extreme care during and specialised environments for handling. These include fluorosilane, used in step 5, and cyanide and hydrofluoric acid used in step 10. Although not generally reported, it is likely that the yield or production rates (patterned area produced per time) are low, particularly when it takes up to a day to create one layer thick patterns. Combining all these points together, it appears that soft lithography would score very low on sustainability or green chemistry ratings. To balance this, it is very important to consider this environmental cost or burden against the benefits offered, to make an educated prediction of its value. This is important because sometimes the ability to pattern at desired length scales offers a wider benefit over a longer term.

5.3.3 Templated synthesis: surely sustainable?

Similar to the methods discussed above, in the case of templated materials, e.g. mesoporous silica considered in section 4.6.6, there remain several issues with these materials, both in terms of the common requirement for hydrothermal synthesis conditions and for the need to remove the template materials prior to use. Although it has been shown to be feasible to modify the synthetic conditions such that hydrothermal methods are not necessary, mild template removal methods are still in their infancy. Therefore, there has been significant research into either templates that are designed for non-destructive removal, or more efficient removal methods—either by weakening the surface–template interaction or increasing the solubility of surfactant in another way such that irradiation, oxidative conditions, or solvent reflux—are capable of purifying the material [8]. Despite the improvements found by these studies, issues with the quality of materials resulting from these sustainability-minded compromises remain, either in terms of residual template contamination or inferior properties. As a result, no such methods have progressed to larger than lab-scale syntheses.

5.4 Adopting green chemistry for nanomaterials

It seems that most methods that we considered revolutionary have given little attention to their sustainability. Despite their preparation being very wasteful and environmentally damaging, most attention aside from material performance has only focused on the toxicity of nanomaterials [9]. However, existing methods for manufacturing nanomaterials suffer from issues pertaining to sustainability. These issues dictate high consumption of water and energy, creating an enormous burden on the environment and result in unsustainable manufacturing, yet the environmental burden of the synthesis/manufacturing process remains largely ignored [1, 10]. These factors clearly stress the urgent need for developing fundamentally new production methods for nanomaterials that are green and sustainable.

Adopting the principles of green chemistry and engineering has the potential to develop greener routes (see figure 5.2) [1]. A solution suggested to overcome the sustainability issues is to adopt syntheses procedures that reduce waste and that operate at room temperature and in aqueous solutions, which in turn can lower the cost of production [9]. A recent white paper produced by the Royal Society of

Figure 5.2. Illustration of how the 12 principles of green chemistry can be used to design new strategies for sustainable methods for the synthesis and manufacturing of nanomaterials. This figure has been adapted from [13] with permission from The Royal Society of Chemistry.

Chemistry has identified 'green materials and processes' as a priority area for future research, in order to deliver new efficient large-scale industrial processes [11].

A World Technology Evaluation Center report [2] on research directions in nanotechnology explicitly recommended that achieving green manufacturing by 2020 is the 'holy grail'. This report suggested that future research should focus on emulating natural designs to develop scalable processes for manufacturing nanomaterials. Further, it has been highlighted that for bioinspired synthesis, future research is required to investigate scale-up, advanced processing and precise control of nanomaterials properties; their implementation into industrial manufacturing will lower the costs and broaden the applicability of nanomaterials [13]. As such, the next chapters will cover how biology makes nanomaterials (chapter 6), and what can be learnt from this, as well as the emerging rules for developing bioinspired syntheses (chapter 7). This will be followed by two case studies. Given that we will be focusing on and discussing a range of biological and biochemical concepts, in the next section we aim to introduce the relevant terminology as a simple guide, while for details, we will refer to specialised texts.

5.5 Biological and biochemical terminology and methods

5.5.1 Molecular biology component

5.5.1.1 Amino acids, proteins and enzymes

Some of the key components of biological machinery are proteins and nucleic acids. Proteins are composed of amino acids (AAs), which are also called residues. AAs have a common structure, as shown in figure 5.3 (top). There are 20 naturally

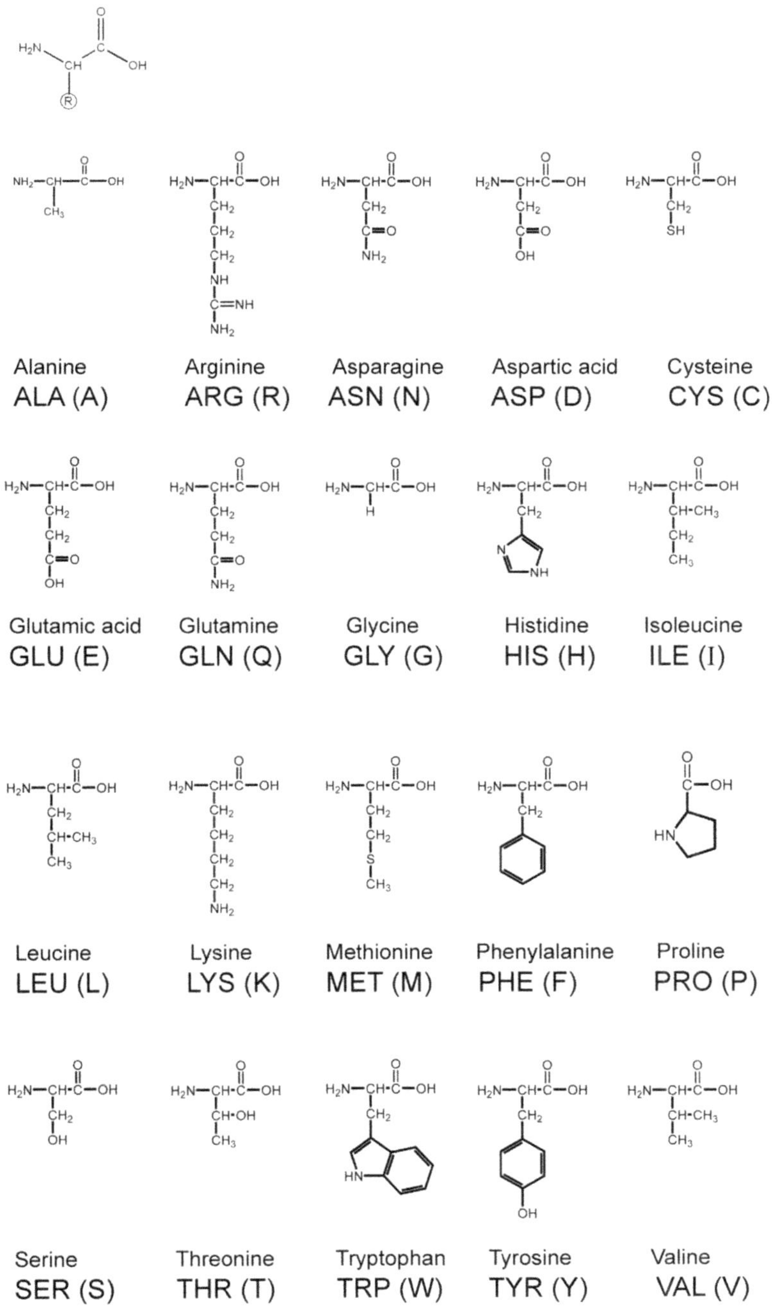

Figure 5.3. Top: generic structure of amino acids with *R* shown as the side/functional group. Bottom: a list of amino acids, their structures, and their three-letter and one-letter abbreviations. Image adapted from [15] with permission from The Royal Society of Chemistry.

occurring AAs, which differ from each other in the side group (shown as *R* in figure 5.3) [14]. AAs are abbreviated to either a three letter code, or a one letter code, as shown in the list in figure 5.3. For example, the AA leucine is denoted as L or LEU. The chemistry of the side group *R* provides specific properties to a given AA. These groups can be categorised as hydrophobic (aliphatic, cyclic or aromatic) or hydrophilic (acidic, basic or uncharged). Histidine, methionine and cysteine are special kinds of AAs, due to the presence of either a heterocyclic side group (histidine) or the presence of sulphur.

The amine group ($-NH_2$) from one AA reacts with the carboxylic hydroxyl group (–OH) from another AA to form a peptide bond. Further addition of AAs results in larger molecules (or polymers), known as proteins. The sequence of AAs in a protein, called their primary structure, ultimately determines their properties and function. Proteins fold into higher order structures as shown in figure 5.4. The secondary structure of a protein arises from the intramolecular assembly of a protein molecule, and commonly results in α-helix, β-sheet, random coil, or β-turns configurations. Non-covalent weaker bonds such as hydrogen bonding and van der Waal's forces are generally responsible for helping form these structures. Within a single protein molecule, a number of such secondary structures are common; their arrangement with respect to each other gives a protein its tertiary structure. In some cases, a number of protein molecules assemble together via intermolecular bonding to create a quaternary structure.

The structure of a protein is very important as it controls the protein function, ranging from its solubility to its recognition by other cellular components. Their ability to recognise certain targets is key in their function, an ability also exploited for controlling nanomaterials recognition as discussed in the next few chapters. Denaturation, breaking protein structure, leads to loss of function and in some cases to detrimental effects. In order to form a stable structure, there needs to be a large (> approx. 30) number of AAs. In the case of peptides, which have a much lower number of AAs, hierarchical structures are not readily possible. However, specific peptide sequences can be designed to self-assemble into supramolecular structures.

Enzymes are a special class of proteins which possess catalytic activities towards certain reactions. Enzymes are highly specific towards reactants (also called substrates) and highly selective in their function. The enzyme's structure and its catalytic 'pocket' renders its function. The pocket provides a selective site for the substrate to bind and react, the change from the substrate to the product results in it being misfit in the pocket and leads to its release.

5.5.1.2 Nucleic acids

Deoxyribonucleic acid (DNA) and ribonucleic acid (RNA) are polymers of nucleotides. There are four types of nucleotides—adenine (A), thymine (T), guanine (G), and cytosine (C)—and their sequence carries genetic information (note that these single letter notations are confusing combined with those used for AAs). As shown in figure 5.5, a nucleotide consists of a sugar and a phosphate (not shown for brevity), which form the backbone of nucleic acids. Nucleotides also contain a base

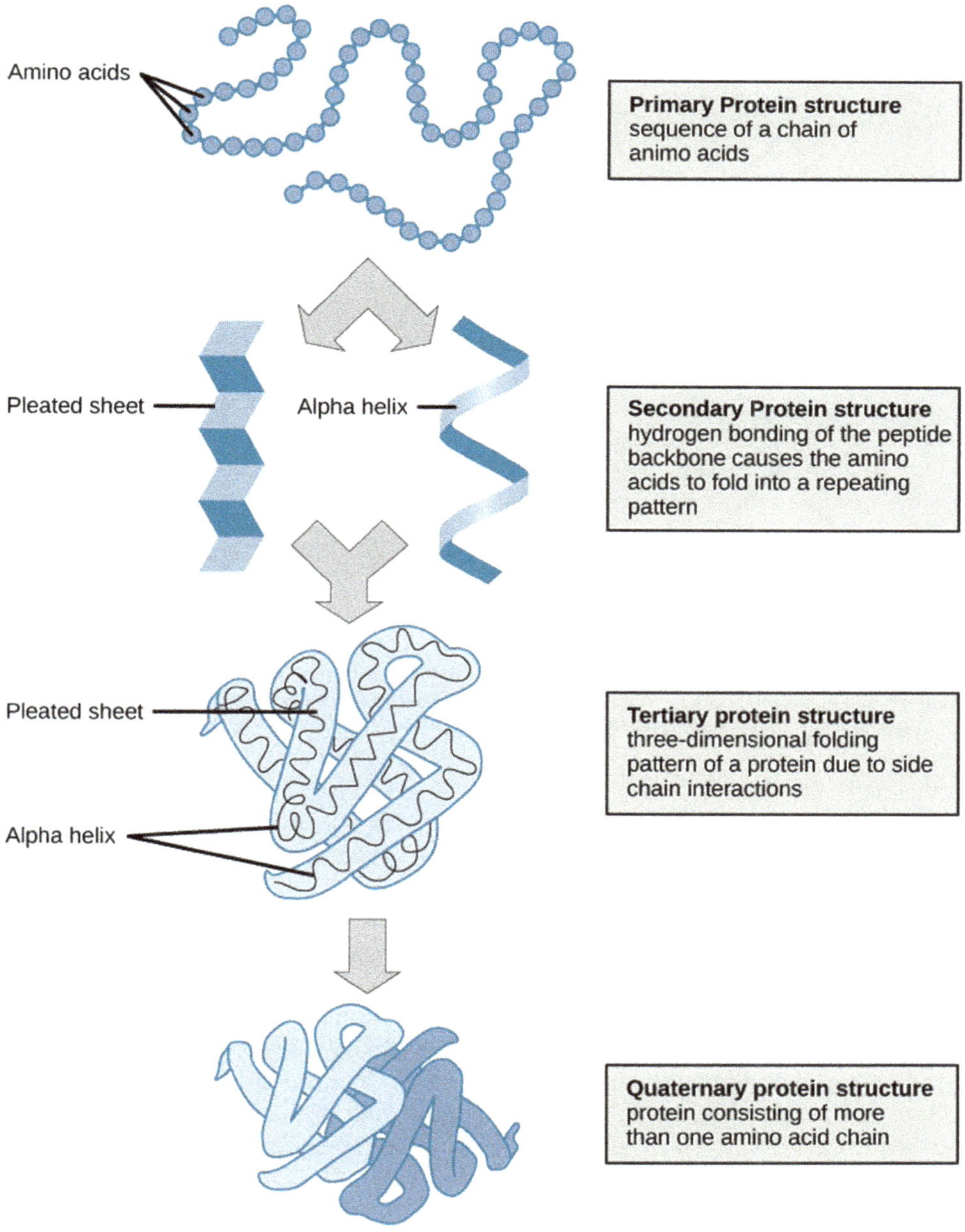

Figure 5.4. Schematic depiction of different levels of protein structure as (a) primary, (b) secondary, (c) tertiary and (d) quaternary. Source: CNX OpenStax—Wikimedia.

(see figure 5.5); there are four types of base, resulting in four types of nucleotides. Hydrogen bonding between A and T, and G and C results in the helical structure of DNA. The sequence of the nucleotides consists of the genetic code, which is

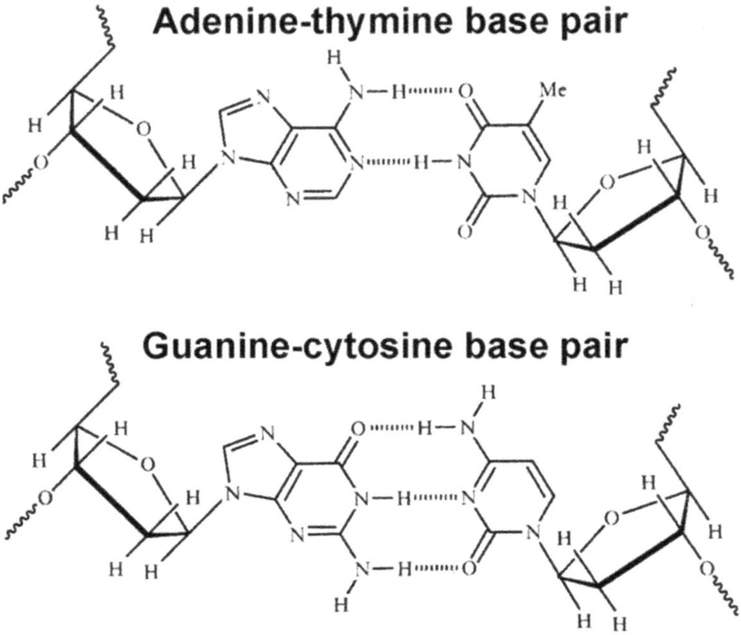

Figure 5.5. Complimentary pairs in DNA with each adenine (A) base perfectly matching a thymine (T) base to link (top), and guanine (G) and cytosine (C) match (bottom) permitting hydrogen bonding. Image taken from [17], copyright 2007 John Wiley & Sons.

'translated' into AAs and then proteins. A group of three nucleotides form a codon, which codes for a single AA. The DNA code is 'read' or transcribed by RNA, which is like DNA except that it is single stranded, and the thymine base is de-methylated (missing the CH_3), so is called uracil (U) instead of (T). When transcription needs to take place the DNA unwinds and opens up, so RNA forms the complementary sequence to the DNA. This is then fed into the ribosome, which is the cell machinery for synthesising proteins. In this way, the DNA sequence codes for a protein sequence. For example, singled stranded RNA codons UCC, GUG and CCA will translate to the AAs serine, valine and proline (S-V-P). It should be noted that as there are more combinations of 3 of the 4 bases than the 20 natural AAs, several codons will make the same AA. There are other codons that dictate at which point the sequence should start and end being read (they could be considered as punctuation codon).

5.5.1.3 Other biomolecules

There are many other types of biomolecules that proteins can produce that are key to biological functions. Those relevant to this book include phospholipids and polysaccharides, among others. Phospholipids are surfactant-like molecules with a hydrophilic phosphate head group with two aliphatic long chains, while polysaccharides are polymers composed of sugar molecules as monomers, e.g. starch and cellulose. Both of these are used for compartmentalisation, among other things, and

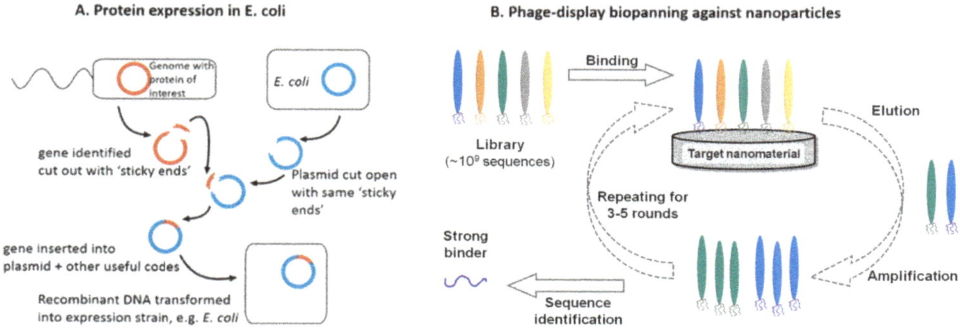

Figure 5.6. Schematics of molecular biological techniques. (A) Protein expression in a bacterial cell such as *E. coli* (reproduced with permission from [18]), and (B) phage display biopanning against nanoparticles.

are discussed specifically with respect to biomineralisation in chapter 6 (sections 6.4.2.1 and 6.4.2.2).

5.5.2 Molecular biological techniques

5.5.2.1 Protein expression in another host

Microbiological techniques have developed methods of using host cells (such as bacteria or yeast) to produce large amounts of protein from a different source. This uses the fact that many cells not only have a genome, but also have smaller loops of DNA code called plasmids. These can be easily extracted from a bacteria, manipulated and reinserted. As such the DNA code of a protein of interest can be inserted into a plasmid. This plasmid can be specially designed to contain protein purification tags and a marker gene (e.g. encode a protein that will change the colour of the cell) as well. The plasmid can then be inserted into a host cell (most commonly *E. coli* bacteria). The bacteria will then produce (or 'express') the protein of choice when it is induced, which is again dictated by the function designed into the plasmid (figure 5.6(A)). Using this method synthetic protein can be designed and expressed in a host organism such as *E. coli.*

5.5.2.2 Protein mutagenesis

Naturally occurring genes can be modified chemically or biochemically to result in a new (or slightly modified) sequence. This will result in translation of a protein with a modified sequence. This can be achieved in a number of ways: 1) disrupting a key base will then read the whole section incorrectly, as there will be a 'read frame shift'(shift how the bases are grouped into 3 so read completely different AA). 2) Whole gene or whole protein sequences can be 'cut out', and silent DNA 'pasted' back in. This genetic engineering strategy can be used to produce DNA which in turn will produce proteins of mutated sequences, or cells with specific proteins missing, known as knock-out mutants. Knock-out mutagenesis is a very important strategy for understanding which proteins are critical to which functions in a cell. In this technique, a knock-out mutant (denoted ΔgeneX) is a cell with that particular gene X missing, and thus it will not produce the corresponding protein X. The cells can then be studied to see if they are different to the native cells (or wild type, as they are often called).

5.5.2.3 Biopanning

A combinatorial approach can be used to identify novel peptides and proteins that strongly bind a target surface/material (also known as 'biopanning'). In this approach, a library of millions to tens of billions of peptides with random sequences is created. Biopanning is basically a fishing exercise. The library is exposed to a target material and the sequences that stick and bind best are separated with the material target while everything else is washed away (figure 5.6(B)). In the context of this book the targets are inorganic nanomaterials, and this is a relatively new concept as the technique was conceived more for biological targets to develop artificial antibodies. Each sequence is attached to an outer surface exposed protein of either a cell (cell display) or bacteriophage (phage display) in the library. Using a cell or phage in this way is the crux of the biopanning technique, as both the protein and the DNA which encodes it are physically linked together. This means that the recovered best candidate protein or peptides can be isolated through pH elution, and the cells or phage amplified, and thus more protein or peptide can be obtained. The process is repeated several times to evolve the strongest binding candidates that can outcompete everything else (figure 5.6(B)). Upon a number of rounds of binding, elution and amplification, tightly bound peptides or proteins can then be sequenced and expressed and their binding motifs studied.

5.6 Summary: key lessons about sustainability nanomaterials production

Based on environmental assessment of methods used for nanomaterials synthesis, we have seen that most existing and emerging methods are unsustainable. The reasons behind this include large amounts of waste produced due to low yield, sequential processing, high energy demands and the need for specialised reagents/environments. This highlights the need for a change in our perceptions and objectivity when it comes to assessing and progressing a new method for nanomaterials. We identified that biologically inspired methods have the potential to design green methods. To this end we have explored the central dogma and basic components of molecular biology and the tool kit it offers for biologically inspired nanomaterial synthesis. In the following chapters, we look for inspiration in biological mineral formation to identify rules and strategies for inventing green methods.

References

[1] Hutchison J E 2008 *ACS Nano* **2** 395

[2] Roco M C, Mirkin C A and Hersam M C 2010 Nanotechnology research directions for societal needs in 2020: retrospective and outlook *Panel Report* (Lancaster, PA: WTEC) https://www.nano.gov/sites/default/files/pub_resource/wtec_nano2_report.pdf

[3] Sengul H, Theis T L and Ghosh S 2008 *J. Ind. Ecol.* **12** 329
[4] Eckelman M J, Zimmerman J B and Anastas P T 2008 *J. Ind. Ecol.* **12** 316
[5] Dahl J A, Maddux B L S and Hutchison J E 2007 *Chem. Rev.* **107** 2228
[6] Warner J 2016 *Chemistry World* (April) p 35 https://www.chemistryworld.com/opinion/toxicity-is-a-hazardous-waste/9605.article#/
[7] Azad Malik M, Revaprasadu N and O'Brien P 2001 *Chem. Mater.* **13** 913
[8] Calugay R J, Miyashita H, Okamura Y and Matsunaga T 2003 *FEMS Microbiol. Lett.* **218** 371
[9] Patarin J 2004 *Angew Chem. Int. Ed.* **43** 3878
[10] Murphy C J 2008 *J. Mater. Chem.* **18** 2173
[11] Romero-Franco M, Godwin H A, Bilal M and Cohen Y 2017 *Beilstein J. Nanotech.* **8** 989
[12] The Royal Society of Chemistry 2011 A sustainable global society. Available at: https://www.csj.jp/international/cs3-whitepaper2010_summary.pdf
[13] Gilbertson L M, Zimmerman J B, Plata D L, Hutchison J E and Anastas P T 2015 *Chem. Soc. Rev.* **44** 5758
[14] Patwardhan S V, Manning J R H and Chiacchia M 2018 *Curr. Opin. Green Sustain. Chem.* **12** 110–6
[15] Berg J M, Tymoczko J L and Stryer L 2002 *Biochemistry* 5th edn (New York: Freeman)
[16] Patwardhan S V, Patwardhan G and Perry C C 2007 *J. Mater. Chem.* **17** 2875
[17] Currie H A, Patwardhan S V, Perry C C, Roach P and Shirtcliffe N J 2006 Natural and artificial hybrid biomaterials *Hybrid Materials* ed G Kickelbick (New York: Wiley) p 255
[18] Galloway J 2012 Biotemplated arrays of nanomagnets using the biomineralisation protein Mms6 *PhD Thesis* University of Leeds
[19] Rawlings A E, Bramble J P, Tang A A S, Somner L A, Monnington A E, Cooke D J, McPherson M J, Tomlinson D C and Staniland S S 2015 *Chem. Sci.* **6** 5586

IOP Publishing

Green Nanomaterials
From bioinspired synthesis to sustainable manufacturing of inorganic nanomaterials
Siddharth V Patwardhan and Sarah S Staniland

Chapter 6

Biomineralisation: how Nature makes nanomaterials

6.1 Introduction

Biomineralisation is the controlled formation of solid inorganic materials by biological organisms [1–3]. The very idea of biomineralisation may seem counter-intuitive to most. Inorganic minerals, like a stone, are inherently non-living, so the idea that these are intrinsically involved in life may seem strange at first. But then, consider the form of your own body. The structure and motility of your body is down to you having a calcium mineral skeleton (we are vertebrates). Other creatures have external skeletons (exoskeletal) like molluscs (they are invertebrates) that use their calcium mineral shells for protection.

From just these two examples it is clear how useful it is for biological organisms to form hard minerals, which serve a structural/mechanical and protective function. Indeed, the birth of biomineralisation is a hot topic in evolutionary biology and geology, as it appears that the rapid increase in biomineralisation and skeletal organisms occurred at the same time as the 'Cambrian explosion' between 525 and 510 million years ago. The Cambrian explosion[1] describes a time when the evolution of new species and genetic diversity rapidly increased over this time. While the reason for the rapid Cambrian evolutionary period is not yet fully understood [4] and also occurs at the same time as significant climate and sea chemistry change and instability, it is clear that the introduction of hard materials into biology introduces much greater diversity in the functions that organisms can have. For example, the earliest biominerals took the form of hard pointed spines that covered worm-like creatures, as well as very early shells on molluscs [5, 6] that act as protection within

[1] While it is known as the 'Cambrian explosion', researchers in the area now believe it was more of a rapid increase in diversification that should be considered in the broader context, and not an 'explosion' or 'big bang' from nowhere.

doi:10.1088/978-0-7503-1221-9ch6

the rocky ocean environment. Fossils later showed the predator 'attack' functions of biominerals, and this went hand in hand with the development of the need to diversify to protect against predators using hard materials to form protective shells. Thus biomineralisation could be enabling diversification through a predator–prey dynamic, being a key factoring in rapid evolutionary diversification over time.

Biomineralisation occurs across the full range of biological kingdoms, from bacteria to humans, producing a range of minerals and biomineral/organic matrix hybrids which we will explore in this chapter. Furthermore we will see that Nature utilises and optimises the chemistry of inorganic material formation discussed in the last chapter, and is able to take advantage of chemical spatial definition (compartmentalisation), organic structural surfaces and chemistry at the surface, as well as strategically positioned organic functional groups, to enable the formation of complex and highly specific mineral forms and structures, all under ambient environmental temperatures and conditions. The production of superior materials under environmental conditions is a key attraction to understanding and utilising this process, as ambient reaction conditions carry huge significance for industry. While these minerals span all length scales (figure 6.1), at the end of this chapter we summarise the lessons that biology can teach us to help us adapt our current chemical synthesis of nanomaterials to more biological green synthesis of nanomaterials.

Traditionally, biomineralisation is broadly classified as follows:

- **Biologically controlled mineralisation (BCM) (figure 6.2)**: Where minerals are synthesised for a specific biological function. Biomineralisation in this class is strictly under biological (genetic) control. We will examine BCMs in this chapter.
- **Biologically induced mineralisation (BIM) (figure 6.2)**: Where minerals are precipitated due only to the environmental conditions the organism has imposed, usually as an insoluble bi-product (to a biological redox reaction, for example). These biominerals may be useful, detrimental or benign to the organisms producing them (e.g. kidney stones), and may be internal or external to the organism. As these are not precisely controlled, there will be

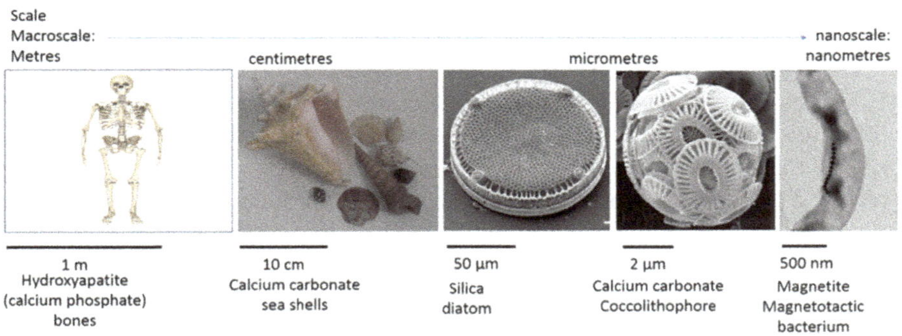

Figure 6.1. A range of biominerals at various scales, from the macro to the nanoscale. Images courtesy of Alex Mit/Shutterstock/Mary Ann Tiffany and [8].

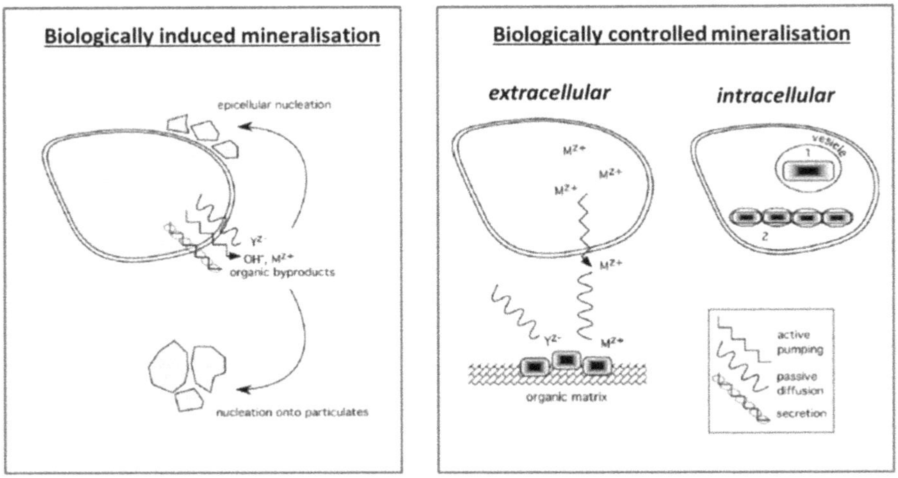

Figure 6.2. Schematic of difference between biologically induced mineralisation compared to biologically controlled mineralisation. Adapted from [2].

no further discussion of these processes in this book, and all subsequent use of the term biomineralisation will refer to BCM only.

While the biomineralisation process varies extensively, generally it involves the uptake of specific elements from the environment by an organism and their deliberate formation/incorporation into a solid mineral structure under precise biological control *in vivo*. Organisms typically accumulate the mineral precursors from their respective environments, extracted from water, soil or food. These precursors can be metal ions (e.g. ions of Ca, Fe and Mg), or even small molecular complexes ($Si(OH)_4$). Organisms are able to transport these precursors from the environment into their cell or cellular matrix. From here the precursors may be stored, then transported to the site of biomineralisation where they are concentrated to form the biominerals. It is the tight genetic control over the process that produces species-specific, ornate unique biomineral structures. Biomaterials are rarely 'pure', as they are deliberately chemically non-stoichiometric and can contain a range of dopants to enhance the functional properties. Most biominerals also have a biological organic material component that can be a molecular or larger organic matrix structure to form a hybrid composite material. Biominerals have very specific and often remarkable properties, unlike those of abiotic component materials. For example, bone has the capacity to regenerate, has massive mechanical strength, is relatively light and has some flexibility, in complete contrast to inorganic calcium phosphate. Such impressive properties evolve over time, in order for a biomineral to perfectly perform their key functions within organisms, e.g. why bone is an ideal structural/mechanical material for its purpose. The origin of these properties is not just the material's composition, indeed the precise hierarchical and often highly intricate ornate structures and morphologies of the materials on multiple length scales are equally responsible for their enhanced properties. Thus, it is critical that biominerals are formed with such highly precise levels of control over composition and structure, to ensure the best function, and this is achieved to such an extent that these complex mineral composite structures are highly reproducible.

In the next section we identify some key biominerals, and explore the link between their required function and their properties and how Nature has optimised:

- the chemical composition of the mineral;
- mineral to organic biomolecules content and interaction;
- structure and morphology;

In order to achieve the specific function. The main specific functions carried out by biominerals can be classified as follows:

- mechanical/structural support;
- protection;
- cutting and grinding;
- sensors (gravity, optical);
- cation storage;
- buoyancy.

6.2 Properties and function of biomineral types

The minerals formed by organisms most commonly utilise Ca, Mg, Si, Fe, and to a lesser degree Ba and Sr ion, as their carbonates, oxalates, sulphates, phosphates, hydroxides and oxides minerals. Relatively rare biominerals are formed from Mn, Au, Ag, Pt, Cu, Zn, Cd and Pb deposited largely in bacteria and often as sulphides or oxides.

Over 60 different biominerals have been identified, with around 50% being calcium-based; of these more than half are calcium phosphates of varying composition. The most abundant biominerals are calcium carbonates and phosphates. Calcium minerals dominate biomineralisation to such an extent that the study of biomineralisation simply used to be called ‘biocalcification’. In terms of structure, 75%–80% of biominerals are crystalline, with the remaining 20%–25% being amorphous in nature, i.e. they do not show structural regularity at atomic scales (e.g. biosilica, amorphous hydrated iron phosphate, amorphous calcium carbonate).

Many biominerals are composite organic–inorganic hybrids and are structurally hierarchical, organised from the nano- to the macroscopic length scale. The organic components of biominerals can include proteins, glycoproteins, polysaccharides and other small organic biomolecules. The organic phase occluded in biominerals may or may not be directly involved in biomineralisation.

6.2.1 Bio-calcium phosphate (hydroxyapatite): mechanical/structural support, motion, cutting/grinding

Calcium phosphate is most commonly found as hydroxyapatite of the pure formula $Ca_{10}(PO_4)_6(OH)_2$. However, it is rarely stoichiometric or pure, so a more accurate and general formula could be:

$$[\mathbf{Ca}_{10-(v+w+x+y+z)}\mathbf{Sr}_v\mathbf{Mg}_w\mathbf{Na}_x(H_2O)_y[]_z]$$
$$[(PO_4)_{6-(a+b+c)}(HPO_4)_a(CO_3)_b(P_2O_7)_c]$$
$$[(OH)_{2-(p+q+r+s+t)}F_pCl_q(H_2O)_rO_s[]_t]$$

showing the full range of dopants and potential vacancies ([ ]).

Bone is the most common material made of hydroxyapatite (figure 6.3). Bone is a hybrid material, with hydroxyapatite as the mineral hard component that forms on an organic matrix which forms the flexible component mainly composed of the protein collagen. The strength and stiffness can both be increased with higher mineral content, whereas it is the toughness that dictates the optimum composition for bones. Thus the ratio of mineral to organic component varies depending on the specific organism, the age of the organism and function of the particular bone. The more collagen present, the more elastic the bone; the more mineral present, the harder it is, showing this variable hybrid composition offers tuneable mechanical properties. Some animals that rely on quick motion, but can compromise on strength (due to the fact the body is supported by a water environment), benefit from having more flexible bones with higher collagen content, e.g. fish such as salmon and mackerel (>50% collagen). However, the larger the animal, the stronger the bones need to be, to support the large body mass. Thus generally the largest animals have the highest mineral content, such as cows, elephants and whales (>75% mineral). Similarly, our bones change with age. When we are born, human bones are mainly composed of cartilage, a flexible organic material which does not fracture easily, enabling us to learn and fail (by falling) without serious consequence. The cartilage is replaced over time as bone mineralises, as our need for support increases with our increasing weight, with optimal bone composition of about 70% mineral occurring around the age of 18–25. After this time the organic component begins to

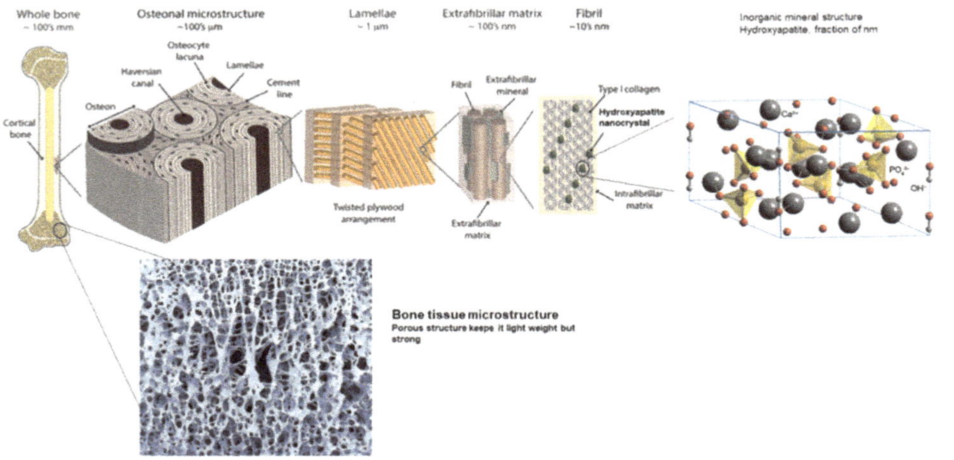

Figure 6.3. The hybrid (organic/inorganic) and hierarchical structure of bone from the atomic structure of hydroxyapatite on the right to the macroscopic structure of a whole bone on the left. Adapted from [7], courtesy of Professor Nick Greeves.

degrade slowly, leading to issues with brittle bones and increased fracturing in older people. Furthermore, bone has an intricate hierarchical arrangement composed of a fine microstructure network, which allows a blood supply, bone cells and proteins to move through the pores. In this sense bone is very much a 'living' biomineral, in that it is constantly remodelling and regenerating in response to environment factors such as physical pressure or signals from hormones. The ornate structure also contributes to enhanced properties for specific functions. Some bone needs to be extra hard, and these are dense and heavy, whereas other bone is required to be light, such as the bones of flying birds, so these are hollow. The properties of a bone can even vary within the same bone, with dense, tough joints and more porous structure in the middle.

Hydroxyapatite is also the hard mineral constituent in teeth. Like bone, teeth have a complex structure, but in simple terms the quantity of mineral increases radially outwards. The organic centre of the tooth is the 'pulp', surrounded by the 'dentine', which has a similar composition and structure to bone, with the hardest 'enamel' forming the outer coating. Enamel is required to be so hard as the purpose of teeth is to cut and grind, while no flexibility is required at all, so enamel is a high density 95% hydroxyapatite biomineral, with dopants such as F^- increasing the hardness of the enamel by reducing the solubility of the mineral, which is why we use fluoride toothpastes.

6.2.2 Bio-calcium carbonate: protection, sensor, buoyancy

Calcium carbonate ($CaCO_3$) has the largest range of biominerals, as there are several different structural forms (polymorphs) used in Nature (figure 6.4). These are

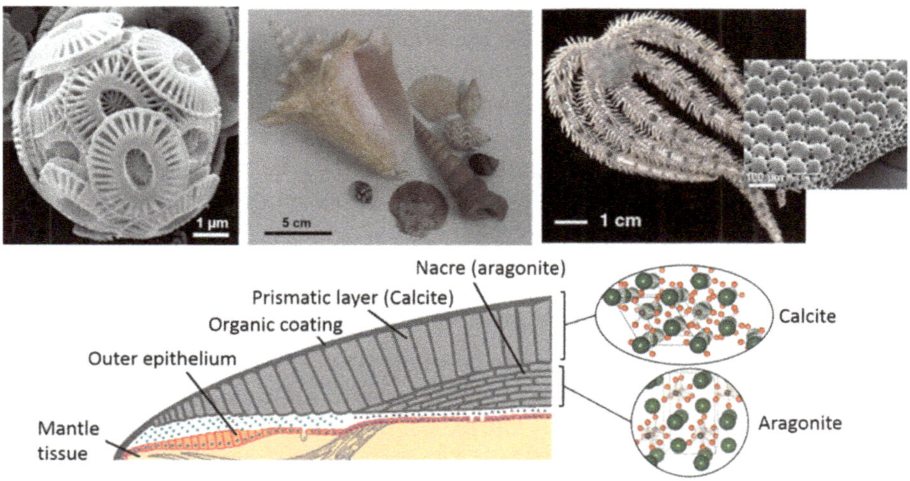

Figure 6.4. Top row: examples of calcium carbonate biomineralisation. Left: a coccolithophore [8]; centre: a collection of shells; right: a brittle star [9] (reproduced by permission of the Royal Society of Chemistry). Bottom row: schematic (adapted from [6]) of the biomineralisation of a sea shell, showing the mantle with the shell formed on top (green triangles show biomineralisation proteins), with the crystal structures of calcite and aragonite shown to the right.

mainly calcite and aragonite crystalline structures (as they are the most stable) but less stable vaterite and amorphous calcium carbonate are also found as biominerals, although are less common, with amorphous calcium carbonate, whewellite/weddellite ($CaC_2O_4 \cdot (2)H_2O$) mainly found in plants or fungi as a mineral storage mechanism for calcium.

Calcite is the most common calcium carbonate biomineral, and is the main component of sea shells, sea urchin spines, bird eggshells and the intricate plates on the very tiny unicellular coccolithophore. All of these biominerals act as a form of protection, so the key requirement is to be hard and tough. It is worth noting that as such they have a much lower organic fraction than bone (usually <5%). Again the materials can be altered with dopants, the most common instance of this is with Mg^{2+}, to make hard urchin spines even harder, for example. Some organisms use either calcite or aragonite (e.g. sponge skeletons), while many sea shells are actually made up of both [6]. Molluscs do not shed their shells, so the shells grow as the creature grows, by increasing the disc-like radius of a scallop/mussel-type shell, or increasing the size and circumference of the spiral turn of a coiled snail type shell. The mineralised shell grows from the shell mantle meaning there is very little organic material (<2%) in the shell mineral, but this biologically controls the formation of platelet aragonite nacre (mother of pearl) on the inside of the shell, and large elongated crystals of calcite form the outer prismatic layer on the exterior. Incredibly, coccolithophore biomineralise their elaborate scales internally within intracellular compartments, using a scaffold made of cellulose and proteins (called a base plate) to support crystal nucleation and growth [10]. Once fully mature, these coccoliths 'burst' out onto its surface and form an exoskeleton around the cell. Although unicellular, the white colour from the scales that coat these algae can be seen from space as huge blooms of them are visible in the ocean [8].

Calcium carbonate biominerals are used for functions other than just protection. Examples include calcite eye-lenses (for extinct trilobites and microlens arrays in brittlestars [9]) and the defensive claws of many mammals and birds. A crystal of calcite in the inner ear also acts as a gravitational sensor, with its movement activating sensory receptors. Aragonite is used as 'love darts' to aid reproduction in gastropods, and some marine molluscs and cephalopods also use aragonite as buoyancy devices.

6.2.3 Bio-silica: mechanical support, transport and protection

Another major family of biominerals is amorphous biosilica ($SiO_2{\cdot}nH_2O$), which again have a mainly structural role. Their amorphous nature can be advantageous; with no set pattern or planes of crystallisation, these materials can be moulded and worked into a range of architectural lace-like structures without a loss in mechanical strength (figure 6.5). The most common example is the beautiful and intricate nano to microscale cell wall/shell structures (exoskeletons, called frustule) of the aqueous single-celled microalgae—diatoms (figures 6.5a and 6.1). Biosilica in diatoms exhibit repeating structural features of ca. 10–40 nm. The patterns are highly sophisticated over multiple length scales and are species-specific with a high degree of

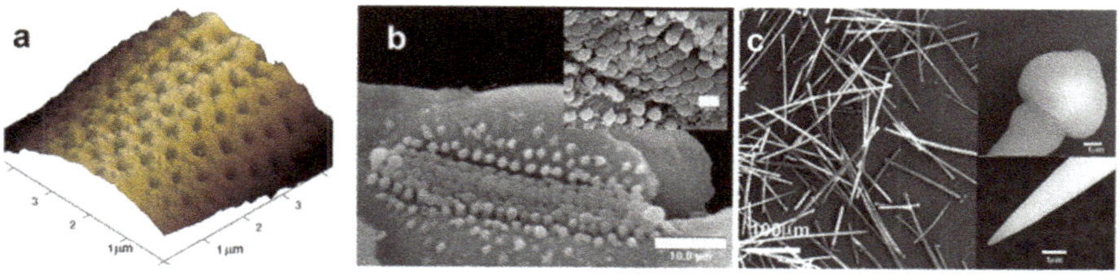

Figure 6.5. Micrographs of ornate biosilica patterns in (a) a diatom, (b) higher plants, and (c) sponge spicule. Image taken from [13], reproduced by permission of The Royal Society of Chemistry.

reproducibility [11]. These structures contain periodic pores which enable the diatom cells to communicate with other cells and the environment. It is interesting to note that diatoms form colonies by 'gluing' cells together—the pores in the silica cell walls allow communication between these cells. In addition, these pores provide a way for nutrient transport into the diatom cells. Silica makes up the biomineral skeletons for the microorganisms radiolarians and some sponges, which are known to form needle-like spicules (figure 6.5c), which are a few tens of microns in diameter and can be as long as a few millimetres. Silica is also found in plants as spines and at cell walls to provide structural support (figure 6.5b). It is the presence of silica in rice that helps the leaves adopt an upright position to maximise incident sunlight. It has been shown that without silica presence in rice plants, the leaves become *floppy* and the crop yields drop significantly due to the lower sunlight absorption. Another role that silica plays in plants is responding to biotic and abiotic stresses. For example, plants can use silica to contain infections occurring on leaves by depositing silica at the infected locations [12].

6.2.4 Bio-magnetite: sensing, cutting/grinding, iron storage

Although not so well known, iron oxides are the other main class of biominerals. There is a large variety of iron oxide minerals, and of these, only a few are biomineralised. These are goethite (α-$Fe^{3+}OOH$), lepidocrocite (γ-$Fe^{3+}OOH$) and ferrihydrite ($Fe^{3+}{}_{10}O_{14}(OH)_2$), which are all ferric minerals, and as such precipitate out of a solution readily. Magnetite ($Fe^{3+}{}_2Fe^{2+}O_4$) is the final remaining biomineralised iron oxide, which is the only mixed valence example, meaning it requires more control from the organism to form it. Iron oxides are generally very hard and dense and can be magnetic (in the case of magnetite), and thus are not deposited in organisms for structural purposes; these characteristics are more commonly utilised to function as teeth and magnetic sensors. As such, in most cases iron-oxide biominerals do not contain organic/bio molecules. Limpet teeth, used for scraping and grinding, are biomineralised goethite, while chiton teeth are formed from lepidocrocite or magnetite, with an underlying core of ferrihydrite, which is proposed to be the precursor phase to the harder mature surface layer of magnetite.

Magnetite nanoparticles and larger structures occur in a wide variety of higher organisms such as honeybees, pigeons, pelagic fish, bats, rodents and humans, with approximately 5 million crystals per gram of human brain tissue. Although the

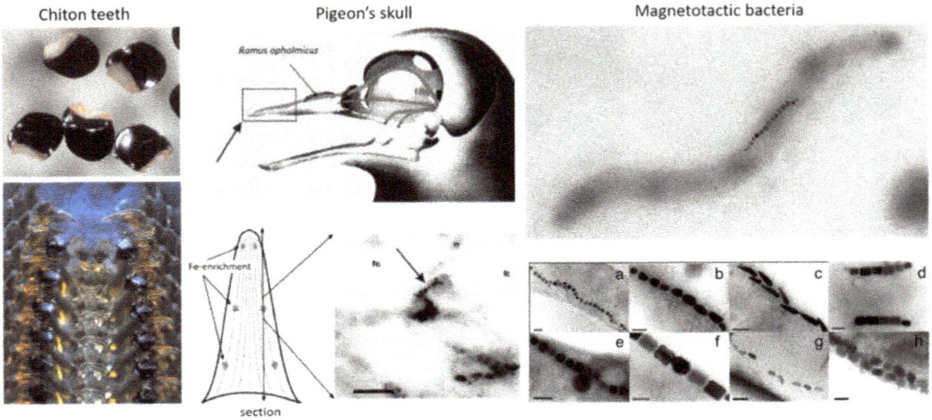

Figure 6.6. Three examples of biomineralisation of magnetite. Left-hand side: chiton teeth coating in magnetite [3] (© IOP Publishing, reproduced with permission. all rights reserved) (top); teeth along the radula showing magentite coating in the forefront (bottom of the image) and without in the immature teeth (at the top of the iamge) (bottom) (from [14], reprinted by permission from Macmillan Publishers Ltd). Centre: presence of nanoparticles of magnetite within beak of a pigeon ([15], reprinted by permission from Macmillan Publishers Ltd). Right-hand side: magnetite nanoparticles biomineralised within magnetotactic bacteria (top); comparison of the range of different morphologies of these nanoparticles (bottom) (reprinted from [16], copyright 2012 with permission from Elsevier.

purpose of these particles is not yet fully understood in all cases, the majority are proposed to be for sensing/navigational use, acting as a natural compass, helping bees and pigeons (figure 6.6(B)) find their way home, for example. However, an excess of magnetite in the human brain has been linked to neurodegenerative diseases such as Alzheimer's and Parkinson's. Magnetite is formed on the surface of chiton teeth to provide a hard material for scraping food from rocks (figure 6.6(Ai)). This is an excellent example to study, as the teeth are constantly being produced and replaced along the tongue-like radula the chiton uses for feeding, with the immature teeth at the back being developed in time to mature to take the place of worn teeth at the front. As such the teeth are at different stages of development along the radula row (figure 6.6(Aii)), so the mineral can be probed over its complete formation from just one sample [14] and shows that the magnetite is formed from a ferrihydrite precursor. The occurrence of intracellular magnetite in lower organisms is limited to a few bacteria, which are magnetic. Magnetotactic bacteria are motile and live in aqueous environments (fresh and salt water) (figure 6.6(Ci)). They take ferrous ions up from the water and use this to biomineralise highly monodispersed nano-crystals of magnetite in liposomes (biological membrane vesicles) within their cells (called magnetosomes) arranged in a linear chain. Remarkably, there is a range of magnetite shapes and sizes across different species, but the shape and size is tightly conserved within each strain, showing the high level of biological control (figure 6.6(Cii)). While these magnets align the bacteria to the Earth's magnetic field, and a navigational function is commonly accepted, there could be other reasons for biomineralising magnetosomes, such as a metabolical function, or detoxifying a harmful iron radical.

Iron oxides can also be deposited primarily for the storing of large quantities of essential iron such as ferrihydrite, which can then be redissolved when required. This is performed within a protein called ferritin, which assembles into a large symmetric hollow protein ball. This protein imports iron ions that precipitate as ferrihydrite within the core. Ferritin is very common across the full range of organisms from bacteria to humans.

6.3 Mineral formation controlling strategies in biomineralisation

The most incredible feature of biomineralisation is the biological control and regulation of the entire process, from the sequestering of ions and molecules from surroundings to the deposition of stunningly beautiful, organised and often multi-component complex structures that are described in the last section. It is this that clearly separates biominerals from chemically synthesised inorganic materials. While there is a diverse range of biominerals and an even more diverse range of properties and functions that these minerals can offer, the process of biomineralisation can be simplified to consider three essential generic steps that are universal throughout (figure 6.7). These are:

1. The sequestering of the reagents required for biomineralisation, and the concentration of these.
2. The nucleation of the specific inorganic mineral.
3. Control of the crystal growth, specifically with respect to the morphology of the final mineral.

The key to biomineralisation is how Nature uses biological components to control and regulate the chemical processes that occur in these steps.

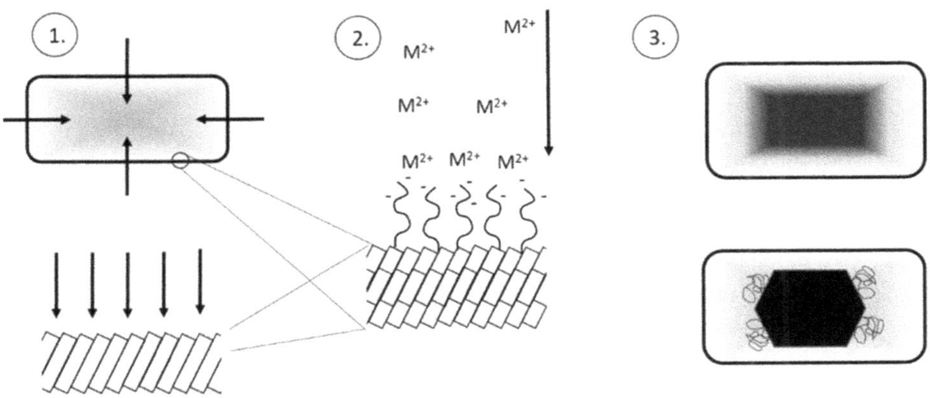

Figure 6.7. Schematic of a generic biomineralisation process. Copyright (2006) National Academy of Sciences [17].

6.3.1 The universal biomineralisation process

(1) While the minerals and thus the reagents vary, the principles of sequestering the cation are the same and basically rely on reagent concentration so precipitation can occur. Confinement is required if you want to concentrate a particular species in that solution in a defined space, compared to the bulk solution. Thus the concentrated cations need to be segregated before biomineralisation occurs, so they can be released at a specific location for mineralisation to take place.

This generally follows one of two processes depending on the size of the mineral. On the larger scale (>internal space within a cell) mineralisation occurs on an extracellular organic matrix, while smaller particles can be mineralised intracellularly within discrete vesicles (on the nanometre scale), while vesicles can also be used in extracellular biomineralisation to concentrate the mineralising species and traffic them to the mineralisation sites where they are deposited, allowing the larger extracellular mineralisation to occur. The cations required are sequestered from the environment often taken up into vesicles. Extracellular mineralisation can occur without the need for vesicle trafficking, if the concentrations of the mineral ions in the environment the mineralisation site is exposed to are high enough to simply precipitate out of the solution onto a nucleation site on the template or surface (see below (2)). In this case, larger scale compartmentalisation occurs in the form of restriction of crystal growth and thus size by features such as cell walls and/or large organic template boundaries (see below (3)).

(2) Biominerals form on biological surfaces (i.e. heterogeneous nucleation) as this has a lower formation energy barrier than in homogeneous nucleation. Hence the formation of biominerals on surfaces is favoured. Most biominerals are precipitated easily from solution and are very thermodynamically stable. This is not a coincidence. Biology will tend to utilise the most accessible chemistry. Thus in many cases simply concentrating the reagents on a surface will result in the correct biomineral being formed. In fact, many calcium minerals precipitate readily and our human biology very carefully regulates the amount of calcium that is absorbed into the blood, and requires biomineralisation *inhibitor* proteins to prevent unwanted mineralisation in the wrong places. We may be familiar with hardened arteries through unwanted calcification, leading to heart disease. With this in mind, the purpose of a biomolecular nucleation site is not simply to induce nucleation, but to induce nucleation preferentially at the correct site, to grow in the correct orientation and of the correct mineral. There are also two aspects to the mineral type: the chemical composition and the structural polymorph (e.g. within the same environment, shells produce both aragonite and calcite using different proteins to direct the formation of different polymorphs). Typically these proteins will preferentially bind the required cations such as Ca^{2+} or Fe^{3+} and this will direct the formation of a particular mineral phase and even orientation. Nucleation proteins are

thus found on the surface of the mineralising site either on the extracellular matrix or embedded in a vesicle membrane. Sometimes the mineralising surface intrinsically has this nucleating functionality so that an additional protein or other biomacromolecule is not required.

(3) There are several ways in which biology achieves various morphologies of the final biominerals. The first is the use of a template. There are two types of template. One is a mould, to grow the biomineral in a confined space of a specific morphology, like ferrihydrite inside the confined ferritin core. The second is to mineralise onto a biological structure as a template, so the mineral forms a shell of the same shape as biological matrix. A further way to control mineral morphology is to control crystal growth by inhibiting growth of specific crystal facets, leading to these facets dominating the final morphology as the fast growing planes grow out to extinctions. Growth of certain crystallographic planes can be inhibited by selective biomolecules binding. Finally, the organism can use chemical control to convert a modellable amorphous phase into a crystalline functional phase once the morphology is achieved, and hence phase transformation is performed *in vivo*. For example, spicule formation in sea urchins occurs via the initial deposition of amorphous calcium carbonate that is 'moulded' into the final required morphology, before being converted into the crystalline form—calcite.

6.4 Roles and types of organic biological components required for biomineralisation

In this section we will discuss the characteristics and properties required to perform functional roles, and the types of biological components that fulfil these in Nature.

6.4.1 Roles of organic biological components

From the three steps outlined above we can clearly identify the roles of biomolecules/organic components/matrices that are required. Broadly, these can be simplified as:

- confinement, concentrating (trafficking) of reagents;
- nucleating;
- controlling morphology: templating onto a matrix surface;
- controlling morphology: crystal growth control (through confinement or epitaxial growth).

6.4.1.1 Reagent confinement/concentrating trafficking

The role of biological components in this step is to form a barrier to divide up the physical and chemical space, to enable reagents and precursors to be isolated, concentrated and moved to other sites. Nanoscale vesicles can feature in both macro and nanoscale process. For intracellular biomineralisation, the vesicle serves as a miniature nanoscale reaction vessel: a place for the reagent to concentrate and mix. Therefore the vesicle must be formed of non-permeable membrane, such as phospholipid, or a fibrous membrane with a specific ion transporter protein embedded, that selectively transports the mineralising reagents in (concentrating them), and the by-produce out, to ensure the correct chemistry is maintained and there is a kinetic driver for precipitation to occur.

6.4.1.2 Nucleating

Typically these are biomolecules and biomacromolecules, with highly acidic side-chains such as proteins with multiple aspartic and/or glutamic acid residues, or sugars, with multiple carboxylic acids and polar hydroxyl groups that will bind the required cations. Note that repeating acidic motifs are the most common biomolecules for the biomineralisation of calcium and iron minerals, whereas the chemistry of silica precipitation differs and thus polar residues such as cysteine, threonine and serine are frequently included, and proteins dominated by basic (cationic) residues, such as lysine and arginine, are utilised in this system to promote the precipitation of silica from silicic acid.

Multiple binding sites are required to give a charged surface to nucleate the formation of a mineral (rather than just binding discrete metal ions), so a protein/polymeric macromolecule must have multiple bind sites or smaller nucleation proteins will aggregate together, and this will direct the formation of a particular mineral phase. The location of these proteins can also orientate the crystals to grow in a specific direction or orientation. Interestingly, these proteins tend to be intrinsically disordered proteins, meaning they are unstructured and flexible, which is advantageous as they will be able to accommodate the growing crystal while still being bound.

6.4.1.3 Controlling morphology: templating onto a matrix surface

Biomineralising on a biological matrix template is used to control the shape of biominerals over a vast range of sizes, from micro (coccoliths) to metres (large mammal bones). Templating biological components are usually tough fibrous macromolecules that self-aggregate to form larger matrices. Often these are made up of multiple organic molecules in complex hierarchies. In a coccolithophore for example, an organic base plate serves to direct the formation of the initial calcite minerals into a ring at the edge of the base plate (proto-coccolith). These structural template matrices are usually multifunctional, being a structural rigid template but also having active nucleating functional groups on their surface, such as aspartic acids presenting negatively charged sites for metal ion binding, so separate nucleation proteins are not required.

6.4.1.4 Controlling morphology: crystal growth control (through confinement or epitaxial growth)

This is mainly controlled by confinement. i.e. a biological component will form a confined space for mineralisation to occur, and once filled no more regents can be added and mineral growth stops. The confining biological components are standardly vesicles on the nanoscale and cell wall or other extracellular confining compartments at sizes typically over a micron, such as aragonite mineralisation envelopes in nacre. However, epitaxial crystal growth can be controlled by small protein and biomolecules that have a binding surface with functional groups, and spacing that complementarily matches the surface of specific mineral crystal planes and steps, to bind strongly and selectively to these surfaces, inhibiting growth. As this sort of shape control relies on inhibiting selective crystal growth, it can be difficult to identify these proteins from general biomineralisation inhibitory proteins. For example, cationic lysine residues will bind strongly to most oxide surfaces, but this is also common to proteins that help inhibit the growth of kidney stones (calcium oxalate). The smaller the morphological-controlling molecules and proteins are, the fewer complications there are with surface matching and mobility in reaching the growing crystal. However, some systems use large molecular crystal matching organic matrices to both nucleate or terminate a specific epitaxially growing crystal.

6.4.2 Types of organic biological components

Some organic components are listed below with their functions (figure 6.8).

- Liposomes: concentrating/confinement, trafficking, templating.
- Polysaccharides: confinement, templating.
- Large proteins: templating, nucleating, mechanical properties.
- Small acidic proteins and macromolecules (anionic): nucleating Ca and Fe minerals.
- Small basic proteins and macromolecules (cationic): nucleating silica, crystal growth regulating.
- Cage proteins, e.g. ferritin: concentrating/confinement, templating, nucleating.

It is clear that many type of biomolecules/components with similar properties can perform the same role, while one organic biomineralisation component can carry out one, several or all of the functions above. In this section we will consider some examples of organic components.

6.4.2.1 Lipid membranes: liposomes (figure 6.8)

Lipid membranes are the primary barrier in biology and as such have a wide variety of functions in biomineralisation. Lipids are amphiphilic molecules, which means they have a long organic hydrophobic 'tail' end, and a hydrophophilic, more polar or charged 'head' group. In a polar solvent like water, lipids self-assemble into a bi-layer membrane, where two 'sheets' of lipids assemble to sandwich the hydrophobic tails between them, shielding them from the water, leaving the hydrophilic head groups exposed. Biological membranes are mainly made up of phospholipids which means that the head group is a phosphate. In eukaryotic cells there is also a substantial amount (approximately 30% of animal cells) of sterols (e.g. cholesterol), which have a hydroxyl polar head group attached to bulky cyclic hydrocarbons leading to the tail groups which can bend the membrane. There is also a small amount (a few %) of glycolipids, which have sugar head groups (monosaccharide or oligosaccharide). Such membranes are the main constituent of cell membranes across biology, both intracellular and at the cell periphery, forming cell walls. On the

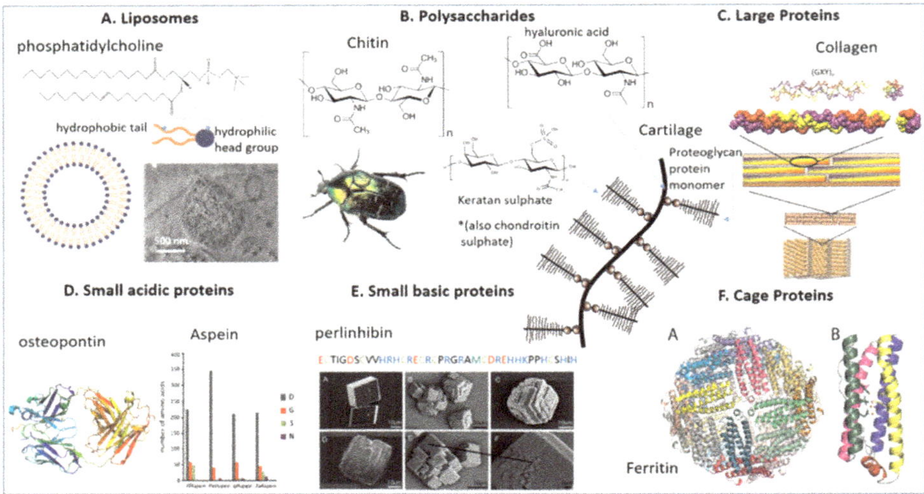

Figure 6.8. Examples of the organic components in biomineralisation. (A) Liposomes. Top: the chemical structure of a typical amphiphilic lipid molecule, phosphatidylcholine, with an uncharged hydrophobic tail on the left and depicted in orange in the cartoon representation, and the charged hydrophilic, phosphate-containing head group on the right-hand side of the molecule, depicted in blue on the cartoon representation. Below: a cartoon showing how the lipids self-assemble into a liposome in water alongside a TEM image of a calcium transporting liposomes for biomineralisation in a sea urchin larvae (reproduced from [18]). (B) Polysaccharides. Left: chitin showing the chemical structure of the monomer unit, and below this a jewel scarab beetle. Right: two further polysaccharides; the top shows the chemical structure of the monomer unit of hyaluronic acid and below the chemical structure of the monomer unit of keratan sulphate. Both of these polysaccharides are building blocks in the cartilage shown to the right. Cartilage is a biological material made up of both polysaccharides and large proteoglycan proteins. (C) Large proteins. The example of collagen is shown from the top, the triple amino acid chain structure is shown and the hierarchical structure is built up below (composed using images from [19] and [7]). (D) Small acidic proteins. Left: the protein crystal structure of osteopontin (from pdb file). Right: a graph to show the huge dominance of the acidic residues in Aspein (reproduced from [20, copyright 2012 with permission of Springer]). (E) Perlinhibin: showing the amino acid sequence highlighting the basic residues in blue, the acidic in red and the polar in green. Below: the affect perlinhibin has on calcium carbonate crystallisation and growth. (A) shows crystals in the absence of protein, with (B)–(F) in the presence of protein (reproduced from [21], copyright 2007 with permission from Elsevier). (F) Cage proteins: showing the protein crystal structure of ferritin (reproduced from [22] with permission from ASBMB).

nanoscale, lipid membranes self-assemble into spherical vesicles called liposomes. Liposomes are well utilised in biomineralisation. First they are used as compartments for sequestering, concentrating and trafficking metal ions such as calcium or iron in extracellular biomineralisation. They can also be used to traffic amorphous calcium phosphate and calcium carbonate precursors to the site of mineralisation to mature into the final mineral once deposited. This has been seen in fish bone formation and sea urchin larval spicules. They can also be used as the site of biomineralisation itself, in the case of intracellular biomineralisation. In this case the liposome serves as a miniature nanoscale reaction vessel: a place for the reagent to concentrate and mix. The liposome membrane has specific ion transporter proteins embedded, which transport the mineralising reagents in (concentrating them), and the by-produce out, to create a chemical gradient across the membrane to ensure that the correct chemistry is maintained and that there is a kinetic driver for precipitation to occur. The CAX family of membrane proteins is a well know calcium ion transporter. These are antiporters, that take Ca^{2+} into the liposomes and export H^+ out. CAX3 is critical in the uptake of Ca^{2+} into liposomes in coccolith scale biomineralisation. MamM and MamB are other liposome membrane protein antiporter systems for the uptake of iron ions into the magnetosome (and the export of H^+). Biomineralising vesicles (such as magnetosomes (specifically discussed further in chapter 8)) also contain a suite of proteins with nucleating site (see below) redox proteins to maintain the chemistry required for mineral formation and shape controlling proteins. Collectively, liposomes can also form an external boundary template, with the lace-like structures of silica forming between liposomes in diatoms.

6.4.2.2 Polysaccharides (figure 6.8)

A saccharide or monosaccharide is the single heterocyclic hydrocarbon unit of sugars. They are a ring composed of carbon with one oxygen, with multiple hydroxyl groups. Polysaccharides or oligosaccharides are composed of chains of these monosaccharide units, polymerised to form a range of sugars and carbohydrates. These large polymer macromolecules are well utilised in biomineralisation due to the way in which: (1) they can assemble into strong yet soft structures and matrices (such as chitin) so can have a templating function, and (2) the hydroxyl and deprotonated acid hydroxyl can coordinate to the mineralising metal ions (such as Ca^{2+} and Fe^{3+}) to offer repeating nucleation sites. Furthermore, these fibrous structural components conjugate to other organic components, readily displaying liposomes and smaller nucleation proteins, while also contributing to many more complex hierarchical organic systems (see section 6.4.2.3). Chitin is commonly used in biomineralisation as a template. It forms the flexible matrix within which to form the iron oxide—goethite needle crystals. This exceptionally tough hybrid organic/inorganic material is used to form limpet teeth. Similarly, the organic matrix of the cuttlefish bone and the templating structure of nacre in sea shells is comprised of chitin. A critical biomineralisation platform is the base plate where the coccolith is assembled in coccolithophores. This is composed of mainly the polysaccharide cellulose, where several proteins attached are responsible for binding calcium ions.

6.4.2.3 Large assembly fibrous proteins (figure 6.8)

Most large assembly proteins have a mainly structural function in biomineralisation, forming a template or scaffold for the mineral to form on or within. However, they also usually have functional amino acids that aid mineralisation upon them. One example is the glycoproteins called frustulins, which aid the formation of the silica diatom shell. These have large areas of hydrophobic amino acid residues (repeats of glycine with numerous proline and tryptophan residues) making them form insoluble specific structures (see collagen, below) identifying the proteins to have a dominant structural framework function. However, an acidic/polar region is also present that identifies as interacting with the forming mineral, with a repeating motif of [Cys-Glu-Gly-Asp-Cys-Asp]. A mix of polar and acidic residues are important for the interaction and mediation of silica biomineralisation. Another class of proteins (HEP) closely associated with frustulins also shows a high quality of polar and acidic residues, with serine and threonine making up a quarter of its sequence, and a fifth of the sequence being glutamic and aspartic acid.

There are also several large proteins that assemble into insoluble fibrils for extracellular templating for biomineralisation. The most common structural protein of this type is collagen. Collagen is a large (>1000 amino acids) protein made up of a repeating sequence of glycine, proline and other amino acids, to such an extent that every third amino acid is glycine, with proline making up a sixth of the total residue composition. This makes the protein hydrophobic and the combination of proline and glycine bends the protein backbone giving the protein a regular twisted structure that assembles into trimers of three twisted strands: tropocollagen, akin to a rope 280 nm long and 1.5 nm wide. Interestingly, collagen contains a large quantity of proline with a hydroxylation post translational modification (9% of the amino acids in collagen are hydroxyproline) and also some lysine. It is the addition of these hydroxyl groups that helps these protein rope units to assemble yet further by cross-linking them in a staggered fashion into long insoluble collagen fibrils. This hierarchical assembly gives the fibrils a building-block unit structure, with gaps between the ends of each trimer twisted strand and structural grooves between them. These have huge importance as the defined pockets (known as the 'hole zones') of nucleation, making up regular nucleation sites along the fibril. Specifically, hydroxyapatite crystallises only within the hole zones initially and the crystals grow along the grooves, so the crystal growth is preferentially orientated. This is a clear example of how organic hierarchical structures can template materials with both short- and long-range structural order.

An example that combines both larger proteins (proteoglycan) and polysaccharides (hyaluronic acid, glycosaminoglycan) to form a complex organic material that interacts and interweaves with collagen is cartilage. Cartilage is the main component of bone when we are born and as such forms the macro-level template of the bone shape. It is tough, smooth and bouncy, so ideal for light youngsters prone to accidents and falling, who don't yet require fully mineralised bones. Collagen and thus mineralised bones form on the cartilage templates over time, so once the bone matures, it entombs the cartilage inside, which then degrades away to leave a hollow void that more bone structures can form in. Cartilage does remain to cover the surface of the bone to protect it against impact, particularly at the joints. The structure of cartilage is very complex and resembles the hierarchical structure of a fern frond: it has a long chain polysaccharide (hyaluronic acid) core (like the central stem), from which numerous proteoglycans extend out perpendicularly at regular 30 nm intervals (as the pinna would). More polysaccharides (around 80) are perpendicularly linked along the length of this protein (resembling pinnules). The repeating units of these carry a negative charge resulting in a large amount of hydration (H-bonding to water) in cartilage, as well as good anionic sites for interaction with collagen and for Ca^{2+} binding to nucleate mineralisation.

6.4.2.4 Small acidic proteins and macromolecules (anionic) (figure 6.8)

There is a vast range of acidic mineralisation proteins, and most of these are nucleation proteins that bind metal ions into a surface to drive spontaneous heterogeneous nucleation. Examples include osteopontin and osteonectin, which are both nucleation proteins in bone that are very acidic residue rich, with large amounts of glutamic and aspartic acid in osteonectin, while osteopontin has a 9 aspartic acid repeat sequence. Aspartic acid repeat sequences are particularly prevalent in biomineralisation nucleation proteins, in a range of systems from bone to a full range of sea shells such as barnacles and mussels. For example, the C-terminal of the calcite nucleation protein aspein is almost entirely aspartic acids. Combined glutamic and aspartic acid rich motifs and repeating sequences are commonplace, and where the sequence is not repeated, nucleation proteins contain motifs to enable aggregation so the acidic motifs are repeated across the aggregate surface. One example is Mms6, from the magnetosome in magnetic bacteria, which nucleates magnetite by binding both ferrous and ferric ions on an Asp-Glu-Glu-Val-Glu motif. This also has a Leu-Gly repeating hydrophobic motif, which promotes the aggregation of the protein through a bulk–non-bulk group—a knob and hole type interlocking interaction and this is further described in chapter 8. Another method of obtaining repetition for a small acidic nucleation protein is to assemble at specific sites on a regular organic surface, such as another protein (like the regular β-sheet structured silk-fibroin-like protein in nacre biomineralisation, or the polysaccharide chitin which occurs in limpet teeth formation (see section 6.4.2.2)).

Further to the nucleation function, small acidic proteins can also be used to control the crystal habit and epitaxial growth of biomineral crystals. This relies on the charge on the surface mapping onto the complementary charged surface of the mineral. Acidic biomineralisation proteins from adult sea urchins interact specifically with the prismatic face of Mg-calcite, while small acidic abalone nacre protein (AP8) directly control calcite morphology to be asymmetrically curved when added to chemical precipitation of this mineral *in vitro*.

6.4.2.5 Small basic proteins and macromolecules (cationic) (figure 6.8)

While it is more uncommon to find cationic proteins or protein motifs in biomineralisation (compared to acidic proteins and motifs), there are examples of basic amino acids controlling the formation of particular crystal phases or morphologies, interacting more strongly with the forming mineral face rather than the precursor ions. Two examples are the cationic/polar region at: (1) the C-terminus of the Japanese pearl oyster shell matrix protein, prismalin 14, which is found at the inner side of the outer face of the shell mantle, and (2) the N-terminus of the nacre associated sea snail shell protein, perlinhibin. Both proteins bind very strongly to the forming mineral, regulating the crystal formation of the prismatic layer of the shell. Both proteins also ‘inhibit’ mineralisation *in vitro* which may demonstrate a preferential crystallisation function. A third example of basic proteins are the highly cationic proteins found in diatoms containing arginine and lysine residues. These proteins also contain propylamine modifications on various residues, making them even more basic. They strongly interact during the nucleation, particle formation, growth and assembly stages in biosilica formation—this is further detailed in chapter 9.

Silaffins are a group of small basic proteins found in silica biomineralisation of diatoms. Again these are enriched for the cationic residues lysine and arginine, along with hydroxyl polar residues serine and threonine. These functional groups tend to have regular repeating motifs and as such the proteins interact with silicic acid and bind strongly to the forming silica, suggesting a crystallisation role.

6.4.2.6 Cage proteins (figure 6.8)

The cage protein ferritin is a remarkable example of a single protein that performs all the functions of biomineralisation for the purpose of iron storage and slow release. Its multi-functionality stretches from sequestering iron ions into its confinement/templating core, nucleating the mineral ferrihydrite and regulating the size of this mineral to the interior core of the protein. Ferritin is a ubiquitous cage protein found in almost all living organisms. Iron-mineral-free mammalian ferritin (apoferritin) consists of 24 subunits, which are each made up of four helix bundle assembles. These subunits assemble to form a dodecameric cage 12–14 nm in diameter. The symmetric structure of this cage gives six pores into the hollow core which act as ion channels, importing the soluble ferrous iron ions. Within the interior of the protein the ferrous ion is oxidised to ferric iron ions at a ferroxidase site on some of the helices. The now higher charge ferric ions are then electrostatically attracted to a highly negatively charged binding pocket, accessed on the cage interior between the four helices of each subunit bundle. This pocket nucleation site is lined with five glutamic acids, a histidine and glutamine to bind two iron ions and a water molecule. The reagents then cross-link to form the ferrihydrite mineral. This grows from each of the nucleation sites until the ferritin is filled with ferrihydrite. The cage protein ferritin is an excellent example of all biomineralisation components in one protein. It encompasses compartmentalisation, behaving as nanoreactor which can sequester the iron ions and then nucleate and direct the mineral growth. Such cage proteins and their mimics show themselves to be very useful in bioinspired green synthesis of nanomaterials seen in chapter 7.

6.5 Summary: key lessons from biomineralisation for the green synthesis of nanomaterials

It is clear that biomineralisation can occur over the whole range of length scales. While it may seem that micro- and macro-level examples of biomineralisation may not be relevant to the formation of nanoscale materials, this is simply not the case, as common trends and features across the whole range of sizes can be applied to bioinspired nanomaterial synthesis. Furthermore, macro-biominerals are hierarchical across the length scales, so have nanoscale intricacies and precision. The common themes that can aid the design of bioinspired approaches to making nanomaterials are the ways in which biology controls:

1. The chemistry of the environment on the nanoscale which affects the nanomaterials formed. This is controlled by ion pumps and redox proteins in biomineralisation, but could utilise other approaches synthetically.
2. The confinement of crystal growth which directs the shape and size of the materials produced, such as formation within liposomes.
3. Organic molecules are adept at forming into a full variety of shapes and architectures at all length scales. However, these are always fundamentally controlled at the molecular level, for example by protein sequence features which introduce bulky amino acid residues that cause a bend in the protein shape. These form intricate scaffolds to template the formation of very specific shaped materials.
4. Patterned arrays of positively charged functional groups can nucleate a specific material by binding metal ions to such an extent they can control the formation of a specific crystal phase and can even direct growth through nucleation of a specific crystal face.
5. Small soluble charged proteins and biomolecules that bind to the forming mineral at specific steps and faces inhibit the growth of these sites, and thus control the morphology of the resulting crystal at the nanoscale.
6. Mixtures of organic and inorganic (hybrid) hierarchical materials can show superior physical properties, which could be utilised when designing new nanomaterials.

References

[1] Behrens P and Baeuerlein E 2007 *Handbook of Biomineralization: Biological Aspects and Structure Formation* (Weinheim: Wiley), p 412
Mann S 2001 *Biomineralization: Principles and Concepts in Bioinorganic Materials Chemistry* (Oxford: Oxford University Press), p 240

[2] Weiner S and Dove P M 2003 *Rev. Mineral. Geochem.* **54** 1

[3] Addadi L and Weiner S 2014 Biomineralization: mineral formation by organisms *Phys. Scr.* **89** 098003

[4] Fox D 2016 What sparked the Cambrian explosion? *Nature* **530** 268–70
Tarhan L G, Droser M L, Cole D B and Gehling J G 2018 *Integr. Comp. Biol.* **58** 688

[5] Maloof A C, Porter S M, Moore J L, Dudás F Ö, Bowring S A, Higgins J A, Fike D A and Eddy M P 2010 *GSA Bull.* **122** 1731

[6] Kocot K M, Aguilera F, McDougall C, Jackson D J and Degnan B M 2016 *Front. Zool.* **13** 23

[7] Zimmermann E *et al* 2016 Intrinsic mechanical behavior of femoral cortical bone in young, osteoporotic and bisphosphonate-treated individuals in low- and high energy fracture conditions *Sci. Rep.* **6** 21072

[8] Von Dassow P, Díaz F, Bendif E M, Gaitán-Espitia J-D, Mella-Flores D, Rokitta S, John U and Torres R 2017 Over-calcified forms of the coccolithophore Emiliania huxleyi in high-CO2 waters are not preadapted to ocean acidification *Biogeosciences* **15** 1515–34

[9] Aizenberg J and Hendler G 2004 *J. Mater. Chem.* **14** 2066

[10] Taylor A R, Brownlee C and Wheeler G 2017 *Annu. Rev. Marine Sci.* **9** 283

[11] Round F E, Crawford R M and Mann D G 1990 *The Diatoms: Biology & Morphology of the Genera* (Cambridge: Cambridge University Press), p 747

[12] Kauss H, Seehaus K, Franke R, Gilbert S, Dietrich R A and Kroger N 2003 *Plant J.* **33** 87

[13] Patwardhan S V, Clarson S J and Perry C C 2005 *Chem. Commun.* **9** 1113

[14] Gordon L M and Joester D 2011 *Nature* **469** 194

[15] Hanzlik M, Heunemann C, Holtkamp-Rötzler E, Winklhofer M, Petersen N and Fleissner G 2000 Superparamagnetic magnetite in the upper beak tissue of homing pigeons *Biometals* **13** 325–31

[16] Staniland S S, Tolosa J, Wilson O, Garcia-Martinez J C, Binns C, Rawlings A E and Bramble J P 2014 *Nanomagnetism: Fundamentals and Applications* ed C Binns (Amsterdam: Elsevier)

[17] Buehler M J *et al* 2006 Nature designs tough collagen: Explaining the nanostructure of collagen fibrils *PNAS* **103** 12285–90

[18] Vidavsky N, Addadi S, Schertel A, Ben-Ezra D, Shpigel M, Addadi L and Weiner S 2016 *Proc. Natl. Acad. Sci.* **113** 12637

[19] Bella J 2016 *Biochem. J.* **473** 1001

[20] Isowa Y, Sarashina I, Setiamarga D H E and Endo K 2012 *J. Mol. Evol.* **75** 11

[21] Mann K, Siedler F, Treccani L, Heinemann F and Fritz M 2007 *Biophys. J.* **93** 1246

[22] Pfaffen S, Abdulqadir R, Le Brun N E and Murphy M E P 2013 *J. Biol. Chem.* **288** 14917

IOP Publishing

Green Nanomaterials

From bioinspired synthesis to sustainable manufacturing of inorganic nanomaterials

Siddharth V Patwardhan and Sarah S Staniland

Chapter 7

Bioinspired 'green' synthesis of nanomaterials

7.1 From biological to bioinspired synthesis

From chapter 6 we have learnt that one of the most interesting facets of biological nanomaterials formation is the involvement of organic biomolecules (proteins, polysaccharides, small molecules) in controlling every stage of biominerals formation, including precursor uptake, precursor/cluster solubility, transport of precursors and mineral formation. It is worth noting the mild conditions used (e.g. ambient temperatures and the use of eco-friendly reagents) [1]. A tangible example we have seen for controlled 'green' synthesis is the formation of bones. This is in striking contrast to the synthetic processes currently used both in industry and in research labs, as discussed in chapter 4. The differences between biological and abiotic routes are on a range of points, as listed below:

- The precursors used in biomineralisation are dissolved ions at low and non-toxic concentrations, versus the hazardous chemical derivatives used in lab/industrial processes.
- Water is the natural solvent for biomineralisation, while a large number of synthetic methods for nanomaterials use non-aqueous solvents or co-solvents.
- Another key difference is the temperature—biominerals are formed at physiological temperatures while abiotic methods invariably need higher temperatures.

A simple example of porous silica can illustrate these differences. As we have seen in chapter 6, diatoms deposit silica that is porous at multiple length scales, due to the hierarchical structure of the constituent particles. Yet, this deposition occurs at ambient temperatures and pressures, in water, and without needing any hazardous chemicals (it simply uses dissolved rocks as precursors). However, in order to make synthetic porous silicas to match biosilica features (e.g. to make mesoporous silica or silica with multiple porposities), hydrothermal conditions are often required (e.g. elevated temperatures between 60 °C and 150 °C and higher pressures), over long durations (hours to days). Those syntheses also commonly utilise hazardous precursors such as alkoxysilanes and organic co-solvents such as alcohols, dimethylformamide and tetrahydrofuran.

So, the question is, how can we translate the learning from biomineralisation, in order to design biologically inspired greener/sustainable routes to nanomaterials?

We note here that the purpose of designing bioinspired strategies is to create sustainable production technologies for desired nanomaterials. It is not necessarily the aim to mimic the biominerals in their structure or form, rather it is to mimic the process/principles. The main reasons for focusing on learning the principles instead of copying the structures of biominerals are twofold. First, with some exceptions, biominerals generally do not have a direct use in commercial applications, or cannot be harvested at the scale required for commercial use. It is thus important to learn the principles and build the methods to produce large quantities of nanomaterials which are desired for applications. Second, learning the principles is a powerful strategy, because it is transferrable to a diverse range of materials/chemistries, even to those that are not found in biominerals. An example of this is titania, a technologically important material in pigments and photocatalysis, which is not biomineralised. However, learning from biosilica formation has led to discovering bioinspired routes to titania synthesis [2]. Hence the use of the term *bioinspired* is appropriate over *biomimetic*. For a more detailed discussion of the terminology of biomimetic, versus biokleptic, versus bioinspired, see [3].

In figure 7.1 we can see an overview of a strategic pipeline that has been successfully used to translate the learning from biomineralisation to develop bioinspired methods. We note that this strategy could be used in an iterative approach in order to refine the learning obtained. This chapter is built and structured around this scheme as a central concept.

In this pipeline, **point 1: Biominerals and Biomineralisation** is covered in chapter 6. We have learnt about *what* biominerals are and their functions. This also includes some aspects of **point 2: Mechanistic Understanding** that biomineralisation is under strict control, where the mechanisms underpinning and the key processes involved

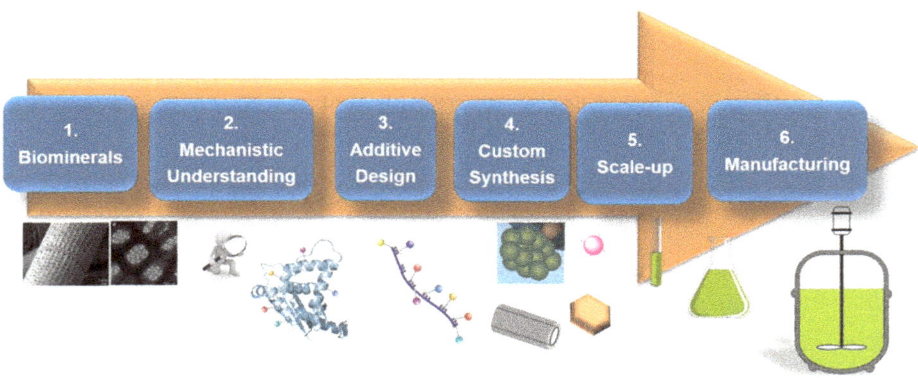

Figure 7.1. Schematic representation of a pathway from biomineralisation to bioinspired synthesis of nanomaterials.

are biologically regulated. The biological machinery exhibits physical, chemical and temporal control over the entire process. The physical and chemical control is relatively easy to mimic *in vitro*, while the temporal control can require highly sophisticated automated systems. Biochemical and biological research has revealed that biomineral formation involves organic biomolecules such as proteins and peptides in crucial stages (see chapter 6).

7.2 Mechanistic understanding

This mechanistic understanding (**point 2** in figure 7.1), once obtained, can be used to develop strategies for bioinspired synthesis. Hence developing an understanding of *how* biology produces high quality nanomaterials is crucial in moving forward. As we have seen, one of the key controlling features of biomineralisation is the use of biomolecules, so it is important to understand the roles that such biomolecules play in the entire process. As such, we will consider two main strategies for developing this 'nano–bio' interaction: (1) investigations of biomolecules that are directly involved in biomineralisation, and (2) investigations of *other* biomolecules that are not associated with biomineralisation but show strong interactions with (bio) minerals (e.g. peptides showing affinity towards synthetic minerals). Below, we describe the overview of both strategies and summarise the common features and learning. In section 7.3, we will elaborate these principles with a detailed example. This learning will be further enhanced with in-depth case studies in chapters 8 and 9.

7.2.1 Biomineralising biomolecules

From the numerous examples given in chapter 6, we can see that there has been an incredible amount of research undertaken to identify and characterise the biomolecules responsible for biological mineral formation. These include isolation via purification of specific proteins or enzymes and understanding their structure and function. Although the characterisation of these biomolecules has provided a significant amount of information, ideally, it is highly valuable to see the molecules in action in order to fully understand their function. Given the complexities of biological systems and the commensurate advances needed in the analytical sciences, it is extremely difficult to watch *live* biomineral formation with a resolution at molecular scales. This either slows the process of designing bioinspired synthesis or creates large barriers.

In order to address these challenges, scientists can (and have) focused on research using genetic engineering or model systems *in vitro*. Genetic engineering strategies have been able to understand the biomineral formation pathways, and identify key genes controlling biomineralisation and the roles of a range of biomolecules *in vivo*. Generally this has been done by making knock-out mutants of a biomineralising organism which results in the organism not making a specific protein and to see if missing that protein affects biomineralisations. Despite this progress, the translation of knowledge to develop bioinspired technologies that are suitable for commercial uptake has been difficult, because these studies do not provide a way to design synthetic/industrial processes. Direct production of biominerals may not be useful, as biominerals are not always the desired products for commercial applications. Secondly, using biomineralisation at commercial scales (tonnes) could be uneconomical. Further, purification of biominerals from cells/organisms is very challenging and costly.

On the other hand, research using model systems has been fruitful in the identification of mechanisms that can be utilised in developing novel manufacturing technologies. Such studies include, for example, probing the interactions between isolated and purified biomolecules and mineral precursors/mineral formation in the lab. As will be further elaborated in chapters 8 and 9 for the specific examples of Si and Fe minerals, such a strategy offers a much improved ability to monitor the mineral formation process via suitable sampling or *in situ* analytical instrumentation. It further helps understand the assembly and organisation of organic as well as inorganic building blocks during mineral formation. In all these cases, it is important to note the key length scales that are common to both nanomaterials and biomolecules, where interesting 'nano–bio' action takes place (see figure 2.1). It is this length scale that includes individual proteins and enzymes (~1–10s nm) and their self-assembled structures (10s–100s nm), as well as precursor clusters (~1–5 nm) that produce nanoparticles (10s–100s nm). Further, it is clear that the molecular interactions between nanomaterials and biomolecules are critically important.

Focusing on *in vitro* investigations and the key length scales, numerous researchers have identified key motifs responsible for binding, catalysis, and facilitating biomineral formation.

Probing these interesting features for a range of biominerals has helped develop deeper knowledge of the mechanisms that underpin the roles of biomolecules. Some of the commonly found mechanisms are given below; it is noted that in many cases, a combination of these mechanisms are also regularly observed. Some biomolecules bind the precursors (figure 7.2(a)), in order to change local concentrations, reactivities and special arrangements, which lead to a specific reaction pathway or the formation of *pre-determined* structures of nanomaterial product. An example where this is documented is of the soluble silicon 'pools' in diatoms. It has been reported that supersaturated levels of silicon are stabilised by biomolecules and

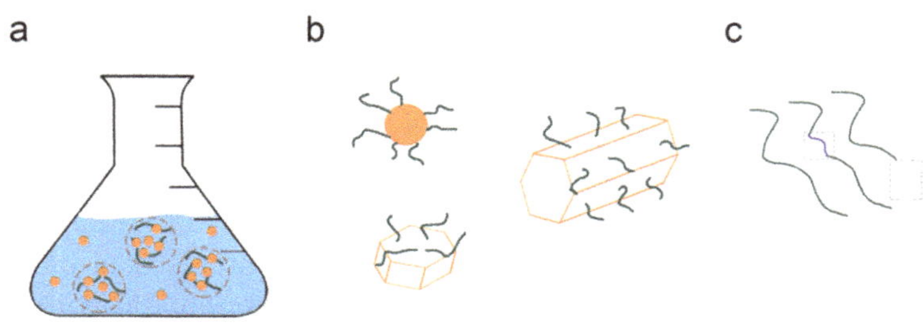

Figure 7.2. (a, b) Schematic representation of commonly observed roles of biomolecules in biomineralisation. (a) Biomolecules creating concentrated pools of precursor, and (b) biomolecules acting as capping agents. (c) One example of using site-specific mutation of a peptide sequence, either by replacing or removing specific parts of the peptide.

transported to the mineral formation sites within the organism. Other mechanisms include where the biomolecules are acting as reducing agents (in the case of metal nanoparticle formation), or capping agents (which stabilise a particular final crystal structure or particle size), see figure 7.2(b). Magnetite formation in magnetotactic bacteria is an important example of this mechanism, where the shapes and sizes of the magnetic nanoparticles are controlled by liposome surface-bound proteins (see chapter 8 for more details).

More specific details of the mechanisms can be further investigated by using sophisticated tools, such as site-specific mutations of the biomolecules, or the use of truncated biomolecules (figure 7.2(c)). The former involves *silencing* (or removing) selected sections of the biomolecules, or replacing them with other motifs, and then monitoring the effects on mineral formation. The latter focuses on a specific *functional* part of the biomolecules, which can be synthetically or biologically produced and tested for its performance in mineral formation. An example of this approach includes the use of a shorter peptide that contains the *functional* part, to develop understanding of how a matrix protein lustrin controls the calcium carbonate formation in shell and pearl nacre [4].

A correlation between the chemical structure of these biomolecules, their function and the properties of the thus formed materials can be created using the strategies explained above. Although the actual '**rules**' vary from system to system, studies have revealed interesting features that are commonly found in biomineralising systems:

- unique catalytic or binding ***sites*** in enzymes or proteins,
- specific (and sometimes repeating) ***sequences*** of key amino acids,
- particular ***self-assembly*** of these biomolecules (intra- and inter-molecular) and
- peculiar ***chemical structures***, modifications or motifs.

7.2.2 Abiotic peptides and proteins from biopanning

Biopanning (explained in section 5.5.2.3) has been highly successful for a range of materials in identifying peptide sequences that show specific recognition for material surfaces. Importantly, this methodology can be used to investigate interactions of peptide and proteins with abiotic minerals just as readily as those biomineralised, allowing us to move away from biomineralisation towards bioinspired for any inorganic nanomaterial. The materials investigated using this approach range from metals and alloys (e.g. Ag, Au, Pd, Gd, Ti, Co/Pt and Fe/Pt), to oxides (e.g. of Si, Fe, Ti, Zn, Sn, Ge, Mn, Cr, Co, Pb), zeolites, sulfides, selenides, arsenides, salts, $CaCO_3$, and even organic materials such as carbons (e.g. fullerenes, graphene and nanotubes) and polymers [5]. A range of peptide sequences identified for selected materials are listed in table 7.1, along with exemplar applications emerging from exploiting these nano–bio interactions.

The key features of peptide binding on surfaces and the benefits of using these peptides, as classified into two broad areas, are given below [6]:

- Selectivity and high affinity binding.
- Ability to mediate synthesis, assembly and/or functionalisation.

Table 7.1. Sequences of nanomaterial-binding peptides for selected systems and their exemplar applications. Table adapted from [5].

Target nanomaterial	Peptide sequence	Applications
	Metals	
Gold	MHGKTQATSGTIQS WALRRSIRRQSY WAGAKRLVLRRE LKAHLPPSRLPS VSGSSPDS	Biocompatibility and synthesis/assembly
Palladium	TSNAVHPTLRHL	Nanomaterial synthesis and controlled catalysis
Platinum	PTSTGQA TLTTLTN SSFPQPN CSQSVTSTKSC	Nanostructure synthesis and biocompatibility
Cobalt platinum	KTHEIHSPLLHK [7]	Inserted into modified ferritin cages
Iron platinum	HNKHLPSTQPLA [8] SVSVGMKPSPRP [8] VISNHRESSRPL [8] KSLSRHDHIHHH [8]	For $L1_0$ phase
Silver	AYSSGAPPMPPF NPSSLFRYLPSD	Biomaterial synthesis
	Metal oxides	
Iron oxide	QKFVPKSTN [9] IKKKKKYKY [9] RRTVKHHVN	Cubic shape controlling Biomolecule immobilisation
Lanthanide oxide	ACTARSPWICG	Biocompatibility
Silica	MSPHPHPRHHHT KSLSRHDHIHHH [10] SSKKSGSYSGSKGSKRRIL	Nanostructure synthesis

	HPPMNASHPHMH RKLPDA	
Titania	RPRENRGRERGL RKLPDA	Biocompatibility and nanostructure synthesis
Quartz	PPPWLPYMPPWS	Biocompatibility
Zinc oxide	EAHVMHKVAPRP RPHRK	Controlling crystal types and morphologies [11]
	Minerals	
Calcium phosphate	KDVVVGVPGGQD	Biomolecule immobilisation; biomaterial synthesis
Hydroxyapatite	NPYHPTIPQSVH	Biomaterial synthesis
Zeolites	VKTQATSREEPPRLPSKHRPG	Biomolecule immobilisation
	Carbon materials	
Graphene [12]	EPLQLKM, QQQLSTH TMGFTAPRFPHY YHRMPQALSAME GAMHLPWHMGTL	Biomolecule immobilisation
Single-walled carbon nanotubes	HSSYWYAFNNKT DYFSSPYYEQLF DSPHTELP	Biocompatibility and multi-material fabrication
	Semiconductors	
Cadmium sulphide	CTYSRKHKC	Multi-material fabrication and nanomaterial synthesis
Gallium arsenide	AQNPSDNNTHTH	SBP binding studies
Zinc sulphide	CGPAGDSSGVDSRSVGPC CNNPMHQNC LRRSSEAHNSIV	Biomolecule immobilisation; nanomaterial synthesis

These points are also illustrated in section 7.3 with selected examples from the literature. The selectivity observed with peptides is remarkable, and is often viewed as the ability of a peptide to *recognise* materials. A number of peptides have been identified (as shown in table 7.1) that recognise surface features of materials, and are able to distinguish small changes in the surface chemistry of the same material (e.g. ionisation) and/or crystal faces/planes. This remarkable property has enabled the modulation of solubilities and stabilities of precursors, intermediates and/or final nano-structures (see examples below in section 7.3).

This recognition ability further helps in controlling nanomaterials synthesis, assembly and even functionalisation. This can be achieved by reducing surface energies upon selective binding of peptides, and thus providing thermodynamic driving force for nucleation and growth of this one preferred phase or face (see section 4.4.3). An example of this is shown in figure 7.3 (top) in the case of ZnO crystals, where peptide binding can modulate the crystal size and shapes. As a result,

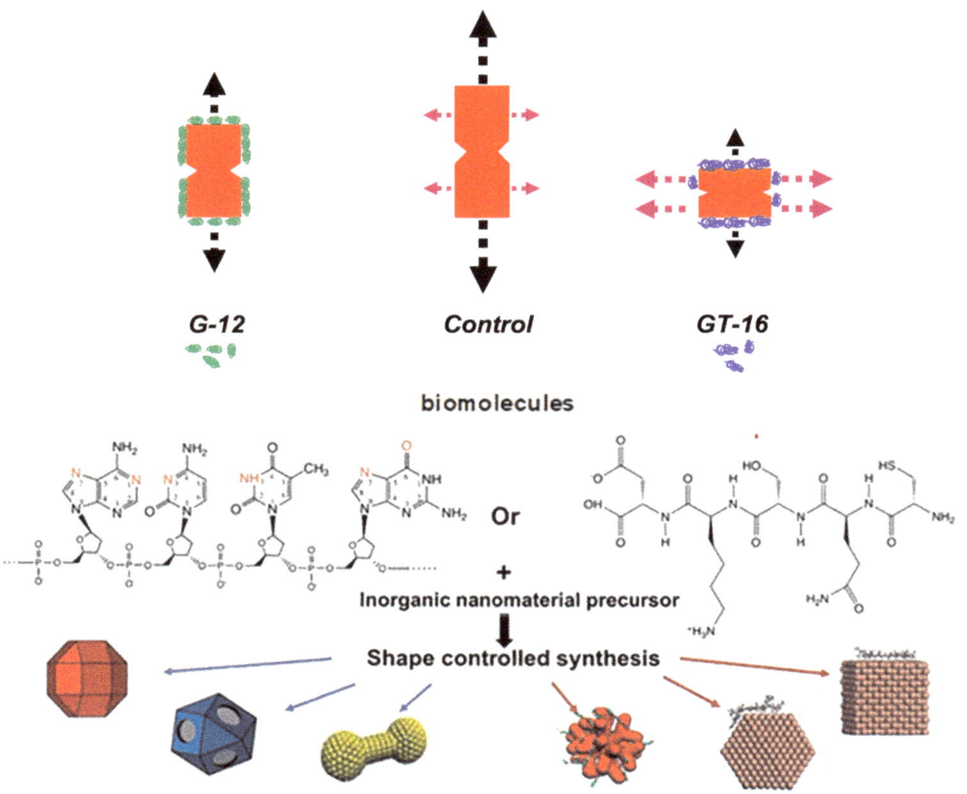

Figure 7.3. Top: schematic representation of two peptides (G-12 and GT-16) that preferentially bind different surfaces of a crystal, thus controlling growth of those faces, resulting in different morphologies. Image adapted from [11], by permission of the Royal Society of Chemistry. Bottom: the toolkit of biomolecules can be used to produce a range of nanomaterials. Image taken from [13], with permission from Elsevier.

peptides not only *catalyse* the formation of nanomaterials by reducing the energy barriers, but they can also select the particle formation pathways. In some cases, these interactions have resulted in reaching new structures that were previously not possible to achieve (see figure 7.3 and examples below).

This approach *only* provides the information on the sequences of tight binding peptides, and on its own does not shed light on the binding mechanisms, e.g. binding selectivity, affinity, rates and energies, material surface properties and peptide structure and conformation, which require further analyses. Hence additional research is required for developing the understanding of the physicochemical properties of peptides and materials using a range of experimental approaches coupled with bioinformatics and molecular modelling and simulations. A number of such experimental and computational tools have been employed successfully to probe nano–bio interactions and they are summarised in table 7.2, while extensive information on a range of techniques used can be found in [14].

7.3 An illustration of exploiting the knowledge of nano–bio interactions

In order to illustrate the key points discussed above, we will use an example of Pd nanoparticles from the literature (extensively reviewed in [15]). The aim is to follow the journey from the identification of strong binding peptides, developing the fundamental nano–bio understanding, all the way to their use in applications. In the case of Pd nanoparticles, Knecht and co-workers initially identified strong binding peptides against commercially available Pd nanoparticles [16]. The most promising peptide was Pd4 (see table 7.3 for the sequence). Using various analyses of the peptide sequence, previous knowledge of metal-binding ligands and, crucially, computational analysis, they identified histidine at positions 6 and 11 as the key residues for Pd binding. Indeed, nitrogen-containing ligands are known to bind metals, in particular those with imidazole rings as found in histidine. In order to confirm this hypothesis, they created a range of *mutant* peptides using the strategy shown in figure 7.2c, where one or both histidine residues were replaced by alanine or cysteine residues (as shown in table 7.3). Detailed experimental investigations of the peptide binding, Pd nanoparticle synthesis using these peptides, and the analysis of peptide structure and confirmation were performed.

The Pd nanoparticles synthesised using these peptides were further subjected to in-depth structural analysis at the atomic scale using x-ray diffraction, x-ray absorption spectroscopy and molecular dynamics simulations. The interpretation of results using reverse Monte Carlo analysis provided atomically resolved mechanistic details of the nano–bio interactions. Importantly, it was revealed that the Pd nanoparticles synthesised using these peptides contained an ordered core and a disordered surface (figure 7.4). Interestingly, the disordering of Pd atoms on the surface depended on the peptide (sequence and structure). These predicted structures of Pd nanoparticles strongly correlated with those observed using a high resolution TEM.

Table 7.2. Experimental and computational tools employed to probe *nano-bio* interactions. Table adapted from [6].

Tools	Affinity	Selectivity	Kinetics and thermodynamics	Surface coverage	Peptide structure and conformation	Surface properties
Fluorescence microscopy	✔	✔				
Site-directed mutagenesis	✔	✔			✔	
Surface plasmon resonance	✔	✔	✔	✔		
Quartz crystal microbalance	✔	✔	✔	✔		
Isothermal titration calorimetry	✔	✔	✔	✔		
Atomic force microscopy	✔	✔		✔		✔
Circular dichroism spectroscopy					✔	
Nuclear magnetic resonance					✔	
Fourier-transform infrared spectroscopy	✔				✔	
Computer models and simulations	✔	✔	✔		✔	✔
Zeta potential	✔		✔			✔
Raman spectroscopy	✔					✔
X-ray photoelectron spectroscopy				✔		✔

This information was used to construct systems containing Pd nanoparticles and strong binding peptides for performing molecular dynamics simulations. The results from these MD simulations were in turn used to predict the reactivity of these Pd nanoparticles in a Stille coupling reaction. In particular, the simulations computed the ease of abstraction of a surface Pd atom; such abstraction is believed to be a key

Table 7.3. Sequences of strong binders for Pd and mutant peptides. Table adapted from [16, 17].

Peptide	Sequence
Pd4	1 6 12 TSNAVHPTLRHL
A6	TSNAVAPTLRHL
A11	TSNAVHPTLRAL
A6,11	TSNAVAPTLRAL
C6	TSNAVCPTLRHL
C11	TSNAVHPTLRCL
C6,11	TSNAVCPTLRCL
C6,A11	TSNAVCPTLRAL
A6,C11	TSNAVAPTLRCL

step in Pd-catalysed carbon–carbon coupling reactions. Remarkably, the predicted reactivities (or ease of Pd abstraction) matched the observed experimental values [17].

This large body of work provided crucial **insights into peptide–nanoparticle binding**, as follows:

- Peptide **sequence** is important and biopanning is a powerful tool in identifying strong binders. Altering a single amino acid can lead to dramatically different binding/synthesis outcomes.
- The sequence provides a specific **structure and conformation** for the peptide, which is important in maximising interactions with surfaces.
- The flexibility/rigidity of peptide sequences is also important, as many peptides re-structure upon binding, leading to **cooperative assembly**.
- Due to the **uniqueness of peptide sequences/conformation**, they yield nanoparticles that are different from each other. This has implications for their performance (discussed below).

What we have learnt from this example, as summarised in figure 7.5, is that by learning about and utilising the nano–bio interactions, bioinspired synthesis of nanomaterials can lead to new or enhanced applications. In particular, the ability of peptides to 'recognise' different surfaces has led to the design of a range of nanocatalysts. The crucial aspect of completing this journey is gaining a molecular- or atomic-level understanding. As such, this knowledge is now being used to predictively design Pd-catalysts for a range of reactions.

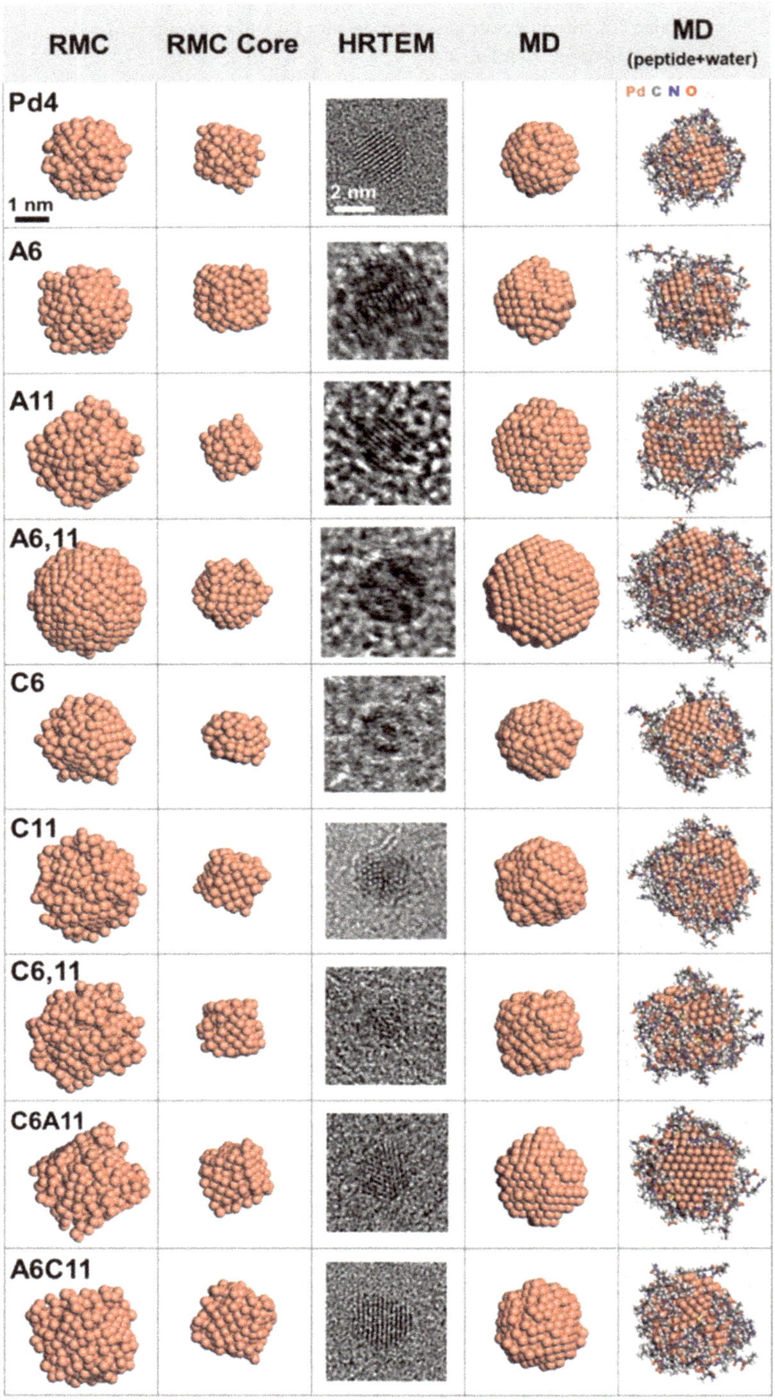

Figure 7.4. Morphologies of Pd nanoparticles synthesised using the peptides given in table 7.3. RMC = reverse Monte Carlo (for surface), RMC Core = RMC for the core of the nanoparticles, HRTEM = high resolution TEM, MD = molecular dynamics. Image taken with permission from from [17], copyright 2015 American Chemical Society.

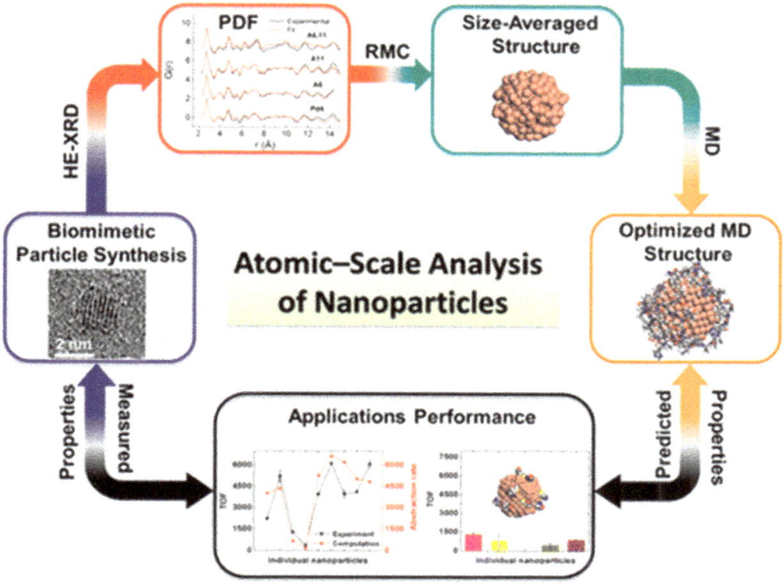

Figure 7.5. Correlation between experimentally measured reactivity of Pd nanoparticles in a Stille coupling reaction with the computed ease of surface Pd abstraction. Image taken with permission from [17], copyright 2015 American Chemical Society.

7.4 Additives as the mimics of biomineral forming biomolecules

7.4.1 The need for additives

The mechanistic understanding of the interactions between biomolecules and nanomaterials is a powerful basis for designing bioinspired syntheses of functional materials. Now the questions to consider in order to move forward are: how do we *apply* this knowledge, and what are the criteria for this *application*? The rationale behind taking this knowledge further is to enable a sustainable/green/economical route to existing materials, to enable the formation of new materials/chemistries that were not accessible previously, and/or to enable new functionalities for materials for specific applications. These needs imply that bioinspired methods provide cost-effective and eco-friendly alternatives, thereby enhancing their accessibility as well as sustainability. Further, the implication is that the methods are reproducible and scalable in order to achieve high quality products, at scale. Given these requirements, when moving from mechanistic understanding to *engineering* bioinspired products and processes, the use of specialised biomolecules (e.g. extracted proteins from organisms or peptides from biopanning) becomes problematic, for the reasons discussed below.

The extraction or genetic production of biomineralising biomolecules is not always viable. These procedures are quite complex, multistep, and require extensive purification. As such, only small quantities of biomolecules are available, typically only to those with specialised expertise, thereby **limiting access** for a wider academic and industrial community. This creates significant limitations associated with using biomolecule-based nanomaterials in sourcing sufficient materials for large-scale manufacture. Genetic engineering can provide ways for larger scale production of such molecules, however, the genetic toolkit again remains with a small group of experts. Given the *application* of the mechanistic knowledge gained requires a multidisciplinary approach, limited access to the specialised biomolecules is a key reason for the stifling of much needed further research and development. Another factor to consider is that the extracts are typically precious, delicate, and **unstable** when isolated from their native environments. This results in either the loss of their function (partly or entirely degrading over time), limiting their usage, or the need for a range of 'co-factors', thus making these biomolecules ineffective for use *in vitro*. These and other factors can result in the use of such biomolecules being **uneconomical and unsustainable** (and sometimes unsafe) for manufacturing. The added value to products or the benefits to a process can significantly diminish due to the issues discussed above, and the knowledge gained from biomineralisation is unable to meet its potential. It is therefore important to focus on mimics of biomolecules ('additives') that can provide the benefits that extracted biomolecules can, yet without the associated issues when it comes to translating the knowledge to the development of new materials, products and processes.

7.4.2 The design of additives and custom synthesis

A successful strategy involves the use of synthetic molecules (defined as '**additives**'), which can mimic the function of biomolecules, enabling custom synthesis of nanomaterials in a sustainable fashion. This strategy is detailed below in a generic sense, and is further expanded with examples of two case studies in the last two chapters of the book.

The general challenge is how to identify/select potentially active additives. We have seen from section 7.2 that the study of biomolecules and their interactions with nanomaterials has provided a range of rules and insights. These features include both the chemical and physical properties of the biomolecules and the materials. For example, the sequence of peptides as well as the chemical composition of the materials are important. In addition, the biomolecule conformation and co-operative assembly with the nanomaterials (or their precursors/intermediates) is crucial. This knowledge can be used to define the key features needed in an additive (**point 3** in figure 7.1) for green synthesis of nanomaterials. It is possible to design synthetic molecules which contain the key chemical functionality required for the synthesis and assembly of nanomaterials. One can also engineer simple (bio) molecules, e.g. self-assembling peptides and polymers, which form scaffolds for nanomaterials formation/assembly. These additives can exhibit one or more features that are associated with their biomolecule counterparts. It is important to understand how these additives interact with the entire process of nanomaterials formation (starting from precursors to the final materials), and their physical and chemical properties (e.g. self-assembly and ionisation). These bioinspired synthesis strategies (point 4 in figure 7.1) can draw from the significant progress made in the

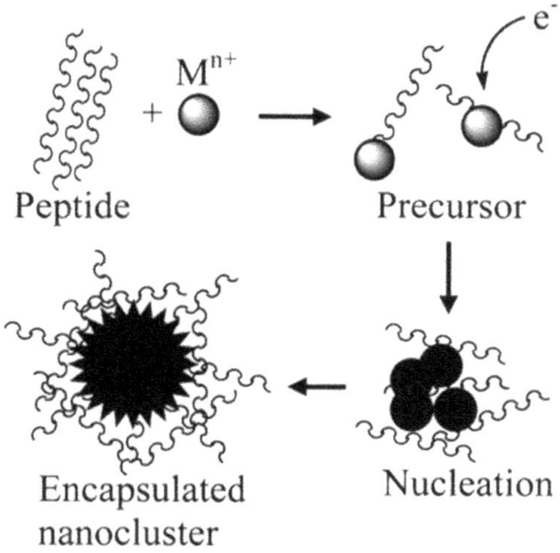

Figure 7.6. Schematic representation the roles of additives in metal nanoparticle formation. Image taken with permission from [19], copyright 2003 American Chemical Society.

synthesis of nanomaterials using bottom-up approaches, which employ sophisticated techniques such as self-assembly, nano–bio interactions, and templated synthesis [18]. These features are outlined in figure 7.6 and illustrated below with selected examples.

In section 7.2.1 and chapter 6 we have seen that some amino acids or chemical functionalities play crucial roles in biomineralisation. Examples include acidic amino acids for calcium carbonate type biominerals, and certain metal hydroxides/oxides or amines for a range of oxides and metals. One example is that of noble metal nanoparticles. A range of bacteria and fungi produce such metallic nanoparticles through the use of specialised proteins that have specific recognition against metal ions/clusters. Further, the chelation of metals with biological ligands is well known in bio-inorganic chemistry (e.g. chelation of Fe ions in heme protein). Through understanding these systems and biopanning (see section 7.2.2), it has emerged that imidazole groups of histidine and thiol groups of cysteine are key in chelating with metals. Using this knowledge, a range of synthesis strategies have been reported for metal nanoparticles (e.g. gold, silver, platinum, and copper) [19]. Using amino acids, peptides or other additives containing imidazole- or thiol-rich motifs, synthesis of these metal nanoparticles has been achieved. The roles of the additives include chelation or ion binding, followed by the formation of cluster/nuclei by lowering the surface energy, and finally controlling the growth and stabilisation of nanoparticles (see figure 7.6). In the growth stages, lattice matching and stereochemical recognition control the morphology of the final materials produced [20]. Knowing these specific (bio)chemical interactions between the additive and metals, one can develop customised bioinspired synthesis, as illustrated with examples below.

In the case of metal nanoparticle synthesis, in an effort to search for greener routes, a number of small molecules have been used, such as polyphenolic antioxidants from tea and vitamins, cellulose and winery waste [21]. For example, gold and platinum have been synthesised using vitamin B2 in water, within minutes [22]. Unlike the conventional methods described in section 4.6.5, these methods occur at room temperature in water, without the need for a separate reducing agent.

There has been extensive research reported on the design and use of a range additives in the formation of calcium-based crystals/nanoparticles [23]. These additives include biological and synthetic polymers (including copolymers), small molecules and inorganic ions. These additives are inspired by calcium carbonate biomineralisation, where proteins rich in carboxylate moieties arising from aspartic acid and glutamic acid residues have been shown to strongly bind the biomineral and regulate their formation and stability [24]. When polypeptides of these residues were investigated *in vitro*, there was clear evidence of preferential binding of these polypeptides with calcium carbonate crystals as well as their effect on crystal habits [25]. Encouraged by these results, a range of acidic additives have been investigated, such as polyacrylic acid, polystyrene sulfonate, and phytic acid (a phosphorylated sugar ring), and each have shown direct control over crystallisation. In the case of polystyrene sulfonate, for example, the polymer was found to bind free calcium ions in the solution, thus altering saturation, and hence the formation pathways. Further, the polymer was also able to preferentially bind the (011) face, causing stabilisation of this face, inhibiting its growth and leading to significant changes to the final morphology [26]. Another study using the same polymer reported that at high concentrations of polystyrene sulfonate, a less stable vaterite phase of calcium carbonate is formed instead of a more stable calcite [27].

In terms of the roles that these additives play, some areas are not well understood. For example, in calcite formation and growth, it is unclear whether the additives influence the nucleation and growth stages [23] or only the crystal growth, thus limiting control only over the morphology [28]. In addition to facilitating nucleation and growth, some additives can inhibit crystallisation via colloidal stabilisation of intermediate phases such as amorphous precursors to crystals. Indeed, in the case of calcium oxalate, the use of citrate has clearly shown that citrate chelates with the early stages of the calcium oxalate species, thus delaying or inhibiting crystallization [29]. These strategies are now being applied widely to many other crystalline nanomaterials. In chapters 8 and 9, we will look more closely at the use of designer additives in the formation of amorphous silica and crystalline magnetite.

In the above examples, the roles of additives have been limited to molecular scale interactions via charge-matching/neutralising and/or reducing the metals. Self-assembly of the additives was rarely utilised. We now focus on examples where the additive also acts as a template by assembling into supramolecular structures (see background on self-assembly in chapter 4). A classic example of this approach is the use of double stranded DNA molecules to programme the assembly of nanoparticles (figure 7.7 (top)) [30]. In this example, DNA-functionalised nanoparticles were prepared and using a complementary DNA strand(s), these nanoparticles were assembled into a desired array/aggregate. This principle has been extended to a range of biological and synthetic molecules in order to assemble nanomaterials. Another example includes the use of block co-polymers, which spontaneously

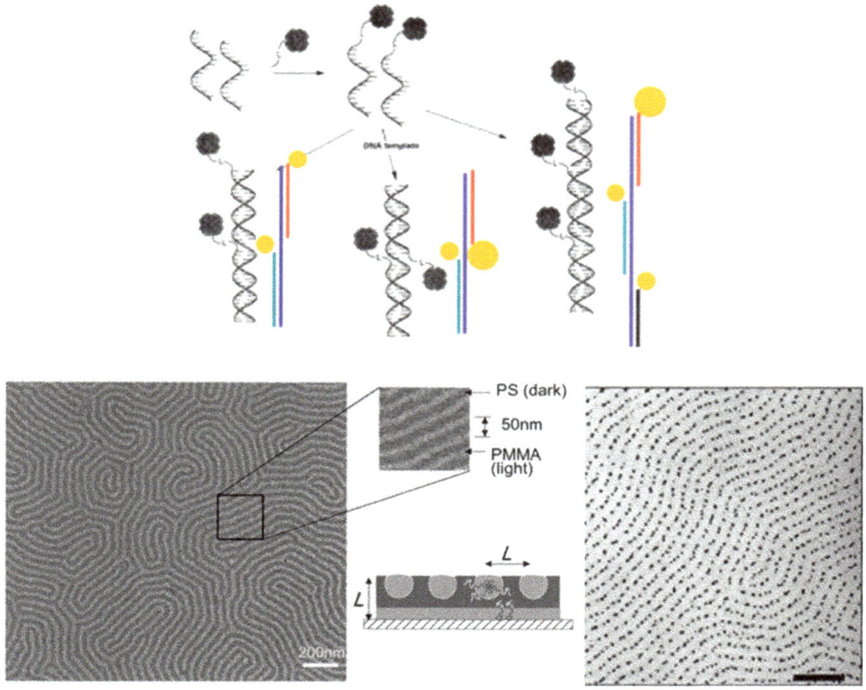

Figure 7.7. Selected examples of using chemical and physical features of (bio)molecules to assemble nanoparticles. Top: the use of DNA to programme the assembly of nanoparticles (shown as black or yellow circles). Image adapted from [30], with permission from Springer, copyright 1996. Bottom: the use of self-assembling polystyrene-*b*-polymethyl methacrylate (PS-*b*-PMMMA) block copolymers to assemble/localise nanoparticles by utilising selective chemical wetting of polystyrene block by gold nanoparticles—images show the self-assembled polymer before (left) and after (right) the addition of nanoparticles and annealing (scale bar = 200 nm). Images taken from [31], with permission from Springer, copyright 2001.

self-assemble due to localised phase separation between *immiscible* polymer blocks (figure 7.7 (bottom)) [31]. The ability of polystyrene to selectively 'wet' gold nanoparticles was exploited by using a polystyrene–poly(methyl methacrylate) co-polymer. These two blocks self-assemble, while the gold nanoparticles selectively assemble with the polystyrene block.

Another example includes the use of polyglutamic acid for the synthesis of porous alumina, where both the chemical and physical interactions are important, as depicted in figure 7.8 [32]. Under the synthetic conditions, polyglutamic acid self assembles into helices, with overall negative charge. Aluminium precursor forms cationic intermediates (hydroxide) in the aqueous synthesis system. This leads to strong ionic interactions between the additive and the inorganic species, leading to the self-assembly of aluminium hydroxide around helical polyglutamic acid. Further condensation reactions lead to the formation of alumina. A similar strategy was used in the case of $BaCO_3$ precipitation in the presence of polyelectrolytes [33]. The ability of polyelectrolytes such as poly(sodium 4-styrenesulfonate) and poly

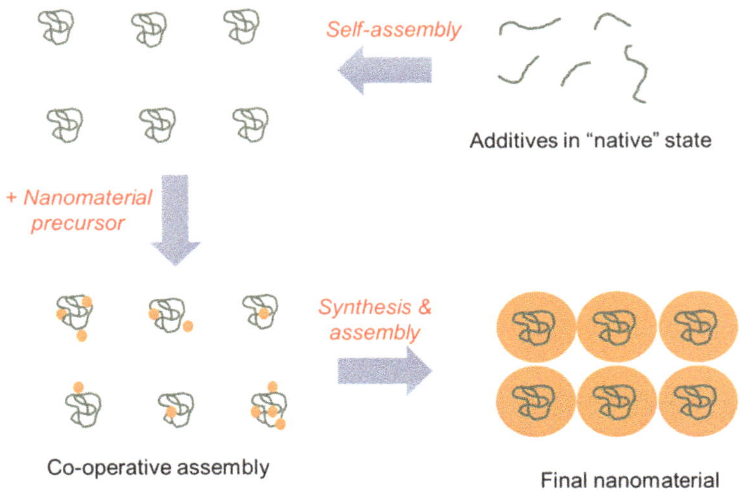

Figure 7.8. Schematic representation of how specific chemical and physical features of additives (shown in green) can be utilised in the synthesis and/or assembly of nanomaterials (shown in orange). Image adapted from [34], reproduced by permission of The Royal Society of Chemistry.

(allylamine hydrochloride) to self-assemble as well as chelate/bind mineral and/or their precursors was used to produce suprastructures of barium carbonates.

In this section, we have seen the need for more accessible additives and the benefits from using them. Using synthetic additives allows us to tailor their chemical properties to understand nanomaterials formation and design specific additives for the custom synthesis of desired nanomaterials. The knowledge of biomineralisation and the roles of organic components enable the design of such additives, which is a crucial first step towards developing sustainable bioinspired methods.

7.5 Compartmentalisation, templating and patterning

Another strategy for developing bioinspired synthesis involves the use of compartmentalisation or confinement. As seen in chapter 6, this is often the first step in biomineralisation. A compartment differentiates an area/volume to concentrate precursors, and controls the location and define the chemistry, thus controlling the material formation. Confinement or compartmentalisation offers various advantages over *homogeneous* particle formation in solutions. This strategy results in very different chemical environments compared to outside of the compartment, enabling formation and control over very specific materials. The compartment is also a template. It provides a mould for the shape and size of the final material, as well as the specific location of the material formation. Both of these factors control the material's chemical and physical features at nanoscales, while the location of the compartment can define the location of the final nanomaterial (e.g. on a surface).

The use of 'nanoreactors' (nanoscale confinement) for compartmentalisation is powerful in producing particles of precise shapes and sizes such that their aggregation can be avoided. Although synthetic nanoreactors have started to appear in the literature, most of the research to date has used biological 'compartments'. These include: within cells (e.g. magnetosomes), protein cages (ferritin), and a range of virus 'cages' [35]. Synthetic systems include self-assembled surfactant-stabilised nano-emulsions (e.g. water in oil), liposome and polymersome.

7.5.1 Confinement in a simple protein template

7.5.1.1 Ferritin

The iron storage protein ferritin has been introduced in chapter 6 (section 6.4.2.6) and is the natural place to begin when considering the biokleptic use of self-assembled biological template 'nano-reactors' for biomineralisation. This is because it is only a small re-purposing proposition. Ferritin is a simple self-assembled cage protein compartment, which already performs biomineralisation of ferrihydrite within its defined 8 nm diameter core. It is thus a simple step to use apoferritin to promote the synthesis of a range of monodisperse nanomaterials within the protein core. Ferritin is ideal for these purposes as it is inherently a biocompatible, precisely folded compartment for precision templating (of spherical particles of 6–8 nm diameter in size), with the mineralisation aided by ion channels situated at the junctions of each subunit, to enable the transport of metal ions into its core. The nanomaterials produced can be used for a wide range of applications [36]. The outer shell of ferritin can also be functionalised chemically or genetically, enabling the ferritin protein shell to act as a multivalent scaffold. For example, functionalisation of the ferritin protein can enable the particles to be soluble in organic solvents, adding more versatility to the applications for which they can be used [37, 38].

The inorganic core can be formed by directed mineral precipitation within the ferritin nanoreactor core space. The ion channels that allow iron ions to enter the core under normal conditions are generic enough to allow a full range of metal ions into the centre, from manganese to gold. The first alternative minerals to be formed within ferritin were iron sulfide and magnesium oxide in the early 90s [39], followed by magnetic nanoparticles such as magnetite [40] and CoPt [41]. Such tight control over the size and dispersity of CoPt makes very attractive particles for recording media. Indeed, the CoPt particles formed in ferritin were tested for their data recording properties [41]. Enhanced clinically applicable gadolinium (III) oxide hydroxide nanomaterials have been synthesised within ferritin [42]. Furthermore, nanotechnologically applicable cadmium sulphide quantum dots (QDs) [43], palladium [44], and gold [45] nanoparticles have also been formed within ferritin's core.

The chemistry of these systems needs to be elegantly designed. It is not simply a case of adding ferritin to a nanoparticle precipitation reaction, as this would result in a mixed population of some particle precipitating within the ferritin, but with nothing to stop particles also precipitating directly in solution. Thus, a multistep process of concentrating the ion, washing, nucleation, and then growth, can usually be adopted. For example, in the case of the Au–ferritin particles, the gold salt was included with ferritin to allow sufficient gold ions to enter the ferritin cavity (figure 7.9A). The solution was then run down a desalting column to remove the exterior gold ions, so that when the reducing agent was added only the gold ions within the ferritin could nucleate a gold nanocluster; these were then grown to fill the cavity by adding more gold salt in the presence of ascorbic acid [45]. A similar multistep process was used for the CoPt example, but in addition they also then heated the particles to anneal them as a final step to form the $L1_0$ phase of CoPt that is preferable for data storage.

7.5.1.2 Further cage protein and virus capsid templates

While ferritin is clearly applicable due to the fact it already biomineralises the nanoparticle in Nature, it is also clear that we can expand this notion to the use of a full range of self-assembled natural protein cage structures as bio–nanoreactors. MjHsp is a well-studied small and stable heat shock protein that forms a cage structure, assembled from 24 subunits with large 3 nm pores allowing reagents such as metal ions to enter the 6 nm diameter core [46]. It is stable up to 70 °C and in a pH range of 5–11, making it ideally robust for chemical synthesis conditions. MjHsp has been used to encapsulate ferrihydrite through air oxidation of Fe(II) in the presence of the protein cage, analogous to ferritin [47]. PepA is another small cage protein with a tetrahedral structure, with a 6 nm interior cavity which can be fed reagents through four 4 nm wide channels at the faces of the tetrahedron, and four 1 nm wide channels at the edges [48]. PepA has been used to mediate the growth of 1.1–2.8 nm sized CoPt magnetic nanoparticles inside the protein cavity under ambient reaction conditions, with magnetic coercivity increasing with increased particle size [48]. It is thought the positively charged metal cation precursors preferentially accumulate and nucleate in the negatively charged interior of the cage [49].

All the examples so far have had spherical cores of approximately 6 nm in size. While these are an excellent template to obtain spherical nanoparticles of this size, there is no scope for different sizes or structure with these proteins. Thus the exploitation of virus protein cages as nanoreactors for bioinspired synthesis of nanomaterials is an obvious next step. A virus is simply nucleatic acid molecules within a self-assembled protein coating/shell, which can only be replicated in a host cell using the machinery of another living organism. Once the replicated molecules in the inside are removed, the protein shells offer a family of biological structures so extensive that they offer a vast range of pre-determined compartment sizes and even a range of shapes, from very simple to ornate and complex, vastly expanding the scope for using biological cages as nanomaterial templates. Furthermore, the variety of pore sizes between the subunits offers further size-selectivity with respect to reagents entering the core and thus the nanomaterials produced.

Each virus template can be utilised in multiple ways, mineralising either within the core of the cage or even the exterior, coating the virus in a mineralised outer shell.

Internal mineralisation

Cowpea chlorotic mottle virus (CCMV) and cowpea mosaic virus (CMPV) are both icosahedral 28 nm sized viruses that infect the cowpea (black-eyed pea) plant, made up of 180 and 60 protein subunits respectively. These are both excellent examples of virus shells where the inherent properties of the structures can be utilised for

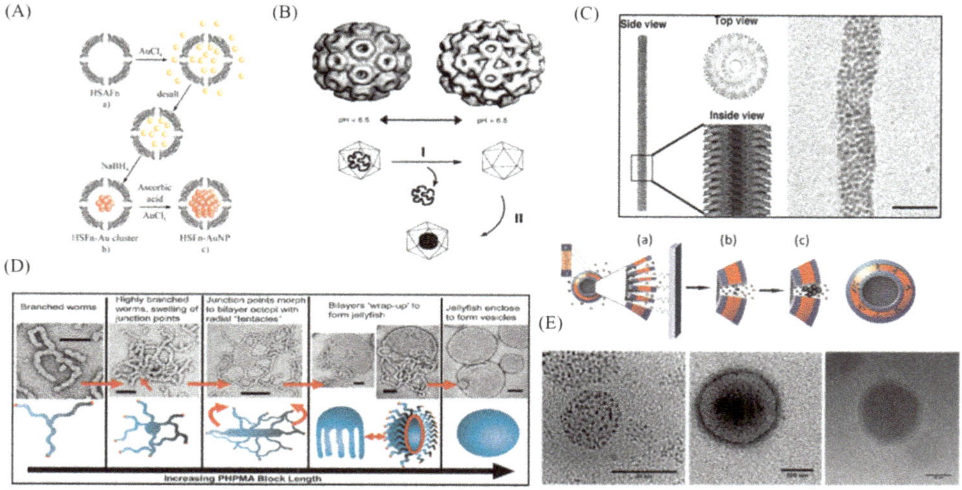

Figure 7.9. Examples of external and internal organic templates for mineralisation. (A) Use of ferritin as a nanoreactor to biomineralise precisely sized gold nanoparticles within the protein cage (reproduced with permission from [45], copyright 2010 John Wiley & Sons). (B) Example of a virus capsid that could be used as a nanoreactor for similar biomineralisation and shows the expansion of cowpea chlorotic mottle virus (CCMV) using pH conditions, demonstrating how this can aid mineralisation reactions (reproduced with permission from [52]). (C) Example of a different shaped template. Cryo-electron micrograph and reconstruction of tobacco mosaic virus (TMV) are shown with its elongated structure, and its use as template for the mineralisation of Pt nanoparticles. Scale bars 20 nm (reproduced with permission from [58]). (D) Schematic and electron microscopy interpretation of architectures achievable during polymersome synthesis demonstrating the potential for control of the formation of a polymeric nanoreactor (reprinted with permission from [71], copyright 2011 American Chemical Society). (E) Schematic and electron microscopy of the formation of magnetopolymersomes. (Ei) Schematic of how electroporation is used to open pores in the membrane to transport in soluble iron (reproduced from [78], with permission of Springer, copyright 2015). (ii) Micrographs of three examples of magnetopolymersomes: (ii) PDB–PEO magnetopolymersomes (reproduced with permission from [78], with permission of Springer, copyright 2015); (iii) modified asymmetric (carboxylate interior) PEG–PHPMA/PHMPA–PMPC magnetopolymersomes (reproduced from [79]); (iv) artificial magnetosome formed from a POPC liposome (reproduced with permission from [69]).

sophisticated bioinspired nanoparticle synthesis (figure 7.9B). CCMV has an uncharged exterior, repelling nucleation on the surface, it also has intrinsic pH-dependent pores that effectively gate the ion channels, and a positively charged inner membrane which provides an ideal environment for the nucleation and precipitation of both paratungstate and decavanadate [50, 51]. The viral capsid undergoes a reversible 10% swelling above pH 6.5. This opens up multiple pores on the virus capsid which can enable the release of the viral RNA from inside, but also allows the empty capsid to uptake precursor through these channels for nanoparticle synthesis (figure 7.9B) [50, 52]. The reversible pH switching of this process allows precise elegant control over uptake and thus precipitation of the nanomaterial through pH. CPMV has also been used as a nanoreactor to form gold [53] and silica [54], again taking advantage of their pH switchable ion transport. It should be emphasised that electrostatically induced, pH-controlled encapsulation and nucleation at a range of template sizes is something that would be extremely difficult to replicate synthetically. The tobacco mosaic virus (TMV) is a non-spherical rod-like structured virus that infects tobacco plants. It contains 2130 coat proteins that arrange into a helix around the RNA, forming a rod 300 nm long and 18 nm wide (with an internal diameter of 4 nm). The central channel of the virus has been used to template both $FePt_3$ and CoPt nanowires [55], as well as Ag nanoparticles and QDs along the internal cavity of the virus [56].

External mineralisation
Biomineralisation on the exterior of CPMV has been performed by modifying the surface to be negatively charged to enable the nucleation of cobalt and iron oxide, resulting in nanoshell materials of a very small size distribution [57]. Similarly, the exterior of the TMV has been used as a template for the synthesis of Pt nanoparticles forming nanoparticle decorated nanowires. Such structures have increased Pt surface area and the overall stability of the Pt nanostructures compared to more conventional particles, enhancing their catalytic ability [58] (figure 7.9C).

7.5.2 Confinement in modified cage protein templates

Confined bio–nanoreactors can also be combined with the use of biomolecules or additives (already discussed in sections 7.2 and 7.4). The functionalisation of the inner (and outer) walls of these nanoreactors with biomolecules derived from biomineral forming systems or their bioinspired counterparts (the additives) can be vital in controlling localisation as well as material properties. This provides the sites for heterogeneous nucleation, thereby reducing the barriers for particle formation. Further, these functional sites help control the chemical and physical properties of the final nanomaterials (e.g. chemical composition and particle shape/crystallinity). We have already considered a very generic chemical modification of the CPMV above (section 7.5.1.2), to add negative charge to the surface to enable nucleation to that surface. Here we will consider more specific modifications to direct specific mineral phase formations through nucleation with very specific biomolecules that direct this phase forming.

For example, exposed surface amino acids have been used to attach specific peptide sequences to the exterior of CPMV. The sequences (HNKHLPSTQPLA) and (CNAGDHANC) were identified to selectively bind to FePt and CoPt respectively from biopanning (section 7.2.2 and table 7.1), and it was found these sequences enhanced the formation of CoPt and FePt respectively on the surface of the virus [57, 59].

Modification of the protein cage can be directed to almost any amino acid site in the protein, so the peptide sequences can be specifically directed towards the exterior (as is the case in the examples above) or to be displayed in the protein core interior. For example, the heatshock protein MjHsp has been modified at the N-terminus (so displayed within the core) with a peptide sequence (KTHEIHSPLLHK) that shows specificity for the $L1_0$ phase of CoPt [60]. This genetically modified MjHsp produced enhanced 6.5 nm CoPt nanoparticles inside the protein cages with increased magnetic coercivity (showing some $L1_0$ character) [60].

There is also potential for engineering the nanoreactor walls such that they can enable active transport of the precursors into the reactors. Recently, this was demonstrated where the proteins responsible for active transport of silicon into diatoms were tethered on the surface of nanoreactors, which pumped in the precursor, while the 'additives' present inside the reactor facilitated the formation of silica particles [61].

7.5.3 Biomimetic compartmentalisation

Natural biomineralisation and biokleptic protein cage templates such as viral capsids are limited by what Nature provides. These bionanoreactors are also designed in the most part to operate under ambient conditions, so may not necessarily be adapted for the more robust reaction conditions required for some inorganic nanoparticle synthesis. Taking a bottom-up approach, artificial biological compartments such as artificial membrane vesicles can be produced with the synthetic requirements designed in, i.e. any new specification arising outside the biological parameters, be they more extreme in chemistry or differing in size. It cannot be underestimated how powerful it is to take components, and knowledge of both chemistry and Nature, to design and engineer in the exact requirements for nanoparticle mineralisation. This flexibility is much less achievable with biological components such as protein cages, whose evolution has been optimised by Nature for their specific natural role, and not the new role imposed upon them.

Liposomes

Liposomes are the obvious starting point, as they are the biological component most commonly used in Nature to aid biomineralisation through trafficking and concentrating precursors, as well as the nanoreactor itself for nano mineral formation (see section 6.4.2.1). Because of their intrinsic and extensive occurrence in Nature and particular relevance to medicine, liposomes have been extensively researched and characterised. There are a number of routes to form liposomes, including but not limited to: sonication, electroformation extrusion, inkjetting, and microfluidics [62]. These different preparation methods allow for multiple modes of encapsulation and in some cases encapsulation of multiple reagents. More complicated preparation methods can allow for the independent design of the inner and outer leaflet of a liposomal nanoreactor, such as the inverted emulsion method developed by Pautot *et al* [63]. Inverted emulsion allows for production of vesicles capable of withstanding two contrasting external and internal environments, widening the scope of reaction type that liposomal nanoreactors can be subjected to, particularly in cases of immiscible solutions [63]. Here we have only been considering smaller unilamellar liposomes, which are composed of just one bilayer membrane in the nanometer size range. However, it should be noted that biology utilises a full range of membrane structure, from giant unilamellar vesicles (in the micrometre range) to more complex multivesicular structures. It is thus worth considering that there is plenty of scope to be inspired by biology to expand the sizes and shapes of templates used to more complex and elaborate membrane compartments. One such example of this is the use of a membrane tubulation (BAR) protein to create elongated membrane tubular structures from giant unilamellar vesicles. For example, giant vesicles containing QDs have been elongated with a BAR protein to make functional, biocompatible QD nanowires [64].

While we can design in more elaborate and elegant features into artificial liposomes, there is also much to be learnt from Nature about using them as a nanoreactor. A key aspect is how to transport the reagents (metal ions) across the membrane. There have been several strategies employed.

1. Encapsulate the metal ion precursors during liposome formation, and then rely on osmotic effects to increase diffusion to change the pH within the liposome to drive mineralisation. This technique has been used to form silver oxide [65] and iron oxide [66] nanoparticles within phospholipid vesicles, as well as CdS, ZnCdS and HgCdS [67] nanocrystal QDs within the core of liposomes, with a controlled one particle per liposome. However, interestingly, these particles do not fill the vesicle core. This is because the metal precursors available are finite and limited by the concentration which is encapsulated. Solid precipitated condensed mineral will always be more dense and thus smaller than the volume of the same quantity of metal ion in a solution.
2. In Nature liposomes have ion channel, specific metal uptake proteins, that provide a supply of metal ions so that the minerals can form and the crystals continue growing to fill the liposome. To mimic Nature, we could also adopt some sort of transport system for the metal ions into the liposome. This has been achieved by the incorporation of Ca^{2+} ionophore A23187 into liposomes and the loading of both Fe^{2+} and Ba^{2+} into preformed liposomes has been seen [68].
3. The ionophore is still biological. We can mimic the action of an ion channel by mechanically forming pores in the liposome membrane. This can be achieved using a process called electroporation. This is where a voltage is passed across a sample containing liposomes and this electric current induces the membrane to form temporary pores. This technique is readily used by molecular biologists as a means of inserting DNA into bacteria. Here the electroporation opens pores in the liposomes, to allow metals ions into the core. In this set up the core contains a basic (high pH) solution, so when the metal ions enter the nanoreactor liposome, they can precipitate as the oxide. This has been demonstrated for a system that mimics the magnetosome in magnetotactic bacteria, to make magnetite nanoparticles that now completely fill the liposome (figure 7.9 Eiv) [69].

Polymersomes

Liposomes are comprised of lipid amphiphilic molecules which can be quite constricting. As we are mimicking natural processes, there is nothing to stop us considering creating our vesicle membranes out of the full range of amphiphilic molecules available to us. We can thus tailor the vesicle functions even more specifically. The largest family of adaptable amphiphilic molecules are block co-polymers. Within this family there is an almost unlimited range of tunable block-lengths, functionalisation (including biocompatibility), hydrophobicity and thus architectures achievable [70, 71]. Polymersome synthesis can occur via numerous methods such as film rehydration, electroformation [72], RAFT synthesis [71, 73], ring opening polymerization [74], layer-by-layer polyelectrolyte formation utilising charge, and more complex methods such as shell cross-linked nanoparticles [75]. The diversity observed in the formation methods of polymersomes is reflected in the almost endless combination of properties that can be incorporated into a polymersome by careful selection and design of the building-block polymer materials. Just like the BAR tubulation example for liposomes, there are multiple, more complex structures that can be achieved with polymer self-assembly by controlled synthesis. For example, the structure that can be achieved with just one block co-polymer (poly (glycerol monomethacrylate)–poly(2-hydroxypropylmethacrylate) (PGMA–PHMPA)) varies from worms to nano jellyfish to vesicles, simply by increasing the length of the PHPMA block (figure 7.9D) [71]. Researchers have successfully incorporated biological components into polymer membranes [76], for example protein channels and enzymes to enable ion transport [77]. Furthermore, electroporation as a mechanical means of opening pores in liposome membranes also works on polymer membranes, which means the same technique can be used to import ions into polymersomes. This has been demonstrated using iron ions to form tiny (2–3 nm) magnetic iron oxide nanoparticles within the membrane itself. This seems to be due to the basic in the interior meeting the iron precursors at the membrane within the pore (figure 7.9 Ei) [78]. This does not mimic the magnetosomes as well as the previous lipid example of the same system (compare figure 7.9 Eii with Eiv). However, due to the tunability of the polymersomes, their inner surface can be modified to aid biomineralising within the core (figure 7.9 Eiii). This is discussed in more detail in chapter 8 (8.6.3) [79]. Polymersomes used as nanoreactors for bioinspired formation of inorganic nanomaterials is a relative new field of research, but owing to the almost unlimited range of possible polymersomes that can be designed and engineered, its potential to grow is certain.

7.5.4 Localisation and patterning on surfaces

It has already been noted that liposomes (section 6.4.2.1) are not only used as bio-nanoreaction vessel for mineralisation but are also key to trafficking and defining the site of mineralisation. To this regard, we can be inspired by nature to use biomolecules to position the site of mineralisation. So in this section the biomolecule has a dual (or even triplicate) role: It directions the formation of the inorganic material (as well as in the case of ferritin control it size) and also locates the materials on a surface in s specific pattern.

Ferritin (section 6.4.2.6 and 7.5.1.1) has been successfully immobilised on a soft lithographically patterned surface. Here micro-contact printing (see section 4.6.1) was used to pattern hydrophobic self-assembled monolayer strips alternating with positively charged 3-aminopropyltriethoxysilane (APTES) stripes on a silica surface. The negatively charged ferritin readily bound to the APTES electrostatically forming a surface of patterned ferritin [80]. In this study they heat their surfaces at 500°C in O_2 to remove the ferritin protein cage reveal just the iron-oxide particles patterned on the surface.

As we learnt in section 2.4.2.2, a key bottle neck for developing high density data storage for the next generation of HDD is finding a way to precisely pattern monodisperse hard magnetic nanoparticles on a surface. Bit patterned media requires each particle to hold a bit of information in an array. Towards this goal 7 nm CoPt nanoparticles were formed in apoferritin (as described in 7.5.1.1) and these were assembled on a silica surface [81]. The particles were not the $L1_0$ preferred crystal form for data storage so this assembly was also subjected to high-temperature (500°C–650°C) annealing to obtain the $L1_0$ phase, again removing the ferritin. In this study they also performed drag testing to see if magnetic information could be written and read on this biomediated surface.

We have already seen how ferritin can be modified with peptide earlier is section 7.5.1. Such modifications have been used to mediate the binding of ferritin to a specific surface. A specific binding sequence for titanium was inserted into ferritin. This modified ferritin cage still maintain the capability to biomineralize iron oxide, but was also capable of selectively binding to a titanium surface. Interestingly, with a higher affinity than the titanium-binding peptide in the absence of the ferritin scaffold [82].

Modifying a biomineralizing biomolecule to enable it to bind specifically to a surface has also been carried out with nanomaterial mediating biomolecules other than ferritin. There are many proteins and peptide that mediate the formation of different nanomaterials and conjugating these to surface material specific peptide to make a 'dual affinity' peptide or protein has proved to be a very successful technique. We will consider two examples, one of a protein that forms magnetite and one of a peptide that forms CoPt.

The protein Mms6 is a magnetite nucleation protein from magnetic bacteria which is described in detail in chapter 8. The unmodified protein has been attached to a patterned micro-contact printed self-assembled monolayer. Magnetite has been synthesised on this surface to make a microscale patterned array of magnetite nanoparticles [83]. The protein has also been modified with a Cys residue to bind directly to gold on a patterned surface, which again can produced a magnetite nanoparticle patterned surface when chemical synthesis is performed on the surface [84] and this has been patterned right down to the nanoscale [85].

The peptide KTHEIHSPLLHK has been used to modified ferritin to template CoPt in section 7.5.2 as it promotes the formation of L10 CoPt. This has been conjugated to a silica binding peptide to form a dual affinity peptide that will both bind to the silica surface and aid the formation of CoPt. [86] The peptide was micro-contact printed onto a silica surface and CoPt was mineralised in a pattern in a similar process to the examples above. Furthermore the KTHEIHSPLLHK has been modified with a Cys residue to bind directly to a patterned gold surface in the same way as the Cys-Mms6/magnetite example above. This has been achieved with patterns as narrow as 226 nm wide, showing a real potential for bioinspired data storage solution [87].

7.6 Scalability of bioinspired syntheses

We have learnt that biosynthesis of nanomaterials occurs across the full range of life forms, from micro-organisms to humans [88], using complex protein- and enzyme-systems. How these 'mild' processes can be mimicked using highly specific designer biomolecules is discussed in this chapter [89]. Indeed, to date, lab-scale biomolecule-mediated syntheses of ~50 inorganic nanomaterials have been reported [90], however, they are expensive and can produce only μg–mg quantities. Since such processes heavily depend on costly biomolecules [91], they are uneconomical to scale-up. Although these biological methods can be eco-friendly, they are expensive, inefficient and/or not scalable to industrial production. On the other hand, results from experiments and process calculations have suggested that the bioinspired routes, which use additives, are potentially scalable and cheaper when manufactured at large scale [91, 92]. This can also help reduce the manufacturing carbon footprint, thus providing a significant cost benefit. As such, we have seen how additives can be identified/selected to meet the desired function for developing bioinspired syntheses. However, the inability to translate these discoveries into economical and reproducible manufacturing is a critical barrier which needs to be addressed in the future [93].

Scale-up, often dubbed as *bucket chemistry*, is an essential stage in order to establish the viability for an industrial-scale production of nanomaterials using bioinspired synthesis [94]. However, bioinspired methods are currently limited to lab-scale batches [95] (there are no reports of larger-scale investigations [96]), hence their true potential has not been realised. It is important to highlight here that scale-up is not trivial, because the transport properties (e.g. mixing and heating) change non-linearly with the production scale [97, 98]. This means that although chemical rates/kinetics do not change with the production scale, the physical parameters, particularly transport properties, do [99]. As a result, the reaction pathways and resultant outcomes (e.g. the properties of nanomaterials) change with scale-up. In most cases it is unknown how this change takes place, i.e. the correlations between the production scale, the transport properties and product properties are unknown. Below we briefly list key steps that can be taken to scale-up bioinspired synthesis of nanomaterials.

An initial step involves a techno-economic assessment of the method, where it is initially assumed that the process will scale-up, and the cost of industrial manufacturing is estimated. Parameters such as the availability of resources and downstream processing are also evaluated. Such analyses can help understand the benefits of bioinspired synthesis and identify any challenges in scale-up, e.g. cost–benefit ratio and efficiency of manufacturing. These challenges are better addressed at the discovery and design stage, due to the ease and cost-effectiveness compared to that at pilot or industrial scale.

Simultaneously, it is important to perform scalability evaluations. These include the experimental and computational approaches to develop the correlations between transport properties, the reaction timescales and the production scales. Established methods, which can be adopted by chemistry researchers, can help assess the scalability and develop pathways to scale-up [100]. Another way forward is to identify and optimise practical ways and equipment configurations [97]. Although some of these stages may seem too applied, the challenges faced during scalability analyses require further fundamental research with inputs from manufacturers. An example of this is the downstream processing (separation or purification of materials and effluent treatment), which is commonly overlooked. A simple centrifugation in the lab for the separation of materials from solvents could prove practically difficult or highly energy demanding at industrial scale. Similarly, washing performed at lab-scale can, at industrial scales, cause large quantities of waste, leading to unsustainable production. Part of the solution to these challenges could be alteration to the synthesis or even the additives used—this requires fundamental research. It is also important to be mindful of the fate of the additives used. Will they remain in the final materials? Will they be removed? If so, will they be discarded (causing pollution) or recycled? If recycled, is there an extra energy demand or cost for purification and reuse? Again, it is beneficial to consider these factors at an early (discovery) stage in order to enable swift and successful scale-up, leading to a rapid *lab-to-market* progress. Although there are not many (or any) examples of scale-up of bioinspired nanomaterials, we hope that this section has provided an awareness, a flavour of the challenges and a potential strategy to address them. With a case study of silica, chapter 9 illustrates how these principles can be applied to the scale-up of bioinspired synthesis.

7.7 Summary: key lessons about the journey towards bioinspired synthesis

One of the main purposes of developing bioinspired synthesis for nanomaterials is to create sustainable production technologies for desired products. In this chapter, we have learnt the principles of how to translate the knowledge of biomineralisation to designing biologically inspired routes. Important lessons learnt are listed below:

- A molecular level understanding of how biology produces high quality nanomaterials is crucial. One of the key controlling features of biomineralisation is the use of biomolecules. Hence understanding the roles that such biomolecules play in the entire process of biomineralisation is extremely important.
- These biomolecules have unique catalytic or binding sites that offer recognition (selectivity and high affinity), and their chemical properties (e.g. amino acid sequence in proteins or peculiar chemical structures, modifications or motifs) are important in this recognition. The structure and conformation of these biomolecules are also important because this leads to particular self-assembly (intra- and inter-molecular) and cooperative assembly with inorganic species.

- These features together enable controlled synthesis, assembly and/or functionalisation of nanomaterials.
- The direct use of biomolecules causes serious barriers to advancing green synthesis. It is therefore important to design 'additives' that can provide the benefits that extracted biomolecules can, yet without the associated issues when it comes to translating the knowledge to the development of new materials, products and processes.
- Confinement for nanomaterials synthesis, particularly when combined with the use of additives, can be powerful in controlling size, shape and localisation as well as material properties.
- It is important to be aware that scale-up is not trivial, because transport properties change non-linearly with the production scale. This means that the reaction pathways and resultant outcomes change with scale-up, and are typically unpredictable for new syntheses.

References

[1] Round F E, Crawford R M and Mann D G 1990 *The Diatoms: Biology & Morphology of the Genera* (Cambridge: Cambridge University Press)
Vrieling E G, Gieskes W W C and Beelen T P M 1999 *J. Phycol.* **35** 548

[2] Sumerel J L, Yang W J, Kisailus D, Weaver J C, Choi J H and Morse D E 2003 *Chem. Mater.* **15** 4804
Kroger N, Dickerson M B, Ahmad G, Cai Y, Haluska M S, Sandhage K H, Poulsen N and Sheppard V C 2006 *Angew. Chem. Int. Ed.* **45** 7239
Sewell S L and Wright D W 2006 *Chem. Mater.* **18** 3108

[3] Rawlings A E, Bramble J P and Staniland S S 2012 *Soft Matter* **8** 6675

[4] Kim I W, Collino S, Morse D E and Evans J S 2006 A crystal modulating protein from molluscan nacre that limits the growth of calcite *in vitro Cryst. Growth Des.* **6** 1078–82

[5] Patwardhan S V, Patwardhan G and Perry C C 2007 *J. Mater. Chem.* **17** 2875

[6] Care A, Bergquist P L and Sunna A 2015 *Trends Biotechnol.* **33** 259

[7] Klem M T, Willits D, Solis D J, Belcher A M, Young M and Douglas T 2005 *Adv. Funct. Mater.* **15** 1489

[8] Reiss B D, Mao C, Solis D J, Ryan K S, Thomson T and Belcher A M 2004 *Nano Lett.* **4** 1127

[9] Rawlings A E, Bramble J P, Tang A A S, Somner L A, Monnington A E, Cooke D J, McPherson M J, Tomlinson D C and Staniland S S 2015 *Chem. Sci.* **6** 5586

[10] Patwardhan S V, Emami F S, Berry R J, Jones S E, Naik R R, Deschaume O, Heinz H and Perry C C 2012 *J. Am. Chem. Soc.* **134** 6244

[11] Liang M-K, Deschaume O, Patwardhan S V and Perry C C 2011 *J. Mater. Chem.* **21** 80

[12] Cui Y, Kim S N, Jones S E, Wissler L L, Naik R R and McAlpine M C 2010 *Nano Lett.* **10** 4559

[13] Wang Y, Satyavolu N S R and Lu Y 2018 *Curr. Opin. Colloid Interface Sci.* **38** 158

[14] Limo M J, Sola-Rabada A, Boix E, Thota V, Westcott Z C, Puddu V and Perry C C 2018 *Chem. Rev.* **118** 11118

[15] Walsh T R and Knecht M R 2017 *Chem. Rev.* **117** 12641

[16] Pacardo D B, Sethi M, Jones S E, Naik R R and Knecht M R 2009 *ACS Nano* **3** 1288

[17] Bedford N M, Ramezani-Dakhel H, Slocik J M, Briggs B D, Ren Y, Frenkel A I, Petkov V, Heinz H, Naik R R and Knecht M R 2015 *ACS Nano* **9** 5082

[18] Weller M T, Overton T, Rourke J and Armstrong F A 2014 *Inorganic Chemistry* 6th edn (Oxford: Oxford University Press)

Ozin G A, Arsenault A C and Cademartiri L 2009 *Nanochemistry: A Chemical Approach to Nanomaterials* 2nd edn (Cambridge: Royal Society of Chemistry)

Whitesides G M and Grzybowski B 2002 *Science* **295** 5564

[19] Slocik J M and Wright D W 2003 *Biomacromolecules* **4** 1135

[20] Mann S, Archibald D D, Didymus J M, Douglas T, Heywood B R, Meldrum F C and Reeves N J 1993 *Science* **261** 1286

[21] Varma R S 2012 *Curr. Opin. Chem. Eng.* **1** 123

[22] Nadagouda M N and Varma R S 2006 *Green Chem.* **8** 516

[23] Xu A-W, Ma Y and Cölfen H 2007 *J. Mater. Chem.* **17** 415

[24] Lowenstam H A 1981 *Science* **211** 1126

Lowenstam H A and Weiner S 1989 *On Biomineralization* (New York: Oxford University Press)

[25] Gower L A and Tirrell D A 1998 *J. Cryst. Growth* **191** 153

Kato T, Suzuki T, Amamiya T, Irie T and Komiyama N 1998 *Supramol. Sci.* **5** 411

[26] Wang T, Cölfen H and Antonietti M 2005 *J. Am. Chem. Soc.* **127** 3246

[27] Jada A and Verraes A 2003 *Colloids Surf.* A **219** 7

[28] Kim Y-Y *et al* 2017 *Angew. Chem. Int. Ed.* **56** 11885

[29] Ruiz-Agudo E, Burgos-Cara A, Ruiz-Agudo C, Ibañez-Velasco A, Cölfen H and Rodriguez-Navarro C 2017 *Nat. Commun.* **8** 768

[30] Alivisatos A P, Johnsson K P, Peng X, Wilson T E, Loweth C J, Bruchez M P and Schultz P G 1996 *Nature* **382** 609

Loweth C J, Brett Caldwell W, Peng X, Paul Alivisatos A and Schultz P G 1999 *Angew. Chem. Int. Ed.* **38** 1808

[31] Lopes W A and Jaeger H M 2001 *Nature* **414** 735

[32] Jan J S and Shantz D F 2005 *Chem. Commun.* **16** 2137

[33] Yu S-H, Cölfen H, Xu A-W and Dong W 2004 *Cryst. Growth Des.* **4** 33

[34] Patwardhan S V, Mukherjee N, Steinitz-Kannan M and Clarson S J 2003 *Chem. Commun.* **10** 1122

[35] Rong J, Niu Z, Lee L A and Wang Q 2011 *Curr. Opin. Colloid Interface Sci.* **16** 441

[36] Jutz G, van Rijn P, Miranda B S and Boker A 2015 *Chem. Rev.* **115** 1653

[37] Wong K K W, Colfen H, Whilton N T, Douglas T and Mann S 1999 *J. Inorg. Biochem.* **76** 187

[38] Sengonul M, Ruzicka J, Attygalle A B and Libera M 2007 *Polymer* **48** 3632

[39] Meldrum F C, Wade V J, Nimmo D L, Heywood B R and Mann S 1991 *Nature* **349** 684

[40] Meldrum F C, Heywood B R and Mann S 1992 *Science* **257** 522

[41] Mayes E, Bewick A, Gleeson D, Hoinville J, Jones R, Kasyutich O, Nartowski A, Warne B, Wiggins J and Wong K K W 2003 *IEEE Trans. Magn.* **39** 624

[42] Sanchez P, Valero E, Galvez N, Dominguez-Vera J M, Marinone M, Poletti G, Corti M and Lascialfari A 2009 *Dalton Trans.* **5** 800

[43] Wong K K W and Mann S 1996 *Adv. Mater.* **8** 928

[44] Ueno T, Suzuki M, Goto T, Matsumoto T, Nagayama K and Watanabe Y 2004 *Angew. Chem.* **116** 2581

[45] Fan R, Chew S W, Cheong V V and Orner B P 2010 *Small* **6** 1483
[46] Kim K K, Kim R and Kim S-H 1998 *Nature* **394** 595
[47] Flenniken M L, Willits D A, Brumfield S, Young M J and Douglas T 2003 *ACS Nano Lett.* **3** 1573
[48] San B H, Lee S, Moh S H, Park J-G, Lee J H, Hwang H-Y and Kim K K 2013 *J. Mater. Chem.* B **1** 1453
[49] Douglas T, Strable E, Willits D, Aitouchen A, Libera M and Young M 2002 *Adv. Mater.* **14** 415
[50] Douglas T and Young M 1998 *Nature* **393** 152
[51] Johnson J E and Speir J A 1997 *J. Mol. Biol.* **269** 665
[52] Speir J A, Munshi S, Wang G, Baker T S and Johnson J E 1995 *Structure* **3** 63
[53] Aljabali A A A, Lomonossoff G P and Evans D J 2011 *Biomacromolecules* **12** 2723
[54] Steinmetz N F, Shah S N, Barclay J E, Rallapalli G, Lomonossoff G P and Evans D 2009 *Small* **5** 813
[55] Tsukamoto R, Muraoka M, Seki M, Tabata H and Yamashita I 2007 *Chem. Mater.* **19** 2389
[56] Dujardin E, Peet C, Stubbs G, Culver J N and Mann S 2003 *Nano Lett.* **3** 413
[57] Aljabali A A A, Barclay J E, Cespedes O, Rashid A, Staniland S S, Lomonossoff G P and Evans D J 2011 *Adv. Funct. Mater.* **21** 4137
[58] Górzny M Ł, Walton A S, Wnęk M, Stockley P G and Evans S D 2008 *Nanotechnology* **19** 165704
Górzny M Ł, Walton A S and Evans S D 2010 *Adv. Funct. Mater.* **20** 1295
[59] Aljabali A A A, Shah S N, Evans-Gowing R, Lomonossoff G P and Evans D J 2011 *Integr. Biol.* **3** 119
[60] Klem M T, Willits D, Solis D J, Belcher A M, Young M and Douglas T 2005 *Adv. Funct. Mater.* **15** 1489
[61] Knight M J, Senior L, Nancolas B, Ratcliffe S and Curnow P 2016 *Nat. Commun.* **7** 11926
[62] Immordino M L, Dosio F and Cattel L 2006 *Int. J. Nanomed.* **1** 297
Maherani B, Arab-Tehrany E, Mozafari M R, Gaiani C and Linder M 2011 *Curr. Nanosci.* **7** 436
Walde P, Cosentino K, Engel H and Stano P 2010 *Chembiochem.* **11** 848
[63] Pautot S, Frisken B J and Weitz D A 2003 *PNAS* **100** 10718
Pautot S, Frisken B J and Weitz D A 2003 *Langmuir* **19** 2870
[64] Tanaka M, Critchley K, Matsunaga T, Evans S D and Staniland S S 2012 *Small* **8** 1590
[65] Mann S and Williams R J P 1983 *J. Chem. Soc., Dalton Trans.* **2** 311
[66] Mann S, Hannington J P and Williams R J P 1986 *Nature* **324** 565
[67] Kennedy M T, Korgel B A, Monbouquette H G and Zasadzinski J A 1998 *Chem. Mater.* **10** 2116
Korgel B A and Monbouquette H G 2000 *Langmuir* **16** 3588
[68] Chakrabarti A C, Veiro J A, Wong N S, Wheeler J J and Cullis P R 1992 *Biochim. Biophys. Acta Biomembr.* **1108** 233
[69] Bakhshi P K, Bain J, Gul M O, Stride E, Edirisinghe M and Staniland S S 2016 *Macromol. Biosci.* **16** 1555–61
[70] Mai Y and Eisenberg A 2012 *Acc. Chem. Res.* **45** 1657–66
[71] Blanazs A, Madsen J, Battaglia G, Ryan A J and Armes S P 2011 *J. Am. Chem. Soc.* **133** 16581–7

[72] Discher B M, Won Y-Y, Ege D S, Lee J C-M, Bates F S, Discher D E and Hammer D A 1999 *Science* **284** 1143
Discher D E and Eisenberg A 2002 *Science* **297** 967
[73] Blanazs A, Armes S P and Ryan A J 2009 *Macromol. Rapid Commun.* **30** 267
[74] Zhang F, Smolen J A, Zhang S, Li R, Shah P N, Cho S, Wang H, Raymond J E, Cannon C L and Wooley K L 2015 *Nanoscale* **7** 2265
Lutz J-F 2010 *Nat. Chem.* **2** 84
[75] Zhang Q, Remsen E E and Wooley K L 2000 *J. Am. Chem. Soc.* **122** 3642
[76] Meier W, Nardin C and Winterhalter M 2000 *Angew. Chem. Int. Ed.* **39** 4599
[77] Onaca O, Nallani M, Ihle S, Schenk A and Schwaneberg U 2006 *Biotechnol. J.* **1** 795
[78] Bain J, Ruiz-Perez L, Kennerley A J, Muench S P, Thompson R, Battaglia G and Staniland S S 2015 *Sci. Rep.* **5** 14311
[79] Bain J, Legge C J, Beattie D L, Sahota A, Dirks C, Lovett J R and Staniland S S 2019 *Nanoscale* **11** 11617
[80] Martinez R V, Chiesa M and Garcia R 2011 Nanopatterning of ferritin molecules and the controlled size reduction of their magnetic cores *Small* **7** 2914–20
[81] Mayes E, Bewick A, Gleeson D, Hoinville J, Jones R, Kasyutich O, Nartowski A, Warne B, Wiggins J and Wong K K W 2003 Biologically derived nanomagnets in self-organized patterned media *IEEE Transactions on Magnetics* **39** 624–7
[82] Sano K-I, Ajima K, Iwahori K, Yudasaka M, Iijima S, Yamashita I and Shiba K 2005 Endowing a ferritin-like cage protein with high affinity and selectivity for certain inorganic materials *Small* **1** 826–32
[83] Galloway J M, Bramble J P, Rawlings A E, Burnell G, Evans S D and Staniland S S 2012 Biotemplated magnetic nanoparticle arrays *Small* **8** 204–8
[84] Bird S M, Galloway J M, Rawlings A E, Bramble J P and Staniland S S 2015 Taking a hard line with biotemplating: cobalt-doped magnetite magnetic nanoparticle arrays *Nanoscale* **7** 7340–51
[85] Bird S M, El-Zubir O, Rawlings A E, Leggett G J and Staniland S S 2016 A novel design strategy for nanoparticles on nanopatterns: interferometric lithographic patterning of Mms6 biotemplated magnetic nanoparticles *J. Mater. Chem.* C
[86] Galloway J M, Talbot J E, Critchley K, Miles J J and Bramble J P 2015 Developing biotemplated data storage: room temperature biomineralization of L10 CoPt magnetic nanoparticles *Adv Funct Mater* **25** 4590–600
[87] Galloway J M, Bird S M, Talbot J E, Shepley P M, Bradley R C, El-Zubir O, Allwood D A, Leggett G J, Miles J J and Staniland S S *et al* 2016 Nano- and micro-patterning biotemplated magnetic CoPt arrays. *Nanoscale* **8** 11738–47
[88] Staniland S, Williams W, Telling N, Van der Laan G, Harrison A and Ward B 2008 *Nat. Nanotechnol.* **3** 158
[89] Galloway J M, Bramble J P and Staniland S S 2013 *Chem. Eur. J.* **19** 8710
Patwardhan S V, Emami F S, Berry R J, Jones S E, Naik R R, Deschaume O, Heinz H and Perry C C 2012 *J. Am. Chem. Soc.* **134** 6244
Rawlings A E, Bramble J P, Tang A A S, Somner L A, Monnington A E, Cooke D J, McPherson M J, Tomlinson D C and Staniland S S 2015 *Chem. Sci.* **6** 5586
[90] Patwardhan S V, Patwardhan G and Perry C C 2007 *J. Mater. Chem.* **17** 2875
Dickerson M B, Sandhage K H and Naik R R 2008 *Chem. Rev.* **108** 4935
[91] Drummond C, McCann R and Patwardhan S V 2014 *Chem. Eng. J.* **244** 483

[92] Patwardhan S V and Perry C C 2010 *Silicon* **2** 33
[93] Roco M C, Mirkin C A and Hersam M C 2010 Nanotechnology research directions for societal needs in 2020: retrospective and outlook *Panel Report* (Lancaster, PA: WTEC) https://www.nano.gov/sites/default/files/pub_resource/wtec_nano2_report.pdf
Patwardhan S V, Manning J R H and Chiacchia M 2018 *Curr. Opin. Green Sustain. Chem.* **12** 110–6
[94] Van Gerven T and Stankiewicz A 2009 *Ind. Eng. Chem. Res.* **48** 2465
[95] Dahl J A, Maddux B L S and Hutchison J E 2007 *Chem. Rev.* **107** 2228
[96] Kharissova O V, Dias H V R, Kharisov B I, Perez B O and Perez V M J 2013 *Trends Biotechnol.* **31** 240
[97] Yates J A 1997 *Pilot Plants and Scale-up of Chemical Processes* vol 1 ed W Hoyle (Cambridge: The Royal Society of Chemistry), p 3
[98] Sharratt P N 1997 *Pilot Plants and Scale-up of Chemical Processes* vol 1 ed W Hoyle (Cambridge: The Royal Society of Chemistry), p 13
[99] Atherton J H 1999 *Pilot Plants and Scale-up of Chemical Processes* vol 2 ed W Hoyle (Cambridge: The Royal Society of Chemistry), p 19
[100] Bourne J R and Bałdyga J 1999 *Turbulent Mixing & Chemical Reactions* (New York: Wiley)
Marchisio D L, Omegna F, Barresi A A and Bowen P 2008 *Ind. Eng. Chem. Res.* **47** 7202
Bourne J R 2003 *Org. Process Res. Dev.* **7** 471

Section IV

Case studies

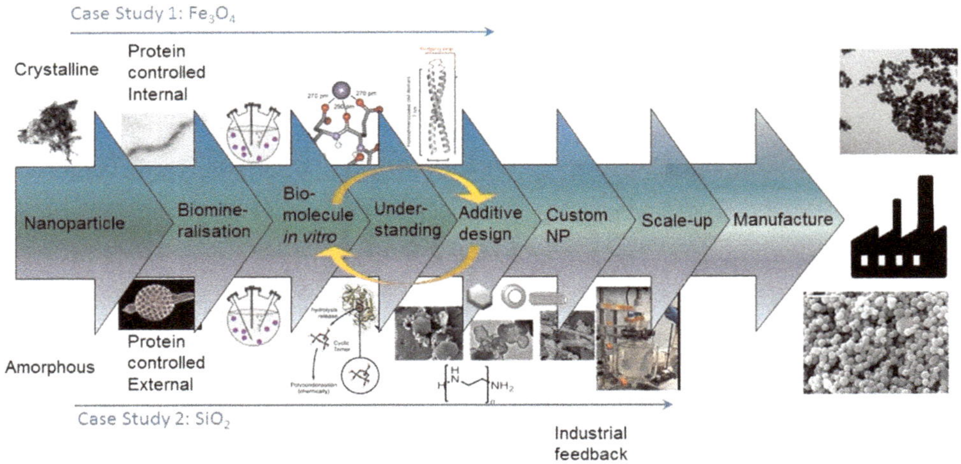

Through the course of this book we have learnt what nanomaterials are, and the real need for them in many varied applications in modern and future contexts. We have explored traditional methods of making nanomaterials, and then compared these to those in Nature. This book has then asked you, the reader, to seek inspiration from Nature, to use its tools and ingenious methods to develop green bioinspired methods of producing nanomaterials. This is beneficial with respect to reduced waste and toxicity of waste, but also goes beyond environmental advantages, potentially providing cheaper methods to create more precise and tailored nanomaterials. It is a learning pipeline, as depicted above. Exploring and learning the fundamentals and concepts behind how Nature makes biominerals in biomineralisation, and how we can understand this process and use them in bioinspired designed synthesis, are covered in section 2, which builds up in steps along this pipeline. However, clarity is often gained, and learning cemented, with an example or two. These examples will run through the processes presented, which are more generally described in chapters 5–7.

The final two chapters plot the journey of a specific material (one for each chapter) through the principles identified throughout this book, from biomineral-isation to bioinspired green synthesised nanomaterial manufacturing, to offer this clarity. The two examples chosen are magnetite nanoparticles and nanoporous silica. They are selected to offer contrasting examples, as they are quite different in several ways. Magnetite is a nano-particular crystalline, well-ordered material, whereas silica is nano-porous and amorphous, lacking any order at the atomic level. Additionally, magnetite is biomineralised internally within magnetic bacteria in nature, whereas silica is biomineralised externally on the walls of diatoms. While both are a synthesised chemical through precipitation techniques, the chemistry is also different. Furthermore, industrially, silica is one of the most widely used materials in applications, with multimillions of tonnes manufactured commercially, whereas a market for precise nanoparticulate magnetite is just emerging with the advent of nanomedicine. As such, both are at different stages of development on the manufacturing journey. In case study 1, magnetite is only just beginning its journey down this bioinspired synthesis pipeline, and research is expected to increase in intensity in the later steps in the future. In contrast, bioinspired silica is further along the process, with much significant understanding already in place to begin exploring scale and manufacture sooner. We will start with magnetite, where the emphasis is on bioinspired production at the lab scale, not having progressed as far down the pipeline as silica. This case study will cover the scientific understanding of the biomineralisation process in depth, and how a bioinspired additive can be developed from this understanding, with a future perspective of where this might lead to with respect to manufacturing. The second case study on silica demonstrates how much further along the journey it is. Research on biosilica, bioinspired silica and synthetic silica has seen tremendous advances recently, and hence this case study extends further into the realms of scale-up and provides a strong guide to developing green manufacturing routes for nanomaterials in the future.

What both examples have in common can be clearly seen from these case studies: we can learn from the natural process, understand the natural process, develop/design and understand the natural process mixed with a synthetic process in the form of bioinspired green nanomaterial synthesis, and develop these additive and methods for scale-up and manufacture.

IOP Publishing

Green Nanomaterials

From bioinspired synthesis to sustainable manufacturing of inorganic nanomaterials

Siddharth V Patwardhan and Sarah S Staniland

Chapter 8

Case study 1: magnetite nanoparticles

8.1 Magnetite biomineralisation in magnetotactic bacteria

Magnetotactic bacteria (figure 8.1(I)) were first identified as early as 1963 by an Italian student who reported their strange magnetotactic behaviour in his thesis [1]. They were then rediscovered and more widely reported by Blakemore in 1975 [2]. While there have been reports of a range of magnetic bacteria, archaea, and multicellular microbes [3], the majority of them fall into the category of a gram-negative aquatic prokaryotic microbe that uses flagellae to propel themselves around the waterways they inhabit. They tend to be microaerobic and thus inhabit the very specific oxic–anoxic transition zone (where oxygen in the water depletes as the depth increases) in their aquatic environments, where an oxygen concentration of about 1% is optimum for them to survive [3]. This does unfortunately make them very difficult for researchers to cultivate and manipulate in the lab. Magnetotactic bacteria biomineralise magnetite [4], although some more exotic strains do also biomineralise greigite [5, 6]. As the ability to biomineralise nanomagnets has now been identified across a vast range of micro-organisms, it has been proposed that the molecular biological machinery required to biomineralise nanomagnets is transferred through horizontal gene transfer. However, there is also another hypothesis that it is a very ancient trait, with all strains originating from a common ancestor from when the Earth's atmosphere was more iron-rich, and had lower oxygen concentrations. In this theory, the scattered microbes that biomineralise nanomagnets are the only ones remaining with this ancient ability, while the rest of the microbes have lost the genes required for biomineralisation over millenia [7].

Magnetotactic bacteria possess liposomes within their cells called magnetosomes, in which the magnetite is biomineralised. Each magnetosome consists of a lipid membrane containing biomineralisation proteins, surrounding a magnetite nanoparticle of precise morphology dictated by the bacterial strain (figure 8.1(II)) [8–10]. The nanoparticles are in a size range of 35–120 nm in diameter, which coincides with the range for single domain magnetism for magnetite, giving them the maximum

doi:10.1088/978-0-7503-1221-9ch8

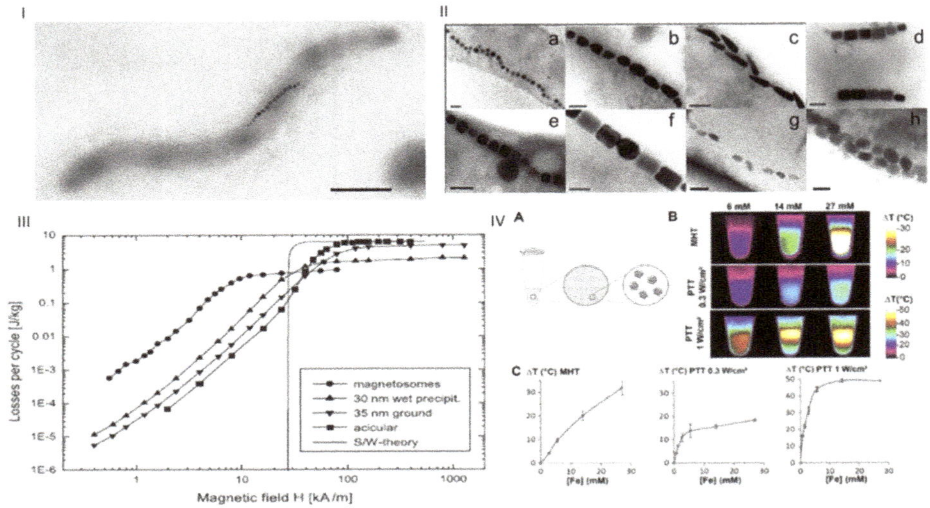

Figure 8.1. (I) TEM micrograph of a magnetotactic bacterium *Magnetospirillum magnetotacticum* MS-1 (scale bar 1 μm), reproduced from [24], copyright 2009 John Wiley & Sons. (II) TEM images of magnetosomes from different strains of magnetotactic bacteria, showing various morphologies: (a)–(g) magnetite particles and (h) greigite particles. (a) and (e) Show cubo-octahedral magnetosomes, (a) from *M. gryphiswaldense* MSR 1, (e) from *M. magnetotacticum* MS-1. (b) Bullet-shaped magnetosomes and (c) tooth-shaped magnetosomes. (d) and (f) Are elongated magnetosomes from (d) *M. coccus* MC 1 and (f) *M. vibros* MV-1. (g) and (h) Irregular bullet-shaped magnetosomes from (g) *D. magneticus* RS 1 and (h) uncharacterised greigite-producing bacterium. Scale bars are 100 nm. Reproduced from [24], with kind permission from D Schüler and T Matsunaga for some of the images. (III) Graph showing dependence of magnetic hysteresis heat loss on the amplitude of the magnetic field for a range of different single-domain magnetite nanoparticles and the theoretical value. Reprinted from [22], copyright IOP Publishing. Reproduced with permission. All rights reserved. (IV) Heating efficiency of magnetosome in water. (A) Schematic of the particle suspension. (B) IR images of magnetosome suspensions at three different concentrations generating heat through magnetic hyperthermia (MHT) and two different powers of phototherima (PTT). (C) Graphical representation of (B). Reproduced from [23], copyright 2018 with permission from Elsevier.

magnetic moment possible [11]. There are typically between 20 and 40 magnetosomes per cell, and they are arranged in long chains, which behave like a compass needle inside the bacteria, allowing them to respond to a magnetic field. In Nature this means that they align to the Earth's magnetic field [12]. A response to a magnetic field is called magnetotaxis [13, 14]. The reason why magnetotactic bacteria display magnetotaxis is not understood. It is widely believed that they use it for orientation to navigate (possibly using a mixture of aerotaxis and magnetotaxis), as they live most optimally in a very narrow chemically defined region in the aquatic environment. However, other functions have been suggested, such as detoxification, catalysis and metabolism. It is not straightforward, and no one hypothesis has yet emerged as the most likely.

8.2 Magnetosome use in applications: advantages and drawbacks

Advantages

Magnetosomes offer an excellent nanomaterial for multiple applications. They are high crystalline, are extremely homogeneous, and are coated in an intrinsic biological lipid membrane, which can be readily functioned and decorated by attaching drugs or biomarkers, etc, making them exceptionally attractive for a range of therapeutic and diagnostic biomedical applications [15].

The lipid coating also means that magnetosomes disperse exceptionally well in aqueous solution, meaning they do not clump or aggregate during biomedical use. Similar quality synthetically produced particles need to be coated post-production in a lipid to bring about similar properties.

Iron oxide magnetic nanoparticles can be used as a carriers for drugs, DNA, proteins or other relevant molecules to facilitate magnetically targeted treatments. For magnetosomes this is fairly easy, as the majority of the lipids in the membrane are amine terminated, allowing easy attachment of biotin, which can then be used to couple to streptavidin, which can in turn be linked to the functional molecule [16]. Another approach to functionalisation is to make genetically engineered magnetotactic bacteria to produce biotinylated proteins anchored to the magnetite nanoparticles, displayed on the surface of the magnetosome membrane, which can then be tagged with streptavidin linked molecules, as before [17]. These methods are in addition to direct covalent attachment to the lipid membrane. The magnetite nanoparticles can then be magnetically guided to a particular site for localized drug treatment or gene therapy, as outlined in chapter 2.

In one study, doxorubicin (DOX), a well-known anti-cancer agent, was attached to magnetosomes and applied to mice bearing a tumour [18]. The attached magnetosome allows the drug to be guided to and accumulated at the site of the tumour, giving a high tumour kill rate with reduced side effects. Such targeted treatments allow gene or drug delivery to be carefully directed to the best location for a successful outcome, and often means lower doses can be used.

Magnetosomes also display superior properties when utilised as contrast agents in magnetic resonance imaging (MRI). The high magnetization of magnetosomes over synthetic nanoparticles gives improved contrast, allowing a more detailed picture of disease states to be observed [19]. Similarly, magnetosomes show exceptional properties with respect to hyperthermic therapies (where an alternating magnetic field is used to heat the magnetic nanoparticles). Magnetosomes have the largest specific heating power compared to comparable synthetic magnetite nanoparticles (up to 5× greater!), making them well suited to this promising treatment area (figure 8.1(III)) [20–22]. Although it is not clear why this is the case, it has been proposed that it is due to the high degree of crystallinity within magnetosomes, or conversely to the subtly different, slightly reduced form of magnetite the magnetosomes produce, which cannot be emulated synthetically. Furthermore, magnetosomes have recently shown exceptional promise as nanomaterials for photothermic therapies, showing even better heating than magnetic hyperthermia (figure 8.1(IV)) [23].

Disadvantages

While magnetosomes offer highly desirable qualities for biomedical applications, they are not currently used commercially, as there are two main drawbacks of the biosynthetic method. The first is that the slow growth rate and low growth density of magnetic bacterial cells give low yields of magnetosomes. Several research groups have developed improved fermentation methods with improved yields (0.32 g l^{-1} day^{-1}, giving a magnetosome yield of 39 mg g^{-1} dry cells), by optimizing media nutrient and oxygen concentration [25]. This has been improved further simply by increasing the air aeration/stirring rate, resulting in 2.17 g l^{-1} at a growth rate of 0.868 g l^{-1} day^{-1}, giving a magnetosome yield 2.5 times higher than the previous result [26]. While these are clear improvements, biomineralisation in magnetotactic bacteria, even in an improved fermentation system, is still an inefficient method of producing magnetic nanoparticles compared to other commercial methods, presenting a real barrier to its successfully commercialisation.

The second drawback of the biosynthesis of magnetosomes within cells is the lack of flexibility this methodology offers. Magnetosomes are almost exclusively pure magnetite of highly specific sizes and shapes, due to the high degree of control exerted by the cell. This is in many ways an asset, if this is exactly what is required. However, if you do not require this exact specification of nanoparticle, the natural system offers little. Thus the applicability of natural unaltered magnetosomes is limited to a small range of magnetite materials of very specific shapes and sizes. This is beginning to be addressed: doping the magnetite magnetosomes with other transition metals such as cobalt is possible *in vivo*, simply by adding these metal ions into the growth media [27]. However, only small quantities can be added to the nanocrystal before the non-ferrous metals have a toxic effect. This also means that growth is retarded and thus yield is reduced. The maximum doping of cobalt into magnetosome has been approximately 3% doping (metal content) [28]. Both manganese and copper have also been doped into magnetosomes, with the former showing less uptake, and the latter having increased uptake but also increased toxicity [28]. The cobalt-doped magnetosomes have increased coercivity, which in turn increased hyperthermic heating in mice [29]. However, the route to flexibility is limited by the metals Nature will accommodate (so far only three transition metal dopants achieved) and concentrations are restricted by the organism. Furthermore, these doped particles are obtained in even lower yields due to the poisoning effect of the dopant, reducing the commercial viability further. Thus, it is prudent to also consider a bioinspired approach to making magnetosomes and magnetosome-like magnetite nanoparticles.

8.3 Biomolecules and components controlling magnetosome formation

8.3.1 Magnetosome biomineralisation protein discovery

The previous section highlights many of the advantages of biosynthesis, but also the disadvantages relating to production and flexibly around adapting the material, meaning the route is not currently accessible for scale-up and manufacture. Thus, we seek inspiration from Nature to aim to produce nanoparticles with the best of both worlds: biosynthetic precision, under green ambient formation conditions, but at high yields, with the ability to design and tailor the particles. In order to do this, we must first understand the biomineralisation process and thus discover which biological components are controlling it.

The names of the magnetosome biomineralisation proteins have mainly been annotated as Mam (magnetosome associated membrane) protein (followed by a letter), led by the Schuler research group working on the MSR-1 strain, whereas the Matsunaga group used the nomenclature of Mms (magnetosome membrane specific) proteins (followed by a number denoting the protein size in K Daltons, or a letter) for the AMB-1 strain. Due to the different groups working on similar proteins concurrently (their similarity unknown at the time), several homologous proteins have the same name. E.g., Mms13 in AMB-1 is homologous to MamC in MSR-1. There are many Mam/Mms proteins, where homologous proteins are found across magnetosome producing bacteria and are unique to only these bacteria. These are clearly good magnetosomes biomineralisation candidates to investigate.

8.3.1.1 Magnetosome genes: the magnetosome island

Interestingly, the genes that encode the Mms and Mam proteins unique to magnetic bacteria are found in one region of the magnetotactic bacterial genome (approximately 100 kilobase). This has been called the magnetosome island (MAI), and encodes around 100 proteins. Genes which encode proteins with similar functions, involved in similar processes, or co-dependent on one another, are often found together in the genome in regions called clusters. Major clusters in the MAI are the mamAB cluster, which encodes 17 genes for the production of proteins for iron transport, magnetite formation, and protein sorting, and which on its own is capable of producing a minimal magnetosome [30]. However, this is not equivalent to native magnetosomes. There has been a detailed knock-out mutagenesis (Δ) study on this gene cluster, where a gene of choice is disrupted so the magnetic bacteria no longer produce the protein that gene encodes. If the cells now fail to make magnetosomes, or the quality of the magnetosomes is very poor, it shows the protein has a role in magnetosome formation. A systematic gene knock-out survey has been performed where each gene was deleted in turn to see which were the most important genes for magnetite production (figure 8.2), identifying critical genes and thus proteins [31]. These will be discussed in more detail in later sections [31]. Later research found that four gene clusters from the MAI (along with a couple of gene clusters from elsewhere in the genome) are required to form magnetosomes within different non-magnetic bacterial host cells. These four clusters are mamAB (already mentioned), as well as the mamGDFC cluster (which encodes four genes), the mamXYZ cluster (of three genes), and the mms6 gene cluster (containing five genes). The other two gene clusters are FeoAB1, which contains genes that encode for iron transporter proteins, and an FtsZm cell division gene cluster (figure 8.2) [32]. This research, for the first time produced magnetosomes in a non-magnetic bacterial strain, by inserting these genes into *R. rubrum* bacteria, giving under 35 genes that are necessary/critical for

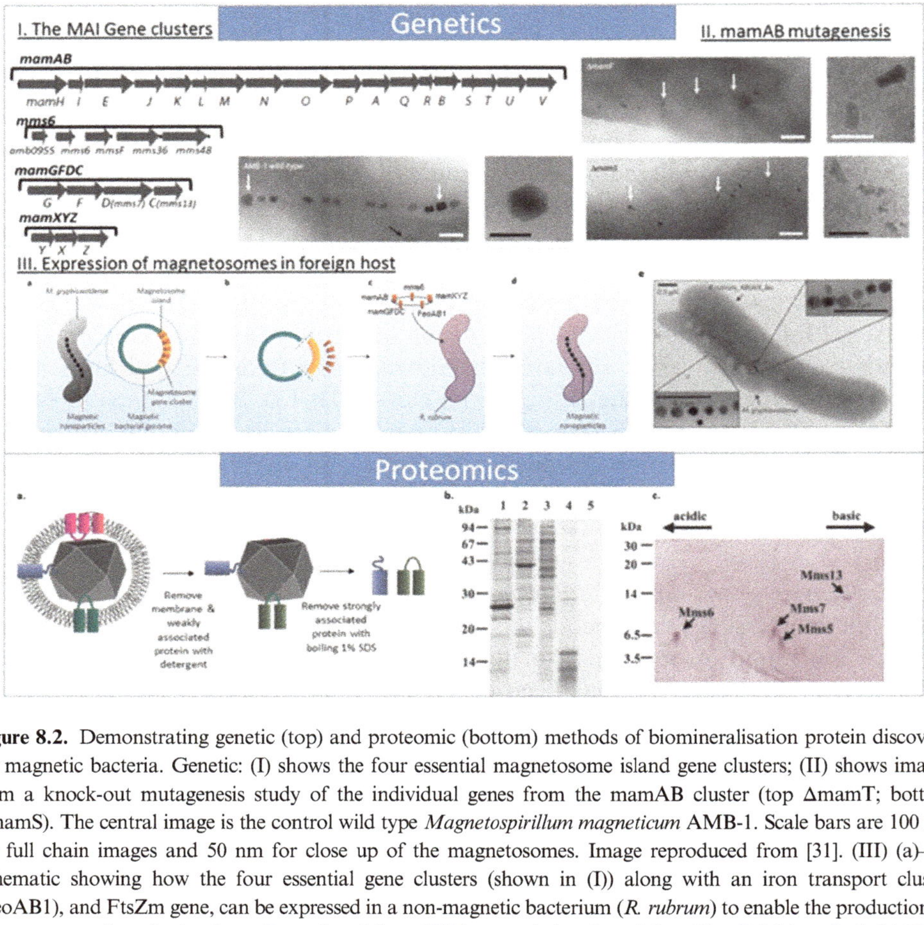

Figure 8.2. Demonstrating genetic (top) and proteomic (bottom) methods of biomineralisation protein discovery for magnetic bacteria. Genetic: (I) shows the four essential magnetosome island gene clusters; (II) shows images from a knock-out mutagenesis study of the individual genes from the mamAB cluster (top ΔmamT; bottom ΔmamS). The central image is the control wild type *Magnetospirillum magneticum* AMB-1. Scale bars are 100 nm for full chain images and 50 nm for close up of the magnetosomes. Image reproduced from [31]. (III) (a)–(d) Schematic showing how the four essential gene clusters (shown in (I)) along with an iron transport cluster (FeoAB1), and FtsZm gene, can be expressed in a non-magnetic bacterium (*R. rubrum*) to enable the production of magnetosomes in a foreign host. Reproduced from [33] by permission from Macmillan Publishers Ltd. (e) TEM image of magnetosome bearing *R. rubrum*, reproduced from [32] by permission from Macmillan Publishers Ltd. Proteomics: (a) schematic showing how the magnetosome membrane containing many proteins can be removed with detergent, leaving only proteins that are strongly bound to the magnetite crystal, which can then be removed for analysis with 1% boiling sodium dodecyl sulfate (SDS). (B) A gel showing the protein masses of *M. magneticum* AMB-1 protein extracted from lane 1. The bacterial magnetosomes: lane 2. The cell membrane: lane 3. The cell cytoplasm: lane 4. The proteins tightly bound to the magnetite crystal: lane 5, show there is no protein inside the magnetite crystal itself. (C) 2D electrophoresis gel of the fraction shown in lane 4 (tightly associated magnetosome proteins). Both (b) and (c) reproduced with permission from [34].

magnetosome production in an alternative host organism [32]. Genetic analysis of this kind allows researchers to recognise the critical genes and thus the proteins they encode for further study and understanding.

8.3.1.2 Magnetosome proteomics

Initially, a proteomics approach was used to identify the proteins specific to the magnetosome membrane, and as such both genetic and proteomic research was performed in parallel. In this process, the bacteria were lysed and the magnetosomes magnetically separated from the rest of the cell. The membrane was then removed with detergents, and the resulting protein mixture then separated and analysed. Some proteins were only loosely associated with the membrane, while others at the other extreme were tightly bound to the magnetite crystal (figure 8.2). Based on sequence homology, many of the proteins belong to well-characterised families, which give a good indication of the likely role they play in the bacteria. However, several of the proteins are unique to magnetotactic bacteria, with no known similar proteins to give indications of function. It is thus likely that these proteins interact to biomineralise the nanoparticle, performing functions such as particle nucleation, maturation, and directing the particles' crystal growth. The function of Mms/Mam biomineralisation proteins found through proteomic routes can then be probed through knock-out mutagenesis, and a detailed analysis of the protein sequence can be performed; comparison to known proteins with known functions can help to understand the function of the new protein.

8.3.2 Bio-components for each step of biomineralisation

Within the magnetosome, proteins control all aspects of magnetite formation. The biomineralisation of magnetite by magnetotactic bacteria can be mapped onto the generic three-step process discussed in chapter 6 (figure 8.3).

(1) **The sequestering of the reagents required for biomineralisation, and concentration of these.**

Iron ions are sequestered from the environment and taken up by the cell (figure 8.3I). Some researchers have proposed that siderophores are used to aid this process. It is indeed observed that the bacteria can scavenge iron, concentrating large quantities of iron even when it is very scarce in the environment. The most important element for concentrating the iron ions is compartmentalisation. In the magnetotactic bacteria the whole biomineralisation process is carried out within the magnetosome bio–nanoreactor. Lipid analysis has shown high similarities in composition between the magnetosome membrane and the inner cell membrane, indicating that the

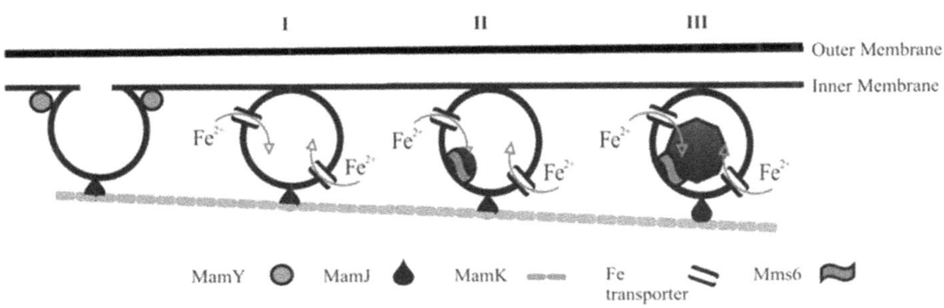

Figure 8.3. Schematic showing the magnetosome formation (adapted from diagram in [35], copyright 2014 with permission from Elsevier) aligned with the generic biomineralisation process. (I) Requesting of ferrous ions; (II) nucleation of magnetite; (III) controlling specific morphological crystal growth.

former is derived directly from the cytoplasmic membrane via a 'budding' process [10]. MamY is the membrane deformation protein (similar to the dynamin and BAR families) which acts to constrict the membrane to initiate vesicle formation [36]. Another protein, Mms16, a GTPase, is thought to hydrolyse GTP to help to form a vesicle-like structure [37]. It is thought that this change in the curvature of the membrane acts to recruit the biomineralisation magnetosome specific protein to the membrane at this point [36], with protein sorting aided by proteins such as MamB. This proposed process is supported by a cryo electron tomographic study, showing that the magnetosomes are attached to the inner membrane, although whether this is the case once the particle has fully matured is not clear [38]. MamK is present as a filament, spanning the long axis of the cell which holds the magnetosomes in the long chains. MamJ anchors the magnetosomes in place by interacting with the MamK fibre. The chain of liposomes serves initially to concentrate the iron ions by importing them across the membrane using iron transport proteins. It is thought the iron is uptaken at all these stages as ferrous ions (Fe^{2+}). The sequestering of the reagents required for biomineralisation, and concentration of these. Specialised iron transporters are capable of transporting iron into the magnetosome against the concentration gradient, from a region of low iron concentration in the cytoplasm to a high concentration within the magnetosome interior. MamB and MamM are found within the magnetosome membrane and are members of the cation diffusion facilitator (CDF) family of membrane transporter proteins, which likely transport iron ions into the magnetosome interior to provide the precursor for magnetite formation [39]. These proteins are antiporters, that transport metal irons in one direction and protons in the other. The formation of magnetite produces protons as a by-product, so these protein drive the precipitation of magnetite not only be providing iron ions, but also be retaining the higher pH required to precipitate magnetite by driving protons out. Magnetotactic bacteria also contain members of the FeoB family of membrane transporters, which are known to transport ferrous iron across membranes [40, 41]. Research which achieved magnetosome formation in a non-magnetic host bacteria, by expressing only the key magnetosome gene, found the magnetosomes formed to be much better quality, larger magnetite particles when the genes encoded FeoB protein were added, showing the importance of good transmembrane iron transport [32].

(2) **The nucleation of the specific inorganic mineral.**

Once a critical iron concentration is reached, chemistry must occur to change the oxidation states of some of the iron ions (convert ferrous (Fe^{2+}) to ferric (Fe^{3+}) ions), to enable magnetite to form (figure 8.3II). A variety of proteins are present in the magnetosome during nucleation and crystal growth, which include MamE, P, and T [42], which are c-like cytochrome redox proteins. These were all also found to be critical in the mamAB knock-out mutagenesis study, suggesting they have a crucial role in controlling the chemistry to allow magnetite to precipitate. Once the chemical conditions are correct, proteins such as Mms6 begin to nucleate the iron into magnetite. Iron continues to be loaded into the magnetosome, allowing the crystal to grow and mature into the correct mineral and with species-specific morphology.

Magnetite nucleation proteins Mms5, 6, 7 and 13.

The proteomic experiment investigating the Mms/Mam protein within the magnetosome membrane found four proteins remain strongly attached to the magnetite crystal, which were named Mms5, Mms6, Mms7, and Mms13, based on their apparent molecular weight (figure 8.2) [34]. All four of these proteins have some common features within their amino acid sequences. All are predicted to be integrated into the membrane due to the regions of their sequences, which are hydrophobic. Mms5, 6, and 7 have one of these regions, whereas Mms13 has two, so is predicted to span the membrane twice, exposing a loop into the centre of the magnetosome to interact with the forming magnetite. Mms5, 6, and 7 all have an $(LG)_n$ amino acid repeating sequence. This leads to a predicted structure that is able to interlock with itself, to self-assemble. Similar repeating sequences have been observed in self-assembly structural scaffold proteins, such as silk protein fibres [43]. Three have an overall isoelectric point (pI) of >8, with the exception of Mms6, which has a pI of approximately 4, giving it a net negative charge at neutral pH. All have a negatively charged (several acidic amino acid residues) region with acidic side chains [34], predicted to be exposed into the centre of the magnetosome to interact with the forming magnetite. For Mms5, 6, and 7 this is the C-terminus end, while it is the central loop in Mms13. Mms6 offers the most acidic region. Several gene knock-out and mutational studies have highlighted the importance of the gene encoding these proteins for the correct formation of the magnetite nanocrystal [44]. The fact that these proteins were tightly bound to the magnetite crystal, combined with what we have learned in chapter 6, concerning proteins offering large arrays of acidic amino acids (in this case through self-assembly), suggest these are nucleation proteins.

(3) **Controlling the crystal growth, specifically with respect to the morphology of the final mineral.**

One of the most remarkable aspects of magnetosome formation is that the morphology is so strictly adhered to (figure 8.3III) within each bacterial strain, but there is so much variety in crystal morphology between strains (see figure 8.1(II)). Unfortunately, however, this is the area of least understanding in the formation process of magnetosomes. The main reason for this is that most of the strains that are culturable and thus readily assessable to work with and research are from the same family (the *Magnetospirillum* family), and as such biomineralise the same cubo-octahedral shaped nanocrystals. The more exotic shaped particles are biomineralised by bacteria that cannot be readily isolated from the environment, let along grown well in a lab or manipulated. The gene sequence and therefore information about predicted proteins is not available. With the advent of extensive rapid and powerful sequencing techniques, there are now some genomes of more elusive magnetotactic bacterial strains with differing particle morphology. However, due to the fact that there is little comparative information to go on, with respect to what sort of sequence patterns we are looking for, it is almost impossible to identify a shape controlling protein from analysis of the genomic sequence alone. So far it has been proposed that MamX is an important protein for controlling crystal growth [45]. A *mms7* knock-out mutation experiment suggests that Mms7 is involved in controlling the crystal morphology [46]. Another protein that is proposed to control crystal morphology is MmsF. MmsF has been described as the master morphological regulator for magnetite biomineralisation *in vivo* [47]. The MmsF protein is encoded by a gene in the same gene cluster as *mms6*. A *mmsF* knock-out mutant has magnetosomes with much smaller, misshapen nanoparticles compared to the control cells. When the *mms6* gene cluster (including the *mmsF* gene) is deleted then a similar but only slightly more severe phenotype is observed. By re-introducing the *mmsF* gene alone back into the $\Delta mms6_{\mathrm{cl}}$ strain, the particle morphology is rescued and a near normal magnetosome is observed (figure 8.4) [47]. These experiments indicate a critical role for MmsF in controlling the growth, size and shape of the formed magnetite crystals. MmsF is predicted to have three transmembrane regions, so will display a loop into the interior of the magnetosome similar to Mms13. This loop sequence has a dense region of amino acids with acidic side-chains, but also has another region of more polar and bulky hydrophobic sidechains.

8.4 Biokleptic use of Mms proteins for magnetite synthesis *in vitro*

Proteins within a biological organism are often aided by the other surrounding proteins and the environment in which they exist. However, taking a purified protein outside of this natural environment and studying how it functions *in vitro* (outside the biological environment) is a very powerful technique for understanding how the protein specifically works. Ultimately, this simplifies the system by removing all the other cell machinery, uncoupling the protein function from the rest of the system. This technique is used across the full range of biology to glean crucial information, from enzyme turn over to drug receptor binding. Further, using protein outside the cell within synthetic set-ups can lead to the production of sustainable catalysis and important biotechnological process. In general, a protein can be produced by using molecular cloning techniques to generate larger quantities of protein. This is achieved by inserting the DNA of the protein required into an *E. coli* bacterium to produce (or over express) the protein for us. The protein can then be extracted and purified from the *E. coli.* for *in vitro* use and analysis.

Magnetosome membrane proteins are not easy to over-express and purify, due their amphiphilic nature (typical of transmembrane proteins) and low charge, leading to insoluble protein aggregates, which means a variety of solubilised

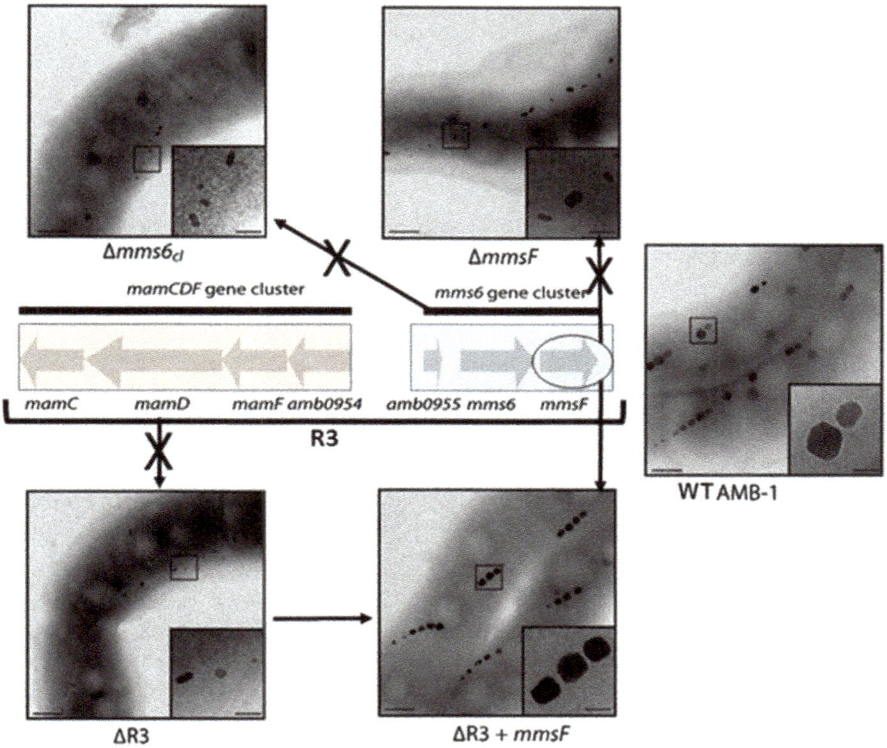

Figure 8.4. Description of the results of the MmsF mutagenesis study. The top left, and right and bottom left TEM images show poor magnetosomes produced, when: the whole mms6 gene cluster is deleted, just mmsF is deleted, and both mamCDFG and mms6 gene clusters (together referred to as R3) are all deleted, respectively. The bottom right image shows the magnetosomes produced in the absence of R3 with the mmsF gene added back in, showing magnetosomes comparable to the wild type (WT) particles (middle right TEM image), showing the MmsF protein alone can result in the activity of the whole R3 region. Scale bars are 100 nm and 50 nm for the inserted images. Adapted from [47], copyright 2012 John Wiley & Sons.

techniques need to be performed. Due to its more acidic nature, Mms6 was the first magnetosome derived protein to be over-expressed in *E. coli* and purified. Mms6 was then added to a room-temperature co-precipitation (RTCP) of magnetite. This reaction is known for its lack of control over the nanoparticles produced, leading to multiple iron-oxide products in a range of sizes and shapes. However, when Mms6 is added to the synthesis, the particles formed are pure magnetite in a regular size of approximately 20 nm with a narrow size distribution [34]. This reaction shows that Mms6 exerts clear control over the particle formation. Analysis has shown the particles are more spherical (similar to cubo octahedral) and also more pure crystalline magnetite compared to control particles [48] (figure 8.5). Furthermore, the ratio of ferrous to ferric ion can be varied, which under normal synthetic conditions directs the formation of alterative iron oxides. However, when Mms6 is added, it can direct the synthesis of magnetite under conditions it would not

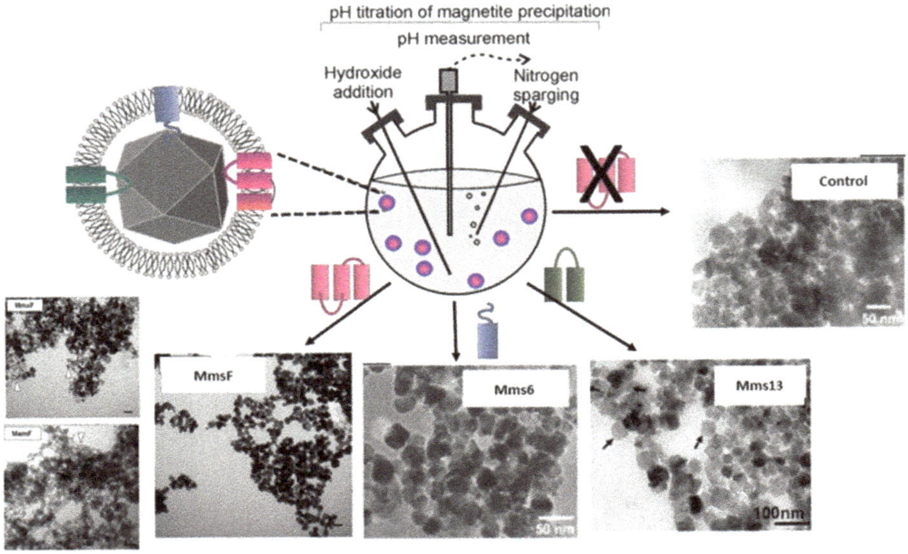

Figure 8.5. Schematic showing how magnetosome membrane (Mms/Mam) proteins can be expressed and purified and added to a chemical co-precipitation of magnetite to control particle formation *in vitro*. Schematic adapted from [49], copyright 2016 John Wiley & Sons, and [52], reprinted by permission from Macmillan Publishers Ltd, copyright 2019. The pink protein represents MmsF, the blue Mms6 and the green Mms13. TEM micrographs show the magnetite particles produced with these protein additives. Scale bars are labelled individually as either 100 or 50 nm. Images adapted from [53], and [54], copyright 2015 with permission of Springer.

normally be formed under in a straightforward chemical synthesis [49]. There are a range of synthetic methods for producing magnetite nanoparticles (see chapter 4). For example, a partial oxidation of ferrous hydroxide (POFHK) produces larger octahedral particles. Interestingly, Mms6 can be added to differing reactions, but Mms6 controls the particles formed to all be approximately 20 nm, regardless of the synthetic method used [48] (table 8.1). Mms6 has also been used in pluronic gel synthesis to slow down diffusion, in an attempt to increase particle size. This was achieved with nanoparticles of around 30 nm being produced, significantly larger than control particles [50, 51].

While more difficult to work with due to its increased hydrophobic nature, Mms13, (also known as MamC), has also been expressed, purified and added to a RTCP reaction (figure 8.5). This also seemed to control the size of the formed particles [54]. Interestingly, the effect was also seen with MamC with a single transmembrane region missing synthesis [54]. When either MamC protein was introduced into a synthetic magnetite precipitation under anaerobic ‘free drift’ conditions over 30 days, the resulting MNP had an increased average size of 30–40 nm. To date Mms7 has yet to be purified and assessed for activity *in vitro*, but MmsF has been expressed and purified and used in an *in vitro* RTCP reaction for form magnetite. When purified, MmsF was added to synthetic RTCP reactions in amounts consistent with previous Mms6 studies and had a striking effect (figure 8.5). It was found to produce very

Table 8.1. Table to summarize the method and characterisation of Magnetite nanoparticles produced via Mms6 protein mediated synthesis.

	Protein type	MNP synthesis method	MNP size (nm)[a]	Size distribution (nm)[a]	Comments	Reference
Mms6 in solution	Mms6	RTCP (1:1)[b]	20–30 (1–100)	—	Tighter size distribution	[34]
	Mms6	RTCP (1:2)[b]	22.3 (23.1)		Negligible size difference	[55]
	Mms6	POFHK	86 (234)		Smaller mean particle size	[55]
	His_6-Mms6	RTCP (1:2)[b]	21.9 (23.1)		Negligible size difference	[55]
	Mms6	POFHK	20 (27.5)	10–30 (10–40)	Smaller mean particle size, Narrower size distribution	[62]
	Mms6	POFHK	20.2 (32.4)	4.0 (9.1)[c]	Narrower size distribution, smaller mean size, cuboidal morphology	[48]
	Mms6	RTCP (1:1)[b]	21.2 (10)	8.3[c]	Larger mean size, cuboidal morphology	[48]
Mms6 in pluronic gels	His_6-Mms6	RTCP (2:1)[b]	30	—	Narrow size distribution,	[51]
	His6-Mms6	RTCP (2:1)[b]	36 (4.6)	11 (1.6)	Increased size	[50]
Mms6 peptide in solution	M6A peptide	POFHK	20–25	10–35	Exhibit spherical morphology	[62]
	GLM6A	POFHK	20	10–30		[62]
Surface-bound Mms6	His_8-Mms6	POFHK	86 (64)	21(26)[c]	Larger mean particle size, narrower size distribution	[57]
	His_8-Mms6	POFHK	90 (69)	15(36)[c]		[58]
	His_8-Mms6	POFHK	87 (60)	19 (21)[c]		[56]
Surface-bound Mms6 peptide	Mms6-pep	POFHK	65 (60)	30 (21)[c]	Negligible effect	[56]

(Continued)

Table 8.1. (*Continued*)

	Protein type	MNP synthesis method	MNP size (nm)[a]	Size distribution (nm)[a]	Comments	Reference
MmsF in solution	MmsF-StrepII	RTCP (1:2)	56 (45)	—	Increased size and magnetite purity	[53]
MamC in solution	MamC-His6	Free drift	30–40 (20–30)	20–80	Increased size, and morphology control	[54]
	MamC-truncated-His6	Free drift	30–40 (20–30)	20–80		

RTCP = room temperature co-precipitation, POFHK = partial oxidation of ferrous hydroxide with potassium hydroxide.
[a] Parentheses indicate values for particles synthesised under identical conditions but without protein additive.
[b] Parentheses indicate ratio of ferric to ferrous ions.
[c] Standard deviation. Table from [59].

regular mono-dispersed particles approximately 50 nm in diameter with a very distinctive and defined morphology, suggesting clear crystal growth control [53]. In the same study, two homologues of MmsF (MamF or MmxF (both with highly similar primary and secondary structure to MmsF)) were also added to a reaction and these showed no control over particle morphology. In fact, they produced particles of lower quality with more alternative iron oxides present than the control particles [53] (figure 8.5). This suggests that potentially small residue changes can have significant effects over the type of reaction products which form.

While this section opened with the motive for studying a protein's function through *in vitro* experiment, we have actually seen that this method goes beyond simply assessing the protein's function. We see the protein is actually a biological additive to control a chemical precipitation of magnetite (see sections 7.2 and 7.4 in chapter 7). This is in fact a very attractive middle ground between biological magnetosome biomineralisation and chemical precipitation. In this biomediated method we add key magnetosome synthesis protein to an *in vitro* chemical precipitation of magnetite to control the crystallization. This is a vast improvement on the chemical method with respect to quality. On the other hand, to make magnetosome without the whole bacterial cell and all the cellular machinery that is not relevant (and in many ways opposed) to nanomaterial development is a clear advantage. In this biokleptic synthetic approach nanoparticle yields are vastly increased compared to magnetosomes, and it also offers an increased level of flexibility (removing the cell also removes any issues of toxicity). For example, using a biokleptic Mms6 *in vitro* RTCP synthesis, cobalt can be doped into the particles in a range of increased concentrations, allowed tunable magnetic coercivity [55]. It also offers a new level of control, allowing a choice of which proteins to add to tailor the particles specifically. However, it is worth considering that membrane proteins are difficult to express and purify and thus this biokleptic method with these magnetosome biomineralisation proteins is a long way from being a commercially viable route to manufacture.

8.5 Understanding Mms proteins *in vitro*

It is evident from section 8.8.4 that Mms6, MmsF, and MamC are able to control, to varying degrees, the formation of magnetite MNPs when added to a chemical precipitation. It is clear that an understanding of how these proteins function is now required (as outlined in section 7.2). The majority of the work has been conducted on Mms6 (which is the subject of a review [59]), as this was the first protein to be isolated and used for *in vitro* mediated synthesis and is also the easier (of the options of Mms5, Mms6, Mms7 and Mms13 initially identified) to purify and work with [34]. Interestingly, all four proteins, Mms6, Mms7, MmsF, and MamC, have two distinct commonalities: self-assembly and iron binding capability (figure 8.6).

Iron binding

The first common feature is the presence of multiple acidic amino acids in a clustered region which should be readily accessible to the iron ion precursors and the forming magnetite nanoparticles. All the proteins have these regions with varying degrees of

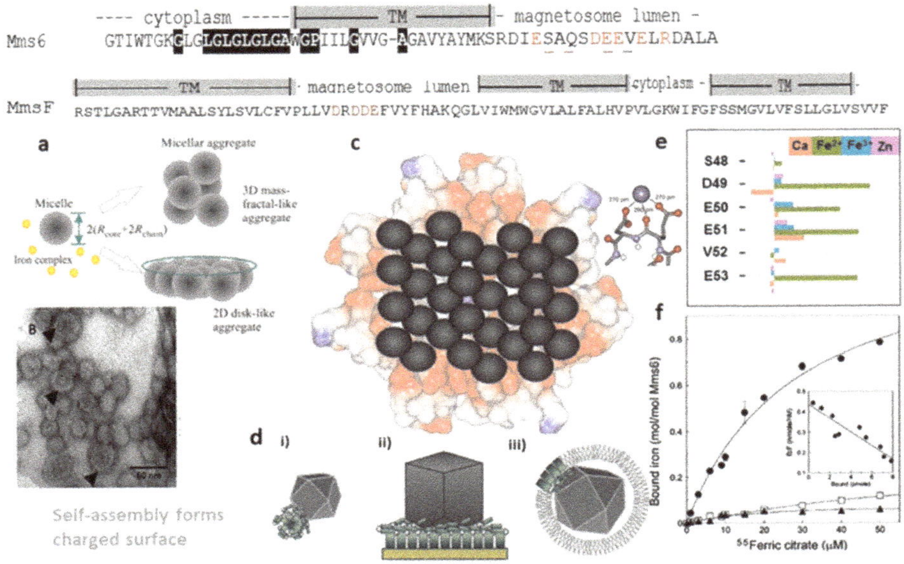

Figure 8.6. Summary of the current understanding of Mms6 and MmsF function. Top: the amino acid sequences for Mms6 and MmsF, annotated with transmembrane region (TM) shown; LG self-assembly repeat section highlighted in black and acidic residues highlighted in red text. (a) and (b) Show the self-assembly feature of Mms6 (micellular) (reprinted with permission from [61], copyright 2015 American Chemical Society) and MmsF (proteinosome) [53] respectively. (c) Bird's eye view schematic of how Mms6 may self-assemble to promote nucleation of magnetite. Right insert shows a computational model of how the acidic residues of Mms6 bind to ferrous ions (reproduced from [49], copyright 2016 John Wiley & Sons). (d) How the curvature of self-assembly may affect particle formation [63], reproduced by permission of The Royal Society of Chemistry. (e) Change in structure of these residues with different metal ions in a 2D NMR [49]. (f) Ferric binding analysis of Mms6 (reprinted with permission [50], copyright 2012 American Chemical Society. Adapted from figure in [59], which in turn uses images from the papers referenced in this caption above.

acidity, with both Mms6 and MmsF having a DEEVE and a DRDDE cluster respectively. Mms6 and MamC (Mms13) have been shown to bind ferric iron ions [34, 43, 50, 54], while Mms6 has been shown to also bind very specifically to ferrous ions [49]. Interestingly, while the looped residue of MmsF can bind iron ions it has a much lower binding affinity than MamC and Mms6 and instead binds much more readily to the magnetite particles.

Self-assembly

All the proteins self-assemble, forming micelles or proteinosomes when purified in aqueous solution. In these structures, the hydrophobic regions of the proteins are believed to be buried within the core of the structure and the hydrophilic, magnetite interaction regions are exposed to the surrounding aqueous environment. When purified these proteins display a secondary structure, suggesting that rather than an amorphous aggregate they actually display a specific folding architecture. This self-assembly is due to the favourable packing of membrane spanning region of the multiple transmembrane region proteins MamC [54] and MmsF [53], and the aggregation through both the amphiphilic nature and the specific repeating LG assembly sequence is responsible for organised self-assembly in the single transmembrane protein Mms6 sequence in the smaller Mms6 and 7. Proteinosome vesicles of MmsF have been observed by TEM [53], while micelles of Mms6 have been visualized [60] and measured with SAXS [61]. Furthermore, the requirement for self-assembly has been tested in a number of experiments. The negative iron interacting region of the protein was used as an isolated peptide, to see if this alone could control particle formation. The peptide without the rest of the protein offered negligible control over particle formation [62]. However, when the iron binding end fused directly to the LG repeating sequence a degree of control was observed, showing how critical the glycine–leucine repeat sections are for self-assembly and thus protein function [62]. In a further similar study, the peptide and the full protein were displayed on a surface to assess if the assembly needed to be controlled or not. In this case again the iron interacting peptide sequence, now randomly assembled on a surface, could not direct mineralisation, but the full-length protein on a surface could [63].

Now we can consider both features of iron ion binding and self-assembly, and extrapolate these findings to inform suggestions about the functions of all these proteins (figure 8.6). For Mms6 and MamC it appears the self-assembly ensures the creation of a proposed iron/magnetite binding surface, which will bind the ferrous and ferric ion to nucleate the formation of magnetite specifically. Interestingly, when particles synthesis is mediated by Mms6 on a surface, the particles formed are much larger. This seems to fit with the hypothesis that Mms6 is a nucleation protein, where the curvature of the surface the magnetite nucleates on directs the size of the particles formed. Indeed, this could explain why the particles formed in solution are so much smaller than magnetosome crystals, as the micelle structure has a very different curvature to the interior of the magnetosome [63]. While MmsF has many of the same features, it is interesting to see that it binds better to the forming crystals and the shape seems to be better controlled with MmsF as an additive. It has been shown that MmsF requires its magnetite interacting region to be constrained in a looped conformation, held in place by two anchored transmembrane sequences. The conformationally free peptide cannot control magnetite synthesis, so we know that structure is critical to the function of MmsF; quite the reverse to Mms6, where the iron interacting/binding end is known to be unstructured. These differences point to MmsF having more of a crystal growth functionality.

8.6 Development and design of additives: emergence of bioinspired magnetite nanoparticle synthesis

8.6.1 Development from biomineralisation proteins: MmsF

We have learnt that MmsF requires a constrained structure, but we also know that Mms proteins are not simple to obtain, and their production is not scalable. This is an excellent example of where we can take the understanding we have to develop to create a bioinspired additive for magnetite synthesis that is much easier to work with

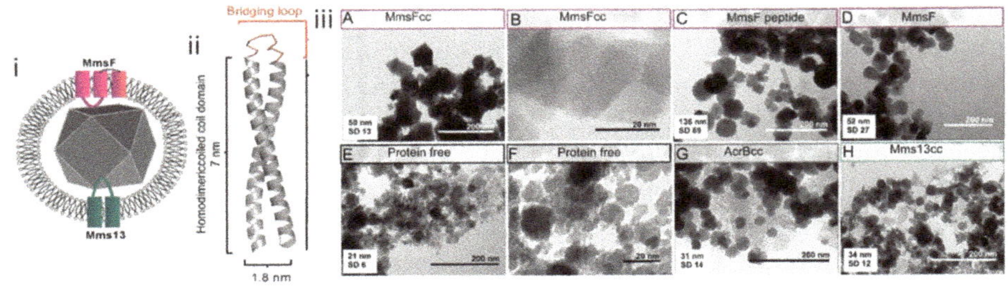

Figure 8.7. (i) Schematic of how MmsF and Mms13 embed in the magnetosome membrane and interact with the magnetite through the loop region. (ii) Structure of the coiled coil (CC) scaffold with the bridging loop region. (iii) TEM images of the particles produced using the CC additive and controls. Adapted from [52] by permission of Macmillan Publishers Ltd.

(section 7.4). Here we know the loop region is required, but the rest of the protein is cumbersome to work with. So, the loop region has been displayed in a constrained loop on a simple 'scaffold' coiled coil protein [52]. This leads to a much more robust, soluble additive that is easier to produce and also more amenable in aqueous chemical reaction conditions. The ease of production and economic benefit of this more soluble additive puts in on the right course for scale-up in the future (figure 8.7).

8.6.2 Screening non-biomineralisation proteins: magnetite interacting proteins

Biopanning (introduced in chapter 5) is a powerful screening technique to discover sequences that bind specifically to minerals (see section 7.2.2). It can be employed to find magnetite specific sequences. However, it can also be employed to find sequences that bind to specific crystal planes of the magnetite. The hypothesis is that biomolecular additives that bind specifically to a specific crystal plan will promote that facet to dominate the morphology, directing towards a particular crystal shape (see sections 4.3 and 4.3.1). To this effect, a protein library of adhirons (displayed on phage) were biopanned against cubic (with [100] crystal faces) magnetite nanoparticles. Adhirons are robust artificial proteins ideal for industry, with a two variable nine amino acid loop region exposed. The hits from this screening were named magnetite interacting adhiron (MIA) proteins. Interestingly, the sequences found to bind most strongly to the [100] magnetite surfaces were very high in basic amino acid residues, with a particular preference for lysine [64]. Adhiron proteins containing both a single loop and two loops showed molecular recognition towards magnetite. This is supported by computational modelling of the amino acids' interaction with magnetite [100] surfaces. Excitingly, these proteins can be used as an additive to direct the formation of cubic nanoparticles when added to RTCP reactions—showing the protein holds the ability to exhibit shape control on the growing crystallization of magnetite nanoparticles [64].

8.6.3 Biomimetic magnetosomes

As we saw in chapter 7, the liposome compartment of magnetosomes can be mimicked in many ways. Indeed, artificial magnetosomes were discussed fully in that section. However, after considering all the work in this chapter, we can use the more detailed understanding of the proteins to improve this method. It was found that while we could use liposomes to form artificial magnetosomes, the use of polymersomes may be more adaptable, flexible, robust, and industrially applicable. We also saw that the whole protein ion channels work effectively to concentrate iron ions into the vesicle; these are very difficult to produce, work with and may not be immediately scalable. A simpler and more effected method of importing iron ions into the polymersomes was to use the physical method of electroporation, which uses electricity to temporary open pores in the polymer membrane. A downside of this is that the reagents (iron ions and base) separated by the membrane immediately mix where they meet at the pore, leading to multiple tiny particles within the membrane itself. Learning from the understanding we have gleaned from the biomineralisation process of magnetosomes and the proteins that perform it, it is clear that a nucleation protein such as Mms6 could aid nucleation within the core of the polymersomes. However, we have also recognised that producing Mms6 is not trivial, and is economically and industrially impractical. Thus we need to mimic the system. We can then consider what we have learnt. We know Mms6 functions by displaying an acidic-rich amino acid surface to nucleate on. Thus, one should design a polymersome with carboxylic acid rich groups covering the interior surface. This was achieved with an asymmetric block co-polymer arrangement of (PEG_{113}–$PHPMA_{400}$) on the outside and ($PMPC_{28}$–$PHPMA_{400}$) directed towards this inside. This asymmetric localisation occurs due to size constraints, with the smaller sized block located on the inner surface and the larger on the outer. PMPC is terminated with the carboxylic group to provide the nucleation surface. When this was electroporated in an iron solution more iron was uptaken and nucleated in the core of the vesicle and thus looks more like a magnetosome than in previous work with non-carboxylated polymersomes [65] (figure 8.8 and both shown for comparison in figure 7.9E).

8.7 Summary: key learning, and the future (towards manufacture)

In this case study we have learnt that there are many ways in which to integrate the biomineralisation of magnetosomes. We can look at the genetics and assess sequences, as well as perform knock-out mutagenesis to investigate which proteins are critical for biomineralisation. We can identify key proteins from their location in the magnetosome membrane, or affinity to the magnetite nanoparticle, and assess these for structure, iron binding ability, etc. Understanding is most powerful when we can use all these techniques to build up a detailed picture of how biomineralisation proteins function to control magnetite nanoparticle growth.

We have also seen how we can use non-biomineralising protein to screen for function that may not be readily seen in Nature. Information extracted from this process can also be used to look for similar sequences in Nature.

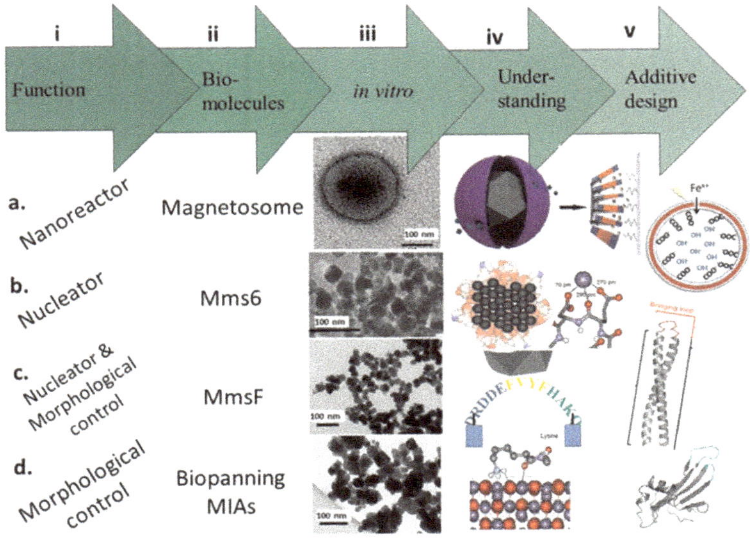

Figure 8.8. Summary figure for this case study. (i)–(v) Show the pipeline so far for magnetite nanoparticles from magnetosomes. Under these headers, an example of all three generic biomineralisation steps are seen. (a) The compartmentalisation is performed by the magnetosome vesicles. Liposome and polymersome artificial examples have been produced. (aiii) A magnetopolymersome *in vitro* that was designed by understanding COOH nucleation sites on Mms6, incorporated COOH sites in the interior (av) and used electroporation (aiv) to import iron ions [65], reproduced by permission of The Royal Society of Chemistry. (b) Nucleation is performed by Mms6 which has been used *in vitro,* and a body or work now describes and understands its function [49], copyright 2016 John Wiley & Sons, and [59]. (c) MmsF seems to control both nucleation and particle shape. It is not fully understood how, but this has too been used *in vitro* as a protein [53] and a more robust additive more suited to scale-up has been produced using a coiled coil scaffold (cv) [52], reprinted by permission from Macmillan Publishers Ltd. (d) The morphological controlling robust MIA protein (dv) was discovered using biopanning. This protein controls the formation of cubic nanoparticles (diii), reproduced by permission of The Royal Society of Chemistry. The positive amino acid residues are thought to be responsible for this by strong interaction with the [100] cubic crystal plane of magnetite (div). Images taken from the references in the caption.

From both of these methods, the trends noted in chapter 6 and the implementation shown in chapter 7 are reinforced. We see some very specific rules occurring, from which we can design future additives.

We see nucleation of magnetite needs an array of acidic amino acids to bind to iron ion. We see with Mms6 that this should be self-assembled in a specific array to obtain maximum effect, but the latest work with polymersomes shows that even just providing this carboxylate charged surface has a nucleation effect.

We see that controlling the crystal growth requires a different set of principles. We see with with both MmsF and the MIA that structural conformation of the biomolecule is essential. Both of these need to be constrained in a loop, perhaps to match the crystal surface for high affinity binding. We also see that basic residues dominate the MIA protein for strong [100] plane cubic magnetite interactions which is demonstrated with computational modelling, whereas the understanding around how the MmsF sequence interaction controls morphology is less clear.

Looking to the future, we can use these principles to design new additives that are more commercially viable. We can adopt these principles into molecules that consider other factors such as robustness to mixing, and other factors of scale-up and manufacture that are not considered in relation to biomineralisation in Nature.

References

[1] Bellini S 1963 Su di un particolare comportamento di batteri d'acqua dolce (About a particular behavior of freshwater bacteria) *PhD Thesis* University of Padua
[2] Blakemore R 1975 Magnetotactic bacteria *Science* **190** 377–9
[3] Schüler D 2007 *Microbiology Monographs* ed A Steinbüchel (Berlin: Springer)
[4] Frankel R B, Blakemore R P and Wolfe R S 1979 Magnetite in freshwater magnetotactic bacteria *Science* **203** 1355–6
[5] Farina M, Esquivel D M S and de Barros H G P 1990 Magnetic iron-sulphur crystals from a magnetotactic microorganism *Nature* **343** 256
[6] Frankel R B, Mann S, Sparks N H C, Jannasch H W and Bazylinski D A 1990 *Nature* **343** 258–61
[7] Lefevre C T *et al* 2013 Monophyletic origin of magnetotaxis and the first magnetosomes *Environ. Microbiol.* **15** 2267–74
[8] Gorby Y A, Beveridge T J and Blakemore R P 1988 Characterization of the bacterial magnetosome membrane *J. Bacteriol.* **170** 834–41
[9] Grunberg K *et al* 2004 Biochemical and proteomic analysis of the magnetosome membrane in Magnetospirillum gryphiswaldense *Appl. Environ. Microbiol.* **70** 1040–50
[10] Tanaka M *et al* 2006 Origin of magnetosome membrane: proteomic analysis of magnetosome membrane and comparison with cytoplasmic membrane *Proteomics* **6** 5234–47
[11] Butler R F and Banerjee S K 1975 Theoretical single-domain grain size range in magnetite and titanomagnetite *J. Geophys. Res.* **80** 4049–58
[12] Dunin-Borkowski R E *et al* 1998 Magnetic microstructure of magnetotactic bacteria by electron holography *Science* **282** 1868–70
[13] Frankel R B, Bazylinski D A, Johnson M S and Taylor B L 1997 Magneto-aerotaxis in marine coccoid bacteria *Biophys. J.* **73** 994–1000
[14] Frankel R, Williams T and Bazylinski D 2007 *Magnetoreception and Magnetosomes in Bacteria* ed D Schüler (Berlin: Springer), 1–24
[15] Xie J, Chen K and Chen X 2009 Production, modification and bio-applications of magnetic nanoparticles gestated by magnetotactic bacteria *Nano Res.* **2** 261–78
[16] Ceyhan B, Alhorn P, Lang C, Schuler D and Niemeyer C M 2006 Semisynthetic biogenic magnetosome nanoparticles for the detection of proteins and nucleic acids *Small* **2** 1251–5
[17] Maeda Y *et al* 2008 Noncovalent immobilization of streptavidin on *in vitro*- and *in vivo*-biotinylated bacterial magnetic particles *Appl. Environ. Microbiol.* **74** 5139–45
[18] Sun J B *et al* 2007 *In vitro* and *in vivo* antitumor effects of doxorubicin loaded with bacterial magnetosomes (DBMs) on H22 cells: The magnetic bio-nanoparticles as drug carriers *Cancer Lett.* **258** 109–17
[19] Hu L L *et al* 2010 Comparison of the $_1$H NMR relaxation enhancement produced by bacterial magnetosomes and synthetic iron oxide nanoparticles for potential use as MR molecular probes *IEEE Trans. Appl. Supercond.* **20** 822–5

[20] Alphandery E *et al* 2011 Heat production by bacterial magnetosomes exposed to an oscillating magnetic field *J. Phys. Chem.* C **115** 18–22

[21] Hergt R *et al* 2005 Magnetic properties of bacterial magnetosomes as potential diagnostic and therapeutic tools *J. Magn. Magn. Mater.* **293** 80–6

[22] Hergt R, Dutz S, Müller R and Zeisberger M 2006 Magnetic particle hyperthermia: nanoparticle magnetism and materials development for cancer therapy *J. Phys. Condens. Matter* **18** S2919–34

[23] Plan Sangnier A *et al* 2018 Targeted thermal therapy with genetically engineered magnetite magnetosomes@RGD: Photothermia is far more efficient than magnetic hyperthermia *J. Control. Release* **279** 271–81

[24] Staniland S S 2009 *Magnetic Nanomaterials* vol 4 ed C Kumar (New York: Wiley), ch 11 p 399

[25] Heyen U and Schuler D 2003 Growth and magnetosome formation by microaerophilic Magnetospirillum strains in an oxygen-controlled fermentor *Appl. Microbiol. Biotechnol.* **61** 536–44

[26] Sun J B *et al* 2008 High-yield growth and magnetosome formation by Magnetospirillum gryphiswaldense MSR-1 in an oxygen-controlled fermentor supplied solely with air *Appl. Microbiol. Biotechnol.* **79** 389–97

[27] Staniland S *et al* 2008 Controlled cobalt doping of magnetosomes *in vivo Nat. Nanotechnol.* **3** 158–62

[28] Tanaka M *et al* 2012 Highest levels of Cu, Mn and Co doped into nanomagnetic magnetosomes through optimized biomineralisation *J. Mater. Chem.* **22** 11919–21

[29] Alphandéry E, Carvallo C, Menguy N and Chebbi I 2011 Chains of cobalt doped magnetosomes extracted from AMB-1 magnetotactic bacteria for application in alternative magnetic field cancer therapy *J. Phys. Chem.* C **115** 11920–4

[30] Losse A *et al* 2011 Functional analysis of the magnetosome island in Magnetospirillum gryphiswaldense: The mamAB operon is sufficient for magnetite biomineralization *PLoS One* **6**

[31] Murat D, Quinlan A, Vali H and Komeili A 2010 Comprehensive genetic dissection of the magnetosome gene island reveals the step-wise assembly of a prokaryotic organelle *Proc. Natl. Acad. Sci. U.S.A.* **107** 5593–8

[32] Kolinko I *et al* 2014 Biosynthesis of magnetic nanostructures in a foreign organism by transfer of bacterial magnetosome gene clusters *Nat. Nanotechnol.* **9** 193–7

[33] Staniland S 2014 An accommodating host *Nat. Nanotechnol.* **9** 163

[34] Arakaki A, Webb J and Matsunaga T 2003 A novel protein tightly bound to bacterial magnetic particles in Magnetospirillum magneticum strain AMB-1 *J. Biol. Chem.* **278** 8745–50

[35] Staniland S S *et al* 2014 *Nanomagnetism: Fundamentals and Applications* ed C Binns (Amsterdam: Elsevier), ch 3, pp 85–129

[36] Tanaka M, Arakaki A and Matsunaga T 2010 Identification and functional characterization of liposome tubulation protein from magnetotactic bacteria *Mol. Microbiol.* **76** 480–8

[37] Okamura Y, Takeyama H and Matsunaga T 2001 A magnetosome-specific GTPase from the magnetic bacterium Magnetospirillum magneticum AMB-1 *J. Biol. Chem.* **276** 48183–8

[38] Komeili A, Li Z, Newman D K and Jensen G J 2006 Magnetosomes are cell membrane invaginations organized by the actin-like protein MamK *Science* **311** 242–5

[39] Uebe R *et al* 2011 The cation diffusion facilitator proteins MamB and MamM of Magnetospirillum gryphiswaldense have distinct and complex functions, and are involved in magnetite biomineralization and magnetosome membrane assembly *Mol. Microbiol.* **82** 818–35
[40] Rong C *et al* 2008 Ferrous iron transport protein B gene (feoB1) plays an accessory role in magnetosome formation in Magnetospirillum gryphiswaldense strain MSR-1 *Res. Microbiol.* **159** 530–6
[41] Rong C *et al* 2012 FeoB2 Functions in magnetosome formation and oxidative stress protection in Magnetospirillum gryphiswaldense strain MSR-1 *J. Bacteriol.* **194** 3972–6
[42] Siponen M I, Adryanczyk G, Ginet N, Arnoux P and Pignol D 2012 Magnetochrome: a c-type cytochrome domain specific to magnetotatic bacteria *Biochem. Soc. Trans.* **40** 1319–23
[43] Feng S *et al* 2013 Integrated self-assembly of the Mms6 magnetosome protein to form an iron-responsive structure *Int. J. Mol. Sci.* **14** 14594–606
[44] Tanaka M, Mazuyama E, Arakaki A and Matsunaga T 2011 MMS6 protein regulates crystal morphology during nano-sized magnetite biomineralization *in vivo J. Biol. Chem.* **286** 6386–92
[45] Yang J *et al* 2013 MamX encoded by the mamXY operon is involved in control of magnetosome maturation in Magnetospirillum gryphiswaldense MSR-1 *BMC Microbiol.* **13** 203
[46] Yamagishi A *et al* 2016 Control of magnetite nanocrystal morphology in magnetotactic bacteria by regulation of mms7 gene expression *Sci. Rep.* **6** 29785
[47] Murat D *et al* 2012 The magnetosome membrane protein, MmsF, is a major regulator of magnetite biomineralization in Magnetospirillum magneticum AMB-1 *Mol. Microbiol.* **85** 684–99
[48] Amemiya Y, Arakaki A, Staniland S S, Tanaka T and Matsunaga T 2007 Controlled formation of magnetite crystal by partial oxidation of ferrous hydroxide in the presence of recombinant magnetotactic bacterial protein Mms6 *Biomaterials* **28** 5381–9
[49] Rawlings A E *et al* 2016 Ferrous iron key to Mms6 magnetite biomineralisation: A mechanistic study to understand magnetite formation using pH titration and NMR *Chem. Eur. J.* **22** 7885–94
[50] Wang L J *et al* 2012 Self-assembly and biphasic iron-binding characteristics of Mms6, a bacterial protein that promotes the formation of superparamagnetic magnetite nanoparticles of uniform size and shape *Biomacromolecules* **13** 98–105
[51] Prozorov T *et al* 2007 Protein-mediated synthesis of uniform superparamagnetic magnetite nanocrystals *Adv. Funct. Mater.* **17** 951–7
[52] Rawlings A E *et al* 2019 Artificial coiled coil biomineralisation protein for the synthesis of magnetic nanoparticles *Nat. Commun.* **10** 2873
[53] Rawlings A E *et al* 2014 Self-assembled MmsF proteinosomes control magnetite nanoparticle formation *in vitro Proc. Natl. Acad. Sci. U.S.A.* **111** 16094–9
[54] Valverde-Tercedor C *et al* 2015 Size control of *in vitro* synthesized magnetite crystals by the MamC protein of Magnetococcus marinus strain MC-1 *Appl. Microbiol. Biotechnol.* **99** 5109–21
[55] Galloway J M *et al* 2011 Magnetic bacterial protein Mms6 controls morphology, crystallinity and magnetism of cobalt-doped magnetite nanoparticles *in vitro J. Mater. Chem.* **21** 15244–54
[56] Bird S M, Rawlings A E, Galloway J M and Staniland S S 2016 Using a biomimetic membrane surface experiment to investigate the activity of the magnetite biomineralisation protein Mms6 *RSC Adv.* **6** 7356–63

[57] Bird S M, El-Zubir O, Rawlings A E, Leggett G J and Staniland S S 2016a A novel design strategy for nanoparticles on nanopatterns: interferometric lithographic patterning of Mms6 biotemplated magnetic nanoparticles J. Mater. Chem. *C* **4** 3948–55

[58] Bird S M, Galloway J M, Rawlings A E, Bramble J P and Staniland S S 2015 Taking a hard line with biotemplating: cobalt-doped magnetite magnetic nanoparticle arrays *Nanoscale* **7** 7340–51

[59] Staniland S S and Rawlings A E 2016 Crystallizing the function of the magnetosome membrane mineralization protein Mms6 *Biochem. Soc. Trans.* **44** 883–90

[60] Kashyap S, Woehl T J, Liu X, Mallapragada S K and Prozorov T 2014 Nucleation of iron oxide nanoparticles mediated by Mms6 protein *in situ ACS Nano* **8** 9097–106

[61] Zhang H *et al* 2015 Morphological transformations in the magnetite biomineralizing protein Mms6 in iron solutions: a small-angle x-ray scattering study *Langmuir* **31** 2818–25

[62] Arakaki A, Masuda F, Amemiya Y, Tanaka T and Matsunaga T 2010 Control of the morphology and size of magnetite particles with peptides mimicking the Mms6 protein from magnetotactic bacteria *J. Colloid Interface Sci.* **343** 65–70

[63] Bird S M, Rawlings A E, Galloway J M and Staniland S S 2016 Using a biomimetic membrane surface experiment to investigate the activity of the magnetite biomineralisation protein Mms6 *RSC Adv.* **6** 7356–63

[64] Rawlings A E *et al* 2015 Phage display selected magnetite interacting Adhirons for shape controlled nanoparticle synthesis *Chem. Sci.* **6** 5586–94

[65] Bain J *et al* 2019 A biomimetic magnetosome: formation of iron oxide within carboxylic acid terminated polymersomes *Nanoscale* **11** 11617–25

IOP Publishing

Green Nanomaterials
From bioinspired synthesis to sustainable manufacturing of inorganic nanomaterials
Siddharth V Patwardhan and Sarah S Staniland

Chapter 9

Case study 2: silica

9.1 Biosilica occurrence and formation

Biological silica has interested scientists for centuries, due to the intriguing structures of biosilica formed, as well as the nature of the entire process of their formation. There have been numerous studies on finding, documenting, and studying ornate biosilica structures. Of particular interest are diatoms, due to their hierarchically-structured silica cell walls, and the variety of structures seen. In fact, in the early days of their study, diatoms were classified based on their shapes, morphologies and structures (see figure 9.1). Biosilica is also found in radiolaria, sponges and many plants. Diatoms and radiolaria form silica-based cell walls, which provide protection and are porous in order to allow transport of nutrients and signals. Sponges use needle-like spicules that offer mechanical support to their bodies, in addition to providing protection against predators. Plants, on the other hand, can produce silica as a by-product of the uptake of silicon from soil. However, this *side reaction* has been used to plants' advantage in many cases. One example is rice, where the leaves become strong due to the presence of silica and hence maximise the absorption of sunlight (this is described in section 6.2.3). Other plants, e.g. cucumber, use biosilica formation in response to a threat or an attack and contain the pathogens in a biosilica *cage*.

In the last 15–20 years, scientists have navigated from morphological observations to molecular and genetic mechanisms. A number of studies on cell biology and molecular biology had made great contributions to our understanding of biological silica formation (figure 9.2). The process of biosilica formation is described briefly here, and readers are referred to the further reading section for references containing detailed descriptions of biosilicification [3–10]. The key stages involved are:

1. uptake of silicon into the living system,
2. its transport inside the system,
3. biochemical transformation into (intermediate and) final form, and
4. deposition into the final location.

doi:10.1088/978-0-7503-1221-9ch9

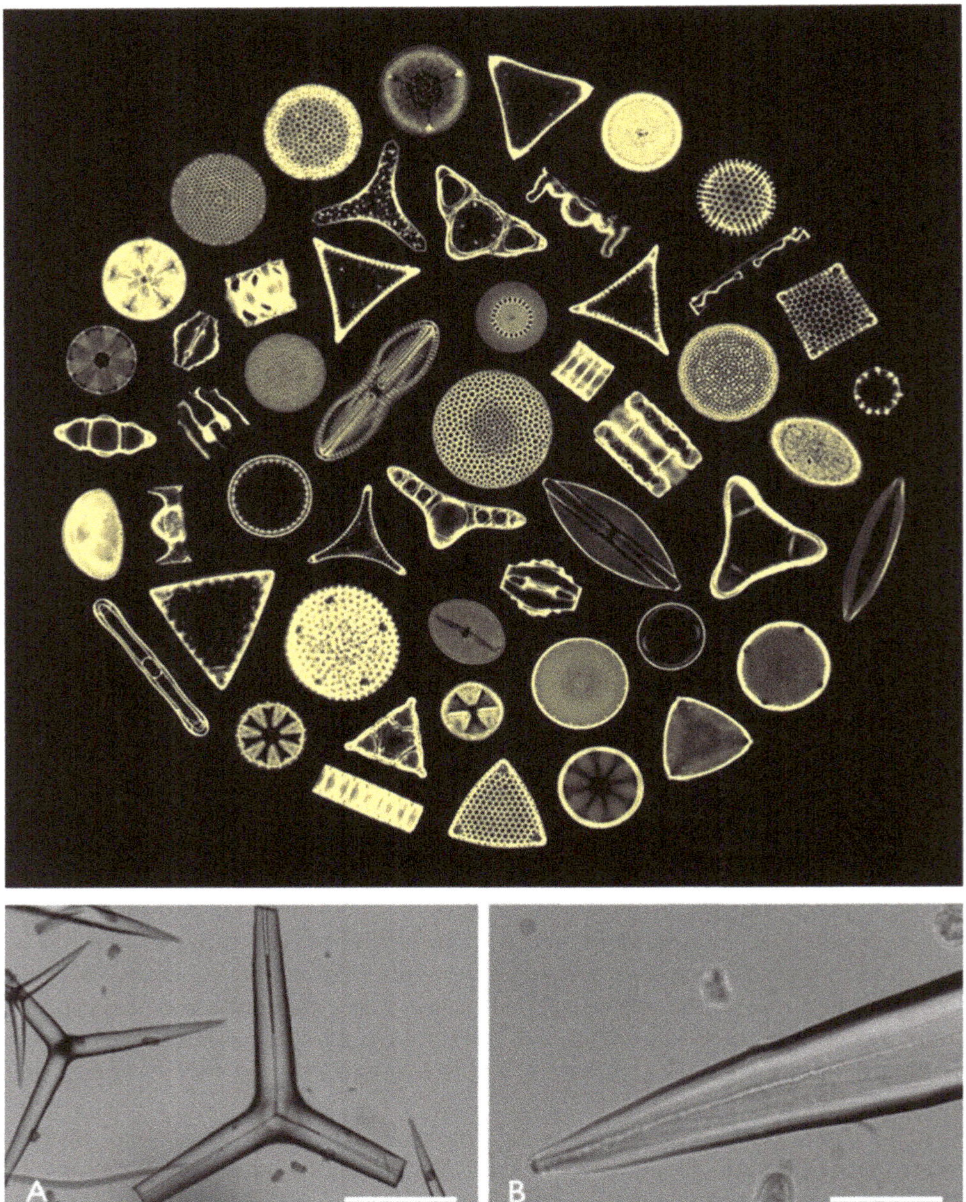

Figure 9.1. Top: morphologies of various diatoms (credit: Randolph Femmer, USGS [1]). Bottom: silica spicules in sponges (image reproduced with permission from [2], copyright 2003 John Wiley & Sons).

This is the general order of the process, however, variations may occur from system to system. For example, it is not fully clear whether the transport within the organisms occurs in the form of orthosilicic acid (a silica monomer), silicic acid oligomers, or as *final* silica. Further, in some examples, steps 3 and 4 can be

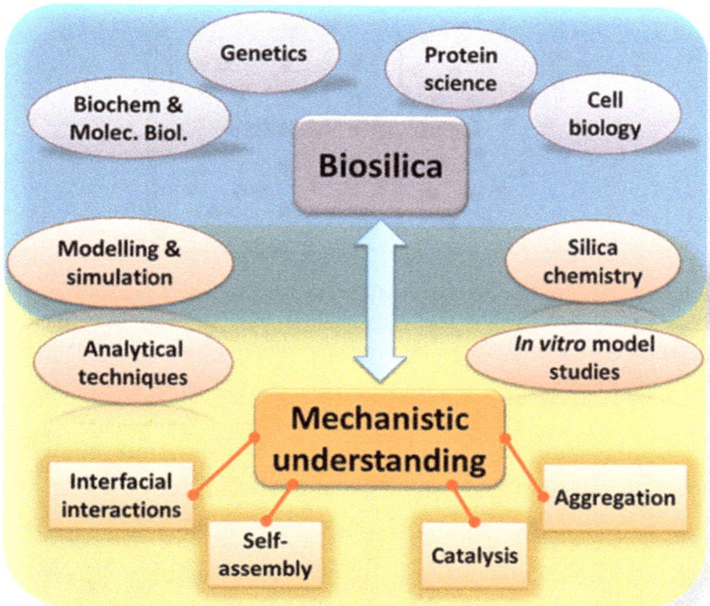

Figure 9.2. Schematic representation of various tools and approaches used to study biosilica and its formation, leading to mechanistic understanding. Image taken from [11], reproduced by permission of The Royal Society of Chemistry.

combined into a single step, where the chemical transformation to silica and deposition occur simultaneously. It is worth noting that each step is under strict biological and genetic control, particularly the order of events and the secretion of the signals/biomolecules governing each step. Given the significant progress made in understanding the processes in diatoms over other biosilica systems, most of the examples below include those from diatoms.

The uptake of silicon in most cases is active, which means that when silicon is needed for growth or cell division, the organism actively and selectively *absorbs* silicon from its environment (figure 9.3). Various organisms have different mechanisms to achieve this, however, they are remarkable in their selectivity. This selectivity allows the organisms to concentrate silicon from a highly unsaturated (~70 μM) extra-cellular concentration to a near-saturated (1–2 mM) or supersaturated intra-cellular concentration (19–340 mM for some diatoms). This is possible due to specific recognition of the soluble monomeric form (silicic acid), as well as the ability to stabilise oligo-/polymeric forms to avoid premature silica deposition. This stabilisation of supersaturated *pools* of silicon allows the organism to effectively supply feedstock to the site of transformation and deposition. This feedstock is then biochemically converted or polymerised into biosilica and deposited in the final form. The final product is typically an organic–inorganic hybrid of biomolecules and silica.

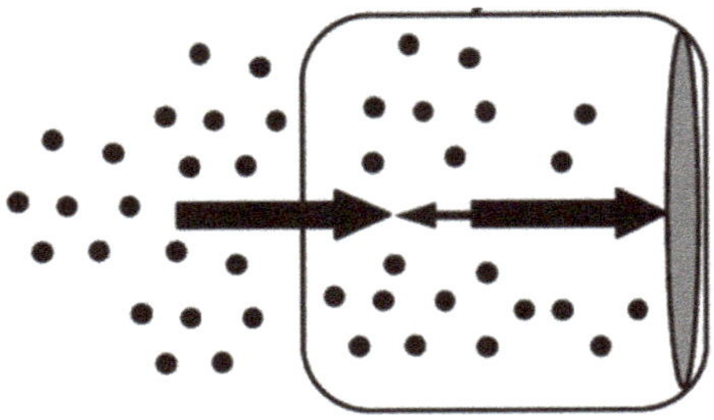

Figure 9.3. Controlled uptake of silicic acid (shown as dots) in diatoms (shown as a box) from outside to inside the cell and to the deposition site (shown as grey elongated oval biosilica). The arrows indicate the direction and quantity (thickness) of transport. Image taken with permission from [8], copyright 2008 American Chemical Society.

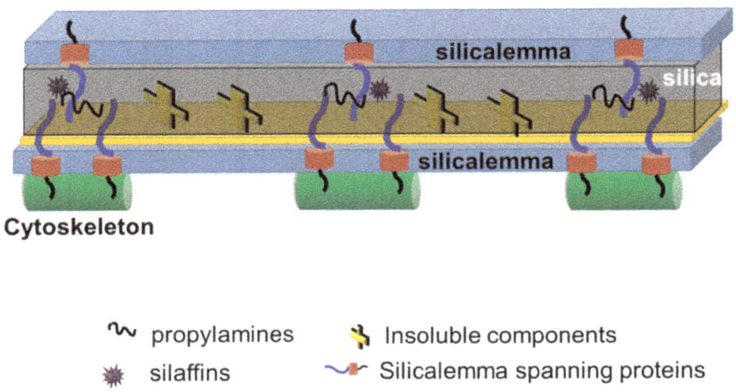

Figure 9.4. A simplified representation showing the proposed localisation of various biomolecules associated with biosilica formation in diatoms. Figure adapted from [40].

9.2 Biomolecules controlling biosilica formation

Each of the stages discussed above involve highly specialised biomolecules—their functions range from recognising monomeric silicon for transport, complexing with intermediate silicate oligomers for stabilisation and catalysing or templating the formation of biosilica from precursors. The molecular/chemical details for specific organisms are discussed in this section. In the late 1990s and early 2000s, the key molecules involved in silicon transport and polymerisation were identified for diatoms, plants and sponges [8, 12–23]. Genomic investigations of diatoms and sponges have provided deeper understanding of biosilicification [19, 24, 25].

A number of biomolecules, including proteins, have been isolated from biosilica of diatoms, sponges and plants. These include, for example, silaffin (a family of silica forming proteins), SiMat proteins, propylamines, silicalemma-associated proteins (SAPs), silacidin, cingulins, and polysaccharides from diatoms [26–32], silicatein and silintaphin proteins from sponges [15, 16, 33–35], and a range of biomolecules from plants [12, 17, 36–38] and choanoflagellates [39]. Through extensive research, the roles of these biomolecules in biosilica formation are becoming clear. Based on the expression, occurrence and function of these organic components, figure 9.4 shows the proposed localisation of various organic molecules associated with biosilica formation in diatoms [40]. Broadly, the functions of these components can be generalised into the following categories:

- framework or scaffold,
- catalytic, and
- regulatory.

The framework components are generally found anchored within a membrane (e.g. the silicalemma in figure 9.4) or the cytoskeleton. These components provide a rigid support for the catalytic and regulatory components to reside on. The catalytic components (e.g. silaffins and propylamines in diatoms, and silicatein in sponges) carry out the chemical transformation of silicic acid (or other form of silica precursors) to silica particles. The regulatory components (e.g. silacidin proteins) help control the biosilica formation (e.g. the amount of silica particles formed). It is proposed that the amine functionalities such as lysine or arginine side groups and propylamines participate in the catalysis of silica polymerisation. Self-assembly of the organic components and organic–inorganic electrostatic interactions also play a key role in biosilica formation. The framework components also control the location for biosilica formation, while other dynamic processes can help move these freshly formed biosilica particles and enable their deposition to the desired sites. The entire process is very complex, with temporal and spatial control over each step.

It is clear that compared to other systems, research on diatom biosilica formation has progressed the furthest. However, silicatein, an enzyme that catalyses silica formation in sponges, has been isolated and studied. Unlike biosilica formation in diatoms, sponges appear to use very few components for biosilica formation, although the chemical mechanisms are quite different, given that diatom biosilica is not formed using enzymes. Silicatein has a specific catalytic site composed of serine, histidine and asparagine.

It is proposed that the association of silicic acid with the hydroxyl group of serine and the imidazole group of histidine is crucial in the enzymatic catalysis. Although the exact details of this mechanism are not fully understood, there are two hypotheses, see figure 9.5. The first mechanism (figure 9.5, top), proposes that silicatein hydrolyses the precursor, thereby activating it for further (non-enzymatic) autocatalytic reaction. The second mechanism is based on biocatalysis of the condensation of fully hydrolysed monomeric silica (silicic acid, as a substrate) to cyclic trimers, which undergoes further non-enzymatic polymerisation. Despite these differences, both mechanisms rely on chemical *recognition* (which could be covalent bonding) between the silica precursor and –OH of serine, and the imidazole N of histidine of the catalytic site of silicatein.

An important reason for investigating biosilicification is to learn from biology how to design sustainable technologies for novel materials for applications. In order to transfer the knowledge obtained from biosilicification, we have seen in chapter 7 that scientists have taken a three step approach:

1. Identify, isolate and characterise biomolecules and genetic machinery controlling biosilica formation.

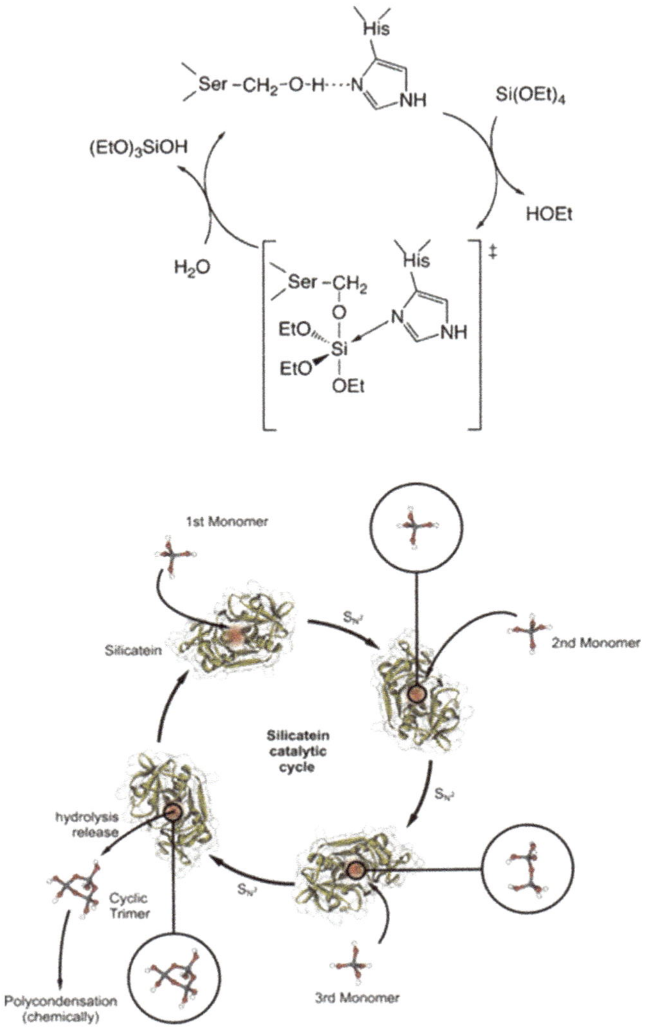

Figure 9.5. Proposed two mechanisms for enzymatic action of silicatein. Images reprinted with permission from [41], copyright 2008, American Chemical Society (top) and [42], copyright 2016, American Chemical Society (bottom).

2. Perform *in vitro* model studies using these bioextracts (and other strong-binding peptides) to understand the roles of various key functionalities.
3. Use synthetic molecules—bioinspired custom-made additives—to mimic the function of bioextracts/peptides and design sustainable processes.

Step 1 is already discussed briefly above, in section 9.2. Below we discuss the next two steps and illustrate them with selected examples.

9.3 Learning from biological silica synthesis: *in vitro* investigation of bioextracts

Once the biomolecules responsible for biosilica formation have been isolated and purified, their roles in the formation can be probed *in vitro*. Such research allows a deeper understanding of the physicochemical properties and function of the bioextracts, and also enables the reconstruction of the biosilicification process (such as that shown in figures 9.4 and 9.5). This can be illustrated with the example of silicateins from sponges.

As shown in figure 9.5, the catalytic triad, responsible for catalysing silica formation, was studied *in vitro*, using a range of conditions and precursors (or substrates). In addition, when the key amino acids from the catalytic site were replaced with other residues, the enzymatic action was lost. Further, the analysis of the crystal structure of the enzyme, molecular modelling, and simulation of the triad and its function provided insights into the catalytic action. It became clear from these studies that the close proximity of serine and asparagine with imidazole N from histidine is important to localise silicic acid via hydrogen bonding, while making it reactive by deprotonating simultaneously. Silintaphin-1 protein is postulated to act as a co-factor to silicatein and enhance the catalytic function. In addition to the chemical or catalytic attributes, silicateins also exhibit unique physical properties—they cooperatively self-assemble with silintaphin-1 into long filaments which form the templates for the hallmark needle-like spicules in sponges. These investigations have clearly identified the key catalytic function of the triad and the importance of self-assembly for templating.

In the case of diatoms, as the extracted biomolecules became available, their roles in silica formation were investigated *in vitro* [29, 43–46]. Initially, these biomolecules were studied individually, however, as additional components were identified and characterised, they were also studied in combination with each other. Similar to the case of sponge silicateins, silaffin (and other) proteins showed unique activities in

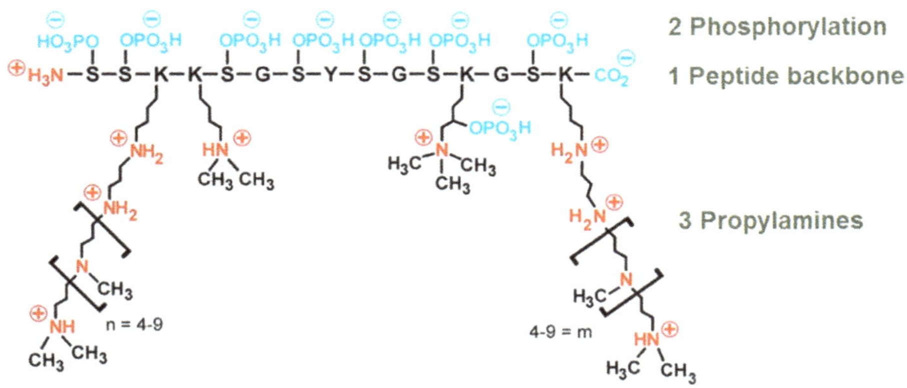

Figure 9.6. Structure of one of the silaffins, showing three key functionalities. Image taken from [29], reprinted with permission from AAAS.

catalysing, regulating, and templating biosilica formation [29, 47]. Scientists deconstructed various key features of those proteins and tested their function individually. An early example of this is the family of silaffin proteins (see figure 9.6), which contain (1) a unique sequence of amino acids in the peptide backbone, and various modifications such as (2) phosphorylation, and (3) propylamine functionalisation.

Each of these three features were studied for their roles in silica formation *in vitro*, using a range of experimental and computational techniques. These included the direct use of the peptide and its mutants (without any modifications) in silica formation, NMR studies of the self-assembly of these peptides (with and without the presence of silicates), and the study of the effects of synthesis conditions (e.g. pH, solvents, counter-ions, and silica precursors) [48]. These studies highlighted various aspects of silaffins that are key to biosilica formation. The amine moieties, the peculiar protonation states and their geometric arrangement lead to a highly cationically charged molecule which creates a favourable micro-environment for silica formation, and also exhibits catalytic action during the condensation reactions. In addition, these biomolecules form self-assembled structures in solution given their specific structure, which template silica formation. The phosphorylation is thought to help anchor the proteins and/or cross-link them with polysaccharides (or other cellular components), in order to exert spatial control over silica formation.

9.4 Emergence of bioinspired synthesis using synthetic 'additives'

The knowledge of the physicochemical properties and functions of the bioextracts have inspired a huge amount of 'synthetic' research on silica formation *in vitro*. These studies can lead to improved understanding of silica (and biosilica) formation in the presence of such additives (and bioextracts). They can further help design sustainable technologies for commercially relevant materials. To achieve these goals, a successful approach taken includes the design and use of synthetic molecules (defined as additives in chapter 7), which mimic the function of these biomolecules. Another approach involves the direct use of bioextracts. The latter provides a great level of control, given that the biomolecules have been optimised via evolution for biosilica formation. However, the issues with their isolation, purification, and stability can limit their use due to significant costs and the limited availability associated with this. To date, none of these biomolecules have been available at quantities of tens or hundreds of grams, while their access is limited to those who can extract them or express recombinantly. Further, it is likely that regulatory approval for their use in manufacturing and consumer products could be problematic. As a result, with industrial use in mind, the former approach of using additives has the potential to be economical, sustainable, scalable, and safe.

Continuing with the examples of silicateins and silaffins from previous sections, let us illustrate how bioinspired silica synthesis was developed, and how the additives were designed. As we have seen in silicateins, both the catalytic residues and the self-assembly of the proteins is essential, and so it could be prudent to design synthetic systems which can offer these features. Moving on to the examples of diatoms, we have seen that certain amino acids (e.g. lysine or arginine) and propylamines provide a highly cationic nature to silaffins. Learning from this, scientists explored the answers to questions pertaining to *in vitro* silica formation, such as [48–57]:

- Which amino acids are important?
- Are the amino acid sequences critical to silica formation?
- Would (homo)polypeptides be sufficient to promote silica formation?
- Do we need peptides or biomolecules?
- Can smaller molecules provide similar activities?
- Is carbon–carbon spacing important in the propylamines?
- Are the number of amines (or propylamine chain length) and amine protonation important?
- ....

There have been numerous investigations addressing these and other questions. Below, we provide a summary of those reports in an attempt to answer most of the questions listed above.

9.4.1 Which amino acids are important?

Many approaches have been taken to further the knowledge of the roles of biomolecules in silica formation. Some investigations focused on the use of individual amino acids in order to identify those that have the ability to catalyse and/or template silica formation. A novel system was designed to address this, in the context of mimicking the activity of silicateins [58]. Two sets of gold nanoparticles were functionalised with imidazole or hydroxyl groups (see figure 9.7). Although this strategy does not directly mimic the catalytic triad, the enhanced silica formation in the presence of these two nanoparticles highlighted the importance of co-operative action by hydroxyl and imidazole functionalities. In other studies, amino acids were also individually used (unbound to any particles) and their roles in silica formation were studied [52, 57]. It emerged that basic amino acids (lysine, arginine and histidine) were able to catalyse the silicic acid condensation reactions.

9.4.2 Would (homo)polypeptides be sufficient to promote silica formation?

The identification of key amino acids inspired the use of respective polypeptides to study their effects on silica formation. As seen in figure 9.8, polypeptides each of these amino acids (lysine, arginine, and histidine) were able to form silica particles within seconds to minutes at ambient conditions, in water at pH 7. It is worth highlighting that until that point, the synthesis of such particles required either heating, alkaline pH, a co-solvent and/or a much longer time.

Polylysine is a peculiar example which combines the physical and chemical properties of biomolecules. In addition to their catalytic function, leading to spherical silica particles, polylysine was able to produce hexagonal silica plates ~50 nm thick and 100s nm to one micron size sides (figure 9.9, top) [59–62]. Such structures had never been reported to form under such mild conditions. When the polylysine conformation, and its dynamic and co-operative self-assembly with silica species in solution were studied, it was found that through charge–charge

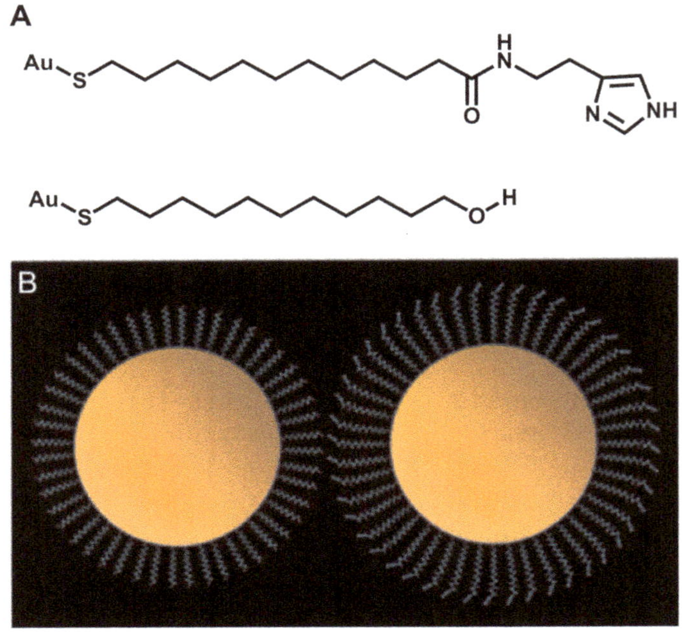

Figure 9.7. The molecules in (a) are used to functionalise two separate sets of gold nanoparticles in (b). Image reproduced with permission from [58], copyright 2005 John Wiley & Sons.

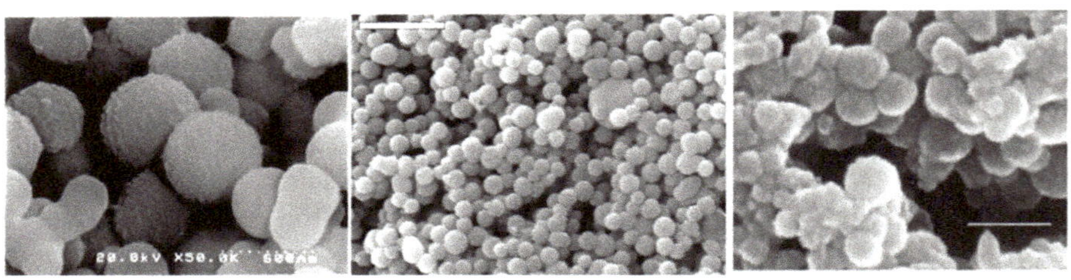

Figure 9.8. SEM images of silica particles formed in the presence of poly-lysine (left), poly-arginine (middle) and poly-histidine. Bars = 600 nm (left), 2 μm (middle) and 300 nm. Images taken from [56] (middle) and [55] (right), copyright 2003 with permission of Springer.

interactions and charge-screening, random coils of polylysine co-assemble with silicate ions into helical conformation (figure 9.9, bottom). A number of polylysine helices pack together into hexagonal single crystals, which template condensing silicates, thus forming hexagonal silica. In another study, a range of block co-polypeptides were designed which contained a catalytic block and a self-assembling block [50]. As expected, these polypeptides provided self-assembled templates for silica formation, which was catalysed by the catalytic blocks. We will expand this strategy later in this chapter. This incredible ability to produce structured materials at ambient conditions is of significant interest for designing high-value materials, as well as for developing responsive materials, such as those that can 'switch' their morphologies [60, 63].

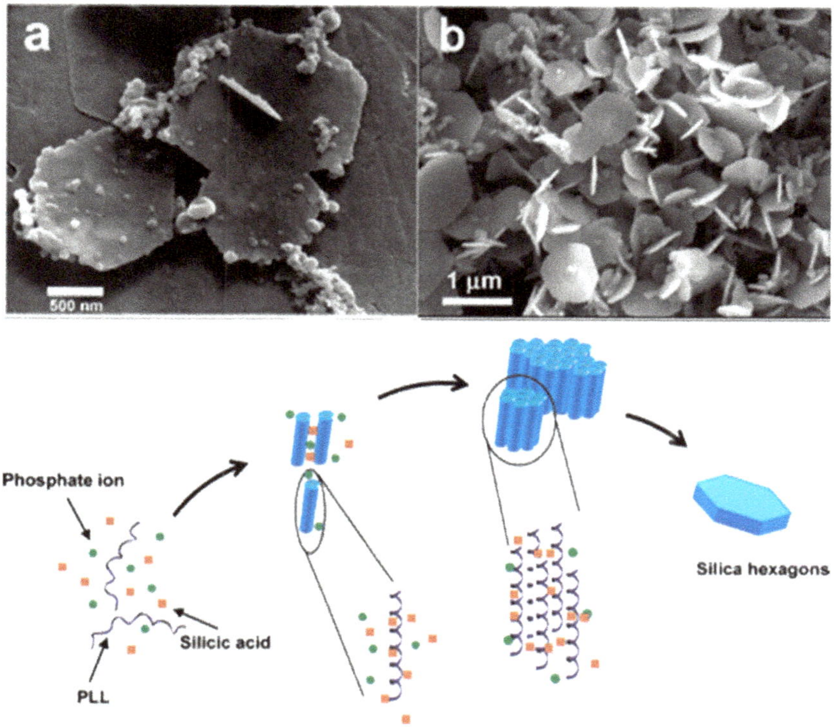

Figure 9.9. Illustration of how additives (polylysine—PLL in this case) can self-assemble into intricate templates and catalyse the formation of interesting bioinspired silica structures. Top: SEM micrographs of hexagonal silica formed using polylysine. Bottom: schematic presentation of the self-assembly, where the blue rods denote helical PLL. Images taken with permission from [60], copyright 2006, (top) and [62], copyright 2005, (bottom), American Chemical Society.

9.4.3 Peptides from biopanning

As discussed in chapter 7 (section 7.2.2 and table 7.1), a number of strong binding peptide sequences were identified against silica surfaces and particles. Although some of these peptides had some (limited) similarity with those isolated from biosilica (e.g. diatoms), there was less than 20% similarity between the peptides reported as strong silica binders. This variation was puzzling, particularly because most other materials, especially metals, have provided consensus in sequences. Therefore, in order to obtain the principles of the peptide–silica interface, further in-depth research was performed [64–67].

In one such study [64], silica particles of three distinct sizes were synthesised using the same method (figure 9.10, top). Using biopanning, strong binding peptides were identified against these particles. Interestingly, there was little similarity between these sequences, as well as with previously reported sequences. Extensive analysis of the surface chemistry using experiments, modelling, and simulation revealed that the particle surfaces have different ionisation. This was attributed to the slight differences in ammonia concentration used during their synthesis as a way to control

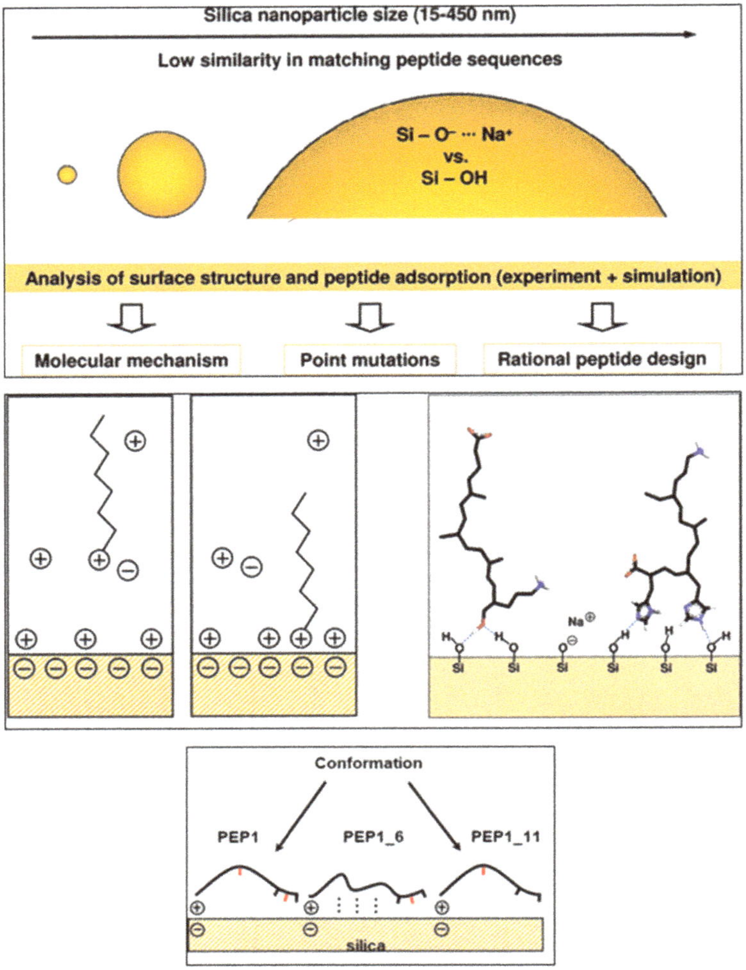

Figure 9.10. Top: an outline of how peptide binding on silica particles of three different sizes was investigated. The middle image shows that the major contributions to peptide adsorption comes from ion pairs (left) and (right) hydrogen bonds. Bottom: replacing a key amino acid (shown in red) leads to the modulation of binding through changes in peptide conformation and available binding sites on the peptide. Images reprinted with permission from [64], copyright 2012 American Chemical Society.

particle sizes. It was striking that such subtle differences were recognised by the peptides. Upon performing a range of peptide binding experiments and simulations, it became clear that the peptides bind via ionic interactions through unoccupied surface charges/sites. Further hydrogen bonds are also formed that provide additional binding strength (figure 9.10, middle). In the next step, the focus was on the role and position of individual amino acids. As we have seen above, histidine has been suggested as a key amino acid in silica binding, while it also consistently appeared in silica-binding peptides. By creating mutant peptides where histidine was replaced with alanine (inactive in silica binding) in either position 6 or 11 of the peptide, the importance of its location was studied. The results showed that histidine at 11th position was important, as replacing that resulted in loss of binding. On the other hand, replacing histidine at 6th position had no detrimental effect, in fact, the binding was improved due to increased conformational flexibility, leading to additional hydrogen bonding (figure 9.10, bottom). These studies provided the following key mechanistic information:

- Different silica sources/synthesis methods exhibit differences in surface properties (ranging from subtle to dramatic).
- Peptides recognise these differences and therefore resultant peptide sequences are different.
- Surface ionisation is an important factor affecting peptide binding—changes in this ionisation can occur from differences in particle sizes or slight changes to synthesis condition or the solvent conditions (pH, ionic strength, and type of buffer).
- The peptide–silica interactions predominantly involved a combination of ion pairing and hydrogen bonds.
- Peptide conformation and its rigidity/flexibility were important. Changes in certain amino acids or at specific location in the sequence led to dramatic changes in binding, even leading to the identification of stronger binders that biopanning did not provide.

9.4.4 Do we need peptides or biomolecules?

This is a question that has been debated widely, not just in the case of silica but also for other minerals. One school of thought is that proteins and peptides are required because they provide molecular recognition, specificity, and selectivity. This is perhaps true *in vivo*, however, *in vitro*, it is worth verifying. As such, the other school of thought is that if the chemical and physical functionalities can be mimicked in synthetic molecules, then there may not be a need for specific peptides or proteins. It is important to recognise here that the ultimate goal is not to mimic biological *structures*, but to mimic the biological *process* in order to develop a bioinspired process. Hence, the need for the peptides/biomolecules may not be so important if synthetic molecules can mimic their function and thus the process. This is supported by some of the examples shown above (the use of polylysine and functionalised gold nanoparticles). Further, it is worth mentioning that bioextracts have been shown to facilitate the formation of non-specific materials. For example, silicatein and silaffin both can catalyse the formation of titania and other materials [68]. Particularly for non-crystalline materials, epitaxial matching between the biomolecules and the materials is of less relevance, further suggesting the possibility of successfully using synthetic molecules.

In order to address this point and other questions listed in section 9.4, a range of synthetic polymers and block co-polymers, small amines and dendrimers have been investigated. The chemistry of these systems was selected based on amine functionalities present either in the backbone (mimicking peptides/proteins) or in the sidechains (mimicking propylamines), see figure 9.6. The advantage of using synthetic molecules is that they are typically well-characterised and their properties readily measured.

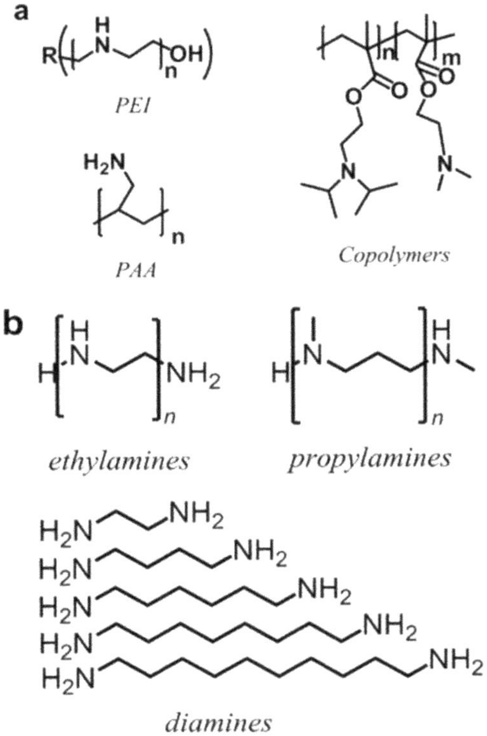

Figure 9.11. Selected examples of the bioinspired additives used in silica formation *in vitro*: (a) synthetic polymers and (b) small molecules [11], reproduced by permission of The Royal Society of Chemistry.

Examples of synthetic amine-containing polymers and block co-polymers include polyethyleneimines (PEI, linear, and branched), polyallylamine (PAA), and various versions of poly(alkylamino methacrylate) copolymers (see figure 9.11(a)). For a wider range of polymeric additives used, see [11]. Each of these polymers were found to catalyse the formation of silica, within seconds to minutes, under ambient conditions. Further, given their chain lengths, these molecules have the propensity to self-assemble into interesting template shapes. For example, in response to the processing conditions (e.g. pH and solvents), a star-shaped PEI assembles into a range of structures. Exploiting this dual catalytic and self-assembling function of many polymers, it is possible to produce a variety of silica morphologies including particles, plates, nanofibers, and coatings (see figure 9.12). This and other research clearly highlights the fact that there is no need to use biological molecules (peptides, proteins, or enzymes) to control silica formation. This is further supported by a large body of literature on the use of synthetic polymers as catalysts and templates for the bioinspired synthesis of wide ranging morphologies of silica (many of which are documented in [11]).

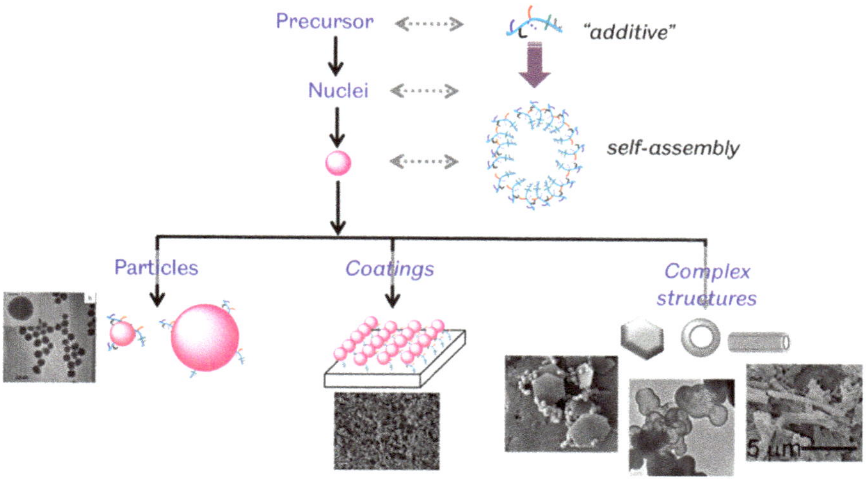

Figure 9.12. Schematic showing precursor to final bioinspired silica structures formed in the presence of additives. The scheme highlights the importance of key chemical moieties of the additives as well as their ability to self-assemble and to interact with a range of silicate species. By controlling these interactions, a range of materials can be produced. Image adapted from [72], copyright 2014 with permission of Springer.

9.4.5 Can smaller molecules provide similar activities?

We have seen that one group of the bioextracts isolated from diatoms is a family of propylamines with variable molecular size (or chain length) and levels of methylation. These propylamines have shown great activity in silica formation, both on their own as well as when mixed with silaffins. These propylamines are much smaller than siliaffins, silicateins and other biomolecules identified from biosilica. This implies that in addition to polymeric additives, smaller additives are likely to be functional. Obviously, the mechanisms of their action are of great interest, and scientists have been curious about the length of these amines, the level of amine methylation, and the size of the carbon spacers between amines. These points were systematically addressed by designing additives, for example those shown in figure 9.11(b) [57, 69–71]. The following key principles emerged in the context of bioinspired silica synthesis [57, 69–71]:

- Additives need to be positively charged throughout silica formation process.
- Additives need to be partially charged under the synthesis conditions, such that they can reversibly undergo protonation and deprotonation.
- When only the alternate amines are charged, the additives seem to provide the best results in terms of silica formation and precipitation.
- The balance between the hydrophilic and hydrophobic parts of the additives controls the catalysis and templating.
- For a given chain length (or number of amine groups), the 3-carbon spacing between the amine groups (as seen in biosilica-associated propylamines) was the best in achieving highest catalysis, compared with 2- or 4-carbon spacing (seen in ethylamines or butylamines, respectively).

- Additives with longer chain lengths (i.e. more amines per molecule) performed better, however, molecules as small as 3–5 amines groups were sufficient.

9.5 Benefits of bioinspired synthesis

The understanding of molecular interactions between additives discussed above and silica has now allowed the rapid synthesis of silica under mild conditions (see figure 9.13). It is worth noting that although organic molecules were used as templates in the formation of silicas and silicates at least a decade before the advent of bioinspired silica (e.g. zeolites and mesoporous silicas), what is new is the bioinspired catalytic ability of these additives. This dual function has enabled the synthesis of silica of desired shapes/morphologies which occurs:

- at room-temperature,
- at neutral pH,
- in water (without the need for co-solvents),
- at ambient pressure,
- very fast (within seconds to minutes), and
- as a single step synthesis.

Further, this method produces comparatively little waste, yet provides superior control of product properties. In contrast, for example, traditional syntheses of mesoporous silica suffer seriously from a range of issues, such as the use of toxic precursors, long synthesis (2–6 days), requirement of hydrothermal conditions (elevated temperature and pressure), and extremes of pH. This is a profound step change in the synthesis of silica-based materials—we now have a green/sustainable method for producing high value silicas. These differences between bioinspired and traditional syntheses of a range of silicas (including commercial products) are summarised in table 9.1. The details of the synthesis procedure are documented in [73], and also in the following videos:

- *Preparation of functional silica using a bioinspired method*: https://www.jove.com/video/57730/preparation-of-functional-silica-using-a-bioinspired-method.
- Small scale synthesis podcast: https://youtu.be/dkwsnrGesbA.

In addition to understanding the *end-product* (i.e. the properties of the silica synthesised), it is also important to understand the *process* by which these materials were formed. Therefore, probing the particle formation pathways and the roles of the additives in this process is of importance. As depicted in figure 9.14, the silica formation process is complex and undergoes multiple stages. During these stages,

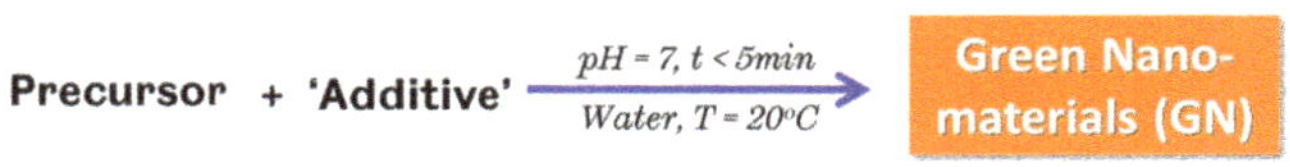

Figure 9.13. Simplified overview of bioinspired synthesis of silica.

Table 9.1. Comparison of selected traditional methods with bioinspired synthesis of silica, adapted from [75].

Silica type	Reagents[a]	Solvents	Reaction conditions t h	T °C	pH	Finishing	Control over
Mesoporous—SBA-15	TEOS, pluronic	Water, ethanol	44	42–100	<7	Calcination at 550 °C (>6 h)	Pore size, space group, wall thickness
Industrial—precipitated	Silicate, H_2SO_4	Water	2.5–3	60–80	<7	Drying	Purity, dispersion
Industrial—gel	Silicate, H_2SO_4	Water	3–5	35–80	ca. 7	Sizing, washing, drying	Pore volume
Colloidal—Stöber	TEOS and ammonia	Water, ethanol	12–24	10–60	ca. 9	Centrifugation, drying	Pore size
Bioinspired	Silicate and additive	Water	0.08	20	7	Centrifugation, drying	Pore size, structure, porosity, particle size, morphology,....[b]

[a] Silicate = sodium silicate or water glass, CTAB = cetyl trimethylammonium bromide, TEOS = tetraethoxysilane, pluronic = a triblock co-polymer of ethylene- and propylene-oxides.

[b] Control over incorporation of foreign material (catalyst, enzymes, drugs, etc) is also possible.

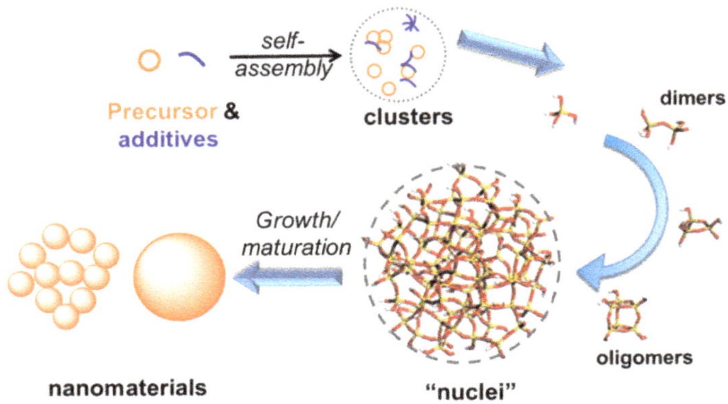

Figure 9.14. Stages in silica formation pathways, showing some of the key molecules/species involved. Adapted from [74].

the concentration of various silicon species (monomer and oligomers in particular) evolve. The additives interact in each stage and play a role in determining the formation pathways taken. Further depth of knowledge created through understanding the formation process in the presence of additives can make it possible to control the synthesis, and hence produce a range of commercially important properties of silica (e.g. porosity, particle sizes, and surface chemistry).

As we have seen in this chapter and chapter 6, biology offers sophistication. In the case of silica, although biosilica from plants and sponges is less ornate, diatoms show a remarkable nanoscale and hierarchical assembly. This spans nearly four orders of magnitudes and ranges from the primary particles (5–10 nm) all the way to the size of the cell (which varies for species between 10 μm to 100 μm). This assembly offers a multiscale structure made up of particles and pores. As this is not just controlled by the biomolecules but also involves other components of cellular machinery and is under strict biological control, such hierarchical assembly is incredibly difficult to obtain on the bench top using bioinspired methods, much less at industrial scales. Better understanding the particle formation process as shown in figure 9.14 is likely to help make advances in obtaining sophisticated features observed in biosilica. Indeed, we have already seen some complex structures being synthesised using bioinspired approaches, such as those shown in figure 9.9; note that biology is not known to produce these kinds of structures. On the other hand, the aim of bioinspired synthesis is not to copy the biomineral structures but to copy the principles and utilise them for producing desirable products. As diatomite or diatomaceous earth, which is essentially fossilised diatoms, is readily available in large quantities, the question to consider is whether we need to produce synthetic 'diatom-like' structures.

9.6 From lab to market

So, using this knowledge, what are we able to do, and what should we do? The ability to control particle formation pathways and the attributes of silica product using a green method is a great and novel outcome. If the goal is to develop industrially viable methods for commercial silicas, then the following attributes are important:

- Purity of the silica produced. This is a really important property, which is generally not given due importance during lab-scale studies. In particular, as the additives (and salt by-products) get occluded in final silicas, the purity of silica should be checked.
- Porosity is the other key property for silica. In most applications, the surface area, pore sizes, and pore volumes govern their performance. These applications include desiccant, adsorption platforms for pollutant removal from air, water and oils, fillers for polymers, cosmetics and toothpastes, and excipients for pharmaceuticals.
- Yield relates to productivity and waste generated. Although the yield is rarely reported in the synthesis literature, the economic viability of the method and the materials largely depend on it.
- Morphology is useful to control, however, in commercial applications, it is not so critical. In contrast, scientific publications put a lot of emphasis on morphology as it can provide information on growth mechanisms.
- Particle size, sphericity, and polydispersity are more related to the handling of solid particles, however, they may also contribute towards the material performance.
- pH and surface silanols provide a good indication of silica compatibility with various formulations (e.g. cosmetics or tyres).

Above, we have only focused on the 'product' attributes. In addition, there are 'process' considerations when it comes to industrial production, such as unit operations and scale-up. Given the focus of this book on materials, the process features are not discussed here. Nevertheless, this feat in controlling bioinspired silica has resulted in a number of patent applications and commercialisation activities. Some of these are listed below, in chronological order of their first disclosure:

- WO2000035993: *Methods, compositions, and biomimetic catalysts for in vitro synthesis of silica, polysilsesquioxane, polysiloxane, and polymetallo-oxanes.* This invention came from the University of California (Santa Barbra), led by Morse. This invention focuses on the use of silicatein and its mimics (polypeptides) in catalysing the formation of a range of silicon-containing materials under 'green' conditions.
- US7169589B2: *Silicatein-mediated synthesis of amorphous silicates and siloxanes and use thereof.* Müller and Schröder (Johannes Gutenberg University, Mainz) disclosed the use of silicatein (natural or recombinant) in the synthesis of silica.
- US7335717B2: *Methods, compositions, and biomimetic catalysts for the synthesis of silica, polysilsesquioxanes, polysiloxanes, non-silicon metalloid-oxygen networks, polymetallo-oxanes, and their organic or hydrido conjugates and derivatives.* This invention also came from the University of California

(Santa Barbra), led by Morse. Similar to their earlier invention, this patent widens to the use of synthetic additives.

- US2003171525: *Structure-directing catalysis for synthesis of metal, non-silicon metalloid and rare earth oxides and nitrides, and their organic or hydrido conjugates and derivatives.* Following from their earlier work, Morse and co-workers extended the use of silicatein (and its mimics) in catalysing the formation of a range of non-silicate materials.
- US7960509B2: *Fibrous protein fusions and use thereof in the formation of advanced organic/inorganic composite materials.* Kaplan (Tufts University) and Naik (Air Force Research Lab) designed a range of fusion proteins that combine silk with silica forming peptides and disclosed their use in the synthesis of hybrid materials, including silica.
- US8383755B2: *Enzyme-medicated cross-linking of silicone polymers.* Zelisko and co-workers (Brock University) developed the use of a range of enzymes, inspired from silicatein and silaffin, for catalysing siloxane bond formation.
- US2008293096: *Enzyme and template-controlled synthesis of silica from non-organic silicon compounds as well as aminosilanes and silazanes and use thereof.* Another invention from the group of Müller and Schröder, which extends their previous invention into the use of a range of proteins, enzymes and peptides for the synthesis of silicates, non-silicates and silanes.
- WO2008022774: *Biosilica-adhesive protein nanocomposite materials: synthesis and application in dentistry.* This invention from the group of Müller and Schröder is on the design of silicatein-silk fibroin fusion proteins and their use in synthesising silica-based materials for dental filling applications.
- WO2010036344: *Compositions, oral care products and methods of making and using the same.* This invention came from the group of Müller and Schröder from Johannes Gutenberg University (Mainz), in collaboration with Grace (a silica manufacturer). It pertains to the direct use of silicatein to deposit protective inorganic coatings on teeth.
- US2011281077: *Silicon derivate layers/films produced by silicatein-mediated templating and process for making the same.* This invention arising from a collaboration between the groups of Dario Pisignano (Pisa University) and Müller and Schröder, developed a method for producing silica coatings using silicatein.
- WO2015150399: *Osteogenic material to be used for treatment of bone defects.* Müller and Schröder further extended the use of silicateins for bone regeneration applications.
- WO2017037460: *Silica synthesis.* This invention, from one of the authors of this book, disclosed a bioinspired green method to synthesise silica and to further purify it by removing the occluded additives at room temperature and in water.

One of the authors of this book has embarked on the journey from lab-based green synthesis to large-scale manufacturing. The highlights of this journey are summarised below, in order to provide a better understanding of the pathway taken

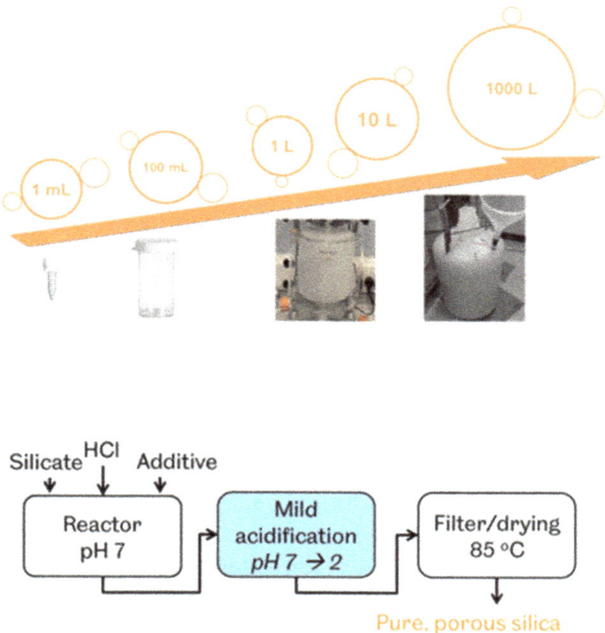

Figure 9.15. Pictorial representation of our scale-up journey from 2 ml scale to 40 L scale (top). The new purification method developed is shown as a block flow diagram (bottom). Images reproduced from [76].

[76]. A systematic approach was taken for process scale-up and process intensification. Initially a techno-economic analysis of a bioinspired method was performed in order to assess the economic feasibility [77]. The results show that using our green methods it is possible to reduce the energy usage of the reaction step by ~95% when compared with a traditional process, and the green nanomaterials (GN) would cost the same as the lowest grade commercial counterparts, yet provide significantly better quality and properties. This is promising and supports further development work (see figure 9.15, top, for a schematic showing the scale-up journey).

It is also worth focusing on aspects of downstream processing, in particular, purification of the products. In order to render porosity and purity for organic-mediated synthesis, calcination is used. However, it is prohibitively energy intensive and strips off any green benefits claimed from the synthesis step. Solvent reflux forms an alternative, but is also energy intensive, as it requires solvent reflux. This means that all the green features of bioinspired synthesis are outweighed by the purification step. Although downstream separations and purifications are generally more energy intensive/expensive, this creates economic barriers to commercialisation and scale-up—the bioinspired method thus cannot add value over existing processes and products. In order to address these issues, a new method for purification was invented, which involves mild acidification, operates at room temperature, and takes a few minutes (figure 9.15, bottom, and WO2017037460A1) [78]. This new method allows a complete removal of organics, with an added possibility of composition and porosity control. Given that it is a non-destructive method, most of the water and additive can be recycled, further improving the sustainability and economics. It is worth noting here that in terms of feedstock, the additives cost significantly more than the silicate precursors. Therefore, the ability to recover and reuse additives is of immense importance—this is not possible with destructive methods such as calcination.

Further, a number of scalability assessments of this method were performed (see figure 9.15, top, and an associated video of reaction in a 5 L reactor: https://youtu.be/S5YvvontRVA). Initially, the synthesis was carried out in lab-scale continuous flow reactors (at small scale: 5–50 ml) [79]. These studies provided information on key process parameters, as well as identifying practical challenges. Learning from these results and refining the process, the scale could be increased from a few mL to 1 L, 5 L, and reaching 40 L. The process was tested in both batch and continuous mode, and in tank and tubular reactors—this is an important step in assessing the scalability of a new method. Key learning from these scale-up trials included that the method is readily scalable and the recoverable yield does not change with scale. Further, as expected, mixing significantly affected the material properties [80]. These findings are currently being utilised in commercialising the process as well as up-scaling to produce kilograms of products.

9.7 Summary: key learning, summary and the future

We have learnt in this chapter that intricate biosilica structures are deposited by a range of living systems, and that they do this via an extremely complex process. Molecular biologists and biochemists have studied these systems extensively over decades in order to reveal the molecular secrets of biosilicification. They have been able to isolate genes, cellular components and biomolecules that control anything from silicon uptake to biosilica deposition. Further, *in vitro* studies of these biomolecules have started to provide key information that can be used to develop green synthesis protocols. These outcomes have spurred on interest in developing synthetic additives that can mimic the function of biomolecules in order to synthesise silica in a controlled and a sustainable fashion (see figure 9.16 also see a summary podcast at https://youtu.be/sDUl7urlsxY). In this journey of bioinspired silica, numerous intriguing features of silica formation and silicate-additive interactions have been unveiled. Some of these new concepts include:

- Cationic molecules that are readily water soluble are generally useful in facilitating silica formation under ambient conditions and neutral pH.
- Additives interact with different (and perhaps selective) stages of silica formation, which leads to the differences in their actions and the features of silica produced.
- Dynamic/reversible protonation of the additives is important.
- Additives can self-assemble or co-assemble with silicates, leading to templating final structures.

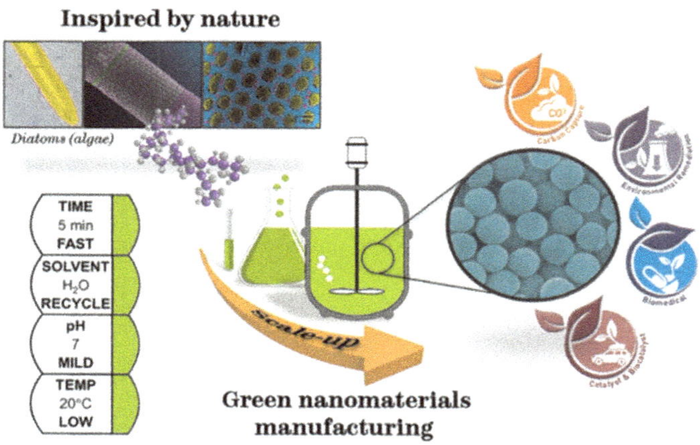

Figure 9.16. An overview of the journey from biosilica to the invention of bioinspired silica, the green credential, scale-up and applications. Image reproduced from [75], copyright 2015 with permission from Elsevier.

- The structure, architecture and amine environment, and the length of the additive play crucial roles in controlling silica synthesis and material properties.

The future focus should be on developing robust science underpinning the correlations between the synthesis–structure–property–performance for these materials, so that they can be easily applied to existing and emerging markets. Scientists should be working with industry to develop these materials for specific applications, and collaborating with engineers to design new sustainable/green manufacturing methods.

References

[1] Femmer R 2016 Diatoms (50 Species) (https://www.usgs.gov/media/images/diatoms-50-species-3)
[2] Pisera A 2003 *Microscop. Res. Tech.* **62** 312
[3] Brook M A 2000 *Silicon in Organic, Organometallic, and Polymer Chemistry* (New York: Wiley)
[4] Simpson T L and Volcani B E 1981 *Silicon and Siliceous Structures in Biological Systems* (New York: Springer)
[5] Evered D and O'Connor M 1986 *Ciba Foundation Symp.* vol 121 (New York: Wiley)
[6] Perry C C and Keeling-Tucker T 2000 *J. Biol. Inorg. Chem.* **5** 537
[7] Hildebrand M 2000 *Biomineralization: From Biology to Biotechnology and Medical Application* ed E Baeuerlein (Weinheim: Wiley), p 171
[8] Hildebrand M 2008 *Chem. Rev.* **108** 4855
[9] Hildebrand M and Wetherbee R 2003 *Prog. Mol. Subcell. Biol.* **33** 11
[10] Perry C C 2009 *Biosilica in Evolution, Morphogenesis, and Nanobiotechnology: Case Study Lake Baikal* vol 47 ed W E G Müller and M A Grachev (Berlin: Springer), p 295
[11] Patwardhan S V 2011 *Chem. Commun.* **47** 7567

[12] Harrison C C 1996 *Phytochemistry* **41** 37
[13] Kröger N, Lehmann G, Rachel R and Sumper M 1997 *Eur. J. Biochem.* **250** 99
[14] Hildebrand M, Volcani B E, Gassmann W and Schroeder J I 1997 *Nature* **385** 688
[15] Shimizu K, Cha J, Stucky G D and Morse D E 1998 *Proc. Natl. Acad. Sci. U.S.A.* **95** 6234
[16] Krasko A, Lorenz B, Batel R, Schröder H C, Müller I M and Müller W E G 2000 *Eur. J. Biochem.* **267** 4878
[17] Perry C C and Keeling-Tucker T 2003 *Colloid Polym. Sci.* **281** 652
[18] Ma J F, Tamai K, Yamaji N, Mitani N, Konishi S, Katsuhara M, Ishiguro M, Murata Y and Yano M 2006 *Nature* **440** 688
[19] Frigeri L G, Radabaugh T R, Haynes P A and Hildebrand M 2006 *Mol. Cell. Proteomics* **5** 182
[20] Matsunaga S, Sakai R, Jimbo M and Kamiya H 2007 *Chembiochem.* **8** 1729
[21] Kroger N and Poulsen N 2008 *Annu. Rev. Genet.* **42** 83
[22] Ehrlich H *et al* 2010 *Nat. Chem.* **2** 1084
[23] Patwardhan S V, Clarson S J and Perry C C 2005 *Chem. Commun.* **1113**
[24] Armbrust E V *et al* 2004 *Science* **306** 79
[25] Bowler C *et al* 2008 *Nature* **456** 239
[26] Kröger N, Deutzmann R, Bergsdorf C and Sumper M 2000 *Proc. Natl. Acad. Sci. U.S.A.* **97** 14133
[27] Kröger N, Deutzmann R and Sumper M 1999 *Science* **286** 1129
[28] Kröger N, Deutzmann R and Sumper M 2001 *J. Biol. Chem.* **276** 26066
[29] Kröger N, Lorenz S, Brunner E and Sumper M 2002 *Science* **298** 584
[30] Poulsen N and Kröger N 2004 *J. Biol. Chem.* **279** 42993
[31] Poulsen N, Sumper M and Kröger N 2003 *Proc. Natl. Acad. Sci. U.S.A.* **100** 12075
[32] Scheffel A, Poulsen N, Shian S and Kröger N 2011 *Proc. Natl. Acad. Sci.* **108** 3175
[33] Cha J N, Shimizu K, Zhou Y, Christiansen S C, Chmelka B F, Stucky G D and Morse D E 1999 *Proc. Natl. Acad. Sci. U.S.A.* **96** 361
[34] Zhou Y, Shimizu K, Cha J N, Stucky G D and Morse D E 1999 *Angew. Chem. Int. Ed.* **38** 780
[35] Schröder Heinz C, Vlad A G, Xiaohong W and Werner E G M 2016 *Bioinspir. Biomim.* **11** 041002
[36] Perry C C and Keeling-Tucker T 1998 *Chem. Commun.* **23** 2587
[37] Currie H A and Perry C C 2007 *Ann. Bot.* **100** 1383
[38] Currie H A and Perry C C 2009 *Phytochemistry* **70** 2089
[39] Gong N, Wiens M, Schroder H C, Mugnaioli E, Kolb U and Muller W E G 2010 *J. Exp. Biol.* **213** 3575
[40] Hildebrand M, Lerch S J L and Shrestha R P 2018 *Front. Mater. Sci.* **5** 125
[41] Brutchey R L and Morse D E 2008 *Chem. Rev.* **108** 4915
[42] Schröder H C, Grebenjuk V A, Wang X and Müller W E G 2016 *Bioinspir. Biomim.* **11** 041002
[43] Murr M M and Morse D E 2005 *Proc. Natl. Acad. Sci. U.S.A.* **102** 11657
[44] Brunner E, Lutz K and Sumper M 2004 *Phys. Chem. Chem. Phys.* **6** 854
[45] Sumper M 2002 *Science* **295** 2430
[46] Patwardhan S V, Holt S A, Kelly S M, Kreiner M, Perry C C and van der Walle C F 2010 *Biomacromolecules* **11** 3126
[47] Groeger C, Lutz K and Brunner E 2008 *Cell Biochem. Biophys.* **50** 23

[48] Knecht M R and Wright D W 2003 *Chem. Commun.* **24** 3038
[49] Mizutani T, Nagase H, Fujiwara N and Ogoshi H 1998 *Bull. Chem. Soc. Jpn.* **71** 2017
[50] Cha J N, Stucky G D, Morse D E and Deming T J 2000 *Nature* **403** 289
[51] Brott L L, Naik R R, Pikas D J, Kirkpatrick S M, Tomlin D W, Whitlock P W, Clarson S J and Stone M O 2001 *Nature* **413** 291
[52] Coradin T and Livage J 2001 *Colloids Surf. B-Biointerfaces* **21** 329
[53] Patwardhan S V, Mukherjee N and Clarson S J 2001 *J. Inorg. Organomet. Polym.* **11** 193
[54] Sudheendra L and Raju A R 2002 *Mater. Res. Bull.* **37** 151
[55] Patwardhan S V and Clarson S J 2003 *J. Inorg. Organomet. Polym.* **13** 49
[56] Patwardhan S V and Clarson S J 2003 *J. Inorg. Organomet. Polym.* **13** 193
[57] Belton D, Paine G, Patwardhan S V and Perry C C 2004 *J. Mater. Chem.* **14** 2231
[58] Kisailus D, Najarian M, Weaver J C and Morse D E 2005 *Adv. Mater.* **17** 1234
[59] Patwardhan S V, Mukherjee N, Steinitz-Kannan M and Clarson S J 2003 *Chem. Commun.* **10** 1122
[60] Patwardhan S V, Maheshwari R, Mukherjee N, Kiick K L and Clarson S J 2006 *Biomacromolecules* **7** 491
[61] Cui H, Krikorian V, Thompson J, Nowak A P, Deming T J and Pochan D J 2005 *Macromolecules* **38** 7371
[62] Tomczak M M, Glawe D D, Drummy L F, Lawrence C G, Stone M O, Perry C C, Pochan D J, Deming T J and Naik R R 2005 *J. Am. Chem. Soc.* **127** 12577
[63] Bellomo E G and Deming T J 2006 *J. Am. Chem. Soc.* **128** 2276
[64] Patwardhan S V, Emami F S, Berry R J, Jones S E, Naik R R, Deschaume O, Heinz H and Perry C C 2012 *J. Am. Chem. Soc.* **134** 6244
[65] Emami F S, Puddu V, Berry R J, Varshney V, Patwardhan S V, Perry C C and Heinz H 2014 *Chem. Mater.* **26** 5725
[66] Emami F S, Puddu V, Berry R J, Varshney V, Patwardhan S V, Perry C C and Heinz H 2014 *Chem. Mater.* **26** 2647
[67] Sola-Rabada A, Michaelis M, Oliver D J, Roe M J, Ciacchi L C, Heinz H and Perry C C 2018 *Langmuir* **34** 8255
[68] Dickerson M B, Sandhage K H and Naik R R 2008 *Chem. Rev.* **108** 4935
[69] Belton D, Patwardhan S V and Perry C C 2005 *Chem. Commun.* **27** 3475
[70] Belton D J, Patwardhan S V, Annenkov V V, Danilovtseva E N and Perry C C 2008 *Proc. Natl. Acad. Sci. U.S.A.* **105** 5963
[71] Belton D J, Patwardhan S V and Perry C C 2005 *J. Mater. Chem.* **15** 4629
[72] Patwardhan S 2014 *Bio-Inspired Nanotechnology: From Surface Analysis to Applications* ed M R Knecht and T R Walsh (New York: Springer), 127
[73] Manning J R H, Routoula E and Patwardhan S V 2018 *J. Vis. Exp.* **138** e57730
[74] Belton D J, Deschaume O, Patwardhan S V and Perry C C 2010 *J. Phys. Chem.* B **114** 9947
[75] Patwardhan S V, Manning J R H and Chiacchia M 2018 *Curr. Opin. Green Sustain. Chem.* **12** 110
[76] Patwardhan S V 2019 *Johnson Matthey Technol. Rev.* **63** 152
[77] Drummond C, McCann R and Patwardhan S V 2014 *Chem. Eng. J.* **244** 483
[78] Manning J R H, Yip T W S, Centi A, Jorge M and Patwardhan S V 2017 *Chem. Sus. Chem.* **10** 1683
[79] Patwardhan S V and Perry C C 2010 *Silicon* **2** 33
[80] Chiacchia M and Patwardhan S V 2018 *16th European Conf. on Mixing (Toulouse, France)*

www.ingramcontent.com/pod-product-compliance
Ingram Content Group UK Ltd.
Pitfield, Milton Keynes, MK11 3LW, UK
UKHW051856100926
8556IPUK00005B/649

Economical, Political, and Social Issues in Water Resources

Economical, Political, and Social Issues in Water Resources

Edited by

Omid Bozorg-Haddad
Department of Irrigation and Reclamation Engineering,
Faculty of Agricultural Engineering and Technology,
College of Agriculture and Natural Resources,
University of Tehran, Karaj, Tehran, Iran.

ELSEVIER

Elsevier
Radarweg 29, PO Box 211, 1000 AE Amsterdam, Netherlands
The Boulevard, Langford Lane, Kidlington, Oxford OX5 1GB, United Kingdom
50 Hampshire Street, 5th Floor, Cambridge, MA 02139, United States

British Library Cataloguing-in-Publication Data
A catalogue record for this book is available from the British Library

Library of Congress Cataloging-in-Publication Data
A catalog record for this book is available from the Library of Congress

ISBN: 978-0-323-90567-1

Publisher: Candice Janco
Acquisitions Editor: Louisa Munro
Editorial Project Manager: Madeline Jones
Production Project Manager: Joy Christel Neumarin Honest Thangiah
Cover Designer: Alan Studholme

Typeset by Aptara, New Delhi, India

Contents

Contributors

Hazi Mohammad Azamathulla Department of Civil and Environmental Engineering, St. Augustine, Trinidad and Tobago

Robert P. Berrens Department of Irrigation & Reclamation Engineering, Faculty of Agriculture Engineering & Technology, College of Agriculture & Natural Resources, University of Tehran, Iran

Robert P. Berrensd Department of Economics, University of New Mexico, Albuquerque, NM, United States

Asim Biswas School of Environmental Sciences, University of Guelph, Guelph, Canada

Daniele Bocchiolac Department of ICA, Politecnico di Milano, Milan, Italy; EVK2CNR Association, Bergamo, Italy

Omid Bozorg-Haddad Department of Irrigation & Reclamation Engineering, Faculty of Agricultural Engineering & Technology, College of Agriculture & Natural Resources, University of Tehran, Karaj, Tehran, Iran

Jessica Budds Department of Economics, University of New Mexico, Albuquerque, NM, United States

Mohammad Delpasand Department of Irrigation & Reclamation Engineering, Faculty of Agricultural Engineering & Technology, College of Agriculture & Natural Resources, University of Tehran, Karaj, Iran

Sina Fazeli Department of Irrigation & Reclamation Engineering, Faculty of Agricultural Engineering & Technology, College of Agriculture & Natural Resources, University of Tehran, Karaj, Tehran, Iran

Seyedeh Habibbeh Hosseiny Department of Range Management, Gorgan University of Agricultural Science and Natural Resources, Gorgan, Iran

Hugo A. Loáiciga Department of Geography, University of California, Santa Barbara, California

Zeinali Masoumeh Department of Irrigation & Reclamation Engineering, Faculty of Agricultural Engineering & Technology, College of Agriculture & Natural Resources, University of Tehran, Karaj, Iran

Assefa M. Melessec Department of Earth and Environment, Florida International University, Modesto A. Maidique Campus, Miami, FL, United States

Fahimeh Mirzaei-Nodoushan Department of Irrigation and Reclamation Engineering, Faculty of Agricultural Engineering and Technology, College of Agriculture and Natural Resources, University of Tehran, Karaj, Iran

Sahar Mohammad-Azari Department of Irrigation & Reclamation Engineering, Faculty of Agricultural Engineering & Technology, College of Agriculture & Natural Resources, University of Tehran, Karaj, Tehran, Iran

Vijay P. Singh Department of Biological and Agricultural Engineering & Zachry Department of Civil Engineering, Texas A&M University, College Station, TX, United States; Distinguished Professor, Regents Professor, Caroline & William N. Lehrer Distinguished Chair in Water Engineering, Department of Biological and Agricultural Engineering & Zachry Department of Civil & Environmental Engineering, Texas A&M University, College Station, Texas, United States

Marcus Randall Center for Agriculture and Financial Big Data Analytics Academia, Bond Business School, Bond University, Queensland, Australia

Alireza Rezaee Department of Irrigation & Reclamation Engineering, Faculty of Agriculture Engineering & Technology, College of Agriculture & Natural Resources, University of Tehran, Karaj, Alborz, Iran

Mohammad-Azari Sahar Department of Irrigation & Reclamation Engineering, Faculty of Agricultural Engineering & Technology, College of Agriculture & Natural Resources, University of Tehran, Karaj, Tehran, Iran

Masoumeh Zeinali Department of Irrigation & Reclamation Engineering, Faculty of Agricultural Engineering & Technology, College of Agriculture & Natural Resources, University of Tehran, Karaj, Iran

Preface

Now, more than ever, one can feel the overwhelming desire by both public and governing bodies to aspire "sustainability" in a society as a whole entity. When it comes to water resources, this notion has coincided with the concept of "integrated water resources planning and management." Plainly speaking, the idea here is to strive for an equilibrium that could not only meet the demands and needs of a community for water resources but also takes into account, and in turn preserve, these natural resources and the environment that surrounds and depends on it as well. On paper, any practices that follow this guideline can be maintained over a long-time horizon. As it turned out, however, achieving this is easier said than done. The main challenge, in a nutshell, is that water management is, by nature, a multi-dimensional process. As such, a truly integrated and sustainable practice must not only account for technical and engineering challenges, but also addresses the socio-economic, legislative, and political aspect of the problem as well.

Having access to water for drinking and sanitation is a well-established right recognized by the United Nations. One tends to forget, however, that for other consuming sectors, water is essentially an economic commodity. Hence, it needs to be treated as one in order to devise a self-sustaining, long-term water resources planning, and management policy. Thus, knowing these economic-oriented principles and their implication from a water resources management perspective can bring us one step closer to our goal of creating a sustainable future. To that end, in this book, we would be exploring the concepts of water accounting, water economics, water pricing, water markets, and water books from a fresh perspective.

A comprehensive water resources policy should also accommodate the role of this natural resource from a social and legislative point-of-view. Often, water is a common, shared good and, to some extent, a limited resource. As such, the socio-political aspect of water resources planning and management can play a permanent role in practical, real-world cases. The sheer importance of this matter became obediently clear when one deals with transboundary water bodies. As such, here we tend to take a deep dive into topics like "rights and laws of transboundary water resources," "water policy and governance," "shared water resources management," "water, culture, civilization, and history," "water quality, hygiene, and health," and "water and society."

The book is structured so that each chapter can be seen as a stand-alone lecture on a specific topic. The interested reader, which can be either an experienced and seasoned researcher or an early-career student who wants to get a foot at the door, can explore the said subject area from a somewhat different and fresh perspective. The book can also be seen as a wholesome lecture, and the reader can, and in fact, we encourage the readers to start from the first chapter and go through the entire book. As such, you would be engaged in an engaging, thought-provocative idea that can bring you a fresh perspective on how you perceive sustainability in terms of water resources planning and management.

Omid Bzorg-Haddad

Acknowledgment

The editor thanks Iran's National Science Foundation (INSF) for its financial support for this research.

Chapter 1

Water accounting

Sahar Mohammad-Azari[a], Omid Bozorg-Haddad[a] and Asim Biswas[b]
[a] *Department of Irrigation & Reclamation Engineering, Faculty of Agricultural Engineering & Technology, College of Agriculture & Natural Resources, University of Tehran, Karaj, Tehran, Iran.* [b] *School of Environmental Sciences, University of Guelph, Guelph, Canada*

1.1 Introduction

High pressure on water resources and uncertainty in water availability in the future would cause complexity in management and planning decisions. In fact, water availability is important not only for basic human and environmental demands but also as a key factor for economic development of different sectors which face conflict in their demands. Nevertheless, water management is based on political, economic and social issues more than only technical aspects. Therefore, the water stakeholders' partnership should be increased in water management and rules should be set for public partnership and clarification in water governance.

In order to institute mentioned requirements, it is necessary to provide high quality information. In this regard, a new approach is required to go further than water balance analysis. In order to achieve this goal, the concept of water accounting has been introduced in water resources management literature.

Water accounting is a process including water problems' recognition, water allocation, and quantification of water related benefits and reporting for decision making in management process. In fact, the main goal of water accounting process is to consider the impact of economic and social factors on water resources management along with hydrological factors. The implementation of this strategy will be started by derivation of water resources and economically important data (Chalmers et al., 2012a).

Generally, appropriate water accounting framework aids to generate accurate standard information about water supplies, consumptions, social and economic aspects of water consumptions in order to answer questions in water resources management. The first question answered through water accounting process is the causes of imbalances in water supply and demands. Furthermore, what are the best opportunities in order to making sustainable and equitable water use and the desirable level of sustainability in current water use? In fact, water accounting framework can find the limitations, problems and opportunities to improve

Economical, Political, and Social Issues in Water Resources. DOI: 10.1016/B978-0-323-90567-1.00001-2

water resources management for future goals through answering mentioned questions.

Water accounting has been one of the most important and remarkable concepts in scientific and professional water related societies. Many national and international organizations focus on this concept and applied in different aspects of water resources management in recent years. Among these organizations, United Nations (UN), Food and Agricultural Organization (FAO), International Water Management Institute (IWMI) and Australian Government are the ones which played an important role in developing water accounting framework (Chalmers, 2012a).

With regard to the importance of water accounting concept, numerous studies have been carried out worldwide considering different aspects of water accounting and its applicability in water resources planning and management.

One of the first studies in this field belongs to Molden and Sakthivadivel. In 1999, they applied water accounting methodology in order to indicate its usefulness in information provision for stakeholders and decision makers' better understanding about water resources use and productivity. They concluded that these types of information can help decision makers to achieve better knowledge in order to improve integrated water resources management.

Peranginangin et al. (2004) developed a water accounting procedure for groundwater and surface water components separately unlike other studies in which both surface and groundwater were considered. This separation helped achieving more realistic estimate of water availability and also applying to evaluate present and future water use performance.

Lange et al. (2007) applied water accounting framework to the Orange River Basin. The accounting process brought an economic perspective to water management for the mentioned river basin. They also linked the water accounting to economic data in order to compute water use and productivity. The results from water accounting and economic data can be applied in future decisions about water allocation, pricing and infrastructure development.

Vardon et al. (2007) was among the most comprehensive studies that were carried out in Australia at a national scale on water accounting related to natural water scarcity crisis. This study introduced water accounting and mentioned the major Australian organizations involved in water related information provision and information use process. The main goal followed in this process was the improvement of the public knowledge on water resources in Australia. It should be noted that the experience of financial accounting in the case of water reporting is presented in this study which is novel approach implemented in line with water accounting and is named Australian Water Accounting Standard. In other word, this approach is similar to financial accounting in reporting and methodology and is based on water accounting standards. In this regard, Australia is one of the countries that made the water accounting standards with the aim to improve water information.

M. Turner (2010) presented a water accounting system (WAS) for strategic long-term water management. This system was compared to other accounting systems and its benefit was reported. The results indicated the potential of WAS for solving tensions between water supply and demands.

Other studies focusing on water footprint accounting and water productivity were carried out in 2011 by Ercin et al. and Perry, respectively. The first study focused on the water footprint accounting and its impact assessment on a hypothetical sugar-containing carbonated beverage. The results revealed that the biggest share at total water footprint of a product is dedicated to agricultural ingredients despite their small fraction in weight of the final product.

In second study, Perry applied water accounting for evaluation of the water productivity in agricultural sector. Some recent researches are presented to indicate the usefulness of water accounting in the scope of water productivity in agricultural sector.

In 2012, Chalmers et al. discussed the generality of water accounting in a review paper. They also evaluated different water accounting systems, their application and differences. In this study they concluded that water accounting can be considered as a useful tool in order to provide high quality information for water management and decision making.

In another study, these authors evaluated the roles of water accounting, water industries and other stakeholders in general-purpose water accounting. Chalmers et al. (2012a) tried to apply regulatory theories which explain financial accounting standards development to present and discuss the national and international future of general-purpose water accounting.

The study of Karimi et al. (2013) is one of the most remarkable studies in 2013. They introduced a new framework to provide explicit spatial information called Water Accounting Plus (WA+). They provided information related to water depletion and withdrawal processes in river basins.

In another study, Hu et al. (2013) discussed about Water Accounting Standard Boards which formed first time in 2009 in order to plan a sustainable water use in Australia and in following, they investigated the impact of stakeholder theories and also public interest on Board. The results revealed that public interests and stakeholders play key role in determining the effectiveness of standard and causes encouragement usage of water reporting standard.

Zhan-Ming and Chen (2013) investigated the virtual water profile of the world in 2004. This investigation was implemented based on a multi-region input–output model. The results can be applied in policy making focusing on water using pattern adjustment and water security enhancement.

In the following, few studies focusing on water accounting, its development and applicability in different water management issues are presented. For example,

Berger et al. (2014) introduced water accounting and vulnerability evaluation model in order to enhance the analysis of water consumption. Results showed the capability of the model in interpretation of water footprint figures and promotion of sustainable use of freshwater resources.

Karimi et al. (2015) evaluated the impact of remote sensing measurements' errors on water accounting and information provided for policy making. They investigated the reliability of WA+ outputs, using a case study. They reported a reliable performance of the majority of WA+ parameters. On the other hand, utilized flow exhibited low reliability and thus, caution must be administrated to use them formulating important policy decisions.

Momblanch et al. (2014) described the usefulness of Australian Water Accounting Standard. However, the applicability of results in integrated water resources management calls for new level of details and the amount of information. Therefore, a different method was presented and compared with the original Australian Water Accounting Standard.

Morillo et al. (2014) proposed joint evaluation of water footprint accounting for high value crops and irrigation management indicators to evaluate the water use efficiency improvement and actions to reduce water use.

Tilmant et al. (2015) presented a dynamic water accounting approach in which the whole basin was considered as a value chain. The results of a hydro-economic model made it possible to achieve effective water accounting.

M.Borrego-Marin et al. (2016) proposed a methodology that used system of environmental-economic accounting for water standard tables in order to assess cost recovery ratio. Results indicated the capability of the method in cost recovery analysis comparability enhancement and knowledge translation between region and countries worldwide.

Pedro-Monzonis (2016) presented a tool that enabled combined use of hydrological and water resources models to fill in the system of environmental-economic accounting tables.

Manzardo et al. (2016) proposed a methodology in order to assess the water footprint accounting of direct water use at an urban level. The methodology was also tested in the municipality in Vicenza, Italy. The results indicated the applicability of the method in water management issues while it requires large amount of data.

Tello et al. (2016) evaluated the superiority and applicability of general-purpose water accounting reports in potential users' viewpoint. They discussed the general-purpose water accounting as a financial technique that can be extended to other areas of natural resources. Users' perceptions indicated the usefulness of general-purpose water accounting.

Burgia et al. (2017) presented a review of water accounting in order to reveal the evolution of researches on water accounting over time in the accounting, management and organization fields.

L. Burritt and L.Christ (2017) focused on the lack of monetary data in water accounting initiatives. The results indicated that considering monetary business in water accounting process help companies to improve water management. On the other hand, developing environmental management accounting along with physical and monetary information in water accounting framework would be the best way for improvement of water management.

Xu et al. (2017) developed an optimization model based on the blue and grey water footprint accounting results of industrial sectors in Dalian City of China. The results indicated the rationality and applicability of optimal model for water management strategy formulation.

Finally, the recent study in this case was carried out by Bazrafshan et al. (2019). They focused on water footprint accounting and virtual water trading of saffron. Valuable information for managers and policy makers were achieved from their findings to extend cultivation area with low economical water footprint, rain-fed cropping and enough precipitation.

1.2 Terms and definitions

As stated before, water accounting is a formation of a mechanism for organization and combination of accumulated information from different sources in order to achieve set of integrated information and link physical water supply and use data with economic data. In fact, the physical data include all the amount of water which enter the domain or exploit from the domain. On the other hand, the economic data can be described as the income and costs of water use in different sectors.

In a classic water management perspective, the hydrologic aspect is the only important issue that has been carried out based on water balance information without considering social, economic and environmental aspects. But water accounting framework goes beyond the water balance, supply and use analysis (FAO, 2017).

Assessments and evaluations during water accounting process are carried out focusing on hydrologic, engineering and monitoring aspects. In hydrology aspects the focus is on understanding physical process governing water flows and storages in various climatic and management conditions. In fact, this aspect in water accounting generally focused on water flow and depletion accounting. Flow accounting methods are based on flows in river, aquifers, lakes, discharges, diversions and abstracts. On the other hand, depletion accounting is emphasized on all the process with the aim of consumptive use. In fact, in water depletion accounting, the focus is on the water depletion causing limited or no access in the future. The process of depletion is categorized in the form of evaporation, apply in production process or water quality degradation (Chalmers, 2012a, b).

On the other hand, engineering aspects consider design, construction and operation of storage infrastructures, water transfer schemes, urban water supply systems, irrigation and drainage schemes and water treatment. In other word, this aspect emphasizes on spatial and temporal water storage management and its delivery from supply source to demand site. In fact, considering both supply and demand sides are the most important evaluation framework in engineering aspect. In this process different aspects of supply and demand sides that play key roles in the water resources management are discussed. The spatial and temporal availability to rainfall, surface and groundwater, unconventional water resources, capacity, operation and maintenance processes of supply, storage and treatment infrastructures will be assessed and evaluated in the supply side. On the other hand, the spatial and temporal demand of various water users, the level of meeting these demands, and spatial and temporal consumptive and non-consumptive patterns are also evaluated in the demand side. The third aspect in water accounting the monitoring aspect focuses on applying water accounting to support management decision or as a means for achieving improvements in policies and practices in both supply and demand sides.

Ultimately all the information which are being collected, qualified and revised during the water accounting process are classified into a number of accounts and analytical indicators. Accounts are sets of data about water asset that reflect the flow status and the available amount of water and the water abstraction for different users. The costs and benefits of water use, and the social and environmental components are also presented in water accounting related tables. Some analytical indicators are also applied in the water accounting framework. These analytical indicators are described in fraction or in productivity form. After deriving accounts, figures and diagrams will be used to reveal the status of the system and present an integrated framework to connect the physical data to economic data to enhance decision making process based on available information (Chalmers, 2012a).

1.3 Basic principles and logics

The basic principle of water accounting is achieving a systematic frame of processes that collect the water related information. The high-quality data will be provided from the collection and analysis of gathered information. This derived information is presented in a report structure in order to meet a category of goals. These goals include water problems and challenges analysis, the reasons of positive and negative outcomes of policies and strategies in water related issues, provision of an integrated and comprehensive data basis for stakeholders, efficient water allocation and conflict resolution, eliminate the bias in viewpoints toward water management and increasing awareness (Godfrey and Chalmers, 2012).

In other word, the social and biophysical plans and strategies required by water user sectors are recognized through the analysis of water related challenges and accumulate the information and documents about policies, plans and measures. Furthermore, recognition and evaluation of different social, political and economic factors with different outcomes in a specified domain can provide a set of information in line with better understanding of the reasons of positive and negative impacts.

On the other hand, access to stated details in a high quality and comprehensive information framework result in improvement and progress in social, political and biophysical plans for fundamental changes in policies and management infrastructure facing the water related challenges. Furthermore, severe scarcity and challenges rising conflict among different water users and stakeholders. Available qualified and reliable data results should be in place for efficient and effective conflict resolution and achieving agreement among users and stakeholders (FAO, 2017; Chalmers, 2012a).

Finally, the main idea in water accounting procedure is awareness rising among different water user groups, planners, managers, policy makers and stakeholders through presenting reports of accurate water related information and outputs and seek for the fast problem solving complicated social, biophysical, economic and environmental challenges.

1.4 Importance and necessity

As mentioned before, high quality information in a standard framework should be collected in water resources planning and management. In fact, unavailability of appropriate information may lead to uninformed discussions as stakeholders do not have access to a basis for evaluating unbiased and incorrect conditions. Consequently, the efficient planning can be also inaccurate and unreliable based on stakeholders' data basis and approximately impossible. The beliefs based information often lacks the scientific aspect of water management and thus have negative impact on land use and water management.

It should be noted that the importance of water accounting arises from the importance of access to high-quality information towards effective decision making related to water management. In fact, water accounting provide access to measurements, monitoring and effective and proper reporting systems which can help build confidence in public and investor sectors for water trading, water extraction, environmental management and other outcomes of public benefit (Godfrey and Chalmers, 2012).

Furthermore, in water accounting framework current status of water resources and per-capita water consumption in different user sectors can be evaluated. The water use and pollution can be recognized in different services and sectors which use water as an input or a good. Finally, a systematic and organized framework will be achieved which results in efficient water governance and sustainability in water resources management.

1.5 Methodology

From water management viewpoint, the water accounting methodologies should be applicable at a river basin scale or at a water resource system scale. The water accounts should contain complete information about relevant water flows and

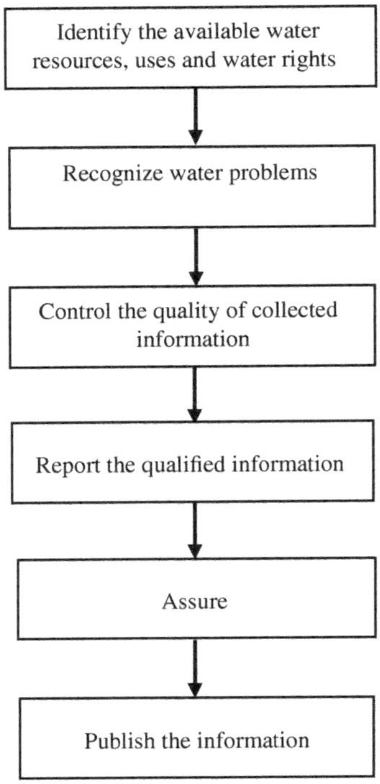

FIGURE 1.1 Flowchart of water accounting process.

storages for water users in the accounting domain. Finally, in order to facilitate generalized use, along with information comparison and transfer, the water accounting methodologies should set up standard procedures for calculating and presenting the water accounts. The flowchart of water accounting process is presented in Fig. 1.1.

Based on different water management strategies, the information and solutions required to meet these goals are different. Consequently, different water accounting systems have been developed for water measuring and reporting at an international scale. These four systems include general purpose water accounting (GPWA), system of environmental-economic accounting for water (SEEAW), international water management institute for water accounting (IWMIWA) and water footprint accounting (WFA) (Godfrey and Chalmers, 2012; FAO, 2017; UNSD, 2012; Tello, 2016).

1.5.1 General purpose water accounting

The GPWA is generally derived of financial accounting focused on general information and reports from a wide range of decision makers. Provision of useful information to water users is one of the main goals of this system along with information for decision making and decision evaluation about water resources allocation. The reports presented in this system include physical turnover statement which similar to statement of cash turnover in financial accounting. This type of statement indicates that how the water availability will change during the reporting period. There are other parts in these reports including financial statements based on common and traditional balance sheet, water income statement, and statement of change in water income (Hu et al., 2013). Water income statement reports the water provision commitment considering rights of stakeholders or any claim about water delivery. The statement of change in water income also indicates the changes in water income during the reporting period. Conceptual statement, sector reporting, additional notes about accounts are other important aspects of GPWA reports.

In fact, considering the application of financial accounting experiences in water reporting resulted in a type of water accounting named standard based water accounting. This is embedded in the framework of GPWA. In other word, GPWA and Standard Based Water Accounting are somehow similar concept to financial accounting in methodology, presentation, and have been formed based on the water accounting standards.

The outstanding aspect of GPWA which distinguishes this approach with other water accounting approaches is the nature of GPWA that emphasizes data preparation for stakeholders. In fact, these data are provided in order to make decisions including evaluating the response to water management and economic, environmental or social allocation outcomes. It should be noted that GPWA is a powerful tool for considering water markets, natural events or reporting on disasters like flood (Chalmers et al., 2012b).

1.5.2 System of environmental-economic accounting for water

The SEEAW was developed by United Nations Statistics Divisions (UNSD) between 2004 and 2007. In fact, this approach brings together experiences and measures of different countries and international organizations in a water accounting framework. In other word, SEEAW is a conceptual framework for organizing water related environmental and economic data. In this approach the key hydrological and economic concepts are explained and presented in a group of tables which reveal the interaction between water, environment and economy. This approach presents data around environmental, economic and social indices like the processes take place in financial accounts.

SEEWA presents a structure including standard tables in order to report hydrologic and economic data. These tables indicate interrelation among water, environment and economy as well as among different economic components. The environmental and economic data are not the only information that is considered in this system. There are other complementary tables which cover social aspects.

The first groups of tables and accounts in SEEWA belong to physical supply and use. Physical supply and use related accounts describe the interaction between environment and economic. In other word, this process focus on (1) water cycle in environment which includes natural flows, surface water resources and groundwater, (2) flows from environment to economy which includes water abstraction from environment by economy for meeting consumption and production goals, (3) inter economy flows which includes water transfer among economy units and (4) return flows from economy to environment which includes water discharge from production economy units to the environment (UNSD, 2012; Pedro-Monzonis, 2016; FAO, 2017).

In order to show the mentioned accounts, some standard tables have been presented in SEEAW frameworks. In these tables, economic activities have been categorized according to international standard industrial classification of all economic activities (ISCI). Based on this classification:

- ISCI 1–3 include agriculture, forestry and fishing
- ISCI 5–33 and 41–43 include mining, manufacturing and construction
- ISCI 35 includes electricity, gas and air-conditioning
- ISCI 36 includes water collecting, treatment and supply
- ISCI 37 includes sewage
- ISCI 38–39 and 45–99 include service industries

General description of physical supply, accounts and flow transfers within economy have been illustrated in Table 1.1 and Table 1.2, respectively.

Another account which plays an important role in SEEAW framework is the water asset accounts. These accounts consist of all the storage status of water resources including surface water, groundwater, and soil water. Furthermore, the increase and decrease in the storage of mentioned resources are stated in these accounts. The details of asset accounts have been described in Table 1.3.

All the information that are presented in SEEAW standard reporting include water resources flow and stocks in the environment, the pressure on environment including water abstraction, wastewater discharge and release to environment, water supply and use in the production process and household sector, water reuse in economic sectors, and all the costs and charges relating to water distribution and services. Payments relating permission of access to water withdrawal or wastewater disposal, financing costs relating to collecting and distribution, hydraulic structures relating to investments and water resources quality are included in the report (UNSD, 2012).

In fact, the SEEAW presents the mentioned accounts along with water resources economic valuation. Furthermore, accounts of water distribution in different regions are considered. On the other hand, water resources storage will be recorded based on the water volume (physical criteria) and water amount and its changes during the reporting period. SEEAW helps information provision at a geographic scale such as local level to river or national and international level (Chalmers et al., 2012a).

TABLE 1.1 Physical (a) supply and (b) use (UNSD, 2012).

		Industries (by ISIC categories)									
		1–3	5–33 41–43	35	36	37	38, 39, 45–99	Total	Households	Rest of the world	Total
From the environment	1. Total abstraction (=1.*a* + 1.*b* = 1.*i* + 1.ii)										
	1.a. Abstraction for own use										
	1.b. Abstraction for distribution										
	1.i. From water resources										
	1.i.1 Surface water										
	1.i.2 Groundwater										
	1.i.3 Soil water										
	1.ii. From other resources										
	1.ii.1 Collection of precipitation										
Within the economy	2. Use of water received from other economic units										
	3. Total use of water (=1 + 2)										
(a)											

TABLE 1.1 (*continued*)

		Industries (by ISIC categories)									
		1–3	5–33 41–43	35	36	37	38, 39 45–99	Total	Households	Rest of the world	Total
Within the economy	4. Supply of water to others economic units										
	Of which:										
	4.a. Reused water										
	4.b. Waste water to sewerage										
To the environment	5. Total return (=5.*a* + 5.b)										
	5.a. To water resources										
	5.a.1 Surface water										
	5.a.2 Groundwater										
	5.a.3 Soil water										
	5.b. To the other sources										
6. Total supply of water (=4 + 5)											
7. Consumption (=3 + 6)											
(b)											

TABLE 1.2 Matrix flows of water within economy (UNSD, 2012).

		Industries (by ISIC categories)								
			5–33				38,39	Households		Supply of water to other
Supplier	**user**	**1–3**	**41–43**	**35**	**36**	**37**	**45–99**	**Total**	**Rest of the world**	**economic units**
Industries	1–3									
(by ISIC categories)	5–33, 41–43									
	35									
	36									
	37									
	38,39, 45–99									
	Total									
Households										
Rest of the world										
Use of water resources from other economic units										

TABLE 1.3 Water asset accounts (UNSD, 2012).

	EA.131 Surface water				EA.132 Groundwater	EA.133 Soil water	Total
	EA.1311 Artificial reservoirs	EA.1312 Lake	EA.1313 River	EA.1314 Snow, Ice and Glaciers			
1. Opening Stocks							
Increase in stocks							
2. Returns							
3. Precipitation							
4. Inflows							
4.a. From upstream territories							
4.b. From other resources							
Decrease in stocks							
5. Abstraction							
6. Evaporation/ Evapotranspiration							
7. Outflows							
7.a. To downstream territories							
7.b. To other resources in the territory							
8. Other changes in volume							
9. Closing Stocks							

In other word, SEEAW framework consists of two groups of accounts. One group includes empirical experience in compiling them in different countries such as physical accounts of supply and uses, economic and hybrid accounts and asset accounts. The second group is rather in the test step. The lack of experience and scientific knowledge cannot make any commitment about their concept or implementation yet. Quality accounts, water valuing, social aspects and impact of natural disasters categorized in the second group (UNSD, 2012). In fact, different steps of implementing the framework of SEEWA include data collection, categorize and organization, summarizing and reporting. These steps should be implemented and evaluated as the interaction among environmental, economic and social sectors.

1.5.3 International water management institute for water accounting

International water management institute (IWMI) plays an important role in water, soil and environment improvement focusing on countries' development. In fact, IWMI has developed a water balance-based water accounting. Components of water balance including inflows and outflows are categorized into water accounting components including net inflow, process use, non-process depletion, committed outflow and uncommitted outflow (Molden and Sakthivadivel, 1999; Karimi et al., 2013). This structure which is known as international water management institute water accounting, here after IWMIWA, is a method in order to determine the outflow per unit of water depleted effectively. It should be noted that IWMIWA can be applied at various scales including basin scale, field scale and service scale. Although, it was originally developed for irrigation schemes at a basin scale, its application has been promoted gradually for basin analysis. In recent years, it is considered as a useful tool for water conservation and efficient productivity solutions. This system provides information about water supply and water use and develops an interactive relationship between water use and economy, available water, use of water in different sectors and value of water use (Karimi, 2013).

In recent years, water accounting plus (WA+) is recommended to develop IWMIWA and updating its concepts. WA+ is a comprehensive tool for assessment of water depletion and productivity in a watershed. In fact, it presents an analytical framework in order to summarize water resources and their management in complex river basins. The structure of WA+ goes beyond the surface water accounting and withdrawal. The unique characteristic of WA+ is incorporation of watershed processes, the role of land use, recognition of manageable flows and differences between gross and net flows. Access to sets of indicators based on information collection process are useful for decision making processes and water resources management solutions.

In fact, the structure of water accounting starts from precipitation and continues with changes in surface and groundwater storages. In WA+ framework, reports are presented about water depletion in a basin using spatial distribution of evapotranspiration rates. Furthermore, information about the part of water depletion and water withdrawal are also presented. Ultimately, WA+ can offer sets of performance indicators. The users also can add excessive indicators based on their personal interests.

On the other hand, WA+ applies satellite measurements as a main data source instead of hydro-climatic measurements. WA+ can also present a strategic viewpoint of possible access to water resources security, resiliency in drought condition and climate change. The WA+ framework also includes, the impact of interactions including water reallocation, reduction in groundwater withdrawal, deficit irrigation, modernize irrigation systems, water storage and retention, wastewater treatment, water productivity improvement, urban extension, changing cropping pattern. All the information provided by WA+ can be presented in 4 sheets in which the state of water management in complex basins can be summarized. The understanding and implementation of these sheets are not complex. The components of these sheets include resource base, evapotranspiration, productivity and withdrawal.

The aims of resource base consist of hydrology, management, utilizable flows, water security and sustainability. In evapotranspiration sheet, the goals of beneficial and non-beneficial flows are accounted. On the other hand, the goals of productivity sheet are biomass returns, carbon sequestration and food security. Eventually, the withdrawal sheet presents the aim of management, regulation, allocations. Generally, WA+ is an efficient and valuable tool for development, planning and management of water resources particularly in the basin which faces water shortage.

1.5.4 Water footprint accounting

As mentioned in many water-related studies, water plays an important role in production procedures of numerous products and goods. The estimation of water amount required for production process is useful information in water resources management. In fact, the amount of water required in production process provide valuable information of hidden water use for water planners, policy makers and managers.

Regarding mentioned issues, a structure is required to quantify the water footprint for a process or product in a particular geographical region in the spatial and temporal dimensions. Water footprint accounting is the mentioned structure which possesses this ability. Water footprint accounting provides a framework to insert final consumers, retailers, various industries into the stage of water governance and accepting the responsibility in planning and policy making. In fact, clarifying the relationship between water use and various products leads to more efficient implementation of water governance strategies and potential of changes for efficient water use (Chalmers et al., 2012a; Godfrey and Chalmers, 2012).

The concept of water footprint was presented in 2002 for the first time. There were lack of practical and scientific experiences about water use in supply chain in the advent of water footprint concept, and in the effect of water use in production process. Gradually, the understanding and data on the hidden water user in production process increase by presenting water footprint concept in the scope of water resources management. In fact, the resulting understanding and knowledge provide a basis for better water management.

In a general viewpoint, different goals can be considered in water footprint accounting according to the scope in which water footprint assessment is important. In other word, the bound of water footprint accounts evaluation and consideration depends on the related goals. For example, in the case of increasing public awareness, global and regional average estimations are enough for products' water footprint. Furthermore, it should be noted that incorporating factors other than water is an important issue in water footprint assessment. As mentioned before, the water footprint is assessed considering various goals in different spatial and temporal levels (Morillo et al., 2014; Manzardo et al., 2016).

In water footprint accounting framework, three levels are considered in information provision. In first level, the spatial and temporal dimension for evaluations is global and annual, respectively. In this level, the sources which provide required data about water use for water accounting are existing studies, database of water use and pollution. This information can be applied in order to increase the knowledge, recognition components related to water footprint and development of the global water use planning.

In the second level, the temporal and spatial dimensions are annual or monthly and national/regional watershed level. The required information in this level are collected similar to the previous level. The only difference is the spatial aspect of data which is national/regional watershed level. Resulting information is applied in order to recognize the spatial variations. On the other hand, this information is a basis for allocation decisions.

Eventually, the third level is carried out in a local or field-scale framework at the monthly or daily temporal scale. Required information in this level are empirical information or the best estimates of water use or pollution that are collected over the year. This information is then applied for water footprint sustainability assessment or to provide strategies for water footprint reduction (Chalmers et al., 2012a; Godfrey and Chalmers, 2012; Manzardo et al., 2016).

Although information provision level is one of the important parts in the water footprint accounting, there is a general classification in water footprint accounting which plays an important role in accounts preparation. This classification includes water footprint accounts related to a process, a product, a consumer and a business.

In the water footprint accounting of a process, the main issue is the focus on blue, green and grey water. In fact, this framework considers not only blue water but also covers grey and green water. In general, the blue water footprint accounting focuses on surface and groundwater consumption in a supply chain of a product. The blue water footprint measures the amount of available water use in a certain period. On the other hand, green water footprint, the volume of green water, is used during a production process. The production process can be agricultural or forestry. In other word, green water footprint can be described as the total amount of evapotranspiration and water used in the production process. The third type of water is considered in water footprint accounting is grey water which indicates the water pollution because of a process.

Other water accounts which embedded in this classification are water footprint of a product. Water footprint of a product is estimated through water consumption and pollution during all the production cycle steps. It is required to specify the production system in order to estimate the water footprint of a product.

Water footprint of a consumer is described as the total volume of fresh water which is used and polluted by consumer for services and goods production. This type of water footprint is calculated through adding direct water footprint of consumers and indirect water footprint of consumers. Direct water footprint is related to the water use and polluted in houses or gardens. On the other hand, indirect water footprint is water use or polluted related to the goods production and services by consumers.

A business water footprint is the last concept considered in water footprint concept. Water footprint of a business is the total volume of fresh water used directly or indirectly for running or supporting of a business. Direct water footprint which is also named operational water footprint of a business is the volume of fresh water used or polluted during an operational process in a business. For example, this part includes water consumption in a product, water consumed or polluted in a washing process or water thermally polluted through cooling process. The water used in kitchens, toilets, gardening or washing clothes is also considered as operational water footprint of a business. Water footprint of supply-chain or indirect water footprint of a business is the volume of fresh water consumed or polluted in order to produce goods or services which is formed through inputs of products. Examples of this type of water footprint of a business include water footprint of infrastructures, materials and energy for general uses and water footprint of ingredients of a good which is bought by a company.

In a general framework, all the parts which have been discussed about water footprint accounting concepts should be linked to input-output modeling and material flow in order to integrate the different concepts of water footprint and related methods together in an analytical framework (Godfrey and Chalmers, 2012; Zhan-Ming and Chen, 2013; Morillo et al., 2014)

1.6 Examples

Water accounting have been developed in many countries. Despite a general framework and structure of water accounting, accounts in each country are

TABLE 1.4 Water consumption in industrial and household sectors (Gigalitres) (Vardon et al., 2007).

	Australia (Total)	NSW/ACT	Vic.	Qld	SA	WA	Tas.	NT
Agriculture	16,660	7322	3725	3454	1302	565	22	70
Forestry	27	4	4	2	1	12	1	–
Mining	401	52	7	109	12	195	21	5
Manufacturing	866	179	249	181	85	83	79	9
Electricity and gas	1,688	59	1536	70	2	19	–	1
Water supply	1,794	676	745	216	24	114	9	9
Other	1,292	453	402	178	40	176	25	21
Household	2,181	679	472	501	181	245	59	45
Total	24,909	9,424	7,140	4,711	1,647	1,409	417	160

different compared to others. In order to better understand the steps of water accounting process and application of different water accounting approaches, the following examples are presented as case studies in order to overcome the water related problems.

1.6.1 Example I

Australia is a pioneer in using water accounting for water resources management. In recent years, due to reduction in precipitations and over consumption in household, industrial and irrigation sectors, severe water shortage was observed in this country. In this regard, provision and access to a mechanism like water accounting can provide a wide view to water shortage modeling scenarios and its management. This can provide a wide range of physical, economic, and environmental information in an integrated framework for planners, managers, engineers, and policy makers.

The establishment of water accounts and information has been carried out by Australian Bureau of Statistics (ABS). Some national sectors including committees, experts and project management teams assisted ABS in this case. The first water accounts presented by ABS restricted to the water supply and use tables. These accounts which presented in the year 2000 were based on the information of 1993–94 and 1996–97 time periods. A sample of water accounts which have been collected by ABS in 2001 has been presented in Table 1.4. In this table the first row and column dedicated to Australian states and territories and user sectors in these states, respectively.

The next generation of water accounts was provided in 2004. These second edition reports had a different framework compared to the first ones and covered more details. The reference time period was 2000–01. In compiling the information, more than one hundred data sources were evaluated. The accounts provided by the ABS helped achieving a broader insight into the water use in Australia and the costs and benefits of this usage. Additionally, it also helped in an efficient water planning and management in severe water shortage situation.

1.6.2 Example II

SEEAW is one of the most applicable water accounting approaches in water related problems. A study carried out in Jucar River Basin in Spain is a good example of water accounting application in water resources management.

Jucar River Basin faced water shortage, persistent drought and overexploitation of groundwater resources. In order to manage the stated challenges, a modeling method based on three computational stages was applied. The first stage was the development of a hydrological model focusing on basin's water resources. Then they obtained hydrological information was applied to simulate water allocations to different users using decision support system (DSS). Finally, water allocations were linked to the economic costs. In all three stages, the asset accounts and physical supply and use information were presented in standard tables based on SEEAW. To provide the standard tables of SEEAW, the water balance model was integrated with rainfall-runoff model's variables using combination of DSS and AQUAACCOUNTS tool. The asset accounts and physical supply, and the tables were derived based on the results of combined model.

The results are presented in Table 1.5, Table 1.6, Table 1.7, and Table 1.8 based on Table 1.2, the main water users in Jucar River Basin were agrarian requirements and energy production based on the amount of water allocation. The surface water was the main sources for these abstractions.

Table 1.6 represents the source of reused water and the volume of return water from different users to water resources. Based on these results, household, and service industries were the main source of reused water. The other results which are obtained in this study were the asset accounts which are presented in Table 1.7. In this table the initial storages of all the water resources in the basin including surface water, groundwater and soil water, the changes in storages (increase or decrease) originating from precipitation, inflow, returns or abstraction, evaporation, evapotranspiration or outflow as well as the final storage in resources are presented. Table 1.8 also indicated the flow cycles between water resources in the environment.

In the final section of this study, some indicators were derived according to the obtained results which are presented in Tables 1.5 through 1.8. Based on the asset accounts table, external renewable water resources (ERWR) were 0 hm^3/year from Table 1.7 (in row 4.a). In fact, ERWR consisted of groundwater transfers and river runoff from other countries. On the other hand, from Table 1.8 (in row EA.133), the Internal Renewable Water Resources (IRWR) obtained in this

TABLE 1.5 Physical use table (Pedro-Monzonis et al., 2016).

Physical use (hm^3)		Industries (by ISIC categories)									
		1–3	5–33 41–43	35	36	37	38,39 45–99	Total	Households	Rest of the world	Total
From the environment	1. Total abstraction	7772 95 6590 114 14,571							376		14,947
	(=1.*a* + 1.*b* = 1.*i* + 1.ii)	7772 95 6590 114 14,571							376		14,947
	1.a. Abstraction for own use	7772 95 6590 114 14,571							376		14,947
	1.b. Abstraction for distribution	1236 0 6590 37 7863							122		7986
	1.i. From inland water resources	1062 95 77 1234							254		1488
	1.i.1 Surface water	5474 5474							3		5474
	1.i.2 Groundwater	1 1									4
	1.i.3 Soil water										
	1.ii. From other resources										
	1.ii.1 Collection of precipitation										
	1.iii Abstraction from the sea										
Within the economy	2. Use of water received from other economic units	150 6 9 165							31	0	196
	2.a. Reused water	109 6 115							31		115
	2.b.Transfers from other territories	41 9 50									81
	3. Total use of water (=1 + 2)	7922 101 6590 0 124 14,736							407	0	15,143

TABLE 1.6 Physical supply table (Pedro-Monzonis et al., 2016).

Physical supply (hm³)		Industries (by ISIC categories)									
		1–3	5–33 41–43	35	36	37	38,39 45–99	Total	Households	Rest of the world	Total
Within the economy	4. Supply of water to others economic units	0 0 0 0 0 27 27							88	81	196
	Of which:	27							88	81	115
	4.a. Reused water	835 81 6590 0 0 91 7598							301	0	81
	4.b. Waste water to sewerage	726 81 6590 0 0 57 7454							187		7898
To the environment	5. Total return (=5.a + 5.b)	271 81 6590 57 6999							187		7641
	5.a. To water resources	455 0 0 455							0		7186
	5.a.1 Surface water	109 35 144							113		455
	5.a.2 Groundwater										257
	5.a.3 Soil water										
	5.b. To the other sources										
6. Total supply of water (=4 + 5)		835 81 6590 0 0 118 7624							389	81	8094
7. Consumption (=3–6)		7087 20 0 0 0 5 7112							18	−81	7049
7.a Losses from evaporation											
7.b Losses in distribution not because of leakages											

TABLE 1.7 Water asset accounts (Pedro-Monzonis et al., 2016).

	EA.131 Surface water						
	EA.1311 Artificial reservoirs	EA.1312 Lake	EA.1313 River	EA.1314 Snow, Ice and Glaciers	EA.132 Groundwater	EA.133 Soil water	Total
1. Opening Stocks	1510 25 0				92,308	1405	95,248
Increase in stocks							
2. Returns	1	54 7389			455		7898
3. Precipitation						20,798	20,798
4. Inflows	4349 303 7283	0			3277	0	15,211
4.a. From upstream territories							
4.b. From other resources	4349 303 7283				3277	0	15,211
Decrease in stocks							
5. Abstraction	49 1 7936				1488	5474	14,947
6. Evaporation/Evapotranspiration	80 66					11,415	11,562
7. Outflows	3768 236 6784	0			2815	3909	17,491
7.a. To downstream territories							0
7.b. To the sea	146 1624				509		2279
7.c. To the resources in territory	3768 89 5139				2306	3909	15,211
8. Other changes in volume							
9. Closing Stocks	1524 27 0				92,290	1405	95,246

TABLE 1.8 Matrix of flows between water resources (Pedro-Monzonis et al., 2016).

	EA.131 Surface water						
	EA.1311 Artificial reservoirs	EA.1312 Lakes	EA.1313 Rivers	EA.1314 Snow, Ice and Glaciers	EA.132 Groundwater	EA.133 Soil water	Outflows to other resources in the territory
EA.1311 Artificial reservoirs	3581				187 3768		
EA.1312 Lakes	89				89		
EA.1313 Rivers	4349 303				487 5139		
EA.1314 Snow, Ice and Glaciers	2306				0		
EA.132 Groundwater	1306				2306		
EA.133 Soil water	4349 303 7283 0				2603 3909		
Inflows from other resources in the territory					3277 0 15,211		

case was 3909 hm^3/year. IRWR was the generated resources from precipitation. Generally, the total natural renewable water resources indicator (TNRWR) was obtained considering IRWR and ERWR. In other word, the sum of IRWR and ERWR results in the amount of TNRWR which was equal to the maximum theoretical amount of available water. Another indicator derived from the results was water exploitation index (WEI) which was applied to assess the abstractions and degree of water stress. The equation which resulted in WEI was the division of mean annual total abstraction of freshwater by mean annual total renewable freshwater resource. In this case study the amount of WEI obtained was 242 percent which reveals a high degree of water stress in the Jucar river basin.

Water consumption index (WCI) indicated the ratio between water consumption and TNRWR which was 40 percent in this case study. Furthermore, ratio of desalinated water, reused water, total water used derived from transfer water were 0.02 percent, 0.8 percent and 0.5 percent, respectively.

The findings indicated the applicability of derived accounts using SEEAW in water management, planning and policy making in water stressed river basins. In fact, the water accounts and water accounting tables provide a report which helped managers and planners to achieve a broad insight into the status of water resources availability and uses in the basin to present efficient solutions and to conquer the water shortage challenges (Pedro-Monzonis et al., 2016).

1.7 Summary

Water accounting is a hierarchical process considering hydrology, water allocations, water management, land use management and eventually quantifying the benefits from water use and reporting relating to evaluating water supply, use and decision making.

On the other hand, the concept of water accounting is not solely focused on presenting current state of water resources but also present information about problems, constraints and opportunities to promote decision and policy making process in water resources management. Furthermore, assembling information resulting from water accounting play a remarkable role in effective investments in water resources issues. This also helps understanding the impact of the water related issues in the future considering the adjustments with climate change impacts and the change in surface and groundwater resources.

This framework integrates the data relating environmental and economic aspects of water supply and use. Therefore, it results in an improvement in water management. In fact, the core of integrated water management has considered environmental, economic, and social aspects. In other word, the basis of integrated water management is considering water as an inseparable part of the ecosystem, natural resources and its role as an economic and social good. The assessment and evaluation of mentioned aspects can be possible in water accounting framework. It should be noted that the water accounting can improve the integrated water management as the only approach that can integrate the economic related accounts with water supply and use for quantitative analysis.

Furthermore, the information that managers use in planning are based on the information that are received from end-users. However, this cannot be applied for economic analysis as end-users related categories rarely considered the economic activities categories which applied for national accounts. In fact, water accounts link water information directly to economic accounts unlike other water information databases.

As a result, without access to reliable information, assessment and discussion around water management problems is impossible. In fact, the key output of water accounting which plays an important role in decision and policy making process is a basis which is acceptable for all stakeholders involving in planning and decision making.

References

Bazrafshan, O., RamezaniEtedali, H., GrkaniNezhadMoshizi, Z., Shamili, M., 2019. Virtual water trade and water footprint accounting of Saffron production in Iran. Agric. Water Manage. 213, 368–374.

Berger, M., Van der Ent, R., Eisner, S., Bach, V., Finkbeiner, M., 2014. Water accounting and vulnerability evaluation (WAVE): considering atmospheric evaporation recycling and the risk of freshwater depletion in water footprinting. Environ. Sci. Technol. 48, 4521–4528.

Borrego-Marin, M., Gutierrez-Martin, C., Berbel, J., 2016. Estimation of cost recovery ration for water services based on the system of environmental-economic accounting for water. Water Resour. Manag. 30, 767–783.

Burgia, I., Contrafatto, M., Rusconi, G., 2017. Water accounting economia aziendale. Electron. J. Manag. 3, 17–27.

Chalmers, K., Godfrey, J., Potter, B., 2012a. Discipline-informed approaches to water accounting. Aust. Account. Rev. 62 (3), 275–285 22.

Chalmers, K., Godfrey, J.M., Lynch, B., 2012b. Regulatory theory insights into the past, present and future of general purpose water accounting standard setting. Account. Audit. Account. J. 25 (6), 1001–1024.

FAO, 2017. Water Accounting and Auditing-A Source Book. Report No. 43, Food and Agriculture Organization of the United Nations, Rome, Italy.

Godfrey, J.M., Chalmers, K., 2012. Water Accounting: International Approaches to Policy and Decision Making. Edward Elgar Publishing, Massachusetts, USA.

Hu, H., Jiang, Y., Jin, Q., Islam, J., 2013. The effects of standardized water accounting: evidence from the water accounting standards in Australia. Life Sci. J. 10 (3), 427–432.

Karimi, P., Bastiaanssen, W.G.M., Molden, D., 2013. Water accounting plus (WA+) a water accounting procedure for complex river basins based on satellite measurements. Hydrol. Earth Syst. Sci. 17, 2459–2472.

Karimi, P., Bastiaanssen, W.G.M., 2015. Spatial evapotranspiration, rainfall and land use data in water accounting- part 1: review of the accuracy of the remote sensing data. Hydrol. Earth Syst. Sci. 19, 507–532.

Lange, G.M., Mungatana, E., Hassan, R., 2007. Water accounting for the orange river basin: an economic perspective on managing a transboundary resource. Ecol. Econ. 61, 660–670.

L. Burritt, R., L.Christ, K., 2017. The need for monetary information within corporate water accounting. J. Environ. Manage. 201, 72–81.

M. Manzardo, A., Loss, A., Fialkiewicz, W., Rauch, W., Scipioni, A., 2016. Methodological proposal to assess the water footprint accounting of direct water use at an urban level: a case study of the Municipality of Vicenza. Ecol. Indic. 69, 165–175.

Molden, D., Sakthivadivel, R., 1999. Water accounting to assess use and productivity of water. Int. J. Water Resour. Dev. 15 (1/2), 55–71.

Momblanch, A., Andreu, J., Paredes-Arquiola, J., Solera, A., Pedro-Monzonis, M., 2014. Adapting water accounting for integrated water resource management. The Jucar water resource system (Spain). J. Hydrol. (Amst.) 519, 3369–3385.

Morillo, J.G., Rodriguez Diaz, J.A., Camacho, E., Montesinos, P., 2014. Linking water footprint accounting with irrigation management in high value crops. J Clean Prod 87, 594–602.

M.Turner, G., M.Bayens, T., C.McInnis, B., 2010. A water accounting system for strategic water management. Water Resour. Manag. 24, 513–545.

Pedro-Monzonis, M., Solera, A., Ferrer, J., Andreu, J., Estrela, T., 2016. Water accounting for stressed river basins based on water resources management models. Sci. Total Environ. 565, 181–190.

Peranginangin, N., Sakthivadivel, R., R.Scott, N., Kendy, E., S.Steenhuis, T., 2004. Water accounting for conjunctive groundwater/surface water management: case of Singkarak-Ombilin River basin, Indonesia. J. Hydrol. (Amst.) 292, 1–22.

Tello, E., Hazelton, J., Cummings, L., 2016. Potential users' perceptions of general purpose water accounting reports. Account. Audit. Account. J. 29 (1), 80–110.

Tilmant, A., Marques, G., Mohamed, Y., 2015. A dynamic water accounting framework based on marginal resource opportunity cost. Hydrol. Earth Syst. Sci. 19, 1457–1467.

UNSD (United Nations Statistic Division), 2012. SEEA-Water System of Environmental-Economic Accounting for Water. United Nations publication, New York, USA.

Vardon, M., Lenzen, M., Peevor, S., Creaser, M., 2007. Water accounting in Ausralia. Ecol. Econ. 61, 650–659.

Xu, M., Li, C., Wang, X., Cai, Y., Yue, W., 2017. Optimal water utilization and allocation in industrial sectors based on water footprint accounting in Dalian City, China. J. Clean. Prod. 176, 1283–1291.

Zhan-Ming, C., Chen, G.Q., 2013. Virtual water accounting for the globalized world economy: national water footprint and international virtual water trade. Ecol. Indic. 28, 142–149.

Chapter 2

Water economics

Sahar Mohammad-Azari[a], Omid Bozorg-Haddad[a] and Asim Biswas[b]
[a] *Department of Irrigation & Reclamation Engineering, Faculty of Agricultural Engineering & Technology, College of Agriculture & Natural Resources, University of Tehran, Karaj, Tehran, Iran.* [b] *School of Environmental Sciences, University of Guelph, Guelph, Canada*

2.1 Introduction

Water, a scarce resource, is not only considered an input to the economic activities including agriculture, industry and hydropower generation, but also plays an important role in ecosystems like rivers and wetlands. However, hydrologic changes, population growth and increasing demand of different user sectors cause problems in water quantity (e.g. shortage) and water quality (e.g. pollution). All these problems as externalities result in negative impact on water users. Thus, water resources management requires consideration of economic fundamentals and instruments. Economics, the science of scarce resources allocation or selection based on limitations, and economic instruments that can balance supply and demand and deal with externalities play key roles in solving challenges in water related issues (C.Griffin, 2006; F.Booker et al., 2012).

In integrated water management, others are also considered in the framework of water economics focusing on water demand management, water instrument applications in water resources management, value of water, efficient water allocation and economic evaluation of water projects. These concepts are discussed in details later this chapter. Some researches considering economic principles and instruments in water resources management scope are also discussed.

A.Ward and Michelsen (2002) evaluated the concepts and policy applications in determination of economic value of water. They reviewed the effective factors of the economic value of water including spatial and temporal factors. They also presented the difference between total, average and incremental water values.

Sharp (2006) evaluated the water demand management in England and Wales and categorized into three groups including the encouragement of users by individual or common needs, considering water as a right or a commodity and the user passiveness or activeness. This study presented the importance of communications associated with demand management.

Economical, Political, and Social Issues in Water Resources. DOI: 10.1016/B978-0-323-90567-1.00007-3

Hutton et al. (2008) applied a cost-benefit analysis to analyze the impact of water supply and sanitation interventions. They documented the cost-beneficial aspects of water supply and sanitation improvement projects.

Moran and Dann (2008) evaluated the water economic value of implementing the Water Framework Directive in Scotland. They considered the national requirement for economic report which emphasized specification of relative value from use.

Similarly, Rigby et al. (2010) estimated the economic value of irrigation water under supply uncertainty in Spain. They indicated the impacts of risk on the preference of farmers towards water allocations.

Kelman and Kelman (2010) focused on the water allocation and availability in arid regions in water shortage condition as a decision making process. In this regard, a water allocation model based on opportunity cost is discussed. Results showed that the proposed model results in water pricing scheme and compensating rationing policy.

A.Ward (2012) focused on the application of cost-benefit analysis in water resources policies and evaluated the recent developments in this method and application of economic principles in water-related decisions.

Alcon et al. (2013) presented a cost-benefit analysis by incorporating non-market benefits of reclaimed water in agriculture from south eastern part of Spain. The results indicated the combination of water resources was the best option in private and social level considering environmental benefits.

Chang and Wang (2013) integrated system analysis with artificial intelligence techniques to achieve a systematical water allocation scheme for water allocation in northern Taiwan. Results indicated that the capability of proposed water allocation scheme.

Dehnhardt (2014) investigated the influence of German policymakers on environmental cost-benefit analysis. He presented an analytical approach based on political economy and policy to show the response of policymakers against the results of cost-benefit analysis.

Tilmant et al. (2014) focused on the storage in multireservoir systems. They proposed a methodology to assess storage economic value in multireservoir systems considering marginal net benefit of storage and hydroeconomic modeling.

Malm et al. (2015) proposed a combined model of cost-benefit and uncertainty analysis to evaluate the leakage from water distribution systems. The results indicated the superiority of combined model compared with the traditional cost benefit analysis as the new model can consider a high degree of uncertainty in the input data. On the other hand, findings showed the cost-effectiveness of leakage reduction through repair than replacing the broken pipes.

Li et al. (2016) developed an efficient irrigation water allocation model to solve the multi-objective problems considering maximum net benefit of the system and irrigation water use under uncertainty. Findings indicated the capability of the model in decision making for irrigation water allocation. On the other hand, applying the model resulted in helping decision makers to present efficient policies for water resources management.

Mulugeta Degefu et al. (2016) applied cooperative bargaining approach for water allocation in transboundary river basins in scarcity condition. They combined bankruptcy theory combined with asymmetric Nash bargaining solution concept and indicated that the proposed method provided useful information about water allocation outcomes under scarcity condition.

Xiao et al. (2016) investigated the applicability of a comprehensive cooperative water allocation framework to evaluate impact of water demand management plan on a water system. The net benefits of the water system were shared among different users and stakeholders through a modified cooperative game. The results showed the impact of cooperation among users on increasing the net benefits. On the other hand, they showed the applicability of demand management in arid and semi-arid regions.

Escriva-Bou et al. (2017) applied a hydro-economic model to assess the economic value of climate change adaptation strategies for water management in Jucar Basin, Spain. Additionally, economic equity of the system was assessed using a new economic loss index. The results indicated the vulnerability of system to long-term global change.

Rey et al. (2017) evaluated the performance of economic instruments in water allocation in five European countries including England, Italy, Spain and Netherlands. Findings can help the improvement of future reforms effectiveness and results in more efficient water allocation.

Porse et al. (2018) applied a comprehensive approach to evaluate economic implications in Los Angeles in the case of shifting to local water reliance. Findings showed that updating economic methods can help to achieve local reliance through institutional reforms, drought-tolerant landscapes and reallocated groundwater rights.

Xiao et al. (2018) applied an agent-based modeling approach to evaluate water users' response to water demand management in a river basin. Results indicated that water demand management strategies can help agents on their economic return and have positive impact on ecological perspectives.

A.Revollo-Fernandez et al. (2019) assessed the economic value of water for industrial sectors in Mexico using the concept of value of marginal products. They considered production elasticity of water as an input to manufacturing industry. Results indicated the need of information on water economic value for investing in new water supply projects and the water resources allocation across sectors.

D.Delorit et al. (2019) developed an agro-economic approach to evaluate the water resources demand management for water rights markets. In this case, the value of water was assessed, and optimal strategies were proposed to add to farm-scale profitability.

Koutiva and Makropoulos (2019) applied an agent-based model to simulate households' water demand behavior. The impacts of water demand management strategies were also analyzed on the households' behavior in drought condition. The results showed the applicability of an agent-based model to help decision makers in understanding water demand management measures.

Mahmoudzadeh Varzi et al. (2019) applied an optimization model for water allocation in agricultural sector under deficit irrigation. The results indicated the impact of crop price on the farmer's income.

Tortajada et al. (2019) evaluated impacts of water demand management strategies including pricing and non-pricing measures on the households' efficient water use in five cities of Spain. Results indicated the greater impact of non-pricing measures on water consumption in comparison with pricing measures. Additionally, users' education, awareness and promotion water conservation technologies played key role in efficient water consumption.

Li et al. (2020) developed a stochastic optimization approach for agricultural water allocation considering uncertainties and risk in Hulan River, China. They allocated the scarce water resources to agricultural sector under different risk levels. Results showed the applicability of developed approach in agricultural water allocation.

Yokomatsu et al. (2020) applied a multi-sector, multi-region economic growth model for investigating the effects of drought conditions on water economic value and water markets. In this case, the water rent was also analyzed through formulating precipitation and river water availability along with agricultural production functions in Pakistan during drought periods. This research introduced a way for evaluating water resources management policies.

2.2 Terms and definitions

Water economics, a branch of economics, is based on economic theory of resources and environment and deals with the water resources related problems and challenges in an institutional structure. In economic theories, the focus is generally on the limitations and useful strategies to deal with them. Water economic can be an appropriate solution to avoid the challenges and limitations (e.g. quantitative and qualitative).

In fact, water-related problems, and challenges are important and remarkable frameworks for application of economic instruments and theories. They also play an important role in better understanding water resources problems and their management and provide powerful insights. At the same time, involving managers and water planners in economics principals beside engineers help improve decision making and management process in water related problems.

The global attitude toward water economics and water as an economic good was a restricted and inefficient attitude before the 1990s. However, based on the Dublin principles after 1990 and consideration of water as an economic good, the attitudes of authorities, responsible organizations and different water related sectors have been changed toward an efficient and new attitude. In fact, before presenting the Dublin principles, the focus of water resources management was on the supply management as a sole solution in such a way that the demand of user sectors was met completely and the allocation was the base of the management strategies institution (ICWE, 1992; Mc Neill, 1998).

However, in implementing water projects, the revenue returns, and social issues were not considered and the political considerations played an important role. On the other hand, the decisions were made without considering economic aspects and water economics did not play any effective role. In contrast, the integrated and comprehensive water management concept was presented and implemented in different water resources issues as part of the Dublin principles, and the attitudes were modified from inefficient to efficient to focus on economic, social and non-structural aspects instead of solely structural aspects.

All these issues lead to include the economic theories and concepts in to the water resources management framework. In water resources economic structure, first stage requires to recognize and define the most important economic policies, problems, and challenges. In this regard, it should be evaluated that the water resources are limited and economic are not. On the other hand, the standards and criteria of water resources management evaluation must be determined.

Generally, water related limitations in water management framework include natural limitations of water resources, financial limitations, human resources limitation, and environmental limitation. For example, water resources may be affected by natural and climatic conditions, inappropriate economic conditions, or lack of access to experienced and educated human resources (Savenije and Van der Zaag, 2002; Olmstead, 2010);

On the other hand, as it is necessary to maintain part of water resources to meet the wildlife and environmental demands, the availability of all the water resources is somehow impossible. The inefficient water resources allocations and inefficient water use challenges necessitate implementation of water economics framework.

In an economic water management framework, the fundamental problems and challenges of water management must be specified and classified after recognizing limitations and introducing them. These classes include the uses, water losses, demands, selection of water projects and their operation. In fact, it should be considered that the main water allocations in different regions are operation management efficient for different supply sources.

Provided all these are implemented appropriately and are managed in an efficient economic framework, they can help to solve the challenges of water resources management. After recognizing the fundamentals of water resources management, standards and criteria should be organized to evaluate the economics of water resources management. These standards include economic efficiency of water, economic efficiency of water related projects, the relationships between planning water projects and economic planning, and water pricing.

Other concepts are evaluated beside general economic concept for integrated water management in the framework of using economic concepts and economic instruments. These concepts that are part of water resources economics include water demand management, water allocation, economic evaluation of water resources development projects, water pricing and economic water valuing (Green, 2000; C.Griffin, 2006; Olmstead, 2010; Cai and Wang, 2006).

As mentioned previously, problems and challenges in water sector were exacerbated gradually and caused to emerge a new concept called water demand management. In fact, engineering solutions focusing on the water supply management cannot be the only solution when water shortage takes place in different water user sectors. Thus, water demand management became widespread among water planners and managers.

This concept also results in evolution of water resources economics. In water demand management, the focus is on the users and sectors which need water for their basic demands. In this regard, the water demand management helps using water efficiently and thus no need to access more water resources.

Water allocation is another concept that is evaluated in water economics framework. Economic instruments play an important role in efficient water allocation. In fact, an efficient allocation will take place in a supply and demand balance. In this case, the marginal cost of water supply (the supply cost of one more unit of water) is equal to marginal benefit of water use (the benefits of presented goods and services by one more unit of water). Therefore, the marginal cost and benefit is equal to the market price in all uses. Although water resources are applied in different economic and social sectors, it is a non-market resource in different aspects. Even in using water as a commercial good, the market-based prices are not specified. In economic viewpoint it requires to distribute water among different sectors based on demand changes. On the other hand, water must be allocated equitable (Babel et al., 2005; Wang et al., 2008).

Water allocation mechanisms are classified in three general groups including public allocation mechanism, market-based allocation mechanism and user-based allocation mechanism. In public water allocation environmental objectives and equitable water distribution are considered as the main goals other than economic efficiency. This mechanism avoids water users' partnership. In market-based allocation mechanism, water is allocated based on prices determined by supply and demand sides in markets. This mechanism results in efficient water use. On the other hand, the user-based allocation is carried out based on recognition and understanding the society demands (Dinar et al., 1997; Babel et al., 2005; Klocke et al., 2006; Wang et al., 2008; Roa-Garcia, 2014; Zilberman and Schoengold, 2013; Read et al., 2014; Hu et al., 2016).

Therefore, focus on water allocation in water economic framework considering economic principles can help modifying allocation mechanism to achieve equity and optimal allocation for different competitive uses.

Finally, some economic principles and indicators must be considered for implementing any water project. In fact, these will be carried out in the framework of water economic concepts along with engineering evaluations. In this regard, economic outcomes must be recognized and categorized. The outcomes will be quantified and changed to a common denominator. The time value of money will then be entered in the quantification process and in some economic indices framework including net present worth, internal rate of return, net benefit-investment ratio, benefit or cost streams, benefit-cost ratio and payback return to present and make decision about the water-related projects.

The water economic value is defined as an economic concept in water resources management and is determined as the influence on the user utilities in the society. The people utility is also determined based on their preference or their willingness to pay for accessing water. As the water is a resource with high conveyance costs, its value will vary based on the spatial aspects. On the other hand, the water price will be the payments by users for water use and is determined based on different factors including costs of water withdrawal, water distribution and willingness to pay. Thus, in determining the economic value of water, the relation between water resources process and the resulting services and goods must be taken into account for their individual values. (Details are presented in next sections).

2.3 Basic principles and logics

Previously for a long time, the focus of water authorities was on the water-related issues through engineering means. Increasing problems and challenges caused the inefficiency of engineering means for solving water-related problems. In this case, entering water economics principles into water resources management framework resulted in a wide and powerful insight to water-related problem solutions. For example, water demand management is one of the most important principles and concepts in water economics framework. In fact, the water demand management is an instrument for achieving adoption and collaboration between water demand and water availability. For water or money scarcity, the water demand management is a vital strategy for economic development and food security (Kampragou et al., 2010; Puri and Maas, 2019).

In other word, application of management systems is required to control the use in different sectors and thus the expansion to new water resources. For costs, the demand water management requires fewer investments and is considered as an incentive for improving efficiency in agricultural and industrial sectors. In contrast, implementing the demand water management without considering rules in an obvious and understandable way leads to negative social risks and negative impacts on economic developments.

From economic viewpoint, two issues must be considered in water demand management including demand more than available supplies and demand which can be met by the available supplies. In first issue, maintaining the demand in the current level is the main goal. On the other hand, the high investment costs of achieving new water resources or increasing supply must be omitted or reduced. In the second issue, the goal is to reduce water demand towards sustainable supply (Kumar, 2005; Maggioni, 2015; Puri and Maas, 2019).

In economic principles, the limitations-related issues are the most remarkable and thus, the strategies related to limitations are followed. Limitations cause the recognition of the value of resources like water. On the other hand, the value of resources opens the efficiency concept as the economic efficiency is the indicator of logic and efficient use of an economic limited resource.

Beside water demand management, achieving efficient water allocation is one of the important issues water economic principles. Using water in different sectors relies on the opportunity costs. Opportunity costs are the benefits which can be achieved from the different uses of water, but these benefits will be ignored for achieving other benefits. The basic economic logic behind the improvement of water allocation is awareness of decision makers and authorities applying economic instruments.

An important aspect in water allocation and development is economic efficiency. Although, efficient water allocation and distribution have been hardly observed economic, economic efficiency analysis is a useful tool for better understanding of the reasons of inefficient allocation and distribution. Economically efficient water allocation refers to the maximum benefits to society considering allocation and distribution of available water. In this case, benefit refers to the economic benefits to the society.

In other word, economically efficient water allocation causes maximum water value in all the economic sectors. Achieving this goal can occur through water allocation and distribution for valuable uses and avoiding fewer valuable uses. The main economic property of optimal and efficient water allocation includes any other supply and allocation which exacerbate the conditions for at least one person in the society (Savenije and Van der Zaag, 2002; Kumar, 2005; Olmstead, 2010).

There are more economic principles other than economic efficiency in water resources management especially in water allocation and distribution framework. Equity, flexibility, marginal cost, predictability and applicability are among other economic principles. In fact, the main reason of inefficient water uses in different economic user sectors that occurs in the point costs is higher than the benefits to the water management authorities. It should be noted that it requires conveying the water between users and sectors based on changes in demand for achieving the flexibility in water allocation. In fact, water allocation should meet the general and political goals of different social groups and changes the inefficient conditions. It should also be applicable and adoptable. The actual marginal cost should be paid, and the outcomes of water allocation and management must be predictable to achieve the best strategy in water allocation and lessen the unreliability.

The stated terms have been considered as the important ones in water economics concept structure to evaluate and investigate different aspects of water resources management and achieve improvements.

2.4 Necessity and importance

The challenges that water resources are facing in recent years and the pressure from the natural factors lead to rising demand and water stresses. Consequently, the competition among different water users has been increased. The traditional water management methods cannot solve these complicated water problems and these require applying new methods and management systems. In this case, economic issues and principles play a remarkable role in water resources management framework and can help better manage supply and demand.

In other word, for scarce resources like water, it requires to consider concepts and policies to increase the water economic efficiency and results in improvement of water resources management. Furthermore, the water allocation must be moved based on the economic principles for different competitive water users to achieve the maximum economic value. On the other hand, increasing the awareness of users about water and its scarcity encourage them to consider the economic principles.

In fact, water economics, including economic efficiency and economic water management in the water resources management helped users to cooperate for efficient and optimal water use. An economic incentive for users' cooperation in planning, management and operation is one of the most cost-effective methods to achieve water resources development goals in the water-related rules framework. According to mentioned issues, considering water economics concept in different allocation and user sectors can result in achieving an integrated sight about water supply and demand management. In fact, economics will be a powerful analytical tool for solving water resources management related challenges and will help to better policy and decision-making processes (C.Griffin, 2006; Olmstead, 2010; Berbel et al., 2017).

2.5 Methodology

There are numerous mechanisms and methods to apply different water economics framework in water resources management. The economic analysis of a project plays an important role in planning, designing and sustainable water resources system management. The decisions, policies and projects in water resources scope are planned to respond to the different user groups' needs. Some of these water-related projects include water reservoir construction, wells, water distribution networks, irrigation and drainage canals, hydropower plant and treatment plant. These projects are implemented to meet the demands in different urban, agricultural, industrial, power generation and flood control sectors.

To verify the economic possibility of implementing the mentioned projects, a cost-benefit analysis framework is needed. The general process of cost-benefit analysis consist of some parts including defining objectives, alternative actions and stakeholders, determining costs and benefits of an action, converting costs and benefits to present time period, considering discount rate, calculating net present value (NPV) and presenting the best action based on the results of analysis. In this framework, each measure, or practice consider as cost if it is directly related to the projects and decreases the benefits. In contrast, every practice which result in increasing the benefit consider as projects' benefit. Costs and benefits refer to the direct and initial economic impacts. In water resources related projects, benefits in urban, agricultural, industry, energy generation, and flood control are increasing the agricultural products, water supplied for urban and industrial uses, the amount of generated power and the boundary under flood control, respectively. To implement the cost-benefit analysis, all the cost and benefits must be converted to the commensurable units (Brouwer and Pearce, 2005; Wiejaczka et al., 2019).

The difference between benefits and costs will be the net value for each alternative. In this case, the most economically suitable alternative is the alternative with the highest NPV. If the NPV is positive, the alternative will be economically accepted. This provides an organized and logical framework for evaluating the policy making and management alternatives. In a mathematic definition NPV for an alternative can be calculated following Eq. (2.1) and its extended form can be presented in Eq. (2.2).

$$NPV = PVB - PVC \tag{2.1}$$

$$NPV = \sum_{t=0}^{0} \frac{B_t}{(1+i)^t} - \sum_{t=0}^{0} \frac{C_t}{(1+i)^t} \tag{2.2}$$

where PVB = the present value of benefits, PVC = the present value of costs, t = time, B_t = benefit in time t, i = interest rate, C_t = cost in time t, and o = the useful time of a project. The calculation of NPV returns all the benefits and costs to the present time step to provide the possibility of their comparison over time. The projects will be feasible economically if NPV is positive. In fact, cost-benefit analysis is a logical and organized framework for policy making and management evaluation (Brouwer and Pearce, 2005).

In the water economic framework, water economic value is another important concept. Valuing water resources is instituted based on the interaction between water resources and infrastructures and the services and commodities resulting water use. In a general insight, values will be categorized in two groups including use values and non-use values. Use values refer to the type of values that are achieved through the direct use of water resources. On the other hand, non-use values achieve the water related services that do not have any direct interaction in this process. The second value group is generally related to the preferences or ethic aspects (Moran and Dann, 2008; Tilmant et al., 2014;E.Garrick, 2017).

Values in the economic framework are estimated using different methods. There are two general groups for estimating water value from economic insights. These valuing methods include indirect and direct approaches. Indirect approaches completely rely on market observations. These approaches include water market transaction, demand functions, travel cost approach, hedonic pricing method, water and variables efficiency calculation, substitution cost approach, and benefit coefficient approach.

In market-based method (water market transaction), the mechanism of valuing is based on observed user behaviors and the direct value of money. In this process, the calculation of shadow price, alleviating the prices and competitive market prices are used. In market mechanism, the results of valuing are based on actual selections and not assumptions.

In demand function approach, the demand function of a user for example a household is derived to estimate the user's willingness to pay for water. In this approach, if the sufficient data is available about the response to changes in water prices, the user willingness to pay and demand will be derived as the user receive the desirable range of water with determined price.

The value of lakes and rivers cannot be estimated easily as one cannot put any price for them. Therefore, the value of these mentioned resources which use for recreational goals is estimated considering the travel cost of people to these recreational places. There two simple methods for travel cost determination. First method is applied to estimate the people recreational demand function. The value is calculated from the area under demand curve. The total value is sum of areas under people's demand curves. In second method, market demand or total demand is determined using standard statistics approaches. The travel cost approach is technically a developed valuing approach.

The hedonic pricing approach is also applied for estimating the value of environmental aspects. In this case, the differences in prices paid by people for the presented commodities in market reveals their preferences for the quality of environmental commodity. In this regard, statistical analysis of prices and the characteristics of commodity are used for estimating the value of environmental quality. Some assumptions must be considered in this approach including efficient and free market and the complete data of people.

The water use in the production process will be determined by the maximum water efficiency approach. This approach considers as a budget analysis method which seek for maximum efficiency related to the water use is calculated from the difference between the total production efficiencies and total costs related to all the inputs except water. The value of product is equal to the value of inputs in the production process. In this approach, two items must be considered including:

- Producers must use determinant data till the marginal product is equal to the marginal price to maximize the benefit.
- Any input can be payable based on its marginal productivity.

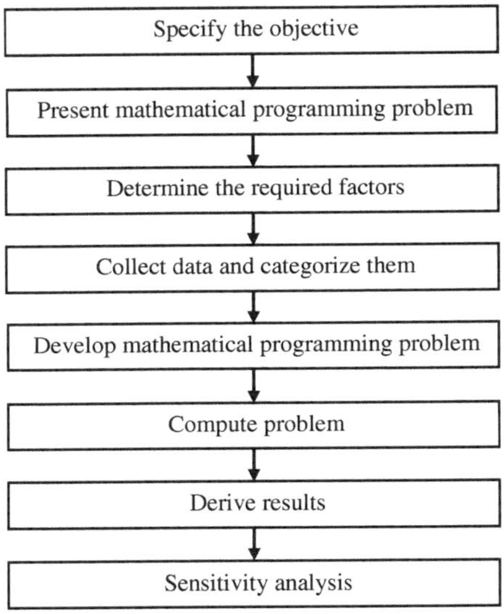

FIGURE 2.1 Flowchart of different stages of mathematical programming application.

In the case of maximization and minimization problems for an economic item, optimization models are used. In solving optimization problems, the economic values of all the inputs are presented. Among all the optimization models, mathematical programming models and dynamic models are the most applicable ones in economic evaluation.

Mathematical programming models seek to optimize an economic factor during a specified time period considering specific constraints. These models can determine the marginal value of water as an input. In this regard, water entered into the mathematical model as a production input and its shadow value or marginal value is calculated from the difference between the desirable value before and after adding one unit of available water for production. The flowchart of different stages of mathematical programming model application is illustrated in Fig. 2.1.

The second group of methods for calculating the economic value is direct methods. These valuing methods determine the priorities based on the willingness to pay of users and include contingent valuation method, contingent ranking, and conjoint methods itself (Tsur and Dinar, 1997; A.Young and B.Loomis, 2005; Brouwer and Pearce, 2005; Olmstead, 2010; Araya et al., 2019).

2.6 Examples

There are a wide range of applications of economic principles in water resources analysis, management, and engineering scopes. All the studies indicate the applicability of economic principle and methods in the improvement of water resources management. In fact, the findings of these studies indicate an integrated insight into the water resources related issues which can result in better understanding of different aspects of water resources management. Two studies are discussed in this section focusing on the applicability of water economics in water-related issues.

2.6.1 Example I

Ahopelto and Vahala (2020) applied cost-benefit analysis to evaluate the impacts of different leakage reduction methods in water supply networks. In this case, as the common process of cost-benefit analysis, NPV was calculated and evaluated for three leakage reduction practices. These three measures included district metering areas, district metering areas with pressure reduction and renovations. For calculating the NPV, the measures' related costs and benefits were also assessed. The value of NPV indicated that renovations were not a cost-effective measure for the leakage reduction. In a general evaluation, the values of NPV were negative but small compared with the median marginal cost of water.

Although, the NPV values were negative for district metering measure, it was the most cost-effective measure. On the other hand, the economic levels of leakage were calculated. In fact, the economic level is the points where costs and benefits of measures were equal. The results indicated that, for the utilities with average leakage levels, water loss management measures are not cost-effective.

2.6.2 Example II

Medellin-Azuara et al. (2010) estimated the economic value of agricultural water. Water economic value was estimated along with the level of spatial aggregation. In this regard, a positive mathematical programming model was applied for water economic value estimation. This method located in the revealed preference methods category. This process was implemented for seven farming regions in North Mexico. Additionally, four scenarios were considered including the impact of changing technology on crop yields change, climate change condition, changing irrigation costs and changing commodity prices.

The applied positive mathematical programming model consisted of three steps. The value of inputs used during the crop production process was calibrated considering a non-linear cost function. The calculation process included the assessment of marginal values in a linear program framework. These values are used in the second step which is a non-linear cost function and a production function. In the final step, cost functions and calibrated parameters were used in a non-linear optimization program.

The mathematical model also included an objective function which was a maximization of farmers' profit subjected to the land area restriction. In this case, the coefficients of each part in the model were obtained by calculating production functions of crops. The calibration process was also estimated based on the difference between base condition and calibrated use of production factors.

The results indicated a high sensitivity of aggregated models to changes in scenarios. On the other hand, the farm-level water economic values were significantly increased with water shortage and the aggregated model was more conservative and the changes based on shortage were moderate. Therefore, aggregated model is a suitable method in the case of water economic value in severe critical condition and new policies.

2.7 Summary

High water demand resulting population growth, over-exploitation of surface and groundwater resources, water resources pollution, lack of water infrastructure and conflict among different water user groups are some of factors negatively impacting the water shortage around the world. In this regard, some instruments and policies are required to overcome the challenges in water resources management. These policies and instruments must focus in a broader bound in water related issues and can solve the challenges and problems more efficient than traditional methods.

After addressing water as an economic good through the Dublin conference, the economic aspects are considered in numerous scopes of water resources management. On the other hand, economic instruments were entered the framework of water resources management. All these mentioned economic concepts have been integrated into a concept named water economics. In fact, this general concept becomes famous in recent years in water resources management literature.

In the framework of water economics, wide range of economic principles and theories are applied to solve water related problems. In the case of water demand management and water allocation, economic policies, and instruments play an important role in deriving demand curve in different user sectors and demand functions. On the other hand, the efficient water allocation can be possible using economic instruments. Additionally, using different methods for water pricing and estimation of water economic value result in finding efficient strategies for water management in critical situations like water shortage.

Finally, the information provided by economic principles help planners, and decision makers to find the source of challenges in water sectors. On the other hand, using economic instruments suggested new ideas for applying different strategies for conflict resolution among different water users and help decision makers to find the best strategies to overcome the challenges in water resources management.

References

Ahopelto, S., Vahala, R., 2020. Cost-benefit analysis of leakage reduction methods in water supply networks. Water (Basel) 12. doi:10.3390/w12010195.

Alcon, F., Martin-Ortega, J., Pedrero, F., Alarcon, J.J., Miguel, M.D., 2013. Incorporating non-market benefits of reclaimed water into cost-benefit analysis: a case study of irrigated mandarin crops in southern Spain. Water Resour. Manage. 27, 1809–1820.

Araya, A., Gowda, P.H., Golden, B., Foster, A.J., Aguilar, J., Currie, R., Ciampitti, I.A., Prasad, P.V.V., 2019. Economic value and water productivity of major irrigated crops in the Ogallala aquifer region. Agric. Water Manage. 214, 55–63.

A.Revollo-Fernandez, D., Roriguez-Tapia, L., Morales-Novelo, J., 2019. Economic value of water in the manufacturing industry located in the valley of Mexico Basin, Mexico. Water Resour. Econ. 30. doi:10.1016/j.wre.2019.01.004.

A.Ward, F., Michelsen, A., 2002. The economic value of water in agriculture: concepts and policy applications. Water Policy 4, 423–446.

A.Ward, F., 2012. Cost-benefit and water resources policy: a survey. Water Policy 14, 250–280.

A.Young, R., B.Loomis, J., 2005. Determining the economic value of water: Concept and Methods. Taylor and Francis, New York, USA.

Babel, M.S., Gupta, A.D., Nayak, D.K., 2005. A model for optimal allocation of water to competing demands. Water Resour. Manage. 19, 693–712.

Berbel, J., Gutierrez-Martin, C., Martin-Ortega, J., 2017. Water economics and policy. Water (Basel) 9. doi:10.3390/w9100801.

Brouwer, R., Pearce, D., 2005. Cost-Benefit Analysis and Water Resources Management. Edward Elgar Publishing, No. 1, Cheltenham, England.

Cai, X., Wang, D., 2006. Calibrating holistic water resources-economic models. J. Water Resour. Plann. Manage. 132 (6), 414–423.

Cheng, F.J., Wang, K.W., 2013. A systematical water allocation scheme for drought mitigation. J. Hydrol. (Amst.) 507, 124–133.

C.Griffin, R., 2006. Water Resources Economics. MIT Press, Cambridge, England.

D.Delorit, J., P.Parker, D., J.Block, P., 2019. An agro-economic approach to framing perennial farm-scale water resources demand management for water rights markets. Agric. Water Manage. 218, 68–81.

Dehnhardt, A., 2014. The influence of interests and benefits on the use of environmental cost-effective analysis in water policy: the case of German policy-makers. Environ. Policy Gov. 24, 391–404.

Dinar, A., Rosengrant, M.W., Meinzen-Dick, R., 1997. Water Allocation Mechanisms: Principles and Examples. World Bank Policy Research Working Paper, Washington DC, USA.

E.Garrick, D., W.Hall, J., Dobson, A., Damania, R., Grafton, R.Q., Hope, R., Hepborn, C., Bark, R., Boltz, F., Destefano, L., O'Donnell, E., Matthews, N., Money, A., 2017. Valuing water for sustainable development. Science 358 (6366), 1003–1005.

Escriva-Bou, A., Pulido-Velazquez, M., Pulido-Velazquez, D., 2017. Economic value of climate change adaptation strategies for water management in Spain's Jucar Basin. J. Water Resour. Plann. Manage. 143 (5). doi:10.1061/(ASCE)WR.1943-5452.0000735.

F.Booker, J., E.Howitt, R., M.Michelsen, A., A.Young, R., 2012. Economics and the modeling of water resources and policies. Nat. Resour. Model. J. 25 (1), 168–218.

Green, C., 2000. If only life were that simple; optimism and pessimism in economics. Phys. Chem. Earth 25 (3), 205–212.

Hu, Z., Chen, Y., Yao, L., Wei, C., Li, C., 2016. Optimal allocation of regional water resources: from a perspective of equity-efficiency tradeoff. Resour. Conserv. Recycl. 109, 102–113.

Hutton, G., Haller, L., Bartram, J., 2008. Global cost-benefit analysis of water supply and sanitation intervention. J. Water Health 5 (4), 481–502 .

ICWE, 1992. The Dublin Statement and report of the conference. In: International conference on water and the environment: development issues for the 21st century 26-31 January. Dublin, Ireland.
Kampragou, E., F.Lekkas, D., Assimacopoulos, D., 2010. Water demand management: implementation principles and indicative case studies. Water Environ. J. 25 (4), 466–476.
Kelman, J., Kelman, R., 2010. Water allocation for economic production in a semi-arid region. Int. J. Water Resour. Dev. 18 (3), 391–407.
Klocke, N.L., Stone, L.R., Clark, G.A., Dumler, T.J., Briggeman, S., 2006. Water allocation model for limited irrigation. Appl. Eng. Agric. 22 (3), 381–389.
Koutiva, I., Makropoulos, C., 2019. Exploring the effects of alternative water demand management strategies using an agent-based model. Water (Basel) 11. doi:10.3390/w11112216.
Kumar, M.D., 2005. Impacts of electricity prices and volumetric water allocation on energy and groundwater demand management: analysis from Western India. Energy Policy 33, 39–51.
Li, M., Guo, P., P.Singh, V., 2016. An efficient irrigation water allocation model under uncertainty. Agric. Syst. 144, 46–57.
Li, M., Fu, Q., P.Singh, V., Liu, D., Gong, X., 2020. Risk-based agricultural water allocation under multiple uncertainties. Agric. Water Manage. 233. doi:10.1016/j.agwat.2020.106105.
Maggioni, E., 2015. Water demand management in times of drought: what matters for water conservation? Water Resour. Res. 51, 125–139.
MahmoudzadehVarzi, M., J.Trout, T., C.Dejonge, K., Oad, R., 2019. Optimal water allocation under deficit irrigation in the context of Colorado water law. J. Irrig. Drain. Eng. 145 (5). doi:10.1061/(ASCE)IR.1943-4774.0001374.
Malm, A., Moberg, F., Rosen, L., J.R.Pettersson, T., 2015. Cost-benefit analysis and uncertainty analysis of water loss reduction measures: case study of the Gothenburg drinking water distribution system. Water Resour. Manage. 29, 5451–5468.
McNeill, D., 1998. Water as an economic good. Nat. Resour. Forum 22 (4), 253–261.
Moran, D., Dann, S., 2008. The economic value of water use: implications for implementing the water framework directive in Scotland. J. Environ. Manage. 87, 484–496.
Medellin-Azuara, J., Harou, J., E.Howitt, R, 2010. Estimating economic value of agricultural water under changing conditions and the effects of spatial aggregation. Sci. Total Environ. 408, 5639–5648.
Mulugeta Degefu, D., He, W., Yuan, L., Zhao, J.H., 2016. Water allocation in transboundary river basins under water scarcity: a cooperative bargaining approach. Water Resour. Manage. 30, 4451–4466.
Olmstead, S.M., 2010. The economics of managing scarce water resources. Rev. Environ. Econ. Policy 4 (2), 179–198.
Porse, E., B.Mika, K., Litvak, E., F.Manago, K., S.Hogue, T., Gold, M., E.Pataki, D., Pincetl, S., 2018. The economic value of local water supplies in Los Angeles. Nat. Sustain. 1, 289–297.
Puri, R., Maas, A., 2019. Evaluating the sensitivity of residential water demand estimation to model specification and instrument choices. Water Resour. Res. 56 (1). doi:10.1029/2019WR026156.
Read, L., Madani, K., Inanloo, B., 2014. Optimality versus stability in water resource allocation. J. Environ. Manage. 133, 343–354 .
Rigby, D., Alcon, F., Burton, M., 2010. Supply uncertainty and the economic value of irrigation water. Eur. Rev. Agric. Econ. 37 (1), 97–117.
Roa-Garcia, M.C., 2014. Equity, efficiency and sustainability in water allocation in the Andes: trade-offs in a full world. Water Alternatives 7 (2), 298–319.

Savenije, H.H.G., Van der Zaag, P., 2002. Water as an economic good and demand management, paradigms with pitfalls. Water International 27 (1), 98–104.

Sharp, L., 2006. Water demand management in England and Wales: constructions of the domestic water user. J. Environ. Plann. Manage. 49 (6), 869–889.

Tilmant, A., Arjoon, D., Fernandes Marques, G., 2014. Economic value of storage in multireservoir systems. J. Water Resour. Plann. Manage. 140 (3), 375–383.

Tortajada, C., Gonzalez-Gomez, F., K.Biswas, A., Buurman, J., 2019. Water demand management strategies for water-scarce cities: the case of Spain. Sustain. Cities and Soc. 45, 649–656.

Tsur, Y., Dinar, A., 1997. On the relative efficiency of alternative methods for pricing irrigation water and their implementation. World Bank Econ. Rev. 11, 243–262.

Wang, L., Fang, L., W.Hipel, K, 2008. Basin-wide cooperative water resources allocation. Eur. J. Oper. Res. 190 (3), 798–817.

Wiejaczka, T., Pirog, D., Fidelus-Orzechowska, J., 2019. Cost-benefit analysis of dam projects: the perspectives of resettled and non-resettled communities. Water Resour. Manag. 34, 343–357. doi:10.1007/s11269-019-02451-0.

Xiao, Y., W.Hipel, K., Fang, L, 2016. Incorporating water demand management into a cooperative water allocation framework. Water Resour. Manag. 30, 2997–3012.

Xiao, Y., Fang, L., W.Hipel, K, 2018. Agent-based modeling approach to investigating the impact of water demand management. J. Water Resour. Plann. Manag. 144 (3). doi:10.1016/(ASCE)WR.1943-5452.0000907.

Yokomatsu, M., Ishiwata, H., Sawada, Y., Suzuki, Y., Koike, T., Naseer, A., Masud Cheema, M.J., 2020. A multi-sector multi-region economic growth model of drought and the value of water: a case study in Pakistan. Int. J. Disaster Risk Reduct. 43. doi:10.1016/j.ijdrr.2019.101368.

Zilberman, D., Schoengold, K., 2013. The use of pricing and markets for water allocation. Can. Water Res. J. 30 (1). doi:10.4296/cwrj300147.

Chapter 3

Water pricing

Sahar Mohammad-Azari[a], **Omid Bozorg-Haddad**[a] **and Asim Biswas**[b]
[a] *Department of Irrigation & Reclamation Engineering, Faculty of Agricultural Engineering & Technology, College of Agriculture & Natural Resources, University of Tehran, Karaj, Tehran, Iran.* [b] *School of Environmental Sciences, University of Guelph, Guelph, Canada*

3.1 Introduction

Water is one of the three most important natural resources beside soil and air. This valuable resource is under extreme threat from the climate change and demand of increasing population in recent decades. As this is an irreplaceable resource, the scarcity has been exacerbated from the high cost to access new water resources. Water is considered as a public good which must be supplied and distributed by governments for different user sectors. In fact, this issue results in more scarcity in water resources.

Water as a public good is used for different purposes including household, agriculture, industry, recreational, tourists, environmental and economic. On the other hand, water have some unique attributes including irreplaceable, renewable, non-renewable, scarcity, exclusive market, and multi supply resource as well as being a social and economic good (Tietenberg, 1992; C.Griffin, 2006).

Considering water as an economic good requires a balance between supply and demand. Achieving the goal of balance between water supply and demand can be possible through efficient pricing strategy in different user sectors. This strategy plays a remarkable role in cost recovery of water provision and operation. On the other hand, it is a useful method in order to encourage the users for efficient water use while considering water productivity.

There are numerous studies and researches that considered water pricing as an economic instrument for efficient water management. These studies focused on the water pricing in different water user sectors. They applied different methods considering natural and economic conditions to compute the price of water.

For example, E.T.Moncur (1987) evaluated the impacts of water pricing on water demand and water conservation in drought conditions for urban water systems. The results indicated that the price, income, household size, and precipitation affected water demand. On the other hand, the increase in water price of less than 40 percent leads to 10 percent reduction in water use during

Economical, Political, and Social Issues in Water Resources. DOI: 10.1016/B978-0-323-90567-1.00011-5

the drought period. Furthermore, a conservation program can slightly mitigate the increase in price.

Dinar and Subramanian (1998) evaluated the water pricing experiences in different regions from 22 countries. The investigations indicated that different countries follow different goal of water pricing including cost recovery, income redistribution, enhanced water allocation and water conservation. Furthermore, the pricing policy is based on average price in developing countries and some developed countries. The response of high-income countries was more positive toward water pricing reforms compared to low income countries.

Varela-Ortega et al. (1998) analyzed the impacts of different water proving policies on agricultural water demand, farmers' income and government revenues. They presented a mathematical programming model where the farmers' behavior and their responses to pricing scenarios were simulated. The results indicated the impact of different pricing policies for agricultural water relied on institutional, regional, and structural situations. On the other hand, water related charges had widespread impacts on efficient water use, farmers' income, and government' revenues.

Ruijs et al. (2008) investigated the effects of water pricing policies in a block pricing model in Sao Paolo. The water demand function was also estimated by an average and marginal price model. A progressive block price can result in a more even income distribution while a trade-off was reported between more even income distribution and incomes of water firms.

Pesic et al. (2012) presented a seasonal water pricing model that used information from meteorological observations of average monthly temperatures and the total monthly precipitations. The obtained results indicated the potential of seasonal water pricing for water conservation which results in covering peak-season supply by water supply companies.

Grafton et al. (2014) applied a stochastic dynamic programming to determine dynamically efficient water prices. The losses in social surplus from premature water supply augmentation were also estimated. Results showed that premature water supply augmentation under the base case caused reduction in net present value of welfare of households.

Ayana et al. (2015) evaluated the impacts of water pricing policy on water management. Findings indicated that high increase in water prices were not desirable for farmers if the water distribution systems and water conveyance are poor and the irrigation systems are insufficient. Nevertheless, strengthening of irrigation research and extension services and improving irrigation system operations can help improve irrigation water management.

Ghimire et al. (2015) determined the price and income elasticity of residential water demand based on uniform water pricing and compared the elasticity of both average pricing considering pre-drought and drought periods. The results provided certain understandings and knowledge about variables affecting water demand. This also indicated that the parcel size, income, and temperature contributed to increase in water demand. On the other hand, water price and household age affected water demand negatively.

Macian-Sorribes et al. (2015) presented a process to design efficient pricing policies. They reported a positive impact of scarcity-based pricing policies on economic efficiency of water use and also allocation of water to high value uses. This policy can be an incentive for water conservation during scarcity periods.

Mamitimin et al. (2015) analyzed farmers' response to increase in water price. More than half of farmers indicated that they will adapt to efficient agriculture if the water prices are increased. On the other hand, water exploitation from groundwater resources will be increased. Eventually it was concluded that the water pricing is not the only strategy which should be implemented in order to achieve water use efficiency in agricultural sector but farmers' awareness about different agronomic methods can increase the impact of water pricing and leads to improve water efficiency.

Renzetti et al. (2015) simulated an across-the-board price increase and an increasing block rate structure and estimated the Stone-Geary and double-log residential water demand functions. They indicated a negative impact of across-the-board price increase on results from Stone-Geary which worsened the inequality of water expenditures. On the other hand, the double-log model indicated no change in the distribution of water expenditures for this price reforms.

Sahin et al. (2015) presented a system dynamic model in which feedback loops in tariff structures, demand levels and financing capacity were added. The presented model indicated the impact of drought pricing regime on lowering frequency and severity of water insecurity and reduction in long-run marginal cost of water supply compared to traditional reactive planning approaches.

Abdelhafidh and Bachta (2016) proposed an alternative price scheme in order to ensure cost recovery and water use efficiency and evaluate its impact on farmers' surplus. The results indicated that annual quantity of water consumption may be decrease if the water price and social surplus increased.

Toan (2016) presented an international review of water pricing policy. The emphasis of this paper was on the alignment of costs with prices. They also discussed the promotion of policy changes for more sustainable irrigation management and environmental protection in the future.

Molinos-Senante and Donoso (2016) presented a water rate model that considered water scarcity value and tried to improve social concerns such as equity and affordability this water. This model followed an increasing block strategy. The results indicated the impact of water rates as an incentive on the improvement of water sustainability and equity.

Ben Zaied et al. (2017) implemented nonlinear progressive water pricing policy using database of two consumption blocks in Tunisia. Finding indicated that the increasing block tariffs can be a useful tool for scarce water management in Tunisia.

Cooper et al. (2017) investigate the farmers' preferences for changing water prices and tariffs structures. The results indicated that the public subsidy of infrastructure in irrigation is not always based on farmers' preference but these preferences are conductive to price reform.

Exposito and Berbel (2017) explored the impact of water pricing on deficit irrigation schemes. The findings indicated that the water demand will be decreased if there is an increase in water prices. It also indicated that water pricing was an ineffective instrument for demand management where farmers face scarcity and adapt to deficit irrigation.

Ratnasiri et al. (2017) compared two conservation-oriented pricing schemes including uniform pricing scheme and an increasing block tariff schemes (IBT) structure to evaluate the effectiveness of water conservation. The results indicated the superiority of IBT pricing scheme compared to uniform pricing scheme for reducing demand. These results help decision makers to find more efficient ways to conserve water.

Sahin et al. (2017) analyzed the response of stakeholders toward pros and cons of new water tariffs to understand the potential of economic and social implications of demand-based tariff structures through a participatory approach. The results based on customers' feedback indicated a positive impact of inclining block tariff on behavioral change and water consumption reduction. On the other hand, the findings can help water planners and policy makers to explore wide range of alternatives for optimal decision making for urban water conservation and management.

Tsitsifli et al. (2017) presented a new method to determine drinking water price. However, they considered social fair for price determination based on Full Water Cost recovery principle and considering non-revenue water.

Kajisa et al. (2018) applied volumetric irrigation pricing instead of area-based pricing in a pilot project framework. The indicated the need for investment in infrastructure for group-level volume measurement and water control. On the other hand, an appropriate system will depend on different environmental and socio-economic background.

Soto Rios et al. (2018) analyzed the impact of water pricing on water security. In this case, they focused on the economic aspects, and water pricing effect on water consumption.

Fuentes-Cortes et al. (2019) presented a multi-objective strategy to define relationships between design and utility policy prices. The applied methodology helps determining cost of utilities according to demand generation dynamics. The results also indicated the advantage of dynamic cost of utilities and the suitable level of subsidies.

Momeni et al. (2019) analyzed the feasibility of increasing water price in agricultural sector in Iran using a hydro-economic approach considering role of water pricing in agricultural water management. The compared, the price of water in agricultural sector in two provinces of Iran and state of California in the United States of America. The results indicated no significant differences between the two case studies in Iran and the one in the USA based on the

Purchasing Power Parity (PPP) index. The authors also suggested institutional reforms which is required before adjusting agricultural water price.

3.2 Terms and definitions

In a general term, price refers to the payments that a consumer pays for receiving a specified bound of a good or a service. Consequently, water price is also the payment toward receiving a unit of water for different uses. The concept of water price includes different definitions and hypotheses and its accurate description is important in policy making. An accurate definition of water pricing forms the discussion framework around this strategy and affects the stakeholders and user insights. An accurate and appropriate water pricing can influence the selection of policies in water sector. Therefore, numerous countries consider water pricing as an important issue in management and policy making (Dinar et al., 2015).

In water pricing, economic and financial sights are important and distinct concepts to achieve sustainable and efficient water management. Economic and financial sights also have different goals. In economic sight, not only water supply cost recovery is considered but also water scarcity costs and response of consumers against water shortage are also covered. In contrast, financial sight only considers cost recovery. There are two other concepts in water pricing framework including supply price and demand price are often focused more than economic and financial sights. Supply price covers all the water supply and distribution related costs. On the other hand, demand price reflects the value of water in consumers' viewpoint.

Nevertheless, management and balance among the different goals in water sector are the main challenges of water pricing. In order to achieve an effective water pricing structure, both supply costs and shortage costs in one hand and water economic value and water demand elasticity on the other hand must be considered during pricing procedure. Water economic value is the marginal value of one additional unit of water for one unit of production in scarcity condition. The water demand elasticity is also water use reduction towards increase in water price (C.Griffin, 2006; Sjodin et al., 2016).

To better understand of water pricing concept, the two main pricing processes are discussed here. Generally, water price determination takes place under two category including administrative pricing and market-based pricing. In administrative pricing mechanism, the prices are determined by the public authorities. These prices can be determined based on supply related costs, opportunity costs or their combination. Opportunity cost is the loss of benefit from an additional unit of water which is not allocated to a productive use. This use is also the most economically optimal option.

Market-based water pricing is the second category in which prices are determined indirectly through a decentralized mechanism. Prices in the market will be decreased or increased based on changes in supplies or demands. In fact, institutions administering this mechanism are complicated because as water possesses some attributes that cannot be considered completely as a private good.

These attributes include the potential of free access and also reuses which are in contrast with the definition of private good (private good refers to a good which is scarce and the usage by a consumer will be restricted if it is used by a consumer). In this mechanism, prices can also be influenced by political and social factors.

While administrative based pricing can cover the costs of water provision and distribution, the market-based pricing has more potential for reflecting and covering opportunity costs and scarcity costs. Furthermore, consideration of social aspects in water pricing results in water availability for meeting the basic human needs (C.Griffin, 2006; Sjodin et al., 2016).

Finally, if the goal of water pricing policy is to balance among conflicting goals of politicians, managers, planners, environmentalists, and civil society, it requires more details other than financial viewpoint, economic viewpoint or economic efficiency. In fact, the sustainability and equity must consider in water pricing process. In a general viewpoint, a pricing policy leads to water sustainability provided it includes consumers' preferences, responses, willingness to pay, cultural and social problems in addition of considering supply and demand.

3.3 Basic principles and logics

Water pricing process gets complicated as the previously mentioned attributes are considered. In this case, a set of principles must be considered for water pricing. These principles include equity, efficient allocation, sufficient and sustainable benefits, cost recovery, resource conservation and storage and social welfare. More than that, consideration of the affordability and willingness to pay is the most important fundamentals in pricing process. Regarding all these principles simultaneously is difficult. Therefore, the priority to consider each mentioned principles is determined according to the sectors in which water price should be computed. For instance, in household sector the priority belongs to equity while in industrial sector, efficient water allocation is the priority.

In order to enhance the pricing structure, consideration of cost recovery is one of the most important principles. In fact, cost recovery is one of the most important parts of sustainable water management. In other word, water sector and authorities must achieve financial benefits and cover the water extraction, treatment, and delivery related costs to promote water services and infrastructure leading towards water sustainability. An efficient and sustainable pricing strategy must provide an incentive for optimal as well as must reveal the minimum investment costs, operation costs and carried out costs based on environmental standards (C.Griffin, 2006).

Equity is another key principle in water pricing structure. Considering the undeniable role of water in human life, enough water must be available for surviving. Therefore, the payments for water use should be affordable for different water user groups. In fact, the payments should cover the costs of utilities in one hand and reveal the actual value of water from users' viewpoint on the other hand to encourage users for efficient use. Furthermore, the water price should be in a range to control some water related challenges including controlling water overexploitation, preventing water pollution, and ensuring enough water availability for future generation.

The price of water should also be in an appropriate range. In other word, if the water price is low, it cannot cover the supply and distribution related cost. In this case, although all the users can afford the price, it will become an inefficient policy resulting in over consumption in users' groups and benefit reduction in the supply side. In this situation, the supply side cannot afford the costs of water-related investments. On the other hand, if the price is in higher bound compared to a specified range, although the supply side benefits and financial capacity will be increased in short-term, the demand side cannot afford the price and the water use will be restricted. In this situation, the supply side will not have sustainable financial capacity for long-term (Sjodin, 2016).

In fact, an average range of these two ranges will result in a sustainable water pricing framework. In this mechanism, users understand the actual water value and the total costs of water withdrawal, water distribution and delivery. In fact, the water pricing based on this structure can help an efficient water use and provide financial capacity for supply side.

3.4 Necessity and importance

As water pricing goals and strategies can be an incentive for efficient water use and conservation of the water quality in the case of appropriate planning and management, they have this potential to overcome the physical and economic water shortage. On the other hand, water pricing can ensure water security through the management of water exploitation, water supply and infrastructural investments. In other word, water prices can cover the costs of promotion and development of water infrastructures. Furthermore, water pricing can ensure the water availability in an affordable price for all the water user sectors around the world.

Water related payments can reflect the water shortage and water use if they are managed and executed efficiently. In fact, the application of water price as an economic instrument for better water management, modification of water use patterns and insights result in a sustainable structure and efficient water use in different sectors.

Additionally, in a sustainability perspective, water pricing plays a notable role in achieving water security. In other word, water security ensures that fresh water resources and ecosystems conservation can improve a sustainable development and everyone can have access to sufficient water resources at an affordable cost. According to this definition, pricing can be considered as a tool for achieving sustainable development as it is an incentive for efficient and sustainable water use and conservation.

3.5 Methodology

As discussed previously, water pricing is considered as one of the most important economic instruments in order to address and solve the management challenges in water-related problems. Application of this policy can lead to a better supply and demand management as well as an efficient water allocation and use.

Numerous methods have been introduced in water pricing literature based on natural, economic and political conditions to assess sector-wise water price especially for urban and agricultural sectors as the two major user categories in majority of countries.

In urban sector, water pricing methods are divided into volumetric water pricing, fixed rates, two-part tariffs and seasonal pricing methods. The pricing structure in this sector includes mostly two parts. Fixed charges are assigned for costs of operation, maintenance, facilities and fixing. The variable charges are related to the volume of water consumption (Dinar and Saleth, 2005; Toan, 2016).

Volumetric method of price determination focuses on the volume of water consumed. In some cases, volumetric pricing includes variable rates for water consumption. The rates may vary as the water consumption exceeds a threshold volume.

In the fixed rates method, the payments for each consumption block are fixed. In this case, if the price is calculated based on the marginal cost, it would result in an economic efficiency, water savings and stability in income. In fact, water can be used efficiently if the benefits from the consumption of its one marginal unit equals to the costs of its achievement. In other word, if the price of water equals to the marginal cost the costs of water consumption and its benefits can be comparable with each other. Lower prices can result in high consumption levels and higher prices will lead to lower consumption levels (Dinar and Saleth, 2005; Griffin, 2006).

The other urban water pricing is seasonal water pricing in which the price is determined based on the weather condition and the water demand. In warmer months the prices are higher in comparison to colder months. The main core of seasonal pricing is focused on social equity. For critical situations and peak use periods, the need for more measurements and capacity increasing is more than the normal situation.

Tariff is another method that is used to determine the price of water in urban sector. This method consists of two sub sets named increasing block tariffs and decreasing block tariffs. In increasing block tariffs, the water price increase is based on the consumption. The limitation of this method is that the pricing in different user groups has not carried out based on the actual water prices. Although this method affects water conservation, it does not necessarily ensure social equity. In this pricing method three goals are considered including affordability, water conservation and economic efficiency. In contrast, for decreasing block tariffs, the relationship water price and consumption is opposite. For higher consumption levels, the price is decreased. The logic of this method is that the water distribution sectors face decreasing unit costs with higher water use. Therefore, decreasing block tariffs convey the cost savings to users. In fact, the users with higher consumption pay subsidies to users with lower consumption (Monteiro, 2005).

In agricultural sector, the water pricing methods have been categorized in three general class including volumetric, non-volumetric and market-based pricing methods. The mechanism of volumetric method is similar to that mentioned in urban pricing parts. In this method the price will be assessed directly based on the measurements of certain flow for example from a reservoir or the fraction of uncertain flow from a river. On the other hand, even if the amount of water consumed is zero a minimum value should be paid.

The second class of water pricing methods is non-volumetric method which is in contrast with volumetric method. In this method the water price is determined based on output (products), input or an area. In fact, this method is applied when there is a lack of sufficient information about the actual volume of water consumption.

In output-based pricing the payments for water consumption is carried out considering the unit of production. On the other hand, the price may be calculated based on the input during the production process. For area-based pricing, charges are determined based on irrigated area considering the kind of crops, type of irrigation system and the withdrawal source. For example, the charges for pumped water or water from river are different.

Finally, the third class of water pricing methods is the market-based methods. In this method, farmers in a market mechanism received suggestions for water price and a price is determined for water use in agricultural sector based on agreement (Dinar and Saleth, 2005; Monteiro, 2005).

3.6 Examples

Generally, a wide range of water pricing methods are applied for water price assessment based on the available information about water consumption and users, natural and economic conditions and the production process. In this regard, numerous researches have been carried out in many countries. Here we discuss the following two examples.

3.6.1 Example I

To investigate the application of different water pricing methods, the study of Ghimire et al. (2015) are evaluated here as an example. This study estimated the water price in the residential sector of Oklahoma City and then evaluated the impact of calculated price by volumetric pricing method on residential water demand. As the rainfall and temperature are the most effective factors on water demand, these two factors were considered during the calculation process. It was expected that the demand is higher in hot and dry condition compared to cold and wet condition.

The current pricing mechanism in consists of a fixed part and a variable part. The fixed part is the base service charge and the variable part is calculated based on the volume of water consumption. The price paid by residential sector is the sum of these two parts. The prices are changed annually and with 6 and 5 percent increase in the fixed and variable part, respectively.

Different stages of price determination and evaluation from different factors and their effect on water demand is completed through regression equation. This equation is a function of household related factors and meteorological factors. These factors include household income, household size, maximum temperature, and total monthly rainfall. Additionally, a factor related to the year of construction of the building is also considered in model. This indicator is 1 if the year of construction was after 1994 and 0 for the building with the construction year before 1994. The reason is that, in the United States of America, all the water related facilities in the houses were changed to an efficient and high-standard one in 1994 and this change influenced the water demand and thus the price for water.

Eq. (3.1) shows the relationship between different effective factors on household water consumption for calculating the average water price (Ghimire et al., 2015).

$$\begin{aligned}\ln\,(\text{water price}) = {} & \alpha_0 + \alpha_1 \ln\,(\text{household income}) + \alpha_2 \ln\,(\text{household size}_i) \\ & + \alpha_3 \ln\,(\text{maximum temperature}_t) + \alpha_4(\text{total precipation}_t) \\ & + \alpha_5\,(\text{year of built}_i) + \alpha_6 \ln\,(\text{gas price}_t) \\ & + \alpha_7 \ln\,(\text{consumer price index}_t) + \varepsilon_{it} \qquad (3.1)\end{aligned}$$

where $i =$ number of household, $t =$ month, $\alpha =$ estimators and $\varepsilon_{it} =$ error indicator. The gas price index and regional consumer price index considered as variable and correlated with the average water price.

In this stage, an average water price was predicted and applied as an explanatory variable in the second stage to evaluate the effect of mentioned factors and calculated price on residential water demand. In the following, the demand function was derived based on price, meteorological factors, household income, parcel size and the year of construction of building. The demand function is shown in Eq. (3.2) (Ghimire et al., 2015).

$$\begin{aligned}& \ln\left(\text{total water consumption}_{it}\right) \\ & = \alpha_0 + \alpha_1 \ln\left(\text{price of first stage}_{it-1}\right) + \alpha_2 \ln\,(\text{income}_i) \\ & \quad + \alpha_3 \ln\,(\text{household size}_i) + \alpha_4 \ln\left(\text{maximum temperature}_t\right) \\ & \quad + \alpha_5\left(\text{total precipitation}_t\right) \\ & \quad + \alpha_6 \ln\left(\text{year of build}_i\right) + \gamma_i + \varepsilon_{it} \qquad (3.2)\end{aligned}$$

where $\gamma_i =$ the random effect term. The equation was established to evaluate the demand changes considering the mentioned factors and water prices for a pre-drought and during drought periods.

The log-log relation between water demand, water price and income of household consisted of coefficients which were the result of model. Obtained results indicated the low impacts of water price on demand reduction especially in during-drought periods and high consumption periods. In fact, in critical conditions the prices did not have positive effect on demand reduction. Specially, in low price level, the changes in residential water demand were negligible (Ghimire et al., 2015).

3.6.2 Example II

One of the water pricing related studies related to water pricing in agricultural sector belongs to Tahamipour et al. (2015). They focused on the agricultural sector of Golestan Province in Iran and determined the water price in this sector using econometric methods. They assessed the demand price of water considering different crops. In the first step, they assessed the production function of crops in Golestan Province. Production function indicates an input-output relation between crop production and effective inputs in the production process.

Water is also considered as an input in the production process. In fact, the value of marginal product of water is considered as an estimator of water demand price. This process is presented in Eqs. (3.3) and (3.4) (Tahamipour et al., 2015).

$$\text{Output quantity} = f\,(\text{seed, fertilizer, pesticide, labor}, \cdots, \text{water}) \quad (3.3)$$

$$\begin{aligned}&\text{Value of marginal product of irrigation water}\\&\quad = \text{output price} \times \text{marginal product of irrigation water}\end{aligned} \quad (3.4)$$

In this case, all the production inputs including amount of seed, number of labors, amount of fertilizer, pesticide etc. must be entered into the production function. These values are specified and multiplied by estimator coefficients in the production functions. Some different and some common production functions were considered based on an economic basis. Then the coefficients in each production function were estimated and all the coefficients of each production function were compared with the obtained coefficients from other production functions. Based on some statistical indicators, the best production function was selected. The coefficients of these selected production functions were derived and inserted into an equation of value of marginal products. Finally, the price of water was calculated and was 1,795 IRR per cubic meter (Tahamipour et al., 2015).

3.7 Summary

Water plays an undeniable role in different aspects of human life, ecosystem, and environment. In most cases, different parts of human life rely on access to water. In recent years, the changes in climate condition and temperature, population growth and increase in the demands of different water user sectors caused various problems in water resources domain. Many countries are facing water shortage and water pollution from water over exploitation. In this case, an efficient policy and strategy is required to overcome the water-related challenges.

Since long, all the water management issues were completely focused on the water supply management and sought the solution of water problems in water supply management framework. However, as the water challenges and problems became complicated and the water supply management cannot be considered as the only solution, the water demand management became widespread among the water experts, managers, and planners.

Water demand management is focused on the water use management and generally tries to manage the current resources without accessing new water resources. In fact, in water demand framework, the users' cooperation is considered along with the supply management to achieve an efficient water allocation and sustainable water management. In order to reach this goal, different instruments are introduced in demand management framework.

Water pricing is one of the most important economic instruments that is defined in the framework of water demand management to handle the water challenges. Using water pricing as a tool for saving water requires consideration of pricing strategies in line with the water management. The water supply and demand can be balanced through pricing and the water resources management goals can be achieved through the improvement of water conservation, water saving and income for public water services. In this case, a group of principles must be considered during pricing process.

Finally, an assessment of actual and accurate water price based on different methods must be considered. The economic and natural and even social and political condition of each user sector and also each region will lead to sustainable and efficient water management and help policy makers and mangers to manage water consumption in different user sectors and encourage the users to understand the real value of water for efficient use.

References

Abdelhafidh, H., Bachta, M.S., 2016. Groundwater pricing for farms and water user association sustainability. Arab. J. Geosci. 9 (8), 525–535.

Ayana, M., Teklay, G., Abate, M., Eshetu, F., Mada, M, 2015. Irrigation water pricing in Awash River Basin of Ethiopia: evaluation of its impact on scheme-level irrigation performances and willingness to pay. Afr. J. Agric. Res. 10 (6), 554–565.

Ben Zaied, Y., Bouzgarrou, H., Ben Cheikh, N., Nguyen, P., 2017. The nonlinear progressive water pricing policy in Tunisia: equity and efficiency. Environ. Econ. 8 (2), 17–27.

C.Griffin, R, 2006. Water Resources Economics. MIT Press, Cambridge, England.

Cooper, B., Crase, L., M.Rose, J., 2017. Cost-reflective pricing: empirical insight into irrigators' preference for water tariffs. Agric. Res. Econ. 62, 256–278.

Dinar, A., Subramanian, A., 1998. Policy implications from water pricing experiences in various countries. Water Policy 1 (2), 239–250.

Dinar, A. and Saleth, R.M. (2005). Water institutional reforms: theory and practices. Water Policy, 7, 1–19.

Dinar, A., Pochat, V., Albiac-Murillo, J., 2015. Water Pricing Experiences and Innovations, Volume 9. Springer.

E.T.Moncur, J, 1987. Urban water pricing and drought management. Water Resour. Res. 23 (3), 393–398.

Exposito, A., Berbel, J., 2017. Why is water pricing ineffective for deficit irrigation schemes? A case study in Southern Spain. Water Resour. Manage. 31, 1047–1059.

Fuentes-Cortes, L.F., Flores-Tlacuahuac, Ponce-Ortega, J.M, 2019. Integrated utility pricing and design of water-energy rural off-grid systems. Energy doi:10.1016/j.energy.2019.04.026.

Ghimire, M., Boyer, T., Chung, C., Q.Moss, J, 2015. Estimation of residential water demand under uniform volumetric water pricing. J. Water Resour. Plann. Manage. 142 (2). doi:10.1061/(ASCE)WR.1943-5452.0000580.

Kajisa, K., Moya, P., Valencia, M.S., San Valentin, M.R., Garcia, L., 2018. Lessons from volumetric water pricing trials at the three surface irrigation systems in Northern Luzon, Philippines. DLSU Business Econ. Rev. 28, 51–69.

Macian-Sorribes, H., Pulido-Velazquez, M., Tilmant, A., 2015. Definition of efficient scarcity-based water pricing policies through stochastic programming. Hydrol. Earth Syst. Sci. 19 (9), 3925–3935.

Mamitimin, Y., Feike, T., Seifert, I., Doluschitz, R., 2015. Irrigation in the Tarim Basin, China: farmers' response to changes in water pricing practices. Environ. Earth Sci. 73 (2), 559–569.

Molinos-Senante, M., Donoso, G., 2016. Water scarcity and affordability in urban water pricing: a case study of Chile. Utilities Policy 43, 107–116.

Momeni, M., Zakeri, Z., Esfandiari, M., Behzadian, K., Zahedi, S., Razavi, V., 2019. Comparative analysis of agricultural water pricing between Azarbaijan Provinces in Iran and the state of California in the US: a hydro-economic approach. Agric. Water Manage. 223. doi:10.1016/j.agwat.2019.105724.

Monteiro, H., 2005. Water Pricing Models: A Survey. Research Centre on Socioeconomic Change Working Paper No.2005/45 doi:10.2139/ssrn.899602.

Pesic, R., Jovanovic, M., Jovanovic, J., 2012. Seasonal water pricing using meteorological data: case study of Belgrade. J Clean Prod 60, 147–151.

Ratnasiri, S., Wilson, C., Athukorala, W., A.Garcia-Valinas, M., Torgler, B., Gifford, R., 2017. Effectiveness of two pricing structures on urban water use and conservation: a quasi-experimental investigation. Environ. Econ. Policy Stud. 547–560. doi:10.1007/s10018-017-0205-6.

Renzetti, S., P.Dupont, D., Chitsinde, T, 2015. An empirical of the distributional impacts of water pricing reforms. Utilities Policy 34, 63–69.

Ruijs, A., Zimmermann, A., Van den Berg, M., 2008. Demand and distributional effects of water pricing policies. Ecol. Econ. 66 (2–3), 506–516.

Sahin, O., Bertone, E., Beal, C.D., 2017. A system approach for assessing water conservation potential through demand-based water tariffs. J. Clean Prod. 148, 773–784.

Sjodin, J., Zaeske, A., Joyce, J., 2016. Pricing Instruments for Sustainable Water Management. Stockholm International Water Institute, Sweden Working Paper No. 28.

Soto Rios, P.C., A. Deen, T., Nagabhatla, N., Ayala, G., 2018. Explaining water pricing through a water security lens. Water (Basel) 10 (9), 1173–1193.

Tahamipour, M., Kavoosi Kalashami, M., Chizari, A.H., 2015. Irrigation water pricing in Iran: the gap between theory and practice. Int. J. Agric. Manag. Develop. 5 (2), 109–116.

Tietenberg, T., 1992. Environmental and Natural Resources Economics, 3rd Edition Harper Collins, New York, USA.

Toan, T.D., 2016. Water pricing policy and subsidies to irrigation: a Review. Environ. Process 3, 1081–1098.

Tsitsifli, S., Gonelas, K., Papadopoulou, A., Kanakoudis, V., Kouziakis, C., Lappos, S., 2017. Socially fair drinking water pricing considering the full water cost recovery principle and the non-revenue water related cost allocation to the end users. Desalination Water Treat. 99, 72–82.
Varela-Ortega, C., M.Sumpsi, J., Garrido, A., Blanco, M., Iglesias, E, 1998. Water pricing policies, public decision making and farmers' response: implications for water policy. Agric. Econ. 19 (1–2), 193–202.

Chapter 4

Water markets

Sina Fazeli[a], Omid Bozorg-Haddad[a], Jessica Budds[b] and Robert P. Berrens[c]
[a] *Department of Irrigation & Reclamation Engineering, Faculty of Agriculture Engineering & Technology, College of Agriculture & Natural Resources, University of Tehran, Iran.* [b] *School of International Development, University of East Anglia, United Kingdom.* [c] *Department of Economics, University of New Mexico, Albuquerque, NM, United States*

4.1 Introduction

Water is a key determinant of economic growth and a critical natural resource that must be carefully stewarded as part of sustainable development. Water scarcity is an important issue in many parts of the world, as demand for water increases for human and economic uses, and water resources are affected by climatic variation. Increased variation in precipitation has increased water scarcity in some regions and catchments. As a result, groundwater extraction has increased in many places, resulting in quantitative and qualitative reductions in aquifers as water tables have fallen and some aquifers have undergone salinization, negatively impacting agricultural land. Increasing water consumption and pollution in water resources has also produced increased competition among different users, resulting in degradation of water sources and environments. Given growing worldwide concern about freshwater supplies and the ability to meet new demands, water security is becoming an increasingly critical issue. Water security can be defined as "the availability of an acceptable quantity and quality of water for health, livelihoods, ecosystems and production, coupled with an acceptable level of water-related risks to people, environments and economies" (Grey and Sadoff, 2007, pg:4); although Jepson et al. (2017) place greater emphasis on the social relations of water management by conceptualizing water security as "a relationship that describes how individuals, households, and communities navigate and transform hydro social relations to access the water that they need and in ways that support the sustained development of human capabilities and wellbeing in their full breadth and scope." Under these circumstances, since at least the 1980s water resource planners in many parts of the world have come to a common understanding that relying solely on water supply enhancement policies is no longer able to bridge expected gaps between supply and demand. This realization has prompted

Economical, Political, and Social Issues in Water Resources. DOI: 10.1016/B978-0-323-90567-1.00003-6

a turn to other approaches, including integrated resource planning, water demand management and the search for greater flexibility (Lazarova et al., 2007).

Recent experiences with the implementation of *water markets* in different regions and countries have demonstrated their potential for improving allocation, but also come with a variety of implementation challenges (Anon, 1995; Howe et al., 1986; Eheart and Lyon, 1983; Lund et al., 1992). While water markets can offer benefits in water-scarce locations where investment in hydraulic infrastructure may not be feasible or desirable water trading also has range of implications, including allocation inequities, environmental impacts (e.g., protecting minimum flows for ecological purpose) legal reform, market and pricing power.

Some countries have introduced economic exchange and market principles to water in order to improve the water allocation system. Considerable efforts have been made to create, strengthen and facilitate water transfers. Prominent examples include some western states in the United States, Australia, Chile, India, and in Spain, where water scarcity has increased the need for reallocation of supplies, whose experience provide important lessons for future practice.

In theory, water trading is based on voluntary agreements between different users, who can trade water rights/entitlements as long as institutional arrangements allow, and transfers are geographically viable. By institutional arrangements, we refer to the rules, both formal and informal, that liberate and constrain choice behaviors in some repeated context (Bromley, 1989). Water markets represent a type of institutional arrangements based on transferrable private property rights. Water trading can both improve efficiency, as well as help meet specific social goals, such as improving environmental flows; however, the potential for success in meeting such goals will depend on the detailed design of institutional arrangements. Broadbent et al. (2012) argue that water market success will often involve including market participants in the design of the market rules.

Improved allocation can be achieved by the formation of a water market, usually with some form of regulatory oversight (government or independent). The role of the government and the definition of water rights and licenses are essential for effective water trading. However, a more fundamental shift is often needed in raising awareness of water market entry and the ability of market forces to improve water management, and encourage the private sector, especially users, to engage.

The objective of this chapter is to introduce the concept of water markets, their requirements, limitations and formation as a prominent water resource management tool. The range of analytical issues involved include the implementation principles and social relationships; the structure of the market, how water prices are formed and varied; and the diversity of water markets formats in different climatic regions and conditions.

This chapter presents and discusses water markets, as a pivotal allocation and exchange mechanism, and how they may help address a variety of water resource management issues. This includes, their potential as an effective mechanism to reallocate water in the context of increasing water scarcity and potentially water conflict (as new demands for water use arise, and water resources approach their limits).

4.2 Definitions and terms

To better understand the chapter, some common terms in this chapter are defined.

- **Market**

  In general, a market is a trading or exchange institution in which buyers and sellers are able to come together and use available information to pursue optimal choices (Zekri and Easter, 2005). A market is efficient when it allocates a good or service in a way that maximizes the net gains from trade (the sum of consumer and producer surplus). A well-functioning market uses price signals to efficiently allocate resources to supply the product at the lowest cost, and equilibrate supply and demand. A market does not exist in a vacuum, but rather within a broader set of institutional arrangements and governance. For example, as a private property regime, well-functioning markets require clear property rights that are exclusive, enforceable and transferrable. Further, trading such rights needs to happen with minimal transactions costs.

- **Water market**

  A water market is an approach that considers water as a private good that can be transferred between users, on the basis of price. As voluntary instruments that form a means by which to transfer water between users efficiently, water market mechanisms are consider as a potentially more effective approach to manage water resources. In general, the water market is a process whereby water rights can be traded legally. A water transfer is a voluntary agreement between buyers and sellers that results in a temporary or permanent change in the type, time or place of water use and / or rights. In this way, water can be transferred between agricultural, urban, industrial and environmental uses. The water market is referred to as a mechanism of allocation of water based on a water rights exchange for consumption that results in some improved allocation of water (Pujol et al., 2005), in the sense of generating net economic benefits and capturing gains from trade. Most basically there is a seller in the market who has a surplus water right to their current need, and a buyer who is facing water scarcity and is prepared to fulfill their need by paying for extra water. So, in principle, both the buyer and the seller can mutually benefit from the transfer. The water market provides water supply for uses where the economic value of water is high, rather than developing new sources of supply.

- **Water Allocation Systems**

Water allocation is the process of sorting the water available between legal applicants in a specific location (e.g. basin, irrigation system) at a specific time (e.g. season). This determines the amount of water available for human use and how it is shared between competing spaces and users. The allocation depends on the volume of water resources available in that season to time period. Water resource allocation methods are generally subject to social and economic factors that play a role in defining methods (including modeling) and indicators. Accordingly, there are several basic allocation methods (Dinar et al., 1997):

1) Marginal price: A marginal cost pricing (MCP) mechanism, in essence, targets a price for water to equal the marginal cost of supplying the last unit of that water. An allocation which equates water's unit price (the marginal value of water) with the marginal cost is considered an economically efficient, or socially optimal, allocation of water resources. The efficiency criterion maximizes the total value of production across all affected sectors of the economy.
2) Public water allocation: Public allocation is more prominent in the allocation of water resources by government bodies. In this approach, water is a state-owned resource that is allocated among the various stakeholders on a non-permanent basis (concession or license), possibly with some delivery or conveyance charge.
3) Water markets: Under this approach water is defined or treated as a private good that is able to be transferred among users, including as a (permanent) private property right.
4) User-Based Allocation: This approach requires collective action institutions that can decide on water rights. Farmer-managed irrigation systems are good examples of user-based water allocation. Studies have shown a wide variation of rules for allocation within such systems; by timed rotation, depth of water, area of land, or shares of the flow.

The six characteristics of an optimal water resource allocation mechanism comprise flexibility, meeting the expectations of stakeholders, reflecting the full opportunity cost of provision, taking into account social values related to water, predictability of allocation and equity mechanisms (Howe et al., 1986). In theory, there are many setting where the market mechanisms better meet these conditions than other allocation methods. In theory, water markets have the potential to generate optimal allocations of water.

- **Optimal Allocation**

Optimal allocation of water refers to that which maximizes the net output of a region, sometimes understood as arising from the intervention of a benevolent social planner. In economics, this is denominated Pareto efficient, which refers to an allocation of resources (including transfers of capital as compensation) that results in no stakeholder being better off without making another else worse off. Another way of conceptualizing the social optimum is through the implementation of trading and technology options that enhance welfare. In economic terms, this results in the equation of the marginal values of water across uses. The social optimum is referred to as "efficient" because water is allocated to the users that most value it (Chong and Sunding, 2006).

- **Water transfer**

As defined by the National Research Council (1992, Pg:2), Water transfer is defined as "any change in the point of diversion or a change in the type or location of use, ranging from simple internal adjustments to actual sales". Reallocation of water involves the transfer of a water right and/or the use of water under a water right from one person or entity to another, and any associated changes to one or more elements of the transferred water right. Such transfers depend on the rules in place (e.g. transfers between sectors). A transfer from an agricultural user to a municipality usually involves changes to all the elements of a water right. A change from agriculture to instream flow (e.g. environmental flow) requires a change in place of use, purpose of use, and elimination of the point of diversion.

- **Transactions**

Reallocation of water from one person or entity to another also includes different forms of transaction. The term transaction refers to the form of the agreement between the owner of the water right and the party receiving the water right. The type of transaction plays an important role in the willingness of water right holders and those seeking water to participate in a market. The permanent sale of a water right results in the transfer of both the water and the water right. This would also be the ultimate outcome if an option to buy a water right were ultimately exercised. Temporary transfers by lease result in the water being transferred for the term of the lease and the water right being retained by the water right holder. Leases can be short or long term (e.g. a single season, or multiple years). Leases can also be for split season whereby a user uses his or her water right for the first half of the season and a party, for example one seeking water for instream flow (e.g., to provide recreational benefits, or protect and endangered fish), leases the water for the second half of the season during periods of low stream flow.

- **Marginal value**

The change in total value for a unit change in some variable or amount of resource, is defined as the marginal value. If two water users have different marginal values for water, the users should be able to benefit from trade, whereby the user with the lower marginal value gives up a unit of water to the other user in exchange for payment. In a linear program, the dual price of a constraint is the same as the marginal value of the resource associated with the constraint.

4.3 Basic principles

The exchange of goods and services between buyers and sellers is conceptualized by the laws of supply and demand. Competitive markets can succeed in maximizing efficiency and the gains from trade. However imperfect markets can fail to maximize social efficiency due to a variety of reasons, including market power (and monopoly pricing), hidden information issues, discrimination over access, and the presence of external effects (e.g., when significant social costs due to environmental contamination are not incorporated in prices. Even if highly competitive markets may produce results that are very inequitable.

4.3.1 Water as a commodity

Unquestionably, water is unique in many dimensions and is necessary for life. Water has been increasingly considered as an "economic good", or commodity, for drinking, agricultural, and industrial uses, with its use being valued in terms of full economic (as opposed to subsidized) cost. Transferrable, valuable and scarce commodities with clear property rights have prices based on the interaction of supply and demand. Therefore, due to its scarcity and exchangeability, water has also been reframed as a commodity that can be subject to market signals and prices. In this regard, the principle of ownership of water (water rights) becomes meaningful.

As mentioned above, water can be seen as a commodity, allocated via private rights. The formulation of these rights and their inconsistency require monitoring of behaviors, especially as water use sometimes has negative external consequences including, undermining the rights of others, groundwater depletion, ecological degradation including through the discharge of sewage and pollutants, implying that monitoring of rights and consumption is essential.

4.3.2 Essential components for forming a water market

- Secure property rights to water
- Low transaction costs
- Presence and balance of buyers and sellers in the market (indicating how close the market is to monopoly power)
- Consideration of all the costs and benefits of market participants
- Minimum barriers for market entry
- Consideration of the interests of third parties and the environment

- Possibility of market formation and activity according to hydrological conditions

4.3.3 Environmental flows

A common concern often raised about water trading is the potential to exacerbate environmental degradation, especially in the case of an unregulated water market. For example, for riverine ecosystems it may be ecologically important to protect minimum environmental flows for ecological purposes (and in some cases maximum flows, such as with connected wetland or overbank regeneration). Flows may also be important from a water or hydrological quality perspective. Flowing water can provide an environmental service to dilute pollution (Chong and L. Sunding, 2006; Colby, 1991; Dinar and Zilberman, 1991). Flowing water can also be important to a variety of recreational activities. As noted by the National Water Commission (2011, Pg:115), "Water trading can cause adverse environmental impacts if it leads to more water being used in absolute terms, more water being used in inappropriate areas, or reduced river flows (for example, where trade increases the amount of water extracted for irrigation upstream in a river catchment and reduces the amount that would otherwise flow down the river to be used by irrigators further downstream)."

Most basically, environmental or instream flows can be protected in two basic ways: (i) through some form of government regulation, such as through some for protection of minimum instream flows (Berrens et al., 1998b); and (ii) through directly and legally recognized instream flows as tradable water right. Of note, (i) and (ii) are not mutually exclusive, and combinations exist. Further, with respect to (i), markets often function with a variety of social and environmental constraints, and the same can happen in water markets. From a non-market perspective, there is considerable evidence that instream flows can have considerable economic value (e.g., Berrens et al., 1996 and 1998a). In practice, one can find a wide variety of environmental flow laws and policies, and cases where explicit tradable rights to instream flows have enhanced environmental protection; Gillilan and Brown (1997) review the political/legal and historical context from instream flow protection across the various western U.S. states.

4.3.4 Conditions to establish water markets

According to the water market approach in different regions, it is necessary to define a set of conditions and requirements to establish it. The water market is formed by using a specific framework and by imposing some requirements. It should be noted that in the management of the water market, the government must play an effective role as enabler, and potentially regulator.

In the following, these requirements are described in various dimensions.

4.3.4.1 Water shortages

As seen in water markets worldwide, the prerequisite for the establishment of water markets is water scarcity, and a viable bargaining "space" (i.e., the presence of potential buyers whose maximum willingness to pay is greater than the minimum willingness to accept of potential sellers). For example, in a setting where high-value urban and industrial users have increasing water demand, and there are willing agricultural sellers with modest commercial returns. Meeting such increasing demand becomes problematic when supply constraints exist; therefore, when there is a shortage of water to meet demand, providing water through the transfer moves water from relatively lower to higher economic value uses (Zekri and Easter, 2005).

4.3.4.2 Separating water entitlements from land

According to the experiences observed in different parts of the world, trading water along with land is one of the constraints to strengthening and expanding water markets. These experiences show that one of the most important steps to strengthen the water markets is the separation of water entitlements from land. Separation of water ownership from land is deemed to facilitate trade in the water market.

4.3.4.3 Clear definition of water rights

The amount of water belonging to each water right based on the specific unit, type of consumption, point of delivery, should be clearly defined. As is shown in all water markets worldwide, this is necessary to facilitate trade and increase water market efficiency. Therefore, one of the most important factors in establishing a water market is defining the water right precisely.

4.3.4.4 Initial allocation of water rights

Different water markets have different criteria for the initial allocation of water rights, some of which are influenced by the cultural and customary structure of those contexts. The initial allocation mechanism plays an important role in market efficiency, especially in its monopolization.

4.3.4.5 Transaction costs

Transaction costs refer to the costs of sourcing information, bargaining and purchasing, as well as monitoring and enforcing the rules (Dahlman, 1979). Where all other factors are equal, transaction costs tend to result in a reduction of trade because they lower the net gains from water transactions (Connell et al., 2005), which can be significant (McCann and Easter, 2004). Transaction costs can either be fixed per transaction (which may be the case with legal or administrative fees), or can be relative to the quantity of water traded (as in the case of commission paid to an intermediary). Transaction costs can thus constrain the viability of market mechanism, to the extent that they may render some water markets ineffective. Transaction costs are divided into four categories (Erfani et al., 2014):

a) Establishment of the system, including making legal changes and infrastructure design costs
b) Pre-trade costs of information gathering about quantity and quality, and helping match buyers and sellers
c) Professional and legal contract costs
d) Cost of system administration including monitoring and dispute resolution

Transaction costs are a primary factors affecting water market efficiency. Generally, keeping transaction costs low tends to facilitate trade, and thus capturing the gains from trade and increasing market efficiency. That is the typical Coasian perspective is that the role of government is to provide clarity around property rights, and then lower transactions costs to facilitate trading. However, there may be cases where the inability to internalize all externalities (e.g., in the presence of significant environmental public goods) may be viewed as a pretext for imposing policy-induced transaction costs (e.g., how transfer appeal processes are allowed and who must bear the cost of a transfer appeal (see Colby, 1995).

4.3.4.6 Physical infrastructure suitable for water transfer

Trading cannot occur if water cannot be transported or sources of extraction cannot be transferred. The existence of appropriate physical infrastructure such as canals, as well as reservoirs and irrigation networks, will improve the performance of the water market and its efficiency and development. It is not possible to form a strong and active water market without access to surplus water in a basin, thus the completion and improvement of irrigation networks should lead to capacity in water markets. Of course, encouraging policies in this regard can speed up the process of building these networks. Attracting public participation, especially in the agricultural sector, and availability of cheap credit facilities for initial investment, will help this process.

4.3.4.7 Existence of executive bodies and popular organizations

Relative to idealized economic theory and the notion of a homogenous good with few external effects, water is a highly variable resource that is hard to store, and can seep and flow and evaporate; further the lack of market monitoring can lead to the rights of others being reduced, groundwater depletion and/or contamination, wastewater (including urban and industrial wastewater, can lead to the destruction of public and private rights). These potential consequences of utilization make it necessary to monitor water rights and consumption. This monitoring will also be implemented by enacting laws, establishing executive bodies, forming local courts and water enforcement. Therefore, the executive bodies which, in addition to monitoring the water market, are responsible for water management, maintaining, constructing transmission facilities, exchanging information, ensuring exchange and resolving disputes, and reducing negative externalities, are also important pillars of the water market. In terms of the specific characteristics of the water, the occurrence of conflicts in relation to water rights, as well as the implications, is likely to be resolved with appropriate measures by the water market regulator. It is imperative that water rights holders have a continuous and close relationship with these organizations and institutions, and also be aware of their management activities and decisions (protection and monitoring).

Water resource studies for each catchment to quantify and establish water resource balances, and record information about water uses from surface water and groundwater, is required for further information of regulators, supervisory bodies, water rights holders, and water users. A complete, transparent, and accessible information system from these data is helpful in facilitating water market interactions. This means that institutional arrangements should be in place for collecting continuous assessment of water resources, and rate of average precipitation, surface water flows, infiltration and recharge of aquifers, evaporation for of each basin, and monitoring and reporting to water rights holders, managers, engineers and planners.

4.3.4.8 External effects

Water rights transfers generally have the potential to affect more than just the two parties involved. Non-exclusivity of the water causes interaction and dependence among users. A water transfer from one location to another affects other water users, by changing water quality, flow volumes, or the timing of flows. In local markets, due to the limited scope of the market, the problem of negative externalities is less likely to occur, but in markets with a wide range of potential negative effects, such as third-party effects or environmental impacts, there is a need for specific rules and regulations to reduce these consequences. Trade may also lead to adverse social or equity impacts in some areas (for example, the reduced viability of agricultural communities and the increase in unemployment resulting from the transfer of water from agriculture to industry or urban use), so all of these should be taken into account. In some water markets studied, rules have been established to address the potential for these externalities.

4.3.5 Water market disadvantages

The development of water markets has not been without problems. There are several limitations and disadvantages of market-based solutions to public resource problems: water measurement when the amount of water fluctuates, enforcing water use laws, investing in transportation and transfers, the sale of water by poor farmers due to economic distress, and environmental impacts. Transferring water from agriculture to drinking water supply may reduce the return flow and affect the third consumer. In addition, increased consumption of drinking and industrial wastewater may increase contamination if measures are not required to limit the treatment of wastewater (Dinar et al., 1997).

A number of barriers to water trading have been identified that hamper the full operation and potential benefits of water markets. Some of these principal challenges are as follows (National Water Commission, 2010):

a) Technical issues: for example, high administrative costs involved in gathering the information needed to establish a market and rules of trade
b) Political issues: for example, the prioritization of certain uses and users (especially drinking water)
c) Social issues: for example, community concerns about water source depletion
d) Cultural issues: for example, water is a deeply symbolic resource that may have specific cultural meanings such as religious significance
e) Managerial issues: water resource managers, for example, are often engineers who have no experience of market mechanisms, while economist and lawyers involved in water reforms may lack an appreciation of the complexity of dynamic hydrological processes.

4.3.6 Types of water markets

Water markets can be divided into two categories: formal and informal,

Formal water markets are those markets in which tradable water rights are introduced through government policy, and the facilities and infrastructure required for its implementation are provided by the government. Usually a government agency is responsible for oversight and all transactions are recorded by official authorities. A variety of countries and regions have formalized legal transferrable water rights to better capture the benefits of informal trades that were happening. By moving out of this informal and possibly illegal status, "the potential to sell water rights makes them more valuable and provides an incentive for conserving water and reallocating it to higher-value uses." (Thobani, 1997, Pg: 166). Notably, tradable water rights can also comprise water leasing (for example for one season), sometimes referred to as spot sales, and indeed can facilitate such transactions. Finally, through the initial allocation of water rights, without charge, to existing users or water rights owners, tradable water rights can overcome the political problems associated with rising water prices, and setting no uniform charge, especially for sectors where it is not viable to implement market prices, such as irrigation. Governments can monitor operations and effectively enforce laws and regulations to prevent abuse of monopoly power, in order to ensure that sales do not have a negative impact on the water available to third parties (i.e. problems of return flow) or on environmental protection. The experience of establishing formal water markets exists in countries such as the United States, Chile, Mexico and Australia (Marino and Kemper, 1999).

Informal water markets are those in which the beneficiaries engage in exchange of water rights without government intervention or formal registration. There has been a long history of informal water markets in Iran and other countries such as Spain, India, Pakistan and China. Various studies show that informal markets are established local governments have failed to respond to rapidly changing circumstances (e.g., local shortages). As noted by Thobani (1997, Pg: 165) "Although such informal markets, where users contract water on their own, are technically illegal, governments usually turn a blind eye to them, perhaps because they manage to reallocate water quickly and voluntarily. In a typical transaction, a farmer sells a specified volume of their surplus groundwater or surface water for a season or a specified period to a neighboring farmer, or several farmers collectively sell some of their water to a nearby town."

4.4 Importance and necessity

Until around the 1980s, water management and policy-making around the world focused mainly on supply management through the construction of hydraulic infrastructure to meet growing demand for water. Influenced by this approach, the basis of resource development was always single-purpose. Such an approach did not address the intrinsic relationships and complexities of human-environment interactions. In other words, water resource development plans were merely for the physical control of water for economic gain, with little regard for environmental and social impacts. In addition, there was rarely public participation in the decision-making process. This approach has left undesirable and sometimes irreparable effects on ecosystems, which prompted the promotion of a more integrated approach to water resources management in global water policy. Therefore, since around the 1990s, water management has undergone significant changes. Integrated approaches to meet the growing need for water emphasize water saving and, voluntary mechanisms for water sales and reallocation (water demand management) between uses, rather than exploitation of new resources (water supply management), through improved management and institutional change. Water markets are one option for such institutional change, as they can mitigate water scarcity by facilitating water trading regimes and markets that allow lower-to-higher value reallocation (Howe et al., 1986; Easter et al., 1998, 1999; Saleth and Dinar, 2000). Water markets also provide price signals that can encourage investment in water-use efficiency and indicate the costs of shifting water from consumptive applications to alternatives, including in situ uses.

4.4.1 Impact of water market mechanisms on efficient allocation of water resources

Water market are efficient mechanisms for allocating water due to their flexible and voluntary nature that captures the mutually beneficial gains from trade.

Introducing a water market is seen as a form of decentralization, and a potential means for the optimal allocation of water resources among users, and the reduction of the effects of water scarcity (Gómez-Limón and Martinez, 2006).

Economists argue that when water is regarded as a marketable commodity it can be managed more effectively than when it is managed and allocated by the state (Gomez and Loh, 1996). According to this view, efficient use of water can only be achieved through market mechanisms, because of the benefits of competition (Griffin, 2006; Lee, 1999).

One of the most important issues in water resources management is the appropriate allocation system in each catchment which is expected to lead to optimization with the water market approach. Allocation can be based on prior and/or historic use, through an allocation system (new water rights), or through market transactions.

4.5 Establishing a water market

The requirements to establish a water market are: 1) separation of water and land ownership, 2) regulation of water trading, 3) infrastructure for water storage and transfer that allow upstream and downstream trading.

A water transfer can be local or transboundary, as long as infrastructure to transfer water is in place, or the possibility of transferring the source of water exists. Water rights can be sold or leased, and can be reallocated between agricultural, urban, industrial and environmental uses, depending on the regulations in place.

The capacity of water access and allocation arrangements to allow water trading critically depends on local circumstances, the range of future scenarios (and the extent to which they can be known) and the available regulatory architecture (Maestu and Gómez Ramos, 2013).

In general, describing the dynamics of water markets is challenging, because these depend on the specific characteristics of different countries, regions, basins and sources, and the nature of water usage there. Water prices are influenced by the amount of water available, location of use, time of use, volumetric flow of water, and the type of water source.

Water governance including, public participation, transparency, and collective decision-making by buyers and sellers, is governed by the rules of institutions. Many efforts have been made to identify the factors that strengthen water markets. For example, studies in Florida, USA show that by assigning responsibility to the water rights owners in these areas, better monitoring of surface and groundwater can be achieved (Ainslie, 2002).

Forming a strong water market, requires the establishment of a set of rules and guidelines entrusted to its rightful owners (Landry, 2000).

In forming a water market, issues such as the negative externalities of water consumption, reclaimed water quality, and its importance in maintaining the watershed's system and sustainability and prioritizing certain uses, especially drinking water, are issues to consider. However, the formulation of some rules should not remove the basic principles and ignore market power. Therefore, the explicit definition of rights, the rules governing them, and ability to exchange water under determined conditions are essential to the water market. Water markets also depend on the principle of information transparency. Based on this principle, right holders are informed about the status of water resources and the water market that allows them to make an economic decision as to whether to consume, save or sell surplus water. Water markets are based on an understanding between all the stakeholders and the rightful owners in each watershed. With the establishment of a water market, drinking water companies can officially purchase water from agricultural water holders, for example.

Groundwater, surface water, treated effluent, sewage effluent, storm water, and even rainfall, can be traded. The price difference of each of these resources is determined by the water market and its institutionalization. It must be acknowledged that water has a substitute, and its price and consumption are influenced by the level of supply and demand given the natural constraints and laws of each region, and the water market can respond to this relationship.

What should not be overlooked in this approach is that the government alone cannot address this important issue and therefore needs the assistance and participation of water right holders. In the optimal management of water resources, it is important to define and describe the responsibilities of water authorities to respond to problems, as in countries with advanced institutions. It is generally agreed that water allocation laws and regulations should be within the purview of water and environmental management organizations, and they should try to determine the price of water according to the value of the water according to market forces. In most economic and managerial theories, the water market plays a strong role in price and allocation, with its oversight body playing an important role in governance. Therefore, it can be said that the key characteristics of the water markets should be as follows:

- The transactions are formal and the water rights are clearly defined
- There is monitoring (regulation) of extraction, use, and transfers
- There is freedom to enter and exit the market
- The government has sovereignty in establishing a market, and a regulator (government or independent) oversees water allocation, use, and impacts

4.6 Examples

Some regions and countries have taken significant measures to establish and strengthen water markets. These experiences of these regions and countries can be used by others in shaping the water market.

This section highlights the process of forming and developing several prominent examples.

4.6.1 Australian water market

In the 1980s and 1990s, successive governments of Australia supported water rights trading by eliminating a variety of restrictions (National Water Commission, 2010).

In the 1980s, as water resources became scarce, water authorities set up a committee to come up with a new strategy for the optimal allocation of available resources, which eventually proposed establishing water markets to increase water use efficiency. This led to the establishment of water markets in the states of South Australia in 1982, New South Wales and Queensland in 1989, and Victoria in 1991 (Murray-Darling Basin Commission, 1995). To allow the market to function effectively, the Australian government made institutional changes. They found that they had to improve record-keeping of water rights. There was evidence of drawing drawn down aquifers, so they started licensing groundwater as well as surface water, and they implemented rules requiring the two to be managed in conjunction. The Australian states are relatively independent, so Australia worked to also allow trade between states.

Drought motivated the Australians to make water markets work, especially in the Murray Darling Basin, which has the most active water market in the world (Raffensperger and Milke, 2017). Most Australian trading is in reservoir and other surface water, although, increasingly, there is also trading in groundwater. Water markets in this country are active because the government removes restrictions on trade. Trade can occur in an area even if water is over-allocated; the logic is that water trade will not cause further over-allocation (Murray-Darling Basin Authority n.d.1). Traders must give notice of the traded prices. Regulators may restrict trade due to physical constraints, hydrological conditions and the need to protect the environment. Private brokers trade via internet. The Australian Government publishes significant information about water trading online.

During 2004–05, 0.167 M^3 was traded in the Australian temporary water market, accounting for 3.5 percent of the total volume of water allocated that season. Less than 1 percent of the total permanent entitlement has changed (Australia, 2006b). Trading was different in each state, with the most activity in Victoria where 38 percent of the entitlement was leased in the temporary market, representing 6.7 percent of the total volume allocated to that state. Approximately 3 percent of the total entitlement was permanently transferred in Victoria in 2004–05 (Australia, 2006b).

Some lessons learned from the experience of establishing the water market in Australia are as follows (National Water Commission, 2011):

Water market development is possible even in complex hydrological systems.

If water markets are well designed, there will be significant benefits to systems that have limited access to water resources (i.e., commonly confront scarcity).

There are general physical and economic characteristics that suggest where water trading will be most beneficial, such as where:

a) Resources are fully developed for consumptive use
b) There is variability in seasonal water availability and between connected systems within seasons
c) There are a large number of connected water users
d) Water users have varying demands and degrees of flexibility to respond to water shortages
e) Water users are exposed to the cycles of global agricultural markets
f) Demands for urban and environmental water are increasing
g) There is pressure for change in the existing structure of water-using industries.

Water market design needs to be informed by the specific and historical features of local water resources management.

General prerequisites to establishing an efficient water market include:

a) Setting an effective cap on total sustainable extractions (preferably before scarcity becomes acute)
b) Establishing well-defined entitlements so that users know exactly what they can buy and sell
c) Establishing a legal and governance framework within which water trading take place
d) Implementation of essential elements of water management such as metering and water accounting.

An incremental approach to water market development is appropriate to manage uncertainty and stakeholder concerns, but benefits will be forgone if suboptimal arrangements are left in place too long.

There is a need to measuring and considering the environmental and social outcomes that could be affected by water market.

Market participants learn quickly and make decisions based on the rules that are in place. Any efforts to stifle market development or impede progress inevitably lead to creativity by market players, which may have unintended consequences for property right holders.

As water markets mature and develop, roles and responsibilities influencing market outcomes need to be assigned carefully to avoid conflicts of interest, which can undermine reform objectives.

4.6.2 Chilean water market

Formal water markets were introduced in Chile in 1981 as part of a package of free-market economic reforms (Bauer, 1998). Water rights were defined as private property, which were expected to provide users with the security of tenure that would encourage private investment in infrastructure, and to enable water trading in order to incentivize reallocation according to meet demand. For these purposes, water rights were separated from land, were afforded constitutional protection, and were regulated by private (civil) law. Water rights were required for all uses except rural drinking water (for people and livestock), with surface water and groundwater designated as separate rights (Budds, 2004), and distinction between consumptive (e.g. irrigation, water supply) and non-consumptive (i.e. hydroelectric power) rights. Water rights were allocated in three distinct ways: (1) existing water rights were recognized through a process of regularization; (2) new water rights to sources that were not fully allocated could be requested; and, where new sources were no longer available, (3) water rights could be purchased through the market. The state administers allocation (1 and 2), and records sales (3), but regulation is undertaken by users and their organizations, and the judicial system in the case of disputes that cannot be resolved by the former.

In 2005, some minor reforms were made to water markets to respond to perceived market failures. The most important of these was the speculative acquisition and accumulation of water rights without using them, which was seen to artificially restrict supply and thus the market, and was addressed through a tax on non-use after five years. In addition, environmental regulations were strengthened for groundwater extraction and environmental flows, which had not been considered in the early 1980s (Budds, 2013).

Since 1981, markets in (permanent) water rights have been generally inactive in most of Chile. This has been attributed to geographical barriers to trading water across space, infrastructural barriers to transfer water between users, cultural barriers to the idea that water should be traded, and administrative barriers to selling water rights that may not be properly registered (Bauer, 1998; Budds, 2004; Hearne and Easter, 1997; Prieto, 2016). Social attitudes are also important, as users may be reluctant to sell water that may be needed for future water security, to buy water rights when illegal extraction (especially of groundwater) is viable amid low regulation, or buy water that has been acquired from indigenous groups due to its controversial nature (i.e. mining companies) (Budds, 2004, 2009). However, in some basins, the leasing of (temporary) water volumes has been active, especially where the presence of major storage reservoirs gives an indication of the availability of water ahead of the agricultural season, as in the Limarí valley in northern Chile (Hadjigeorgalis, 2008; Hearne and Easter, 1997; Romano and Leporati, 2002).

While some authors have posited that inactive water markets are due to the 'optimal' original allocation of water rights that has produced few sales (Ríos and Quiroz, 1995), other empirical studies have suggested that water trading has largely occurred from small to large farmers, especially in times of hardship (Hadjigeorgalis, 2008; Romano and Leporati, 2002). Furthermore, peasant farmers may also have less access to new rights, which are likely to be groundwater rights since most surface water was already allocated by 1981, due to less familiarity and/or willingness to engage with state administrative regulations and procedures (Budds, 2004). Moreover, the role of (user-regulated) water markets in managing demand has been questioned, as users have sought to increase supply through other means than the water market (Budds, 2009). This is manifest in the context of the expansion of natural resources industries in semi-arid and arid areas of northern Chile, where responses to acute water scarcity have primarily comprised supply-led hydraulic infrastructure (e.g. storage reservoirs, desalination plants) rather than more active water trading, not least because many water sources no longer yield the supply to fulfill the water right to them (Budds, 2020).

4.6.3 Western US water markets

Water markets of numerous varying details (e.g., for both permanent sales and temporary leases) are commonly found in the western United States (US), and in some cases have existed for many decades (WGA, 2012). The magnitude of transaction costs, and the locus of who bears them, can vary greatly across state (Colby, 1995). Western water rights are primarily prior appropriation rights ("first-in-time, first-in-right" priority systems, and subject to "use-it-or-lose-it"), where the water right holder has a claim in an annual queue. In many cases, a significant proportion of high priority (and thus higher valued) rights were historically concentrated or held by irrigated agricultural uses. Water markets have helped build flexibility into these systems. Agricultural-seller-to-agricultural-buyer trades (especially short-term leases) are common within the numerous irrigation districts found across the western US. However, water markets have also certainly facilitated transfers away from agricultural uses and towards growth in municipal and industrial uses, as well as environmental uses which keep water in the river (see Gillilan and Brown, 1997).

To give a sense of scale, for the 20-year period, 1988–2009, the quantity of water traded across the 11 western states varied in the range of 1–3 million acre feet (MAF) annually (WGA, 2012). High trading frequency states included Colorado (with roughly 2,000 trades over the period). High water volume or quantity traded states included California (with over 13 MAF traded over the period) (see WGA, 2012).While in many cases market transfer have moved permanent water from relatively lower-valued agricultural uses to higher-valued alternatives, in what is referred to as "buy and dry" if it takes agricultural land out of production. However, innovative transfer and lease programs are also pursued with the intent of protecting the integrity of agricultural communities (e.g., including dry-year contingent contracts, and partial season leasing programs) and keep farmers in farming (WGA, 2012).

To give one innovative example of the variety of water markets found, a new market has been recently created for moving leased water from willing individual agricultural sellers to a single federal entity buyer for the purpose of protecting environmental flows in the Middle Rio Grande of the western US state of New Mexico. The market was developed and implemented by the local, quasi-governmental irrigation district (the Middle Rio Grande Conservancy District

[MRGCD]) pursuant to an interagency government agreement and consultation under the provisions of the US Endangered Species Act (ESA) of 1973, as amended. Public preferences have demonstrated significant support for protecting the silvery minnow (Berrens et al., 1996 and 1998a). Leased water is released to protect an endangered fish species, the *silvery minnow*, as part of the species survival plan under a 2016 Biological Opinion (US Fish and Wildlife Service, 2016). The ESA has the power to restrict development and government agency actions that jeopardize an endangered species. Thus, the required water lease market (implemented with a fixed lease price) was one part of broader species protection plan. The market went active in the water-scarce summer of 2020, and was able to meet all initial water quantity targets for signing up willing sellers.

4.7 Summary

Water resources management around the world, especially in arid regions, has increasingly pursued flexible approaches to dealing with scarcity, while engaging or improving public participation. The search for innovative governance and institutional arrangements has only heightened with the evolving challenge of managing natural resources, such as water under a changing climate. In this chapter, water markets were introduced as one approach to adding flexibility into a regional system. Water markets provide benefits by facilitating voluntary transfers among users, and encouraging flexible water management (e.g., for dealing with drought or changing preferences and circumstances). When scarcity and the changing opportunity cost of water are reflected in market transfer prices, water is more efficiently utilized.

However, one of the common challenges to implementing functioning water markets has been establishing the rules or institutional arrangements that regulate water markets. A prominent example is resource monitoring, without which water markets cannot function effectively due to the unseen or negative consequences of water extraction and use. A second example is the consideration of accommodating instream flows for environmental protection, whether as constraints on market transactions or through directly allowing the purchase of instream water rights (Gillilan and Brown, 1997). Therefore, the establishment of clear rules, along with building the governing entities and capacity to manage and monitor the performance of these markets, are critically important requirements. Successfully establishing a functioning water market needs to be based on the specific regional characteristics; i.e., there needs to be process design focused on meeting the needs of the market participants and protecting the water resource base. The challenge for many regions and even countries is that simply rewriting broad laws encouraging water transfers is not enough. The real work is in fully establishing property rights, building governance organizations that, among other duties, implement strong monitoring programs (for understanding water quality and quantity and its movement) and conflict resolution mechanisms, as well as designing clear rules for regular trades and transfers.

References

Ainslie, K.F., 2002. Water management policy in Florida: regional politics and market development. public policy analyst, the Cato Institute. Policy Report.

Anon, 1995. Arizona interstate water transfers issue again on table. Arizona Water Res. 4 (No. 1), 1 January.

Bauer, C., 1998. Against the Current? Privatization, Water Markets and the State in Chile. Boston: Kluwer Academic Publishers.

Berrens, R.P., Bohara, A.K., Jenkins-Smith, H., Silva, C.L., Ganderton, P., Brookshire, D., 1998a. A joint investigation of public support and public values: case of instream flows in New Mexico. Ecol. Econ. 27 (2), 189–203.

Berrens, R.P., Brookshire, D.S., McKee, M., Schmidt, C., 1998b. Implementing the safe minimum standard approach: two case studies from the U.S. endangered species act. Land Econ. 74 (2), 147–161.

Berrens, R., Ganderton, P., Silva, C.L., 1996. Valuing the protection of minimum instream flows in New Mexico. J. Agric. Res. Econ. 21 (2), 294–309.

Broadbent, C., Brookshire, D., Coursey, D., Ganderton, P., Tidwell, V., 2012. Water markets in New Mexico. In: Brookshire, D., Gupta, H., Matthews, O.P. (Eds.), Water Policy in New Mexico. RFF Press, pp. 153–175.

Bromley, D., 1989. Economic Interests and Institutions: The Conceptual Foundations of Public Policy. Blackwell.

Budds, J., 2004. Power, nature and neoliberalism: the political ecology of water in Chile. Singap. J. Trop. Geogr. 25 (3), 322–342.

Budds, J., 2009. The 1981 water code: the impacts of private tradable water rights on peasant and indigenous communities in Northern Chile. In: Alexander, W. (Ed.), Lost in the Long Transition: Struggles for Social Justice in Neoliberal Chile. Lexington, Lanham.

Budds, J., 2013. Water, power, and the production of neoliberalism in Chile, 1973–2005. Environ. Plan. D. 31, 301–318.

Budds, J., 2020. Securing the market: water security and the internal contradictions of Chile's Water Code. Geoforum 113, 165–175.

Chong, H., Sunding, D, 2006. Water markets and trading. Annu. Rev. Environ. Resour. 31, 239–264.

Colby, B.G., 1991. Economic incentives and agricultural drainage problems: the role of water transfers. Econ. Manag. Water Drain. Agric. 56, 803–820.

Colby, B.G., 1995. Regulation, imperfect markets, and transactions costs: the elusive quest for efficiency in water allocation. In: Bromley, D., Vatn, A. (Eds.), The Hndbook of Environmental Economics. Blackwell Press, pp. 475–502.

Connell, D., Dovers, S., Grafton, Q., 2005. A critical analysis of the national water initiative. Australasian J. Nat. Res. Law and Policy 10 (1), 81–107.

Dahlman, C.J., 1979. The problem of externality. J. Law Econ. 22 (1), 141–162.

Dinar, A., Zilberman, D., 1991. The Economics and Management of Water and Drainage in Agriculture. Kluwer Acad, Boston.

Dinar, A., Rosegrant, M.W., Meinzen-Dick, R., 1997. Water Allocation Mechanisms Principles and Examples. World Bank, Agriculture and Natural Resources Department: Policy Research Working Paper, Washington, D.C..

Easter, K.W., Rosegrant, M.W., Dinar, A., 1998. Markets for Water: Potential and Performance. Kluwer Academic Publishers, Boston.

Easter, K.W., Rosegrant, M.W., Dinar, A., 1999. Formal and informal markets for water: institutions, performance and constraints. World Bank Res. Obs. 14 (1), 99–116.

Eheart, J.W., M. Lyon, R., 1983. Alternative structures for water rights markets. Water Resour. Res. 19 (No. 4), 887–894 Aug..

Erfani, T., Binions, O., Harou, J.J., 2014. Simulating water markets with transaction costs. Water Resour. Res. 50 (6), 4726–4745.

Gillilan, D., Brown, T., 1997. Instream Flow Protection: Seeking Balance in Western Water Law. Island Press.

Gomez, S., Loh, P., 1996. Communities and Water Markets: a Review of the Model Water Transfer Act, Vol.4. Hasting West-Northwest Journal of Environmental law and Policy, No. 0 (Fall), San Francisco, California.

Gómez-Limón, J.A., Martínez, Y., 2006. Multi-criteria modelling of irrigation water market at basin level: a Spanish case study. Eur. J. Oper. Res. 173 (1), 313–336.

Grey, D., Sadoff, C.W., 2007. Sink or swim? Water security for growth and development. Water Poliey 9, 545–571.

Griffin, R., 2006. Water Resource Economics: The Analysis of Scarcity, Policy and Projects. Cambridge, Massachusetts, The MIT Press.

Hadjigeorgalis, E., 2008. Distributional impacts of water markets on small farmers: is there a safety net? Water Resour. Res. 44 (10), W10416.

Hearne, R., Easter, K, 1997. The economic and financial gains from water markets in Chile. Agric. Econ. 15 (3), 187–199.

Howe, C.W., Schurmeier, D.R., Shaw, W.D., 1986. Innovative approaches to water allocation: the potential for water markets. Water Resour. Res. 22 (4), 439–445.

Jepson, W., Budds, J., Eichelberger, L., Harris, L., Norman, E., O'Reilly, K., Pearson, A., Shah, S., Shinn, J., Staddon, C., Stoler, J., Wutich, A., Young, S., 2017. Advancing human capabilities for water security: a relational approach. Water Security 1 (1), 46–52.

Landry, C., 2000. A free market solution to groundwater allocation in Texas: a critical assessment of the house natural resources. Committee Interim Report on Groundwater, December.

Lazarova, V., Levine, B., Shack, J., Cirelli, G., Jeffrey, P., Muntau, H., Salgot, M., Brissaud, F., 2007. Role of water reuse for enhancing integrated water management in Europe and Mediterranean countries. Water Sci. Technol. 43 (10), 25–33.

Lee, K.N., 1999. Appraising adaptive management. Conservation Ecology 3. 2, online at http://www.consecol.org.vol3/iss2/art3 .

Lund, J.R., Israel, M., Kanazawa, R., 1992. Recent California Water Transfers: Emerging Options in Water Management. Report RD-38, U.S. Army Corps of Engineers Hydrologic Engineering Center, Davis, CA.

Maestu, J., Gómez Ramos, A., 2013. Conclusions and recommendations for implementing water trading. In: Maestu, J. (Ed.), Water Trading and Global Water Scarcity: International Experiences. RFF Press, New York.

Marino, M., Kemper, K.E. (Eds.), 1999. Institutional Frameworks in Successful Water Markets: Brazil, Spain, and Colorado. USA WORLD BANK TECHNICAL PAPER NO. 427.

McCann, L., Easter, K.W., 2004. A framework for estimating the transaction costs of alternative mechanisms for water exchange and allocation. Water Resour. Res. 40, W09S09. doi:10.1029/2003WR002830.

Murray-Darling Basin Commission, 1995. An Audit of Water Use in Murray-Darling Basin. Murray-Darling Basin Authority, Canberra.

Murray-Darling Basin Authority. (n.d.1). Guidelines for water trading rules, Introduction. http://www.mdba.gov.au/sites/default/files/pubs/01_WTG-REFERENCE_final.pdf. Accessed 30 Apr 2016.

National Research Council of the National Academies, 1992. Water Transfers in the West: Efficiency, Equity, and the Environment. National Research Council of the National Academies, Washington, DC.

National Water Commission, 2010. The impacts of water trading in the southern Murray–Darling Basin. National Water Commission. Commonwealth of Australia. http://www.nwc.gov.au/__data/assets/pdf_file/0019/10783/681-NWC_ImpactsofTrade_web.pdf. Accessed 30 Apr 2016.

National Water Commission, 2011. Water Markets in Australia: a Short History. National Water Commission. Commonwealth of Australia.

Prieto, M., 2016. Bringing water markets down to Chile's Atacama Desert. Water Int. 41 (2), 191–212.

Pujol, J., Raggi, M., Viaggi, D., 2005. The potential impact of markets for irrigation water in Italy and Spain: a comparison of two study areas. Aust. J. Agric. Resour. Econ. 50 (3), 361–380.

Raffensperger, J.F., Milk, M.W., 2017. Smart Markets for Water Resources. Global Issue in Water Policy. Springer Publication doi:10.1007/978-3-319-55008-4.

Ríos, M.A. Quiroz, J.A., 1995. The Market for Water Rights in Chile: Major Issues. Washington DC: World Bank.

Romano, D., Leporati, M., 2002. The distributive impact of the water market in Chile: a case study in Limarí Province, 1981-1997. Q. J. Int. Agric. 41 (1–2), 41–58.

Saleth, R.M., Dinar, A., 2000. Institutional changes in global water sector: trends, patterns and implications. Water Policy 2 (3), 175–199.

Thobani, M., 1997. Formal water markets: why, when, and how to introduce tradable water rights. The World Bank Research Observer, Vol. 12, No. 2 (Aug 1997), pp. 161–179. Published by Oxford University Press.

US Fish and Wildlife Service, 2016. Final Biological and Conference Opinion for Bureau of Reclamation, Bureau of Indian Affairs, and Non-Federal Water Management and Maintenance Activities on the Middle Rio Grande, New Mexico. Consultation Number 02ENNM00-2013-F-0033.

Water Account Australia 2004-05. Canberra: Australian Bureau of Statistics, November 2006b.

Western Governors Association (WGA), 2012. Water Transfers in the West. Western States Water Council. Western Governors Association.

Zekri, S., Easter, E., 2005. Estimating the potential gains from water markets: a case study from Tunisia. Agric. Water Manage. 72, 161–175.

Chapter 5

Introducing water banks: principles and importance

Sina Fazeli[a], Omid Bozorg-Haddad[a], Marcus Randall[b] and Robert P. Berrens[c]

[a] *Department of Irrigation & Reclamation Engineering, Faculty of Agriculture Engineering & Technology, College of Agriculture & Natural Resources, University of Tehran, Karaj, Alborz, Iran.* [b] *Center for Agriculture and Financial Big Data Analytics Academia, Bond Business School, Bond University, Queensland, Australia.* [c] *Department of Economics, University of New Mexico, Albuquerque NM, United States*

5.1 Introduction

Water scarcity can significantly constrain the growth and development of regions and countries (Barbier, 2004). Continuing population growth shifts out the demand for food, and without additional efficiencies in production increases global water withdrawal and consumption. Global warming, has meant that there is an overall reduction in freshwater availability in many regions, as well as more frequent and severe drought periods (Montilla-López et al., 2016). This phenomenon is especially evident in arid and semi-arid regions that rely on irrigated agriculture, and agricultural consumption can commonly account for up to 80 percent of total water use (Montilla-López et al., 2016). Thus, water security and the larger interconnections between food, water and energy systems play an important role in achieving sustainable development (Beck and Walker, 2013). For example, due to the close feedback between water supply and energy production, without water security, energy supply and sufficient food supply are at risk.

The twin stressors of population growth and climate change mean that dealing with increasing water shortages requires a greater focus on building mechanisms for adaptability (Brown et al., 2019). In recent years, the limitations of available water resources have prompted many countries to refine their perspective towards water resource management. However, management approaches to cope with water scarcity commonly remain focused on increasing the total fresh water supplied (increasing reservoir capacity, groundwater withdrawal or reusing of treated wastewater). Evidence shows that these approaches have limited success for both environmental (most water resources have been used currently) and

Economical, Political, and Social Issues in Water Resources. DOI: 10.1016/B978-0-323-90567-1.00009-7

economic (producing additional water supply is expensive) reasons (Montilla-López et al., 2016). This situation has led to a gradual reorientation in water policy to focus more on managing existing water resources (i.e., moving from a primary focus on supply management to demand management and resource reallocation) (Montilla-López et al., 2016). Moving away from a primary focus on augmenting supply requires development of detailed policies, strategies, and step-by-step plans to be more consistent with the environment and institutional context. Such measures include various economic instruments or incentive and exchange mechanisms such as water pricing, water markets, voluntary agreements, and the establishment of water banks (for creating organized systems of deposits and withdrawals).

The formation of water banks and connected water markets don't create additional supply, but they do create regional flexibility in confronting water scarcity. The common goal of a water bank is moving water to where it is needed more or valued more.

Water banks, which collect deposited water rights and facilitate transfers, are not a new phenomenon; they have long been established in various corners of the world. For example, water banks under various titles, such as the California Drought Water Bank or the Idaho Water Supply Bank, have been in the United States for many years. Other countries that utilize water banks, of various scale and scope, include Australia and, Spain. In general, water banking is used effectively in water resources management in many countries and regions. Water banking is used to solve water-related issues by facilitating the re-allocation of water in one catchment to other catchments will conserve water resources so that it can be used in emergencies and droughts. The need for banks is increasing as climate conditions change and water reserves decline.

This chapter examines the structure of water banks and the conditions for their establishment, and provides examples of water banks in the world.

5.2 Definitions and terms

To appreciate the context and contribution of this chapter, it is important to first have a clear understanding of its key concepts and terms. These are given below.

5.2.1 Water bank

The terms "water bank" and "water banking" have been widely used for a variety of water resources management practices. Water banking is usually used in places with significant storage capacity or the ability to easily transfer water. The term water bank is commonly used for two purposes: (1) physical storage of water for later use; and (2) a method to facilitate the transfer of water rights (Dellapenna, 2000; National Research Council, 2004). Or the focus is on (2), or the combination of (1) and (2).

So as not to be confused with notions of simple storage, a water bank and more generally the concept of water banking is used here and defined as the depositing of water or water rights to facilitate transfers (Montilla-López et al., 2016). Or, as (O'Donnell and Colby, 2010. Pg: 2) state "an institutional mechanism designed to facilitate transfer of water on a temporary, intermittent or permanent basis through voluntary exchange and exchange of various types of surface, groundwater, and storage entitlements." Water banking is often viewed as part of a broader regional management approach that allows options for the reallocation of limited water resources. In fact, water banking is a measure of water transfer at specified times. The water bank can also use or store water appropriately for the future for other people's use, and receive money or water correctly instead of the water provided to other people. In other words, the bank lends or sells water to people in need (Murcia, 2019). A water bank facilitates the transfer of water under a set of rules regarding prices and quantities, with the bank functioning as an intermediary in the water market between buyers and sellers (Quesne et al., 2007). A water bank is an entity that is an intermediary between potential buyers and sellers of water rights. This could be for wither temporary (spot) or permanent. By facilitating voluntary exchanges of water and reallocating water rights to more valuable needs as well as creating strategic reserves for the future, the water bank can enhance the water security needed to achieve sustainable development goals.

Cartron et al. (2002) define water banking as a tool for reallocating water rights or transferring the water from the water rights holder to another person and operated by a central institution. The bank not only identifies potential buyers and seller of the water rights, but also the water right holders can deposit their water rights in the bank. In this case, the bank is responsible for maintaining these water rights or transferring them to another person. MacDonnell et al., (1994) introduces the water bank as an intermediary tasked to facilitate water exchange for the allocation of water to new uses. So the water bank, as an intermediary, tried to find buyers and sellers, but with the distinction of being as an institution, it has predefined organizational processes, and in accordance with the laws of the region, it has adopted certain restrictions on its activities. It is important to note that the establishment of the water bank in each country and region is done according to the socioeconomic conditions of the same country and region (O'brin, 1999).

5.2.2 Water rights

According to Clifford et al. (2004, Pg:156) A water right is a legal authorization to use a certain amount of water for a specific beneficial purpose(s), and may also have a timing dimension. Water rights are identified in the form of permits or certificates. A permit is the right to develop a water use on a specified schedule with reasonable progress under specific conditions such as protection of senior water right holders. A certificate is granted once the water right development schedule and all the conditions have been satisfied. Water rights typically specify

- Source of water
- Pointed of water diversion or withdrawal
- Purpose of water use
- Quantity of water that may be used
- Location of use
- Conditions such as seasonal use
- Priority date (if within a prior appropriation system)

5.2.3 Water withdrawal

The amount of water extracted or produced from the source. Extraction may occur through pumping of an aquifer or a surface water diversion (Clifford et al., 2004. Pg: 156).

5.3 Basic principles

As mentioned previously, water banking is defined as "an institutional mechanism designed to facilitate transfer of water temporary, intermittent or permanent basis through voluntary exchange and exchange of various types of surface, groundwater, and storage entitlements." (O'Donnell and Colby, 2010, Pg:2). The bank is responsible for facilitating water transactions. These transactions include the review of leasehold rights, including the registration of liabilities and payments, access to water storage, monitoring of storage adjustments, and permitting water to be delivered to the intended destination.

The most important opportunity that establishing a water bank can achieve, is to create a long-term strategic reserve to increase water security; the bank does this by increasing flexibility for transfers, but not through increasingly supply. Therefore, if successful in enhancing flexibility the bank can greatly increase the probability that a particular target is met. For example, the target may be available drinking water, which is considered the most important water requirement and it can be used in times of severe water shortage and drought.

Using a water bank provides an opportunity for the combination use of surface water and groundwater resources, which is called conjunctive use. In the long run, each region has periods of drought and wet. The bank can store underground water stored in the aquifer during the wet periods and use it during drought periods. A water bank can also play a role in flood management if there is adequate infrastructure. In addition, the bank in a large area, where the concentration of rainfall varies in different parts of the country, can transfer water from the areas where the concentration of rain is higher to the lower water points through the exchanges it performs.

There are great potential economic and environmental benefits by investing in sound water resources management practices. Privatization and markets of water provide the necessary incentives to use water resources efficiently. This will lead to a reduction in expensive investments in the development of water resources. Further many banks pursue their environmental goals in addition to economic objectives and water supply. Some banks try to restore river water to an acceptable level. This will prevent some species from being endangered. For example, the Idaho Water Supply Bank in the United States provides instream flows in the lower Lemhi River during low flow periods (Clifford et al., 2004). Also, each bank can also provide its users with incentives to purify wastewater and drainage of land by setting reasonable prices and criteria for storing water in a specified quality bank. With this policy, the water bank can also play a role in improving water quality. Any bank can along with the quantitative water trading, also deal with qualitative aspects of water trades.

Water banks can issue specific instructions such as (Clifford et al., 2004):

- Determining how the water rights can be deposited in the bank,
- Determine the volume of water transferable in the bank if necessary,
- Identifying who can buy or rent from the water bank,
- Determine the terms of the contract and the price,
- Facilitate regulatory measures.

The structure of the water bank is discussed below.

5.3.1 Market structure of water banking

The interaction between market participants in transactions is determined by the market structure. The market structure also plays a key role in price determination and the publication of market information. Price and market information are two important components in the development of water banks, especially in the early stages of market development. Without sufficient informations about these, participants would find it difficult to decide how to conduct their transactions. Therefore, determining the structure of the bank has a significant role in its success and activity water bank. Following Clifford et al. (2004), there are various types of water bank structures:

Clearing House: A Clearing House is the simplest type of structure (Clifford et al., 2004). In this case primary role of the bank is to provide venue for posting bids of prospective buyers and sellers (i.e., willingness to pay and willingness to accept, respectively). The venue could range from a physical bulletin board in some earlier time, to a website more currently, allows a place for buyers and sellers to find each other. While this may often then involve individual-level negotiations and exchanges with time. Equilibrium prices may emerge via repeat interactions between buyers and sellers. The only role of the bank in this structure is to provide sufficient information to the bank participants; the bank has no role in setting prices. The only role of the bank in this structure is to provide sufficient information to the bank participants; the bank has no role in setting prices.

Fixed Price: Another common structure in water banks is the fixed price structure (Clifford et al., 2004), where fixed prices are posted presumably for a uniform set of water rights (e.g., standard quality and reliability). The fixed price is further assumed to mimic a market clearing price, i.e. the bank manager has enough information to accurately estimate the clearing price in the market. Factors such market and climate conditions, which affect the amount of water that can be exchanged, are considered to be the same (or at least approximately held constant) during all transactions.

Water supply options: A water supply option structure provides banking participants greater flexibility to manage their financial and water supply risks (Clifford et al., 2004). Instead of buying water directly, participants enter constructional agreements to buy or sell options to water, which specify price, quantity, timing (e.g., some future date) and location of delivery. While not yet widely used, as banking and trading grows more complex, more complicated water supply options are more likely to be implemented.

Auctions for water bank: The water bank may use an auction market structure (O'Donnell and Colby, 2009a). Auctions can be structured in a wide variety of ways and represent an approach for allocating scarce goods based on competition (Clifford et al., 2004). Most basically, alternative forms include both (i) procurement auctions where the bank buys or obtains water from willing sellers; and (ii) conventional auctions where the bank auctions water – already held or deposited – to willing buyers (Clifford et al., 2004).

Auctions have a high efficiency so that it can be assured that the product (here water) is dedicated to its most valuable use. While there is a large economics literature about auction design and expected properties, generally auction methods are very effective in determining the market price of water. And auctions always provide information to the bank. For example, depending on the structure of the auction, participants reveal information on willingness to pay (demand) and willingness to accept (supply) through their bids and offers. So, even when no transaction occurs, the bank is obtaining information about relative value.

5.3.2 Purpose of water banking

Generally, the common purpose of a water bank is to provide a system of collecting deposits and organizing withdrawals. But, the goals of any given water bank may be broader (Clifford et al., 2004):

A) Facilitate efficient market exchange mechanisms to allocate resources, and bring disparate buyers and sellers together,

B) Ensuring future water for particular purpose. E.g., human consumption, agricultural and fish,

C) Protecting resources by encouraging rights holders to conserve and deposit their rights into the bank,
D) Creating reliable water supply opportunities during drought periods,
E) Resolving disputes arising out of competition between surface water and groundwater users.

Transaction efficiency is characterized by flexible scheduling, improved water delivery, efficient water storage and flood control. By balancing the priority of rights, the bank can help better meet demand using available resources.

5.3.3 Bank management

Bank management and operations have a considerable impact on the effectiveness of the bank program. Trust is a very important issue and directly affects how users interact with the water bank and accept its operations and policies. A water bank may be in the form of a public or private entity. Depending on the social circumstances of a particular water source, a bank may be organized as a cooperative of local institutions and citizen groups. In some cases, the bank is run by an elected board of directors, which includes citizens who are eligible.

5.3.4 Water banking: questions to answer

Water banks operate in various ways, with differing purposes and locus of control, depending on the region or country. Nevertheless, the common goal of banks is to transfer water to demand centers. This section provides a list of questions that should be considered when establishing a water bank. A summary of this list is given in Table 5.1 (Clifford et al., 2004).

5.3.5 Water banking formats

Water banks can be sorted into three broad categories or formats (Clifford et al., 2004): institutional banking, surface water banking, and groundwater banking. The following explains each of these three types of banks.

- **Institutional banking**
  In institutional water banking, there is a legal and accounting mechanism for exchanging water rights, so there is no physical storage of water in the bank. That is why they are also called the "paper exchanges" within some larger regional accounting system. These structures are provided for areas where physical water is limited or for large geographical areas. As noted by Clifford et al. (2004, Pg:4), such banks often require the flexibility of multi-year deposits: "For example, the Idaho Water Supply Bank requires a minimum term of five years for all water rights deposited in the bank. The Texas water bank is another example of this type of bank."

TABLE 5.1 Implementation framework to establish water bank.

What principles should be considered in determining the most appropriate structure for the water bank?	Approval of new and necessary rules to create banks Limiting transaction costs to allow smaller water rights to be competitive Examine how the water transfer process works
What issues should be addressed in the operating framework of the water bank?	To consider the interests of all stakeholders in the water bank; and policies to attract more participation
	Specify transfer terms in the annual plan (short-term leases or permanent purchases)
	Determine the term of storage, one year (annual or long term)
How does the water bank encourage participation?	Provide a good view of transaction prices from other successful examples
	Possibility of experiencing hypothetical transactions before actual transactions take place
	Explain the bank's benefits to various groups
How does the bank identify its market area/geographical area?	Based on population density and amount of water use
	Bank boundaries are often defined by numbers of potential participants and conveyance abilities (natural and man-made)
Who or what type of water can participate in the bank?	Determination of water resources entering the bank (Surface water, groundwater, reclaimed water)

- **Surface storage water banking**
  Surface storage banks are formed around reservoirs and dams. This type of banking can be stored and exchanged in the bank. By definition, a water exchange is the physical storage of water. Unlike institutional banking, these types of banks have short-term use. Also, unlike institutional banks, surface storage banks are usually more reliable in supply. In general, surface reserve banks operate on an annual basis, where deposits and exchanges are limited to one year. In this type of banks, the exchangeable water rights are determined based on either specific volumes or proportions of storage available (Clifford et al., 2004).

- **Groundwater banking**
  The groundwater water banking provides a mechanism to exchange water rights for water withdrawal in an aquifer. Groundwater banking is a water resource management approach and increasing water resource capability. Important factors in groundwater banking include regulatory laws, ownership of stored water and the amount, timing, location of water withdrawal. Groundwater sources are fed in years of high rainfall and stored then water can be used during drought years. Additionally, groundwater rights holders can use surface water instead of their rights and store their rights in the aquifer and use it in dry years. People can benefit by buying and selling these rights when using groundwater.

Feeding and withdrawal water from aquifers affect groundwater levels. The using of the aquifers as bank depends on several factors, including pumping water by other users, leakage, natural feeding. Generally, by feeding the aquifer or injecting water, the volume of water increases and can be used in the future.

5.3.6 Groundwater banking methods

Conjunctive use or management refers generally to a method of using surface and groundwater resources in combination to increase flexibility and reliability. Generally, when surface water resources increase, this water is used instead of groundwater by water users or feed groundwater and recharge aquifers. Conversely, during period of drought or aridity, when surface water is scarce, groundwater resources are used. Surface water can feed groundwater resources in both physical (direct) and *in-lieu* (indirect) ways. (Semitropic, 2004).

The physical method itself consists of two methods of natural and artificial nutrition, each of which is briefly described.

Natural Nutrition: The result of water infiltration into the aquifer through natural waterways. It contains rain, snow-melt water, or excess water used to irrigate plants.

Artificial Feeding: In this type of feeding, water is stored in ponds (dams, dikes, etc.), which allows water to penetrate the aquifer. Another form of artificial feeding is the direct injection of water into the aquifer through injection wells.

Groundwater users switch to surface water instead of pumping groundwater resources, thereby protecting or enhancing groundwater in the aquifer. Groundwater will remain in the watershed for future use in either the physical or in-lieu method (Christian-Smith, 2013).

5.4 The advantages and disadvantages of groundwater banking

Groundwater banking can be compared to the amount of surface water storage (i.e., reservoirs or tanks). Groundwater banking has advantages over using surface water tanks. Artificially recharging groundwater reservoir may be more attractive than constructing surface water storage and/or delivery systems because compared with surface water reservoirs, groundwater reservoirs have lower evaporation rates, require no construction, often have large storage capacities, generally do not interfere with surface land uses, are less subject to contamination and drought, and typically are more cost effective and efficient (MacDonnell et al., 1994). For example, the cost of these projects is not as high as that of equipment and storage of surface water. The damaging effects of groundwater on the environment are expected to be less (e.g., by reducing land subsistence) than the effects of surface water reservoir (O'Donnell and Colby, 2010).

Groundwater banking, of course, also has its disadvantages. One of the most important limitations of groundwater banking is that the "water rights" are not clearly defined, the "rule of capture" is often used as the means and measure of groundwater ownership (Lueck, 1998). However, effective groundwater banking management necessitate a special allocation system to determine the amount available for transactions. Pumping also has its own capacity constraints and production demand is limited when demand fluctuates throughout the year.

5.5 Water bank categories

In general, the term "water bank" include a wide range of institutional designs. According to Montilla-López et al., 2016 several types of water bank are classified into different categories based on the experience of water banks that have been implemented. They are described below:

Categorized by the nature of the organization responsible for its implementation

A) Public Water Bank: These banks are operated by government agencies, usually there is an expert related the field of water.
B) Private Water Bank: These banks are operated by a private initiative. In general non-profit organizations are responsible for their implementation.

Categorized by type of rights being exchanged

A) Permanent Water Bank: water Rights holders transfer their water rights to the bank permanently. These rights obtained by the bank can later be reallocated to other users (current or new users).
B) Temporary or Spot Water Bank: The activity of these banks is like a permanent water bank, but the transfer water rights is temporary (e.g. an irrigation season). Most of the bank's activity is related to drought periods and purpose to reduce the effects of drought and scarcity.
C) Option Contract Bank: Option contracts give the holder the right (not the obligation) to buy or sell the underlying asset (Cui and Schreider, 2009).

Main goal of option contract is to provide premium. Options contracts allow the buyer to avoid the risk of water shortages for their activity.

Categorized according to their purpose

A) Water Bank for reallocation of resources as a production input: Water banks can reallocate water resources by using water rights exchanges. Reallocation depends potential buyers and potential sellers. Reallocation can also move water from low-value to high-value uses.
B) Water banks for environmental purposes: Although these banks buy water rights, they do not reallocate it because their goal is to solve the environmental problems due water shortages (Montilla-López et al., 2016).
C) Water Bank for Risk Management: The uncertainty of water supply is due to variability in climatic and hydrological parameters that affects on water resources availability. Therefore, users face considerable risk and cannot make decisions that lead to optimal economic results. To minimize the negative impacts of the decisions, the way these banks work is based on negotiating water options contracts. This helps to improve supply security for the buyer of water option contracts (by reducing supply security for the seller of such contracts), thereby an effective transfer of risk between users with different levels of risk aversion (Montilla-López et al., 2016).

Categorized by management strategy

A) Active Water Banks: In these types of banks, managers adopt an active strategy as "market makers" and use the bank budget to buy water rights and then try to sell them to users. For this purpose, the bank attempt that total purchase and sale have not a net cost (the amount paid for purchases should be equal to the income from the sale), or this cost does not exceed from the maximum budget provided by the bank for this aim. Also in these cases, the conditions for buying and selling of right are set by the bank manager, so these banks become a kind of monopoly market through one-sided trading. Active water banks can be used to strengthen market activity and practice more useful control over market performance. (Montilla-López et al., 2016).
B) Passive Water Banks: The activity of such banks is limited to linking buyers and sellers to supply and sell water at any moment according to supply and demand. The bank administer in this case acts as an intermediary for buying and selling (broker), or as clearinghouse or through sealed bid double auctions (Montilla-López et al., 2016).

5.6 Importance and necessity

Water banks have emerged as one important way in which greater flexibility can be developed within a regional water allocation system. Water banking has long been used to solve water problems by allocating water within a basin and consuming it during drought. The necessity of establishing a water bank requires the creation of a realistic, dynamic and effective condition in the interactions between civil society, government and water users in catchments to achieve effective governance in water resources management. In this regard, water resources-based development measures, taking into account the water potential of each basin, will be operationalized with the participation of stakeholders and water rights holders, in a way that strengthens people's support for water resources management. A dialogue is established between the people and the government with an emphasis on conservation, management, and optimal and social use of water. Finally, the idea of a water bank not only can lead to reform of water resources management, but it will also help to transform agricultural policies.

Although water is considered a public good but the use and distribution of water rights may be either rented or sold. Water banks equitably manage how this is done so that water is not in short supply and groundwater sources are sensibly used. Water banks act as exchange markets and work across different regions so that costs are fair to users. Additionally, they have mechanisms that allow users to borrow allocations and give these back later, water laws permitting. Groundwater and groundwater interaction should be considered by careful management of the water resources that are physically banked in the aquifers and virtually traded and used.

Water banks are one of the ways to remove obstacles and challenges. The bank facilitates the transfer of ownership of water in a situation that requires short-term changes in water supply.

Water banks are responsible for providing information about basin water balance, demand and consumption data, registering rights, recording withdraw for various uses, obtaining bid and purchase prices and storing water to further enhance the basin water management. In the bank, all exchanges and the number of shares of the water right holders are known and every market information, is updated. The bank plays an important role in encouraging the private sector to invest in water supply and conservation projects and can communicate between water applicants and investors (Dosi and Easter, 2000).

5.7 How a water bank works?

For each water bank there is some responsible body, whose administrative services can vary greatly. The first step in most banks is to collect the water rights from the sellers who willing to sell and give it to those who are willing to buy. This is done with the encouragement of the water rights holder, which means the water bank encourages the water rights holders to save on water use so that they can store their surplus water in the bank. The water bank enables water rights owners to sell all or part of their rights. In these conditions, the role of the water bank is to bring together those who are willing to buy and sell water rights. Such a bank has several key functions. The bank determines what rights

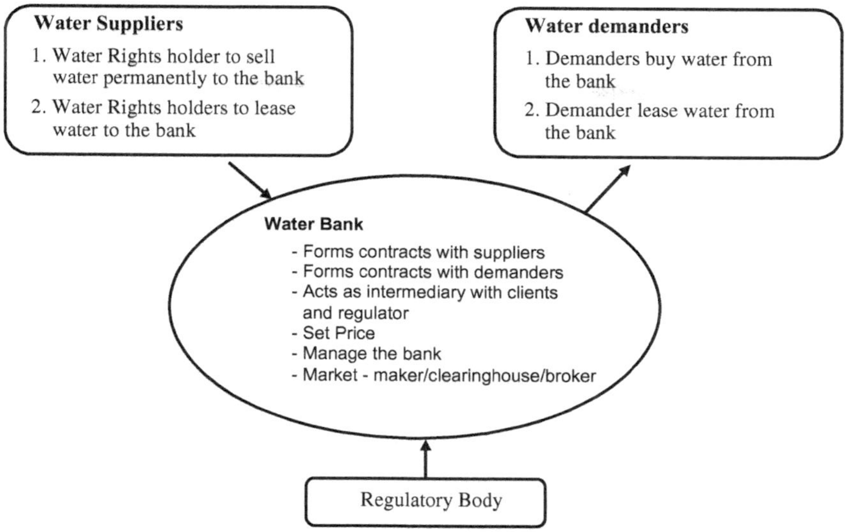

FIGURE 5.1 Water Bank Conceptual Model- Source West Water Research (Source: Clifford et al., 2004).

are subject to market conditions and determines the amount of water associated with each water rights. The bank determines who can buy or rent water, so it sets the rules for rent. The bank determines how much water can be transferred to another without affecting a third party. By enacting rules on the water market, the bank acts as an intermediary that accepts water rights to provide water rights to user tendency to supply the water they need. Banks present completely different designs in terms of participants, types of water and pricing and allocation rules. The conceptual model of water bank show in Fig. 5.1.

5.8 Examples of water banks

There are various experiences of water banks in many parts of the world. This section examines some of the world's water banks. According to the purpose and function of these banks, their experiences can be used to inform the potential of this institution to improve water management in an area or region. Water transfers in the United States has been specially developed in the Western States. So, this section examines water banks in both California and Arizona. In the United States, water banking laws are typically enacted at the state level (but also happen within smaller agricultural irrigation or conservancy districts). Australia's water bank activity is also briefly described. In these arid states and countries, while water rights come in various forms, in order to strengthen water trading, water rights are usually considered separate from land rights.

5.8.1 California

Water banks are increasing common in the Western US, There are a variety of active water banks in the state of California in the U.S. They vary in size and purpose (environmental, drought, dry-year purchase and conjunctive groundwater management). Water banks have also been established in states such as Texas, Colorado, Idaho, Nevada, New Mexico and Arizona. As one prominent example, in 1991, the state of California established a water bank called the California Drought Water Bank. This state example helps us understand the *initiation* of a bank. As noted by lund et al. (1992, pg:8) The 1991 and 1992 California Drought Emergency Water Banks were the first major State-brokered water transfer programs in the nation. The Water Banks were established in response to severe water shortages to provide water for critical municipal, industrial, and agricultural needs, preservation of fish and wildlife, and carryover storage for potential additional dry years. The initial market structure of California Drought Water Bank was a clearing house that pooled water and allocated supplies to critical needs in the state (Clifford et al., 2004; Howitt and Lund, 1999). The main purpose of the Drought Water Bank was to move drought water supplies from the northern part of the state to the southern part of the state through a market-based approach (participation). As a rough average, 75 percent of the rainfall is in northern California, while there is 75 percent water demand in California's central and southern parts of the state. To start, three different methods were used (O'Donnell and Colby, 2010, pg:13) "a) fallowing contracts, whereby the surface irrigation water was sold in lieu of irrigating; b) groundwater contracts, where groundwater would be used instead of surface water and the surface water would be sold; c) stored water contracts for releasing water from reservoirs". In total in 1991, for example, all three types of contracts acquired 1012 Mm^3 of water (Israel and Lund, 1995). Based on the bank's successful experiment, the California Department of Water repeated the program in the 1992, 1994. In 1992, the bank was similar to the 1991 bank with some modifications, but the volume of water reallocated decreased to 235 Mm^3 (Hadjigeorgalis, 2009). Due to increased rainfall in 1992, the demand for water declined and as a result the activity of the water banks decreased.

In 1991, all members purchased water from a single pool but in 1992 six separate pools were operated, each of which was to sell water at a single melded price to members of the pool. Prices for the 1992 water bank could varied from one pool to the next.

After a single wet year in 1993, critically dry conditions again returned in 1994 so a new water bank was set up, which bought 272 mM^3.

The California Drought Water Bank example for the early 1990's shows how a water banking initiation can be an effective way to respond to drought, "helping to improve water supply for users with higher water productivity, and reduce risk" (Jercich, 1997). In various incantations, the state of California has implemented versions (e.g., 2009 and 2018) of the Drought Water Bank to similarly respond to water scarce years. The following provides an example of more detailed timeline, from the nearby state of Arizona.

5.8.2 Arizona

Once water banking is initiated at the state level, it can significantly evolve as the next example shows. The state of Arizona conducts water banking operations through the Arizona Water Banking Authority (AWBA). The AWBA was established in 1996 to store the unused portion of Arizona's annual Colorado River entitlement in Central and Southern Arizona. The AWBA stores water in underground aquifers to earn long-term storage credits. These credits can be recovered (pumped) during a shortage to provide back-up water supplies (known as "firming") for Arizona water users. To help understand this important example of the evolution of a state water banking system, the historical timeline is laid out below (Closely following the Arizona Water Banking Authority).

TimeLine:

1986– The Arizona Legislature established the Underground Water Storage and Recovery Program to allow entities to store water underground and recover (pump) it at a later time (Arizona Water Banking Authority website).

1994– The Legislature enacted the Underground Water Storage, Saving, and Replenishment Act, a comprehensive program, which streamlined and consolidated various types of recharge projects that emerged after 1986. It encouraged the use of renewable water supplies, particularly Arizona's entitlement to Colorado River water, instead of groundwater, through the underground storage, saving and replenishment of water (Arizona Water Banking Authority website).

1996– The Legislature established the AWBA, effective April 30, 1996 (Arizona Water Banking Authority website).

1998– The Arizona Water Banking Study Commission issued a Final Report that describe the finding and recommendation of the committee (Arizona Water Banking Authority website).

1999– The AWBAs enabling legislation is expanded based on the recommendations of the Arizona Water Banking Study Commission's 1998 Final Report. The legislation allows the AWBA to perform water banking services for Arizona, which includes loaning previously developed long term storage credits and/or storing water on their behalf. It also gave AWBA the authority to supplement storage of Central Arizona Project (CAP) water with effluent and/or other surface water supplies (Arizona Water Banking Authority website).

1999– The Secretary of the US Department of Interior adopts regulations providing for "Off-stream storage of Colorado River water Development and Release of Intentionally Created Unused Apportionment in the Lower Division States." This enables interstate water banking in the Lower Colorado River Basin (Arizona Water Banking Authority website).

2001– The AWBA entered into an agreement with Southern Nevada Water Authority (SNWA) and the Colorado River Commission of Nevada (CRCN) for interstate water banking on behalf of Nevada (Arizona Water Banking Authority website).

2002– The AWBA adopted Resolution 2002–1 prioritizing the use of long-term storage credits accrued with state of Arizona General Fund appropriations for purpose of firming Fourth Priority on River municipal and industrial (M&I) users (Arizona Water Banking Authority website).

2004–The Arizona Water Settlement Act (AWSA) was signed into law by the US Congress. The AWSA obligated the State of Arizona to firm delivery of up to 23,724 acre-feet of priority CAP water reallocated to Indian tribes under the AWSA for a 100-year period (until 2108) to the same level that deliveries of CAP M&I priority water is firmed during shortages (Arizona Water Banking Authority website).

2006– The Indian Firming Study Commission (IFSC) issued a Final Report that describes and evaluates the State's (Arizona) obligations under the AWSA. In the same year, House Bill 2863 established an Indian Firming Program based on the recommendations of the IFSC and identified the AWBA as the agent for the State to meet its Indian firming obligations under the AWSA (Arizona Water Banking Authority website).

2008– The AWBA executed Resolution 2008–1 establishing a long-term storage credit replacement account for Fourth Priority mainstream Colorado River M&I users (Arizona Water Banking Authority website).

2010– The AWBA executed Resolution 2010–1 to establish the procedures for on-River contractors to enter into an agreement with the AWBA for firming (Arizona Water Banking Authority website).

2014– The Legislature authorized the AWBA to purchase long-term storage credits created by others. Credits may be purchased after all available excess CAP water supplies have been scheduled for storage (Arizona Water Banking Authority website).

2014– The "Recovery of Water Stored by the Arizona Water Banking Authority: A Joint Plan by AWBA, ADWR and CAP" is released. The recovery plan describes the methods for future recovery of AWBA long-term storage credits (Arizona Water Banking Authority website).

2015– The Legislature extends CAWCDs authority to levy and collect ad valorem water storage tax within its three-county service area beyond 2016 to 2030 (Arizona Water Banking Authority website).

2015– The AWBA and Gila River Indian Community entered into an intergovernmental agreement establishing the annual procedures to ensure the State's firming obligations to the Community are satisfied (Arizona Water Banking Authority website).

2019– Arizona Governor Ducey signed state legislation authorizing Arizona's participation in the Lower Basin Drought Contingency Plan (LBDCP), and Federal legislation was signed authorizing both the Colorado River's Upper Basin and Lower Basin Drought Contingency Plans. The AWBA plays an important role in Arizona's plan for implementing the LBDCP (Arizona Water Banking Authority website).

In 2008, Arizona's total use of Colorado River water is about 2.7 million-acre feet of water (3.3 billion cubic meters) (Guenther, 2008). Th water bank mechanism was used to store this excess water. The Bank of Arizona has stored unused water from the Colorado River in its reservoirs to increase the security of water supply during drought. The bank does not use market mechanisms to facilitate exchanges between different users, but rather is known as a system to facilitate additional water storage in the state. After 2002, however, the bank began interstate water exchanges with the states of Nevada and California. The Bank of Arizona pays for storage and transportation costs each year. The Bank of Arizona provides financing through the costs of removing water from the bank's reserves, taxes on each transaction, government grants, and income from interstate exchanges.

5.8.3 Australian use of water banks

Australia is located in the driest inhabited continent in the world (Murcia, 2019). Australia is also generally known for active water trading (Murcia, 2019). Water banking is widely found in Australia's Murray-Darling Basin. Initially, those rights were traded only on a temporary basis. But since the 1990s, permanent rights have been traded. About 20 percent of the water used in Australia comes from commercial transactions. In this regard, the bulk of Australian water deals are backed by intermediaries including brokers. These intermediaries gained the confidence of consumers of water and some subsequently set up and managed water banks. In fact, local brokers are often perceived as preferable to local transfers, but water banks are well suited for long-distance transfers (Bjornlund and McKay, 2001). The purpose of these private water banks is to reallocate resources, match supply and demand in the short term (temporary transfer of rights) and long term (permanent transfer of rights). The water exchange in these banks is done through two mechanisms: a) via the Internet, using bulletin board where buyers and seller declare their offers and b) via double auctions sealed bids, as in stock markets (Bjornlund, 2003).

5.9 Summary

Reducing the institutional impediments to water transfers across alternatives water uses will be critical in adapting to expected future water shortage due to population growth and climate change (Brown et al., 2019). The creation and implementation of water banks is one way to alleviate such institutional impediments to water transfers. While they come in a variety of forms and sizes, in general water banks facilitate the exchange of water rights (whether permanent or temporary) and the movement of scarce water from relatively low-valued uses to relatively higher valued uses. Water banking approaches do not create new water, but they can add considerable efficiency and flexibility for dealing with water scarcity in a regional water management system, and thus improve regional water security. By enhancing water security, linked food security and energy security are also possible. Such connections are seen in the experiences of a variety of successful water banks around the world. The success of these water banking tools is conditional on the establishment of a capable and flexible legal framework where governance entity has authority to regulate a legal framework based on social, physical, and economic condition. Of course, like any planning setting, water banking will also have its own challenges, and before any major policy making, the impacts on the various economic, social, political, and environmental contexts in the long term should be examined. Further, since water banks are often a new arena of policy making, the geographic area and its potential for water banking should be well studied. One of the basic prerequisites to establishing water banks is having accurately defined water rights, as well as the infrastructure to transfer the water rights traded. Water market tools can also be used to motivate stakeholders to participate in the bank. The analysis of various international experiences with water banks demonstrates the power of facilitated market mechanisms to transfer water towards higher-valued uses, and improve efficiency. In some cases, this many include environmental uses. Water banks can help provide flexibility to regions in responding to increasing scarcity.

In closing, this chapter has provided an overview of the roles and functions of water banks, as well as a summary of the steps required to establish an effective water bank. Our hope is that this review helps spur additional investigation and implementation of water banks.

References

Barbier, E.B., 2004. Water and economic growth. Economic Record 80 (248), 1–16.

Beck, M.B., Walker, R.V., 2013. On water security, sustainability, and the water-food- energy-climate nexus. Frontiers of Environmental Science & Engineering 7 (5), 626–639.

Bjornlund, H., McKay, J., 2001. Operational mechanisms for the efficient working of water markets—Some Australian experiences. In: Proceedings of the Joint Conference of the International Water and Resource Economics Consortium and the Seminar on Environmental and Resource Economics, Girona, Spain, 3–5 June 2001.

Bjornlund, H., 2003. Efficient water market mechanisms to cope with water scarcity. Int. J. Water Resour. Dev. 19 (4), 553–567.

Brown, T.C., Mahat, V., Ramirez, J.A., 2019. Adaptation to future water shortages in the United States Caused by population growth and climate change. Earth's Future 7 (3), 219–234.

Cartron, Dominique, Susan C. Kery, and Bonnie Colby. "*Alternative: bank water*" (2002). Workshop on Water Banking in the Lower Rio Grande, Southern New Mexico New Mexico Water Resources Research Institute.

Clifford, P., Landry, C., Larsen-Hayden, A., 2004. Analysis of Water Banks in the Western States. Washington State Department of Ecology Publication number 04-11-011.

Christian-Smith, J. "Improving Water Management through Groundwater Banking: Kern County and the Rosedale-Rio Bravo Water Storage District" Pacific Institute Farm Water Success Stories: Groundwater Banking, 2010.

Cui, J., Schereider, S., 2009. Modeling of pricing and market impacts for water options. J. Hydrol. (Amst.) 371 (2009), 31–41.

Dellapenna, J.W., 2000. The Importance of Getting Names Right: The Myth of Markets for Water. 25 Wm. & Mary Envtl. L. & Pol'y Rev. 317. Symposium Issue II – Water Rights and Watershed Management: Planning for the Future. Winter 2000.

Dosi, C., Easter, K.W., 2000. Water Scarcity: Economic Approaches to Improving Management. Center for International Food and Agricultural Policy.

Guenther, H.R., 2008. Arizona Water Banking Authority Annual Plan of Operation 2009. Annual reports. Arizona State Government Publication.

Hadjigeorgalis, E., 2009. Place for water markets: performance and challenges. Appl. Econ. Perspect. Policy 31 (1), 50–67.

Howitt, R.E., Lund, J.R., 1999. Measuring the economic impacts of environmental reallocations of water in california. J. Agric. Econ 5, 1268–1272.

Israel, M., Lund, J.R., 1995. Recent California water transfers: implications for water management. Nat. Resour. J. 35, 1–32.

Jercich, S.A., 1997. California's 1995 water bank program: purchasing water supply options. J. Water Res. Plan. Manag. 123 (1), 59–65.

Lueck, D., 1998. First possession. The New Palgrave Dictionary of Economics and the Law. Peter Newman, Ed.

Lund, J.R., Israel, M., Kanazawa, R., 1992. Recent California Water Transfers: Emerging Options in Water Management. Center for Environmental and Water Resources Engineering, Davis, CA, USA.

MacDonnell, L.J., Howe, C.W., Miller, K.A., Rice, T.A., Bates, S.F., 1994. Water banks in the west. Natural Resources Law Center University of Colorado School of Law.

Murcia, M.E.M, 2019. International comparative analysis of regulations for water markets and water banks. Integrated Water Resource Management. https://doi.org/10.1007/978-3-030-16565-9_5.

Montilla-López, N.M., Gutiérrez-Martín, C., Gómez-Limón, J.A., 2016. Water banks: what have we learnt from the international experience? Water J. 8 (10).

National Research Council of the National Academies, 2004. Managing the columbia river: instream flows. Water Withdrawals and Salmon Survival.

O'Brien, M., 1999. Possible solutions: policy tools to achieve flexibility to meet new conditions, preliminary thoughts for coping with future droughts. Nat. Resour. J. 39 (1), 175–177.

O'Donnell, M., Colby, B., 2010. Water Banks: a tool for enhancing water supply reliability. University of Arizona, Department of Agricultural and Resource Economics http://ag.arizona.edu/arec/people/profiles/colby.html. Accessed February 15 (2010): 2011.

O'Donnell, M., Colby, B, 2009a. Water Auction Design for Supply Reliability: Design, Implementation and Evaluation. Arizona Water Institute: Enhancing Water Supply Reliability http://www.azwaterinstitute.org/ewsrpublications.html.

Quesne, T.L., Pegram, G., Heyden, C.V.D., 2007. Allocating scarce water: a primer on water allocation, water rights and water markets. WWF-UK Available at: http://assets.wwf.org.uk/downloads/scarce_water.pdf https://waterbank.az.gov/about-us/history.

Semitropic water storage District-Groundwater Banking. (2004). http://www.semitropic.com/GndwtrBankFAQs.htm, Accessed, June 2009.

Chapter 6

Rights and international laws of transboundary water resources

Zeinali Masoumeh [a], **Omid Bozorg-Haddad** [b] **and Vijay P. Singh** [c]

[a] *Water Resources, Department of Irrigation & Reclamation Engineering, Faculty of Agricultural Engineering & Technology, College of Agriculture & Natural Resources, University of Tehran, Karaj, Iran.* [b] *Department of Irrigation & Reclamation Engineering, Faculty of Agricultural Engineering & Technology, College of Agriculture & Natural Resources, University of Tehran, Karaj, Tehran, Iran.* [c] *Caroline & William N. Lehrer Distinguished Chair in Water Engineering, Department of Biological and Agricultural Engineering & Zachry Department of Civil & Environmental Engineering, Texas A&M University, Texas, United States*

6.1 Introduction

Transboundary water resources themselves face significant challenges due to increased population growth, industrialization, urbanization, environmental degradation, as well as climate change and hydrological changes (Salman, 2007). Due to the scarcity of water resources, continued population growth, and rapid growth rate of water consumption, various statistics and forecasts have been presented so far regarding the increasing water consumption and future water scarcity crisis. For example, population growth, rapid growth in consumption, and industrial revolution have led to a threefold increase in freshwater consumption from 1950 to 1990, a 17-fold increase from the early 20th century, and a 45-fold increase from the previous three centuries (Mokhtari and Ghaderi., 2008). According to some statistics, water consumption has doubled in the last 40 years, and it is predicted that by 2025 at least 0.25 of the world's population will live in countries experiencing severe water shortages (Gardner and Engelman, 1997). According to available statistics, nearly 20 percent of the world's population is currently deprived of access to safe water, and nearly 6.2 billion of Earth's inhabitants are deprived of access to adequate water for sanitation (Mosert, 2003). According to forecasts by international research institutes, by 2025, one billion and eight hundred million people will be living in areas with absolute water shortages and more than two-thirds of the world's population will face moderate to severe water shortages (Ansink, 2009). Some forecasts confirm that in 2030, nearly 55 percent of the world's population is facing a water deficit (Falkenmark, 1990), and by 2025, water shortages will reach over 62 percent.

Economical, Political, and Social Issues in Water Resources. DOI: 10.1016/B978-0-323-90567-1.00013-9

Despite the statistics and forecasts of their organizations and institutions on the future water crisis, some do not consider these statistics realistic and reasonable. For example, Biswas believes that according to statistics from these institutions, future water needs are considered exponentially growing and increasing. While they mistakenly assume that water resources such as coal mines and petroleum resources are disposable, the fact that water is a renewable resource that can be reused over and over again can be reused (Biswaz, 2011). Others believe that the future water crisis will not be due to the scarcity of water resources but due to the lack of proper management as well as the benefit of sufficient investment in the sector (Koutsoyiannis, 2011).

Therefore, due to the crisis of water scarcity and inadequate distribution and increasing consumption of water resources, there are many issues and disagreements in the common catchment water resourcesof the world. It is predicted to be a critical factor in future tensions and conflicts in the world, especially in the West Asian region. The underlying problem that plays a critical role in most of these tensions and conflicts is that the amount of water demand of a country exceeds the amount of water available, which points to the importance of fair and equitable allocation of water resources (Mianabadi, 2013). Therefore, proper familiarity with international water laws and principles and the principles of joint water resources utilization can reduce conflict and tension between riparian states of an international river (Mianabadi 2013).

6.2 Definitions and terms

The terms and topics discussed in this chapter have a clear definition of general and consistent definitions that are acceptable to everyone. This section deals with each of these items.

The concept of "international water resources": The term "international water resources" is used to identify natural water resources for several countries. The term links all water resources (surface, groundwater, and frozen) of international importance (Caponera, 1980).

Definitions of Common Boundary Rivers: Defining the International Boundary River: Grossius and Waterloo first introduced the theory of river internationalization in the nineteenth century. Before that, all rivers were considered to be inland and under the sovereignty of countries adjacent to them, and as a result of any country, regardless of whether the waters in its territory might be the source of river or the source of catchment in another country, it had the right to exploit it at any rate. This theory, entitled "Absolute Territory Sovereignty," was first put forward by the "United States Attorney General" in the year 2008, and had been accepted by most states for decades. Later in the Treaty of Versailles, the term "international rivers" was replaced by the word "multi-territorial rivers" for the first time. According to this treaty, the International River is a river that can be navigated and links more than one country to free waters. An important conference (Barcelona Conference) was held in Barcelona on April 5, 2011. The conference, which was convened at the invitation of the United Nations within the framework of the Treaty of Versailles, played a major role in formulating and expanding the legal regime of river shipping. Under this system, any river that complied with the terms and conditions contained in the above treaty was called international and included in the treaty provisions. Instead of the standard term of International River, the Treaty of Barcelona used the term "internationally important navigable waterways." The International River, however, is a river that crosses the territory of more than one country or is a river that separates the land of two countries from one another, and the sovereignty of the states is shared. According to the international law, rivers are divided into two main groups: national rivers and international rivers.

National Rivers: Rivers within a country that are part of the territories of a country, such as the Sefidroud River in Iran and the Mississippi River in the United States.

International Rivers: 1. Rivers that pass through more than one territory and separate two or more countries, such as the Danube and the Arvand Rood and the Aras. International rivers, in turn, are divided into two categories: 1. Boundary rivers that form the boundary between two countries, such as Arvandrood (Iran and Iraq). 2. Rivers that are double soil or pass by many countries; for example, the Nile, the Rhine, and the Aras. The Nile passes through Sudan, Egypt, Ethiopia, Uganda, Kenya, Tanzania, Zaire, Burundi, and Rwanda, and the Danube River, which crosses the common border between the Czech, Slovenian, Hungarian, Bulgarian, Romanian, Yugoslav, and Romanian states or we can mention the Aras River which lies between the countries of Iran, Turkey, Armenia, and Azerbaijan. The report refers to all rivers in more than one territory, rivers, or transboundary waterways.

Methods of Boundary Determination in International Rivers: Several technical methods are used to determine the boundary limit in international rivers.

Determining the boundaries of two riverbanks and sharing them: Determining the boundaries between the two countries is the determination of the boundaries between the two banks and the shared use of river water. In this situation, the river is governed by two states, and there is no boundary line in the river. However, it is the river itself, which is considered a common territory of the parties (such as the Mappen Treaty between the Netherlands and Prussia).

Choosing one of the two river banks as the border: The sovereignty of the river is transferred to one country, and the other has no control over its periphery; for example, since the border between France and Portuguese Guinea has been designated by the Vincent-Pinezes River, or Under the Treaty of Andrews between Russia and Turkey the Danube seas are given to Russia. Fair Line: A straight line is a line whose distance from each point to the nearest two shores is the same, so a straight line divides the river into two equal parts and is related to the horizontal water level. Most lanes are used in lakes and rivers where navigation is not possible; for example, the Canadian-US borderline between Lake Erie and the Detroit River is specified.

Crossroads or Thalweg: In most rivers, there is a deeper passage, which is the safest way to navigate, and the ship wreckers use the same passage when crossing the river; it is the border between Iran and Iraq.

Direct crossings: Optional directional lines define river crossing, for example, the St. John's River between the United States of America and Canada through reference markings in the middle, or by the river, the borders of the two countries are designated.

How to Set Border through the Dialog Line with the Double Line for Two Submarine Rivers: If the navy is in the dominant interest of the governments, the Dialog is recognized as a boundary, but there may be two divisions in the river (especially at the estuary) where, as governments own their routes, the boundary line between the two channels is delineated and consolidated, such as the "Guadiana" border between Portugal and Spain (1).

Boundary Determination by Combining the above Methods: This method can be created by combining several types of boundary determination systems.

Some Important Rules for the Operation of International Rivers: These rules are generally divided into two groups: Shipping Operations.

In scientific studies, one of the essential requirements of any discussion is to understand the concepts and terms in the relevant field in order to clarify the meaning, intent, and purpose of the word and vocabulary as well as explicit terms in the process of exploration and explanation of the subject.

Since many water-related vocabulary and terms have technical and engineering implications and are not widely used in legal matters, water law is a particular issue that most law professionals have little to do with technical terms. On the other hand, water experts and engineers are also unfamiliar with legal issues. Therefore, it is necessary to discuss some concepts and terms related to water.

Water for different needs and purposes has different categories and types. One of the most critical water classifications that have a fundamental legal basis is the division of water into surface and groundwater. Each of these categories has different components and subsets and therefore needs to be identified for better understanding.

Law: This term is used in two general and specific ways. When this term is used in the cloud of public policy and is intended to be used in a general way, it means all laws and regulations may be made by competent government agencies, such as legislature with the head of state or a member of the executive branch. In a general sense, the law includes all legislative approvals, as well as administrative letters and divisions. However, from a specific point of view, the law refers to rules that are directly ratified by parliament or through referendum as required by the constitution and must be enforced from the date of ratification. On the whole, if we want to have a complete and straightforward definition of law, it can be considered as a set of criteria and rules that are required to establish a clear and orderly framework.

Contract: The word contract is a synonym that is the result of an agreement between one or more people. A contract can be written or oral. Legally, any agreement that arises between different people and as a result of that agreement, a commitment that includes individuals, is called a contract. Under Article 5 of the Civil Code, the contract is the obligation of one or more persons to commit and be acceptable to one or more persons. Accordingly, any contract will be based on two primary conditions. The first is the existence of parties to the contract, meaning that the contract is created by the agreement of one or more persons, and the second is the result of obligation.

Public Shares: Property owned by legal entities in public law such as government and the municipality. These properties are not transferable (Jafari Langroodi, 2007).

Litter: Litter means "sleep" and is the part of a river, creek, or wetland that is in any location according to hydrological statistics, floods, and maximum floods with a 6-year return period from the Ministry of Energy or regional water companies. It determines the Bed and Breakfast Regulations.

Reasonable consumption: Reasonable consumption is the amount of water that is determined in each specific time and place (according to consumer needs and compliance with public needs and facilities) by the provisions of the Fair Distribution Act (Fair Water Distribution Act, 2).

Useful Consumption: The amount of water that is determined by water law at any time and place (according to consumer and public needs and facilities), which can be revised by changing scientific and technical methods (Water Law, 2).

Principles and logic of the subject

- **Water Wars**

More than 45 percent of the world's land is in common international catchments (Wolf et al., 1999). Moreover, the lives of more than 40 percent of the world's population depend on international rivers. In Turkmenistan, this is 100 percent, Egypt 97 percent, the Netherlands 89 percent, Pakistan 80 percent, Sudan 77 percent, and Iraq 66 percent (Swain, 2001). Elsewhere in the world, the crisis in the arid and semi-arid region of Western Asia is far more dangerous and serious than elsewhere. Locating the West Asian region in the arid belt of the world has given the region only 5 percent of the world's fresh water, with only 1 percent of the world's population. These resources are also commonly used in common. More than 50 percent of the population of the Middle East region lives in shared watersheds, leading to competition and disputes over the use of these limited resources (Mokhtari and Ghaderi, 2008).

In general, factors such as being in arid regions, seasonal rivers with highly variable and irregular currents, typical catchments, economic conditions of regional countries, political instability, and political tensions and governments, along with distrust of governments in the region, are among the factors that have made the water crisis in the Middle East more critical than anywhere else in the world, making water a critical factor in the region's tensions and conflicts (Korkutan, 2001).

Since water is no substitute for oil, some analysts have predicted that the next century's war will be over water resources, not over oil reserves (Bulloch and Derwish, 1993; Engleman and Leroy, 1995; Kuchukmohammatlou, 2009; Ohlsson, 1995; Seabright, 1997; Swain, 2001; Zeitoun, 2007). For example, Smith (1995) believes that there is no place in the world that, like Central Asia, is prone to conflict over water resources. However, many believe that water has rarely been a major cause of war and violence in the present century (Barnaby, 2009; Kauffmanet et al., 1997; Swain, 2001; Wolf, 1998; Yoff et al., 2003).

However, there is ample evidence of some intense tensions and conflicts between some governments and countries over water resources. For example, it is said that in the last 50 years, 1831 water disputes have occurred around the world, of which 37 have caused wars or explosions (Akbari et al., 2007). Tensions between Iraq and Syria in 1975 resulted in Syria's exploitation of the al-Thuriya Dam and the reduction of the Euphrates River to a full-fledged war front, and Iraq threatened to bomb al-Thurra. This dispute led the two countries to deploy their frontier forces at the borders, which ultimately ended with the mediation of the former Soviet Union and Saudi Arabia (Korkatun, 2001).

On the one hand, the exploitation of the Arvand Rood border River was one of Saddam's main pretexts for an eight-year military invasion between Iran and Iraq Also, Iraqi Prime Minister Nouri al-Maliki warned in early 2012 that Arab nations could be at war if no immediate action was taken to address water scarcity. In general, water resources may not be the leading cause of conflict and intense tensions between governments and nations, but they may have been the cause of wars and conflicts between governments. In other words, although it is argued that no water war has occurred, the absence of war does not mean that there will be no conflict over shared water issues (53- Zeitoun, 2007).

On the other hand, some analysts see water as a means of participation, not a means of conflict between governments (Biswas, 2011; Moosert, 2003; Swain, 2001; Wolf, 1998). Almost all international water-related relations are participatory (Ansink, 2009). Moreover, water can serve as a partnership tool or as a shared goal for intergovernmental partnerships. Shared water resources not only reduce tensions between the two countries but also facilitate the improvement of relations and increase the level of participation and cooperation of governments in various economic fields. It also works politically. India-Bhutan participation in the construction of Chukha 336 MW hydroelectric dam on Wang Chu border river and joint exploitation of this dam is the Government of India provides one of the best examples of participation in border river operation in recent decades which entails 60 percent grant and 40 percent loan, and as the partnership continued, its electricity generation capacity increased to 370 MW.

The agreement between the two countries was that electricity produced first had to meet Bhutan's domestic demand. Surplus electricity was sold to India at two rates and then at a single rate. This was far below the cost of electricity produced in India. The positive impact of the two countries' cooperation on the joint exploitation of water resources led to the continuation of these partnerships and the construction of electricity projects (1020 MW and 900 MW). Studies show that Bhutan is expected to generate more than $ 100 million in electricity sales to India annually, with the current population of less than 2 million. It could be a notable income for a relatively small country. Accordingly, it is predicted that by 2015, Greece would have the highest per capita GDP rate among South Asian countries (Biswas, 2011). Iran-Turkmenistan cooperation in the joint construction and operation of the Dostami Dam is another successful example of the two countries' participation in the operation of a border river. Dam Dosti is built on the River Harirud (the common border between Iran and Turkmenistan). With the supply of potable water and agriculture in 2005, it reached an operation that, according to the agreements, the water is divided equally between the two countries (Mianabadi, 2013).

Equitable utilization and distribution of water resources are one of the main concerns and concerns of the countries concerned. If this is done with a fair and equitable approach, it will not only reduce the conflict and tension in the exploitation of these resources, cross-border and shared, but also enhance the participation and cooperation of countries in other political, economic, and even cultural fields. There is a fair and equitable allocation of water resources shared between several stakeholders (local, national, and international). It is a complex process that requires not only specialized aspects of water resources redistribution but also requires useful indicators and benchmarks such as political, economic, and social indicators in the fair redistribution of water resources. Overall, it can be concluded that water can serve as a tool or a goal for greater participation between governments and countries in other areas, rather than as a source of conflict and tension. Studies show that the costs of non-participation in the operation of border rivers for the countries concerned are much higher than the benefits of unilateral exploitation of these rivers. Not only will it have to be paid for by governments and politicians, but it will be the primary pressure on the hundreds of millions of poor people whose lives depend on these resources (Biswas 2011).

Political tensions and challenges in the management and utilization of shared water resources.

From the geopolitical point of view, water is not separate from politics. The severe shortage of water resources, the unnecessary increase in per capita water consumption, the uneven distribution of rainfall, and the lack of suitable alternatives have caused controversy, conflict, and political and sometimes military tensions in some parts of the world. It has secured internationally shared water and hydropower issues. It has increased competition for higher utilization of shared water resources between two or more countries, such as high-water countries for higher power and country. Downstream users use power to gain more water (Weiberg, 2008). These political crises and tensions in the countries of West Asia (the Middle East) and North Africa due to different tribes and races and the governments' efforts to satisfy these tribes also provide water resources. There is enough to provide food security and improve livelihoods, which is of paramount importance in large-scale government decisions. This has exacerbated the political tension and crisis in the exploitation of border rivers in West Asia (the Middle East) and the countries of North Africa. Among the prominent cases of water resources tensions around the world are tensions between the three countries of Turkey, Syria, and Iraq in the joint exploitation of the Tigris and Euphrates Rivers, one of the root causes of which is the existence of different ethnicities in the these countries. Governments of these three countries are trying to satisfy these ethnicities by creating food security and work by using more water resources to meet their agricultural needs (Zawahri, 2006). For example, one of Turkey's most important objectives for the implementation of the large-scale Anatolian (GAP) water resources project has been "national security" in the south of the country (Korkutan, 2011). Due to the deprivation and poverty of the southern regions of Turkey and the fact that the principal inhabitants of these weak areas are the Kurdish tribes, the region's backwardness and abuse by the separatist group PKK have caused this problem. After extensive studies, Turkish officials seek the leading solution to increase security in the region by raising the standard of living and Kurdish tribes in the south. With the development of Gap's water and agricultural projects in the south of the country, the excuse is to break away from the PKK separatist groups. On the other hand, the Syrian Shiite government, whose economy is based on agriculture, considers its primary strategy to provide national security for political stability and the satisfaction of its Sunni tribes in providing sufficient water for the agricultural needs of these tribes.

In other words, the U.S. Department of Defense launched the HDD research project in 2009 to enhance stability and security in Helmand Province, Afghanistan. This research project is comprised of experts from the U.S. Army Corps of Engineers (USACE) Engineer Development Research and Engineering Center (ERDC), U,S. Geological Survey (USGS), and the Secretary of Science and Technology Advisory Center.

The U.S. State Department was STAS. After completing its work, the research team held frequent briefings with the U.S. military and civilian decision-makers on the role and importance of water in stabilizing and securing the area (Palmer and Moloney, 2009), the most relevant results of this group's research are:

1. Efforts to secure and stabilize Afghanistan must be in line with the short-term goals of the agricultural sector. The activities and operations undertaken

should emphasize improved policymaking and capacity building in water resources management and increase quantitative and qualitative water efficiency. Excessive extraction of groundwater resources is a concern, and if the harvest exceeds recharge, groundwater reserves and wells are not a sustainable option for water supply.

2. There is no right quantitative and qualitative information to assess the water status of Hilmand province. Therefore, the U.S. military is studying and implementing some plans and infrastructure to make more use of the Hirmand River under the pretext of stabilizing Afghanistan's Helmand province. This may have disproportionate consequences for the situation of border towns in eastern Iran.

Insufficient water resources not only drastically reduce security and stability indicators in border cities and areas facing water crisis, it can also increase crime and insecurity in other cities with rich water resources. For example, the drying up of Lake Hamoun was due to the phenomenon of smuggling in the region.

The continuation of this trend led to the widespread migration of indigenous people to other parts of the country, especially the city of Gorgan. This caused various social and sometimes political problems in the city. Therefore, not providing sufficient water resources in the area may not only undermine security and increase the socio-economic, political, and social impacts of the area but may also affect other social, economic, political, and security indicators of the city. Water resources are in good condition.

Some analysts and experts see food and occupational security as a precondition for achieving "national security"; while the two themselves are dependent on "water security", this issue is much more pronounced in countries and regions where the economy is dependent on agriculture.

In terms of national security, water resources management and planning in the border areas are more important than in other areas. Due to the central role of border guards in the border security of each country, the well-being and comfort of border guards are of paramount importance, which are highly dependent on food, occupational, and economic security. On the other hand, living in many border areas of the country is dependent on agriculture and livestock activities, thus providing sufficient water resources in border areas, and border towns will have a significant impact on border security. To be taken seriously by policymakers and government officials, problems with security in the border areas remain an essential factor affecting the security of border cities (Mianabadi, 2012).

Also, tensions and conflicts between stakeholders in the exploitation of shared water resources of the border rivers are not always those of the upstream countries because of the geographical power. They have rivers and are dominated by rivers, and the downstream countries have always complained about the unfair treatment of water by the upstream states. For example, in the exploitation of the Nile River, which is shared by ten countries, Egypt is the main exploiter of the Nile because of its military power in the region. Although a little more than 9 percent of the Nile catchment is located in Egypt, and 95 percent of the Nile originates outside Egypt, 97 percent of the Nile is exploited by Egypt (Mokhtari and Ghaderi, 2008). Another example that can be mentioned is when Ethiopia was planning to build a large dam on the Blue Nile, Egyptian President Anwar Sadat threatened not to build any dam on Lake Tana and urged to attack and jam (Dinar and Wolf 1994).

Studies have shown that many factors increase vulnerability and increase the likelihood of military tensions in a common catchment area, some of which are rapid changes in the environmental growth and rapid population growth. One-sided water resources projects include economic and military imbalances, water scarcity crises, and lack of friendly relations between coastal states (Hensel et al., 2006; Wolf, 2007), which can be managed properly and these rivers can be operated and sustained by drastic changes.

Principles of Management of Operation of Cross Border Water Resources

Due to the widespread and direct relationship of border water resources management issues to social, economic, political, and security issues, there are different theories regarding the exploitation of these resources. Every country strives to look at this with its own interests and priorities. Creating synergies between countries with shared water resources and equitable allocation of these resources so as not to damage the natural ecosystem is one of the most important challenges of international institutions and laws in the present century (McCaffrey, 2007).

There are four doctrines and general theoretical approaches to how to govern and exploit shared natural resources, especially border rivers, which fall into two categories of extreme theories: 1- Absolute territorial sovereignty (Harmon doctrine); 2- Absolute territorial integrity (integrity) Absolute Rivers and Moderate Theories; 3- Shore States' Common Use of Water Resources of an International River (Common Water Rule); and 4- Integrity Satisfaction or restricted sovereignty (territorial limited sovereignty).

Doctors of Absolute Territory Sovereignty or the Doctrine of Humans: By virtue of this doctrine, the portion of the international rivers flowing into the territory of any country except for the inland waters of that state shall be considered and any country may derive from it. The use of water in anyway is deemed necessary, regardless of consequences for other countries. This doctrine was put forward by Harmon for the use of the Rio Grand River, a joint venture between the United States and Mexico. Because of the benefits of this doctrine to the upstream states of common rivers, it is evident that the upstream states rely on this principle in water tensions and disputes. Because this doctrine is contrary to justice and fairness, it has been rejected by many international organizations and institutions.

Doctrine of Absolute Territorial Integrity: According to this principle, the coastal states of rivers cannot make serious changes in the natural state of their country, causing significant changes and negative effects in other countries. because these changes affect the territorial integrity of the country. The downsides will be affected. This principle is in complete contradiction with the doctrine of ours, and the low-lying countries, as the beneficiaries of this principle, have demanded that this principle be recognized in international water laws. But this principle, like the principle of absolute territorial sovereignty, is not recognized as part of contemporary international water laws.

Coastal States Doctrine of International River Water Use: This doctrine is based on the principle that all states of an international river have a common sovereignty over that river and that an international river is an economic entity whose resources belong to all coastal countries. Since this principle, as an ideal, ignores concepts such as sovereignty and national issues, it has not received widespread acceptance by states because they believed that this principle at hand puts pressure on the group agreement.

Doctrine of Territorial Integrity or Restricted Territorial Sovereignty: According to this principle, every coastal country has the right to use and operate international rivers in its own country provided that such use would not cause damage to the territory of the states. In this doctrine, the principle of the "fair and just use" of shared water resources and the "prohibition of damaging the territory of other countries" is of great importance. Of the four doctrines, only the doctrine of territorial integrity and limited territorial sovereignty formed the basis of modern international water laws.

Principle of fair and equitable use of water: This principle falls under the doctrine of restricted territorial integrity, which is based on the shared sovereignty and equality of rights, which does not necessarily mean an equal sharing of shared water resources. Various indicators and criteria have been proposed by international law to evaluate the fair and reasonable utilization of shared water resources. Although these criteria have been introduced and explained by some international laws, the lack of a clear definition of the relationship between this principle and other principles of border management is one of the main ambiguities of the existing rules between international laws related to water resources. This is one of the most important principles cited by the above states to prove their claim.

Principle of non-material damage to other countries: This principle is also part of the doctrine of territorial integrity. According to this principle, no government is allowed to use its sovereign water resources in international catchments to cause severe damage to other governments and countries. These damages include quantitative, qualitative, or a combination of both. This principle has been used in many conventions, treaties, and international water agreements, and many downstream countries rely on this principle for access to sufficient water resources.

Principle of Data and Information Sharing: The regular exchange of data and information is the first stage of partnership between coastal states and governments (Kibaroglu, 2002). Based on this principle, each coastal state in the catchment is involved. International law is responsible for sharing data and information with other coastal countries for the optimal management and utilization of shared water resources. Paragraphs 29 and 31 of the Helsinki Law recommend that this principle be enforced, while Articles 8 and 9 of the 1997 United Nations Convention make it mandatory.

Principle of Information, Consultation, and Negotiation: According to this principle, although the coastal country in the international catchment area has prior information, consultation that negotiations of other countries' water projects may result in. There will be severe damages to that country. Like other principles, this principle has been recognized by most international conventions, treaties, and agreements. Naturally, the upper states sometimes disagree with this principle.

Principle of Peaceful Settlement of Disputes: This principle states that all states in the international catchment area should seek a peaceful settlement of disputes, even when the governments are not hopeful of reaching a consensus and are worried that they will not be able to reach an appropriate negotiation.

These five principles have been incorporated into many international water conventions, treaties, and agreements, some of which are important: the Helsinki 1996 Act, the 1997 UN Convention, the 2004 Berlin Law on Management Water Resources, 1960 Indus River Treaty between Pakistan and India, 1995 Protocol SADC Joint Water Resources South Africa Development Agreement, and 2002 Sava Basin Agreement (Common to ö Bosnia, Herzegovina, Croatia, Serbia, Montenegro, Slovenia, and Albania). The 1995 Mekong River ICWC Statute Agreement. In general, the lower countries tend to focus on the principle of "prohibition of damage to the territory of other states, whereas the above countries are interested in focusing on the "fair and reasonable use" of international water laws (Mianabadi, 2013).

International Water Laws and Treaties

From time immemorial, the importance of international waters have been recognized as a bridge of communication and trade, as well as a source of agricultural and industrial use. Moreover, over time, the history of humanity shows that many ancient civilizations formed along large rivers such as the Huang Ho, the Tigris, and the Euphrates, the Nile, the Andes, the Ganges, the Yangze, the Yellow, and the Tiber. Moreover, people welcomed having a law governing the use of shared waters in their territory (Caponera, 1980).

Processes created by international law, in general, and especially in the case of global water resources, are many and varied. A starting point (a traditional topic that is still used in many ways) is found in the statute of the International Court of Justice.

Article 38 stipulates that the court must make decisions by international law concerning disputes:

(A) International conventions, whether public or private, shall clearly state the laws recognized by competing countries.
(B) International customs, as evidence of a general practice accepted as law.
(C) The general principles of law recognized by civilized nations.
(D) Judicial decisions and teachings of the most experienced public propagandists of different nations, as a subset, means determining the legal rules (Caponera, 1980).

The most common method of establishing rules of conduct between countries in which international water resources are related is the international agreement referred to in Article 38 of the Statute of the International Court of Justice. International law does not specify a specific form for these agreements. However, rarely, if ever, does it happen that the agreement is non-written. In Article 2 of the Convention on the Law of Treaties (Vienna, 23 May 1969), which came into force on 27 January 1980, the term "treaty" means: "An international agreement concluded between states in writing and concluded under international law, whether a single instrument or two or more of the relevant documents and any specific designation thereof."

There are more than 260 standard basins and 270 common groundwater aquifers that share two or more countries (; Wolf et al., 1999). Out of this large number of international rivers and shared waters, 158 rivers lack any specific structure for stakeholder engagement (Ansink, 2009). Also, of the more than three hundred internationally signed treaties that directly deal with the management and operation of the shared rivers, nearly two-thirds of them are in European and North American countries (Rogers, 1993). In Europe, for example, the four rivers that share between four or more than four countries are signed by 175 treaties—managed and managed. While in Africa, twelve border rivers are shared between four or more than four countries; there are only 34 treaties for the joint operation of these rivers. In Asia, for the five rivers that share between four or more, there are only 31 Essen treaties (Swain, 2001). Also, out of the 145 treaties signed between different countries in the 20th century, nearly 86 percent are bilateral, and many of them have not been considered to be part of the treaties (Jägerskog and Phillips, 2006). For example, bilateral agreements between Syria, Turkey, and Iraq can be signed.

The exploitation of the Tigris and Euphrates Rivers or the agreements between Egypt and Sudan on the joint exploitation of the Nile without the participation of other coastal states has been pointed. Despite this, many signed and ratified treaties and agreements have encountered numerous problems during implementation. This has led to increased political tensions between the countries concerned.

Since the functions of border rivers are different, the conflicts and tensions between their stakeholders in the exploitation of shared water resources can also vary (locally). For example, boundary river functions can include agriculture

(Hirmand, Jordan, and Nile), hydroelectric power (Sindh and Colombia), potable water supply (Jordan rivers, Harirood, and Vinyl), shipping (Columbia, Mekong, and Danube rivers), or fishing (Mekong River). As a result, the type and intensity of tensions that occur between border states of a river can vary. Among all these functions, rivers with agricultural and drinking uses have the highest potential for causing stress. Because the use of shared water resources for drinking and agricultural use is a vital need that cannot be replaced by another, but the use of a typical border river to meet hydro demand can be achieved through alternative methods, including solar power (Iyob, 2010).

6.3 International water law

Water has always been one of the main tools for civilization throughout human history, always causing disputes between nations, governments and even tribes. The expansion of international common water laws goes back thousands of years, but today's newly developed water laws go back two centuries (Dellapenna and Gupta, 2009). The expansion of international water laws began in the 1870s with the establishment of two international organizations, the International Law Agency or Society (ILA) and the International Law Institute (IIL) continues to this day. ILA International Law Institute and IIL International Law Association are two non-governmental scientific organizations active in various fields of international law. The resolutions and rules issued by these two organizations constitute a subsidiary source of international law and are used in the establishment of legal rules. These rules are not legally binding on their own, but the conventions and treaties derived from them have come into force with the signature and ratification of states (Avarideh et al., 2016). The focus of IIL International Law Resolutions is the commitment not to cause significant harm to other watershed countries. The first IIL resolution, known as the Madrid Declaration and adopted in 1911, is an absolute ban on activities that could harm other basin countries. It was in sharp contrast to the Harmon Principle (1895). This approach was limited by the Salzburg Resolution adopted by the IIL in 1961 (Salman, 2007). The Salzburg Resolution continues to emphasize the commitment of States not to harm other countries, but its right to use waters. The common river had restricted access to other basin countries. Therefore, the absolute prohibition of the Madrid Declaration was partly enacted by other resolutions adopted by the IIL in 1979 and 1997, which addressed the environment of international waterways and any action that caused water pollution or caused undesirable damage to the country. The resolutions and rules adopted by the International Law Society or ILA are different from those adopted by the IIL, emphasizing the principle of fair and equitable use of international waterways.

The first related resolution, known as the Dubrovnik Declaration, was issued by the ILA in 1956. According to the statement, each country within its territory has the right to use the relevant international river water, but how it should be used in view of its effects on other watershed countries. An approach beginning with the Dubrovnik Declaration was amended by the New York resolution issued by the ILA in 1958. The resolution states that each of the legitimate watershed governments is in a fair and reasonable proportion of the waters of an international catchment for useful use. In August 1996, the ILA, at its 52nd meeting in Helsinki, Finland, reported the results of its surveys on international river water use, known as Helsinki rules (Salman, 2007).

International laws governing the operation of common rivers are divided into two general categories: 1) International laws and treaties on the common operation of rivers for shipping purposes; and 2) laws and treaties on the joint exploitation of rivers for non-navigable purposes (Mianabadi, 2013).

International Rules and Principles on the Use of Shipping for the Purpose of International Water

Since the ninth century, shipping vessels in Europe have become the most important prospect for the future of global shipping, and for this reason, numerous treaties and agreements have been signed between countries, including the Congress of Vienna (1815). The Barcelona Convention (1) and the Geneva Convention (1) address the development and exploitation laws of the shipping states, and the rights of coastal states. Prior to World War II, shipping on international rivers was free for European countries, but after the war when the countries were divided into West and East Blocs, the right to navigate in the river was granted only to coastal states.

Exploitation for Non-Navigation: With the expansion of non-navigational exploitation of rivers, proprietary maritime laws alone did not meet the new needs. As a result, countries have been trying with each other to reach a legal understanding of non-shipping exploits, with some of the key points of such agreements being briefly mentioned (Fathi, 2016). A number of multilateral conventions are designed to be used in a territorial area defined by a specific geographical area. These are "regional agreements" that aim to create a unified legal structure for a specific geographical area (usually across continent, Asia, Africa, Latin America, Europe) that they refer to.

International Rules and Principles on Non-Shipping Uses of International Waterways

Documents that reflect the legal status of transboundary waters include:

1) Helsinki Rules (1996) and other resolutions of the International Law Institute IIL and the ILA International Law Association concerning the Non-Shipping Uses of International Waterways
2) European Water Convention on the Conservation and Use of Transboundary Waterways and International Lakes (1992)
3) United Nations Convention on the Use of Non-Shipping International Waterways (1997)
4) The Berlin Rules on Water Resources (2004)

So far, various laws, treaties, and resolutions have been adopted for the joint exploitation of shared water resources, but none of these laws, due to existing defects and ambiguities, refer to fair and equitable exploitation of shared water resources. One important reason is the existence of different doctrines and attitudes to environmental law and philosophy. Among the laws and treaties and resolutions adopted around the world, the three Helsinki laws, the Convention on the Rights of the Operator of International Waterways for Shipping and the laws of Berlin are of greater importance and popularity (Mianabadi, 2013).

1) **Helsinki Laws**

The “Helsinki Laws on the Use of International River Water Resources” was adopted by the ILA International Law Society at the Helsinki Summit in 1966, and by virtue of these laws, all states along the river are subject to exploitation. Water resources have equal rights, and no country has the right to use river water if it causes damage to other coastal states. In addition, the Government shall, without notice and without the consent of the other coastal governments, change the route of the international rivers to the detriment of the State, but shall be responsible for all actions taken and any unilateral actions taken by the said Government shall have no right. It does not create a government for that. Although the title refers only to international rivers, Article 1 states that these rules apply to the use of water resources in a catchment. Therefore, these laws can be cited in the combined exploitation of international surface and groundwater resources shared by several countries. The management of international border aquifers by an international legal instrument was first mentioned in these rules.

Helsinki Laws emphasize the “fair and reasonable use” of shared water resources as a fundamental principle of international water law. Subject to Article 5 of these Rules, the criteria to be applied in the “fair and reasonable use” of any water resource shall be as follows: The catchment areas to be considered between several countries are as follows: 1- Geographical conditions of the catchment area; 2- Hydraulic characteristics of the catchment area; 3- Climatic conditions of the catchment area; 4- Rate of use of each country in the past and present time; 5- Economic and social needs of each country; 6- Population of each country; 7- Availability of other natural resources; 8- Avoid unnecessary waste of resources; 9. Water compensation in one or more countries involved in the basin as a tool to resolve the conflict between Kshvr Ha V; and 10- satisfaction and the degree of persuasion Kshvr Ha a country without serious damage to the other. Helsinki laws cover a wide range of issues, including the use of international waters for shipping and non-shipping purposes. Important about these rules is that they do not contain a separate reference to make a commitment not to harm a country, but that the damage that may arise from a country’s use of the river. It identifies an important factor in the fair use of shared water resources. Despite any compulsion to adopt and enforce Helsinki laws, these laws have been used and cited in many international water treaties (Salman, 2007; Correia et al., 1999).

2) **European Water Convention**
United Nations Economic Commission for Europe (UNECE) is one of five UN Regional Committees to monitor and resolve economic, environmental, and industrial issues and governments.
3) **The 1997 United Nations Treaty**
In 1959, Bolivia submitted a proposal to the General Assembly of the United Nations. That organization will address the legal problems related to the operation and use of international rivers. Following the 1963 UN Secretary-General's report, the UN General Assembly recommended that the International Law Commission (ILC) 'reviews the rights of non-shipping use of international waterways. In 1970, the resolution of the General Assembly of the United Nations was adopted, entitled "The Gradual Development and Formulation of the Rules of International Law on International Waterways." According to the resolution, the Commission on International Law started the user on the subject of international waterways. Five Special Rapporteurs have guided the Commission's activities over the past 20 years, and two nongovernmental legal entities, including the ILA International Law Association and the International Law Institute (IIL) have contributed to these investigations. After 20 years, the Commission issued its first draft in 1991 entitled "The Rights of Non-Shipping Uses of International Rivers" on the Two Principles of "Reasonable and Fair Use" of Common Seas and "Obligation," provided no severe damage is done to other countries along the riverbank. The draft was submitted to the UN General Assembly in 1994 and submitted for signature by the UN member states following the 1997 amendments to the Convention on the Use of Non-Shipping International Waterways. The month of 2014 (August 2014) came into force after reaching the quorum of the acceding countries.

The Convention contains 37 articles, which are divided into seven sections; Part I: Introduction, Part II: General Principles; Part Three: Planned Measures; Part Four: Protection, Maintenance, and Management; Part Five: Emergencies; Part Six: Miscellaneous; and Section 7: Arbitration and Dispute Resolution.

The main provisions of the Convention are in sections one, two, three, and four. The first section includes the definition of a relevant international waterway term. Article 2 of this section defines the term "waterway" as "a system consisting of a surface and underground waterway that, with its physical connections, forms a unit and leads to a common point." International waterways define "runoff as waterways that are located in different parts of the country."

Also, Article 3 of the Convention states that prior bilateral or multilateral protocols and agreements between States shall be preferred to the provisions of this Convention.

Article 5 The second paragraph expresses the principle of fair and equitable use and participation in the international waterway, which is the cornerstone of the Convention. According to this principle, countries participate in the use, development, and protection of international waterways fairly and reasonably.

Article 4 Provides factors for determining the fair and reasonable use of States by the international waterway. The conditions of application and weighting of these factors are similar to those in the Helsinki rules, but the factors introduced are somewhat different.

These include 1- Geographical, hydrographic, hydrological, climate, environmental, and other factors relevant to the basin's natural characteristics; 2- Social and economic needs of the concerned countries; 3- Water-dependent population in each basin country; 4- Impacts of using international waterways in one basin country versus other basin countries; 5- Existing and future uses; 6- Maintenance, protection, development, and economics of water resources use and cost of actions; and 7. Availability of alternatives of comparable value to a planned or existing use.

One of the key provisions of the Convention is the requirement for non-serious injury, as set out in Article 7. Paragraph 1 of this article states that States shall take appropriate measures to prevent significant damage to other States of the international water basin. Clause 2 of the same article states that if substantial damage to one of the waterways is to occur in any place, the State causing such damage shall, in case of disagreement for such use, take all appropriate measures while considering the provisions of Articles 5 and 6, in consultation with the State concerned, to eliminate or mitigate the damage suffered and to discuss, where appropriate, the payment of compensation. Although the emphasis is on avoiding harm, it is painful and sometimes very complex and expensive to stop the term or terminate the activity that has begun. If possible, compensation can be paid for damage (McAfee, 2015). Article 10 also states that in the absence of agreement or custom among States, no use shall be inherent over other uses and shall, in the event of a conflict between the different uses, be resolved. Special attention is paid to meet vital human needs.

The third part of the Convention contains nine articles, the principle of informing about planned actions and detailing some of the different aspects of this requirement. The essence of this principle is that if the planned actions are made in the country or other countries on the international waterway, that government is obliged to timely inform its programs to the other countries involved. If those States consider that the said programs are inconsistent with the requirements of Article 5 or 7. The process of consultation is followed, if necessary, by negotiations to reach a fair solution.

Part IV of the Convention deals with the protection, maintenance, and management of international waterways and includes provisions for the protection and maintenance of waterway ecosystems, prevention, reduction, and control of pollution, and advice on international waterway management. It has also emphasized that waterways.

6.4 Laws of Berlin

The laws of Berlin are very comprehensive and consist of 73 articles, divided into 14 chapters, covering various water resources issues not covered by the Helsinki Act and the 1997 UN Convention. The rules were discussed and approved at the seventy-first ILA International Law Society conference in Berlin in August 2004. The former title of these laws, known as the "Revised Laws of the International Law Society for the Sustainable and Fair Use of Water Resources Management," was renamed as "Berlin Laws on Water Resources Management." These laws are applicable to all water management issues, including national and international water management issues. Also, these laws examine the relationship between the principle of "fair and reasonable use" and the principle of "prohibiting damage to the territorial territory of other states."

The main difference between Helsinki law and the UN Convention on the one hand, and Berlin law on the other, is that Helsinki law and the UN Convention emphasize "fair and reasonable use." In contrast, the Berlin law emphasizes the "obligation to manage water resources." The commoner emphasizes fair and reasonable. Another critical difference between these laws is that the laws of Berlin are subject to the need not to inflict severe damages on other coastal states. However, the Helsinki laws only emphasize the need not to damage the indicators and criteria for the rational and fair use of water resources. The 1997 Convention is similar to the Helsinki Laws, except that, unlike the Helsinki Laws, it provides separate legal provisions for "prohibiting damages to other countries." Moreover, the environmental law of the Berlin law is far more comprehensive than any other law (Cooley et al., 2011; Salman, 2007).

6.5 Necessity and importance

Water resources do not pay attention to the boundaries created by human beings and move according to their nature. Understanding this fact and paying attention to the natural movement of these resources will help solve problems related to water consumption and protection at the international level. Therefore, sometimes a water source, such as a lake, a river, etc., may be considered as a border between two or more countries, and then as a result, factors such as soil erosion or river bed erosion may change over time. Either the exploitation of a country with the aim of using the waters in its territory may have bad consequences for the neighboring countries, or one of the parties may want to use the same common waters to meet the demands and exploit its borders. These may occur in all countries that share a common water border, and the consequences may be disputed. Therefore, the existence of international rights and laws is essential for common transboundary waters.

For example, the construction of dams for various purposes in the downstream country may cause flooding in the upstream country. This and other possible consequences may be felt not only on the main river or lake water route but also throughout the natural system in the catchment or international drainage basin. This basic unity of river/lake or drainage system is due to the fact that they almost always have a component that is the same as the hydrological cycle. Any action taken by a country to improve the natural climate may affect other parts of the watershed (Caponera, 1980).

Also, natural erosion that occurs in an upstream country may damage canals, dams, and port facilities in a downstream country on the same river. Or, again, the water used for irrigation in an upstream country may prevent the use of that water in an upstream country for navigation, boating, and industrial purposes.

Similarly, the presence or absence of downstream facilities, or the use of water for irrigation, may deprive the upstream countries of the possibility of using the river for purposes such as navigation or boating. Finally, pollution from the upstream use can have detrimental effects and cost significantly for downstream water treatment (Caponera, 1980). According to the above examples, there may be a conflict of interest between countries with a common international water resources system, such disputes cannot be resolved and the parties will not agree unless government performance laws are in place and are specified in the use of cross-border and shared water resources.

As noted earlier, there are several principles in the use of cross-border resources that many countries consider to be one of the principles in the interest of substantiating their claims. Some of the most important principles cited in water negotiations and disputes are:

- The principle of historical right to exploit the river
- The principle of fair and rational use of water
- The principle of not inflicting severe damage on other countries
- Principle of water requirement (agriculture, drinking, industry, environment)
- The principle of necessity of early notification
- Principles of water law (Helsinki law, UN treaty, and Berlin law)
- The principle of sovereignty and currency integrity

For example, in the dispute over the Nile basin, which is one of eleven countries with Sudan and Egypt, the largest beneficiaries and consumers, Egypt's and Sudan's consumption and utilization rates have been contested by other coastal states. But Egypt, citing the principle of "historical right" and Sudan, citing the principle of "fair and rational exploitation" in the Helsinki Act, defends their efforts to exploit the Nile River, which raises the issue and lack of a comprehensive definition of water law. There is tension and conflict between these countries.

Ambiguity in current international water laws

Another factor contributing to the failure of international law to resolve water disputes is the ambiguity of some of the provisions of the law.

International law is water. For example, Article 5 of the 1997 UN Convention emphasizes the "fair and reasonable" use of the water of a shared river, while Article 7 of the Act refers to "prohibiting severe damage to other states." However, since this treaty does not set out the appropriate criteria for fair and rational use of water resources without causing severe damage to other countries; this has led each country to rely on one of the two articles in its interest.

For example, in the dispute between Turkey, Syria, and Iraq on the joint use of the Tigris and Euphrates Rivers, Turkey cites Article 5 of the 1997 UN Convention on the Fair and Reasonable Use of Water Resources, while implementing the GAP project. According to Article 7 of this treaty, Syria and Iraq claim that Turkey's actions upstream of these rivers have caused severe damage to both countries.

6.6 Example description

For example, the Hirmand River Treaty between Iran and Afghanistan in 1973 can be mentioned. The materials for this Contract are as follows:

Article 1- The Government of Iran and Afghanistan agree that each year the amount of water of the Hirmand River that reaches the Kamal Khan Dam shall be divided between Iran and Afghanistan from Kamal Khan Dam to Balaban.

Article 2- In order not to consume the tonnage of water from the present turret, the Afghan Government undertakes not to construct or even repair any other creek in the distance in question.

Article 3- Authorized officers and minibusses shall unanimously determine in autumn each year the amount of water that reaches the Kamal Khan Dam, and shall determine from the Kamal Khan Dam the amount of water that each Party shall divert from Hirmand, They are considered to be equal in proportion to the two.

They will notify each other after separation if the divisions and names of the mines and the amount of water they have given each other have been notified.

Article 4- The equipment necessary for determining the river water and the flow of the flow shall be provided by the Iranian and Afghan Technical Officers at the expense of the Government of Balochistan and shall be fitted at appropriate locations from the port of Kamal Khan and, if necessary, from the Kohk-e-Siskar dam section.

Article 5- To divide and regulate the water of the Hirmand River from the Kamal Khan Dam referred to in the present Contract, the technical officers of the Parties shall be based on which they deem appropriate, establish permanent divisions in their territories, and the expenses incurred by the Contracting Parties after the acquisition. They will be in Agreement with each other in the design according to the amount of water they take.

Article 6 Until the construction of permanent barriers and gages of geysers, the former shall be constructed and shall pay to the other if the parties need to close the gutters.

Article 7- Because the streams that flow directly from the Hirmand River from the Kamal Khan Dam and flow into the soil of the Parties shall be to prevent excesses and defects in each Party's part, up to four years from the date of signature of this Agreement. Gradually make the lid with the lime bricks each in their own proper order.

Article 8- The Government undertakes not to take any action or action from the Kamal Khan section to the ten Mohammed Khan and Sikh Khars, which is the last subdivision, to cause any part or all to lose its share.

Article 9- The Government of Iran and Afghanistan shall mutually agree that the technical agents and the agents responsible for carrying out the operations outlined in this Contract, as well as the operations of the Clerks, Builders, Cleansers, and Clerks and Subordinate Clerks and Deposits between the Kuh-e-Sikhsar Ward shall act. This Contract has been permitted on the other side without permission.

Note: The materials, equipment, and objects required by the agents to carry out the duties outlined in this Contractor to carry out the duties of their lives when leaving the border with them shall be exempt from customs duties.

Article 10- If the Hirmand River is transferred to a new bed above the Kamal Khan Dam, the first article shall be implemented elsewhere, which constitutes the Kamal Khan Dam in the interest of the present dam, with the Agreement of the Government, shall be designated if it is moved lower than the Kamal Khan Dam. Alternatively, to deepen the current context to prevent irrigation for the lands of Sistan, the Government will agree to grant the parties consent to use the water quota specified in the first subparagraph under separate conditions.

What if the Afghan government is reluctant to enter into water negotiations with Iran for years is a worrisome outcome? The Afghan diplomats worry that the outcome of blue talks may not ultimately benefit the Afghan people. This concern is so severe that despite holding 19 consortium meetings on the Hirmand issue, the Afghan side is still making severe suggestions or demands on the table. The negotiations are out of the question and involve only a reactionary policy. That is to say, in response to Iran's proposed packages, it is consulting for or against. On the other hand, the Afghan public opinion regards the Hirmand River as an inland river whose sole proprietorship and rights belong to Afghanistan. Public study in Afghanistan is based on energy supplies and water swaps instead of oil. In such circumstances, negotiations on the nature of water or the increase of Iran's water share, as guaranteed by the Hirmand Treaty, are not easy for the Iranian side.

The Iranian government is trying to persuade the Afghan authorities to accompany and co-operate with waterfront promotional packages. In the Iranian media environment, it has been repeatedly reported that Afghanistan has not been loyal to its commitments in Hirmand and the barriers above. The ravine has undermined Iran's right. However, the results of the consortium's reports show that Iran has not been successful in proving its claim.

Afghan officials have repeatedly emphasized in official interviews with Iranian diplomats that Afghanistan remains committed to the Helmand Treaty of 1351 and that cooperation is required to implement the agreement and to enter 820 million cubic meters of water at the edge.

What is at stake is the difference in water inputs in the non-normal water years and the months in which the river is almost dry downstream. The Helmand agreement has determined how much water should enter the Iranian soil every month, while insisting that Iran has no right to floods and water surpluses in high-water years. Afghanistan says the water that flows into Iran during the rainy months is often more than Iran's right, and the Iranian government directs and stores it in half.

Iran says, "Hamoon is an international wetland," and nature's right should be considered separate from the political and economic rights of the countries because droughts have harmed the region's ecosystem and are causing increased dust, migration of birds, and people across the region. The Afghans, however, insist that the negotiations of 1972 did not explicitly address but in practice the water requirement of nature and that Iran had to manage its water resources in order to survive.

According to the census of 1355, the population of Sistan and Baluchestan province was 660 thousand people, and now despite extensive migration, the population of the province, according to the latest census, has increased to more than two million and seven hundred thousand people.

6.7 Summary

The primary purpose of this chapter is to review international laws and regulations of transboundary waters. In general, international water laws and regulations on transboundary waters, challenges and policies of shared water resources management, common water use doctrines, ambiguities in laws and treaties, and finally water laws in Iran are examined. In the beginning, concepts and terms are defined in order to determine their general meaning. Then, the basics and logic of the subject were studied, and then the necessity of the existence of comprehensive and general water laws and regulations are examined. To clarify the issue, an example of the Iran-Afghanistan agreement on the Helmand River is given.

References

Akbari, M., Abdi, A., Mokhtari, H., 2007. Evaluation of Consequences of Transformation of Transboundary Rivers into Interbrains: Case Study: Zaab Kouchak. National ConCongress on Border, Opportunities and Challenges, Urmia, Iran.

Ansink, E., 2009. Game-Theoretic Models of Water Allocation in Trans Boundary River Basins. PhD Dissertation. Wageningen University, Wageningen. Retrieved from edepot.wur.nl/10993.

Avarideh, F., Atari, J., Abdollahi, M, 2016. A comparative study of international principles and rules governing water distribution in transboundary rivers. Environ. Science Quarterly Volume 14 (Number 2).

Barnaby, W., 2009. Do nations go to war over water? Nature 458 (7236), 282–283. doi:10.1038/458282a.

Biswas, A.K., 2011. Cooperation or conflict in transboundary water management: case study of South Asia. Hydrol. Sci. J. 56 (4), 662–670. doi:10.1080/02626667.2011.572886.

Bulloch, J., Darwish, A., 1993. Water Wars: Coming Conflicts in the Middle East. Gollancz, London, UK, p. 256.

Caponera, D.A. (1980). The law of international water resources. Some general conventions, declarations and resolutions adopted by governments, international legal institutions and international organizations, on the management of international water resources. Background paper 1/ Rev. 1.

Cooley, H., Christian-Smith, J., Gleick, P.H., Allen, L., Cohen, M.J., 2011. Climatechange and transboundary waters. The World's Water 7, 1–22. doi:10.5822/978-1-59726-228-6_1.

Correia, F.N., Da Silva, J.E., 1999. International framework for the management of transboundary water resources. Water Int. 24 (2), 86–94.

Dellapenna, J.W., Gupta, J., 2009. The evolution of global water law. The Evolution of the Law and Politics of Water. Springer, pp. 3–19.

Dinar, A., Wolf, A.T., 1994. Economic potential and political considerations of regional water trade: the Western Middle East example. Resour. Energy. Econ 16 (4), 335–356.

Engelman, R., Leroy, P., 1995. Sustaining Water: An Update. Population Action International, Washington, DC.

Falkenmark, M., 1990. Global Water Issues Confronting Humanity. J. Peace. Res 27 (2), 177–190. doi:10.1177/0022343390027002007.

Fathi, T., 2016. One River and Several Countries. East http://sharghdaily.ir/.

Gardner-Outlaw, T., Engelman, R., 1997. Sustaining Water, Easing Scarcity – A Second Update. Washington, DC, USA.

Hensel, P.R., McLaughlin Mitchell, S., Sowers, T.E., 2006. Conflict management of riparian disputes. Polit. Geogr. 25 (4), 383–411.

Iyob, B., 2010. Resilience and Adaptability of Transboundary Rivers: The Principle of Equitable Distribution of Benefits and the Institutional Capacity of the Nile Basin. Oregon State University. Oregon State University, Corvallis, Oregon Retrieved from http://scholarsarchive.library.oregonstate.edu/xmlui/handle/1957/17900.

Jafari Langroodi, Mj, 2007. Terminology of Law. Ganj Danesh Publications, Eighteenth Turn, Tehran, Iran.

Jägerskog, A., Phillips, D., 2006. Human development report 2006: managing transboundary waters for human development. Human Development Report Office Occasional. Paper (p. 33). UNDP. Retrieved from http://hdr.undp.org/en/reports/global/hdr2006/papers/jagerskoganders.pdf.

Kaufman, E., Oppenheimer, J., Wolf, A.T., and Dinar, A. (1997). Transboundary fresh water disputes and conflict resolution: planning an integrated approach. water international, 22(1), 37–48. doi:10.1080/02508069708686661.

Kibaroglu, A., 2002. Settling the dispute over the water resources in the Euphrates-Tigrisriver basin. In: Castelein, S. (Ed.), From Conflict to Co-operation in International Water Resources Management: Challenges and Opportunities. Delft, The Netherlands: UNESCO, pp. 263–277.

Korkutan S., (2001), The sources of conflict in the euphrates-tigris basin and its strategicKorkutan, S. (2001). *The sources of conflict in the Euphrates-Tigris Basin and its strategic consequences in the Middle East*. NAVAL POSTGRADUATE SCHOOL MONTEREY CA.

Koutsoyiannis, D., 2011. Scale of water resources development and sustainability: small is beautiful, large is great. Hydrol. Sci. J. 56 (4), 553–575.

Kucukmehmetoglu, M., 2009. A game theoretic approach to assess the impacts of major investments on transboundary water resources: the case of the euphrates and tigris. Water Resour. Manage. 23 (15), 3069–3099. doi:10.1007/s11269-009-9424-3.

Law office and parliamentary affairs, environmental organization. (2006). Environ. Law Collection, Volume I.

McCaffrey, S., 2007. The Law of International Watercourses, 2nd Ed. Oxford, UK, Oxford University Press, p. 550 Retrieved from http://www.oup.com/us/catalog/general/subject/Law/PublicInternationalLaw/GeneralPublicInternationalLaw/?view=usaandci=9780199202539.

Mianabadi, H., 2012. Water, Security and Transboundary Rivers. National Congress on Border Cities and Security; Challenges and Strategies (CBCS 2012). University of Zahedan, Zahedan, Iran.

Mianabadi, H., 2013. An Overview of International Water Law. National Congress of Civil Engineering, Shahid Nikbakht School of Engineering, Zahedan, Iran, May 17-18.

Mokhtari H., and Ghaderi M., 2008. Prospects of Middle East in 2025; Case studies: euphratesMokhtari, H., & Ghaderi, M. (2008). Prospects of Middle East in 2025; case studies: Euphrates and Tigris, Jordan and Nile river basins. *Geopolitics*, *4*(1), 36-74.

Mostert, E., 2003. Conflict and co-operation in international freshwater management: a global review. Int. J. River Basin Management 1 (3), 267–278. doi:10.1080/15715124.2003.9635212.

Ohlsson, L., 1995. Hydropolitics: Conflicts Over Water as a Development Constraint. Middle East. London, Zed Books (London, UK), p. 230.

Rogers, P., 1993. The value of coopcooperation in resolving international river basin disputes. Nat. Resour. Forum 17 (2), 117–131. doi:10.1111/j.1477-8947.1993.tb00167.x.

Salman, S M a, 2007. The Helsinki rules, the UN watercourses convention and the Berlin rules: perspectives on international water law. Int. J. Water Resour. Dev. 23 (4), 625–640.

Seabright, P., 1997. Water: commodity or social institution?. Stockholm, Sweden: Stockholm Environment Institute.

Smith, D.R., 1995. Environmental security and shared water resources in post-soviet Central Asia. Post. Sov. Geogr. 36 (6), 351–370.

Swain, A., 2001. Water wars: fact or fiction? Futures 33 (8–9), 769–781. doi:10.1016/S0016-3287(01)00018-0.

Wegerich, K., 2008. Hydro-hegemony in the. Amu Darya basin. Water Policy 10 (S2), 71–88.

Wolf, A.T., 2007. Shared waters: conflict and Cooperation. Ann. Rev. Environ. Res. 32 (1), 241–269.

Wolf, A.T., Natharius, J.A., Danielson, J.J., Ward, B.S., Pender, J.K., 1999. International river basins of the World. Int. J. Water Resour. Dev. 15 (4), 387–427. doi:10.1080/07900629948682.

Yoffe, S., Wolf, A.T., Giordano, M, 2003. Conflict and cooperation over international freshwater resources: indicators of basins at risk. J. Am. Water Resour. Assoc. 39 (5), 1109–1126. doi:10.1111/j.1752-1688.2003.tb03696.x.

Zawahri, N.A., 2006. Stabilising Iraq's water supply: what the Euphrates and Tigris rivers can learn from the Indus. Third. World. Q. 27 (6), 1041–1058.

Zeitoun, M., 2007. The Conflict vs. Cooperation paradox: fighting over or sharing of Palestinian-Israeli groundwater? Water. Int. 32 (1), 105–120. doi:10.1080/02508060708691968.

Chapter 7

Water policy and governance

Masoumeh Zeinali[a], Omid Bozorg-Haddad[b] and
Hazi Mohammad Azamathulla[c]
[a] *Water Resources, Department of Irrigation & Reclamation Engineering, Faculty of Agricultural Engineering & Technology, College of Agriculture & Natural Resources, University of Tehran, Karaj, Iran.* [b] *Department of Irrigation & Reclamation Engineering, Faculty of Agricultural Engineering & Technology, College of Agriculture & Natural Resources, University of Tehran, Karaj, Tehran, Iran.* [c] *Department of Civil and Environmental Engineering, St. Augustine, Trinidad and Tobago*

7.1 Introduction

The water situation It has received a lot of attention in recent years. Experts and theorists in different sciences have expressed different views on the nature of policy and the water crisis. In other words, there are various interpretations and analyzes of the water crisis and ways to deal with it. The executive bodies in the country's water governance sector only pay attention to water engineering and have neglected to study and research its social subsystems. Meanwhile, some see the declining population as a way to escape the crisis, some suggest importing water and transferring water between basins, segregation of drinking water as another way to shut down agriculture in non-strategic sectors, and realizing water prices for consumers as other solutions. The water crisis has been raised, with some even calling the water crisis a global problem.

In discussing decision-making in the field of water crisis policy and management-making, some have launched various campaigns to provide a solution to save water in the world. Apart from judging these attitudes and analyzes, the vacancy of water crisis policy is felt in a specialized way. Increasing the number of "forbidden" plains, the inability of the water allocation system to harvest more than the capacity of water resources and observing the allowable limit, reducing the quality and salinity of water resources, non-observance of water rights of the ancients and the environment and the emergence of conflicts are signs or consequences. Going beyond the conceptual understanding of water governance and choosing the type of view of the current water governance mechanisms of the country and having basic information from the history of the subject, we can examine the hypothesis that in the 1340s the water governance situation in the country changed dramatically and With the performance of these

Economical, Political, and Social Issues in Water Resources. DOI: 10.1016/B978-0-323-90567-1.00005-X

changes in the water governance process so far, the hand of the government or the government in resolving open affairs and the possibilities of the rest of society in participating in various affairs has been limited. As a result, the country's governance mechanism is largely limited to regulating limited and closed procedures for planning in the centralized state system, the use of judicial procedures, and the use of administrative facilities. In other words, in the face of the problems and challenges that have arisen in recent decades, economic and market measures and agreement, consensus and public participation, have a small share. If this hypothesis is accepted, to what extent can the creation of greater diversity in the country's water governance mechanisms be effective in fundamentally reforming affairs? How can more diversity be provided in the country's water governance mechanisms? What institutional measures should this diversification take? To what extent can a better and more analytical understanding of the country's water governance system and its shortcomings provide the necessary information to confirm and reject this hypothesis and answer the above questions based on a conceptual understanding of the issue of water governance? So far, several studies have been conducted on the state of water policy and governance in various countries. Despite the many similarities in the definition of water governance and its components and factors, the differences in the methods of assessing water governance in countries are significant. Even the goals for assessing water governance vary widely. These include goals such as determining the dimensions and characteristics of a particular mode of governance (pragmatism), comparing countries, strategic planning, identifying the main gaps and how to repair them, and identifying risk areas based on policy and governance components. In different countries, many studies have been conducted on water policy and governance and the existing systems in the water sector. The following is a brief summary of some of these studies. Ruth and Pieper (1994) developed a system-based model for examining the shape of the coast and the Cape of Good Hope in Massachusetts. Due to climate change in the world, this model deals with how seawater progresses in the nose. Nasal and coastal erosion is a dynamic issue, and the system-based model of modeling has greatly helped to simulate changes in this process in the coming years and the policies adopted in this regard. Simonovich et al. (1997) used the system dynamics method to evaluate the long-term water resources and analyze the policies implemented in the Nile River Basin in Egypt. Huerta et al. (2001) used the system dynamics method to manage a catchment in Mexico. The model has been a tool for policy-making and dispute resolution between the five provinces and the National Water Departments over surface water distribution. Hu et al. (2005) used the system dynamics analysis method to examine the effects of water management policies and decisions for South Taiwan. The model offers the increase effect the capacity of water facilities, water treatment and groundwater. Rashid and Hayes (2009) examined the problem of wastewater management by providing a simple and efficient method based on dynamic programming. In this method, they proposed two dynamic models; a collection system model and a wastewater transmission model. They considered the benefits of this approach in policy-making and decision-making for future planning for the development of the sewer system to be among the advantages. In the Indian state of Gujarat, India's government has been forced to replace existing water laws with new ones due to overuse of groundwater resources and salinity. Rajiv and Gapta covered the issue in an article published in 2007. This can be considered as a partnership Public-private partnerships should be considered in conjunction with government agencies, citizens, communities, religious communities, non-governmental organizations and the private sector.

In China, groundwater levels have fallen due to the use of groundwater resources for agricultural use, making it difficult for people to get water. To address this, the Chinese government has set up a government agency in every Chinese village funded by rural taxes, with the task of planning and managing the village's irrigation. Feng described the issue in 2010, and Huang and Shu in 2017. In 2007, Mukherjee and the Shah published an article comparing the rule of water in India and China.

In Spain, a new law was passed in 1985 after previous failures related to groundwater dependence in order to supply water for public needs (for example, the failure of the subterranean aqueduct's water supply system to supply Madrid's water in the 19th century). Groundwater ownership became government.

In 2007, Lamas and Gardiono analyzed water management practices in Spain.

In East Anglia, England, according to studies by the Environmental Protection Agency, there are predictions of future droughts.

A recent study by owners of environmental organizations in Africa found that the Saharan aquifer system in Africa.

7.2 Definitions and concepts

Public policy is a set of rational actions taken in a process that consists of necessarily political actions. These political actions can be seen as a policy-making process and can be seen as a set of interconnected timelines. The policy steps are as follows:

Preparation of the agenda: The elected officials put the problems on the general work order. Many problems will not be addressed at all, and other problems will be addressed after a long time.

Settlement solutions: Officials set replace policies to address a problem. Replace policies will be implemented in the form of executive instructions, court rulings, and legal action.

Adopting a policy: Alternative policy to be adopted with majority support of members of parliament, an agreement between the directors of the organizations, or a court order.

Policy Implementation: A policy is implemented by those executive units that mobilize financial and human resources to pursue policy.

Policy Assessment: Government audit and accounting units will set whether the executive agencies, the legislature, and the courts are subject to the legal requirements of the policy and its goals or no.

These steps indicate current actions that take place over time. Each stage is related to the next stage, and the last stage (policy estimation) is connected to the first stage (preparation of the agenda) and the middle stages, which is a kind of cycle or non-linear cycle of actions.

Water policy is a process that has a special shape and nature in government systems and is a framework of written and unwritten rules and procedures that are generally developed over decades.

Policy Making: Lyndon Bloon describes the policy process as follows: "First, a problem is assessed, then goals are identified, a possible policy maker is identified and considered to achieve the goals, and finally. Raw politics is united (Lind Bloun, 1968). In other words, it refers to the process of policy-making and preparation of various proposals and plans in the field of problem with the supporting years and providing a proposal for approval by considering the proposed options and then implementing it. Monitoring the implementation of the relevant plan and adjusting its shortcomings is the last step in the policy-making process.

In the policy-making process, policymakers both adopt a policy and make specific decisions, but the two concepts are in no way synonymous. Decision-making by decision-making is the solution to an approach with a short-term, day-to-day, case-by-case approach. But policy-making is seen as the basis for subsequent decisions. Approved by members of parliament. In other words, such policies, by defining the objectives in a particular field, provide the necessary frameworks for short-term policies and ultimately decision-making.

The use of the term governance has a historical background. However, today's use of this term has led to various definitions and interpretations and wide-ranging discussions in the field of social sciences. Most importantly, its emerging application has led to the formation of a highly defining perspective on "development management" or, more accurately, the emergence of a current under this heading.

The phenomenon of governance is an interdisciplinary issue between the branches of law, sociology, management, economics, and political science. Researchers in this field are mainly focused on new topics such as the fragmentation of power between governmental and sub-governmental levels, encouraging and encouraging decentralization of the government and increasing reliance on new types of citizen counseling and participation.

Governance often takes a fresh look at limiting traditional hierarchical power (top-down) and changing its direction toward collective decision-making. Researchers in this field believe that the inefficiency of governments in doing new things has created this concept, so by separating different areas of government and distinguishing between the private and public sectors and the distinction between government and civil society, trying to change the traditional structure of power and They give more weight to parallel government institutions in decision-making and policy-making. Therefore, effective and legitimate power is the power that is obtained through the participation and conflict of various institutions and forces at different national, regional and global levels.

Governance in various ways have defined. The Oxford Dictionary defines governance as follows "the activity or method of governance, control or authority over the activities of subordinates and the military of laws and regulations". The United Nations also defines governance as: "Governance is a set of individual, institutional, public, and private measures for joint planning and management, and a continuous process of understanding between different and conflicting interests." "The framework for participatory and adaptive measures is moving and includes official institutions and unofficial arrangements and the social fund of citizens."

Thus, the concept of governance is based on the cooperation of government and civil society. This concept is based on the basic principle that governments, instead of solely taking full responsibility for governing society at all levels, should better be with citizens, the private sector and public institutions, as one of the institutions or agents responsible for governing society. Be considered. With the entry of citizens, or in a more organized sense, civil society, into decision-making and policy-making, the process of governing the country from an authoritarian and authoritarian system will become a democratic, participatory process.

From the perspective the Ottawa Governing Institute, governance is a combination of traditions (rules, customs), institutions and processes that determine how power should be exercised, how citizens have the right to participate and protest, and how to make decisions in the field.

According to the Tokyo Institute of Technology, the meaning of governance recourses to a complex set of values, norms, processes, and institutes by which society formally and informally governs the process of development and development. Governance requires the government, the international community (social and economic activists, community-based institutes and without structure groups, the media, etc.) to make sense at all levels, local, national, regional and global.

Another definition of governance is defined by the World Bank, the Development Cooperation Committee, the Organization for Economic Co-operation and Development, the United Nations Development Program, and international development organizations and institutions. A complex set of interactions between structures, cents, functions, and processes that are characterized by three key values: accountability, transparency, and participation".

There are two points to consider when using the word governance:

First, it is a different kind of government. In fact, governance as a meaning determines that power flows in and out of official authority and government institutions in the form of three main groups of players: the government, the private part, and civil community. Governance is a particular form of action in which there is sufficient perception of power. That is, its logic is based on the concept of power and imposition. But in the realm of government, the state is limited and there is no place left for the use of power, and one-sided relations give way to multilateral ones. The atmosphere of governance is horizontal, and the government puts itself on a par with other institutions. That is, non-state actors are so independent of each other that they cannot impose anything on each other, but at the same time they are so interdependent that they know they cannot act without each other.

Second, governance underscore the process and determines that decisions are made according to the intricate relations between a number of actors with variant priorities and tastes. Governance establishes an agreement between competing and conflicting priorities and tastes, and this is the foundation of the concept of governance.

What has deliberately or inadvertently led to the use of the term "government" by officials and the public today is the extraordinary acceptance and credibility that this concept provides for government action and authority. However, the space of governance is no longer a state-centered space, and the role of government in society is redefined. In this definition, the government must abandon many of its past tenures, but this does not mean that the government institution should be set aside, because the government must continue to maintain the following three monopolies:

1. Monopoly of physical violence
2. Security monopoly
3. General services

In a general division, three concepts of governance can be distinguished:

- Corporate governance
- Good governance
- Global governance

Among these, the concept of good governance is mostly used, which is discussed here.

While the definitions of governance are not relatively controversial, the components of good governance are debatable. Some argue that different groups have different perspectives on "optimal governance" influenced by their ideology. Some see good governance as possible, mainly by removing restrictions on smoothing market action and minimizing the role of government, and poor governance by insufficient markets and large-scale government. Others cite six characteristics such as transparency, accountability, and the right to decide at the lowest levels for good governance.

Good governance is an approach formed as a result of the failure of the policies of the welfare state and the liberal economy since the late 1980s. This approach or theory, after implementing the policies of "structural adjustment" or "Washington consensus" desired by the World Bank and cheering developing countries to privatize, deregulate and other reforms that were mainly economic and with the results of these policies determined And failures in achieving the desired goals were introduced. Good governance is a theory that not only has an economic approach, but also emphasizes the provision of efficient public services, a credible judicial system, and an accountable administrative system. In 1989, the World Bank devoted its annual report to good governance. (Andisheh Tadbir Water Institute, 2014).

So far, in situations where the roles of government, market and society have been balanced, "distributed governance" has been recognized as the superior type of governance (Anoush Nouri Esfandiari, 2012). Good and smooth interaction between the main actors of the society, ie the government, the private part and civil society organizations and local societies, provides the ground for the realization of good governance in various dimensions. Good governance is not just about doing the job, it's about how you do it. Both goals and methods are equally important for good governance, so each country can blend and localize the conditions of good governance with its own cultures, values, and norms.

According to the latest definition of the World Bank, good governance, in adopting clear, forward-looking government policies (indicating the transparency of government activities), transparent bureaucracy, accountability of executive and social bodies for their own activities, public participation, active participation Everyone crystallizes before the law. In general, it can be said that good governance, management practice (political, economic, executive, etc.) are the resources of a country to achieve the set goals. This exercise includes strategies and institutions through which individuals and social groups have the ability to pursue their own legal interests and rights according to their limitations.

Although in some texts good governance is defined as good governance, these two concepts cannot be considered synonymous, because not all institutions of society are summarized in the executive branch and other institutions are in the process of administration. A country contributes as much as it does. But for many thinkers, the idea of a good government is a prerequisite for good governance. The Asian Economic and Social Commission (ESCAP) has explained the important principles of good governance and the factors that strengthen these principles, and that governments can take effective steps in the process by respecting and enforcing these principles. Accordingly, the basics of good governance are:

- Participation
- The rule of law
- Transparency
- Accountability
- Responsibility
- Forming public consensus
- Equal Rights (Justice)
- Effectiveness and efficiency

Good governance is one of the new topics that has attracted the attention of the world's scientific and international circles in the last two decades. Researchers have defined good governance with different approaches (process, liberalism, and socialism), dimensions (political, economic, and human), and goals (economic development, political development, and sustainable human development).

7.3 Fundamentals and logic

Policy-making studies and analyzes the actions of governments. Public policies are in fact the manifestation of government governance in practice, which can be considered as structural and related complexes, consisting of decision-making goals and actions that can be applied to public authority at the local, national and international levels (Karbasian, 2014).

Government policy occurs at two levels: the level of sub-systems and the high level of national government, policy at the level of national government sub-systems, includes procedures and policies that create continuity of political policies. Policy at high levels of national government, includes It is the procedures and policies that can make a significant difference in the direction of past policies, albeit changing them. At these two levels of government policy, there are networks of policymakers. Kikkert, citing the coordinates of policy networks, considers their most important feature to be the sustainable communication between different governmental organizations and semi-private organizations in the world that the policy process takes place (Kickert et al., 1997). Sociologists see policy networks as prominent forms of power (Kinok, 1990). Political scientists have also used the meanings of networks and policy groups to describe the impact of governmental and non-governmental actors on policies with specific decisions (Hazlores, 2001), and have described networks as a framework for the relationship between governmental and non-governmental actors. And efficient, they are unique to their data and information (Alkaniz, 2010).

Overall, policy-making networks are prominent forms of power, such as the bridge that guides governmental and semi-private organizations on a common policy path. Therefore, these networks ensure that, on the one hand, political or intrinsic demands are transferred to the system and, on the other hand, the output or executive aspects of policies are properly implemented.

7.4 General water resources policies

1. Creating a comprehensive management system in the whole water cycle based on the basics of sustainable development and land management in the country's catchment areas.

2. Promoting productivity and paying attention to the economic, security and political value of water in its extraction, supply, maintenance and consumption.
3. Increasing the amount of water extraction and minimizing the natural and unnatural liquids of water in the country by any possible means.
4. Development of a comprehensive plan to observe the appropriateness of the implementation of dam and watershed management and aquifer management projects and irrigation networks and equipping and leveling lands and maintaining water quality and combating drought and preventing floods and recycling and using unconventional waters and improving knowledge. And techniques and strengthening the role of people in extraction and exploitation.
5. Controlling the waters that leave the country and prioritizing the use of common water resources (taken from the set of environmental laws and regulations, 2006).

In an article entitled "Government Management and Management Governance", Mintzberg (1998) examines governance in five models:

A) Governance as a machine: In this model, governance is considered as a machine that is governed by rules and regulations and standards. This model is compatible with the traditional bureaucratic model and its core is "control". Hierarchical relationships and the lack of flexibility and accountability are the hallmarks of this approach.
B) Governance as a communication network: In this way, flexibility and freedom of action have taken the place of comprehensive and hierarchical control, the government is considered as a complex network of relationships that solves problems through temporary communication.
C) Performance-based governance: This model emphasizes the definition of performance indicators and its accurate measurement.
D) Government-based governance at least: In this model, minimizing government interference in the affairs or government of Hudhaqli is considered. Proponents of her case have been working to make the actual transcript of this statement available online.
E) Rule-based governance: This approach is rooted in the values, beliefs and opinions of a group. Mintzberg refers to this method in the book Organizing Five Models of Effectiveness as the "Special Mission." A distinctive feature of this model is "loyalty", which is adapted from the governance style in Japan.

Avon Hughes discusses governance practices in two models of governance based on the welfare state and modern public administration. Clark and Newman use the term managerial government. They believe that in the present age, government structures have undergone drastic changes that are the result of the transition from a welfare state to a managerial state. Various topics such as the

Observer State, the Evaluation Government, the Small Government, the Empowering Government, the Powerful Government, the Contracting Government, and the Axis Government that evoke a managerial state play a key role in modern government management.

Rhodes (1996) discusses six ways of governing in an article entitled "New Governance; Non-Governing Governance" which includes:

1. Governance as a minimal government
2. Governance as the management of companies
3. Governance as a new state management
4. Governance as "good governance"
5. Governance as a system of social cybernetics
6. Governance as self-organized networks

These issues show the dominance of the economic approach in governance methods, and the models discussed are based on economic expectations and seek to maximize economic benefits; even the new state management with the new values it offers is confined to the economic approach.

7.5 Importance and necessity

Based on the unique value and position of water resources in ensuring the welfare and well-being of society, the need for continued access to adequate and safe water for all at all times should be emphasized. Strong links between "sustainable economic development" and "water management" have made the need to reform the country's water management inevitable. In this direction, "Washing the springs" and a new interpretation of government and people's participation in water management must be presented. however, the implementation and establishment of a "comprehensive, participatory and integrated" management in the country's watersheds is a challenging and time-consuming development that must be done with the help of farmers (especially water users). Such a major move (as a "transformation plan") requires a relative consensus of all factors governing water governance, from sovereignty to creditors, in line with a "participatory perspective" that can only be achieved through collective understanding. In other words, for each of the country's water basins, "the key to success in the evolution of water management is to stimulate the collective will of the main actors in the basin in a single direction to achieve good water governance." Of course, the key to achieving this key is the actors' awareness of the nature of water resources, the current state of these resources, and the increasing trend of pressures on these resources (resulting from growing development and climate change). Certainly, the actors' awareness of the consequences of continuing the current trend also plays a key role in this development. Because, despite the differences between the main actors of the basin and each other, this awareness and vigilance of their connecting elements is during the process of transformation. It is obvious that the general culture and sensitization of the people to the unique value of clean water will lead to the emergence of a strong "social capital" (intelligent and firm) to continue the path of transformation.

Activists' access to processes combined product management products and even their replacement or exchange can have a significant impact on policy outcomes and effects in recent decades, some have consciously and unknowingly contributed to the development of the water crisis in the country (Nouri, 1395). Elites believe that the current water crisis is the result of past policies rather than environmental change over the past few decades. The behavior of the natural environment and the social environment is in solving future problems and challenges (Azimi, 2014).

One of the most important problems in water resources policy is engineering. Many officials and managers look at water resources engineering in the same way as they look at water projects in terms of technical issues such as design and design. Profitability considers that social and environmental conditions are not taken into account, as evidenced by the lack of presence and low presence of social scientists and ecologists in water policy, while the social and cultural dimensions of structural measures and water projects are unknown.

Imitating non-native versions and paying little attention to native management

Imitation and imitation of non-indigenous areas in the field of water extraction leads to misguided policies, because each country must be politicized based on an appropriate assessment of its natural and social conditions. Experts consider the ecological approach to be a strategy for the integrated management of water, land, and human resources. Considering the relationship between the various components is practical.

With this approach, the continuity of human-nature relations and the convergence of populations with government agencies and non-governmental organizations provides a good basis for environmental protection and the development of sustainable development. A country whose economy is based on water and soil or on environmental resources in general, and on the other hand is limited in relation to these resources, cannot be managed on the basis of unprofessional thinking and imitation (Froon, 2016).

Recent global studies have shown that water governance shortcomings can disrupt basic human and environmental needs, reduce economic growth, reduce efficiency and effectiveness, discourage investment in the water sector, and frustrate stakeholders in the decision-making process. Face. Indeed, how can this concept of importance be interpreted on this basis? And what are the roots and background of this interpretation in the epistemological system of water development management and management?

Because there are different versions of the concept of "governance" and this can be an important obstacle in the way of evaluating the governance of the country's water and planning to correct it, it was necessary to formulate the issue coherently from a single and specific point of view.

7.6 Water governance

Good governance is enshrined in the United Nations Millennium Development Goals as a key condition for arriving the Millennium Development Purposes. Statements such as this have been used in water management texts for many years. Apparently, the main reason for these statements should be considered in the relationship between the components of governance and the important and lasting political, economic and social consequences. The profound implications of governance shortcomings are detrimental in many respects to the sustainable use of water resources and related services. Recent global studies have shown that water governance shortcomings disrupt basic human and environmental needs, reduce economic growth, reduce efficiency and effectiveness, and undermine investment in the water sector by discouraging and involving farmers in the decision-making process. Water management includes political, economic, social, and institutional processes that provide the necessary arrangements for stakeholders to participate in the planning and policy-making of sustainable water resource management, including transparency and participation. It is popular and accountable.

The issue of water governance has a maximum history of 12 years in the world, which originated from the water crises of the countries. In Iran, despite issues such as population growth and unbalanced distribution, intensification of competition between drinking water, agriculture and industry, environmental issues of wetlands and lakes, overloading in basins, increasing inter-provincial conflicts and even A city in common water resources and the poor condition of groundwater are serious tensions in the water sector. If these issues are not properly managed, it will have adverse consequences for the country, including technical and engineering, environmental, legal, security, political, cultural, social, tourism and economic. The issue of governance can be generalized to different areas, including the water sector (depending on its characteristics).

According to the Global Water Partnership (GWP) definition, "water governance" refers to the totality of systems involved in decision-making on the development and management of water resources. In the language of global water cooperation, "water governance" refers to a range of political, social, economic and ministerial systems that exist for the development and management of water resources and the providing of water services at different levels of community. In water governance, the issue of how to decide on water is discussed: how, by whom, and under what conditions. Both the way decisions are made to allocate and regulate water, as well as the official and unofficial institutions by which power is exercised.

"Water governance" is essentially a political process, in the sense of balancing the interests of a competitor over who deserves what services, how services are provided, who pays and how to balance the interests of competitors, and how decisions are made. Water resources require political choice. Water governance systems usually reflect political facts at the national, provincial, and local levels.

Therefore, it is impossible to discuss the rule of water separately from the rule of society. At the Freshwater Conference, water governance is recognized as one of the three pillars of water security. At the third meeting of the World Water Council in Kyoto in 2003, it was stated that after numerous meetings at the global level, there is a broad agreement on the importance and implications of water governance: governance determines who and what What are the conditions for access to water, how is water quality maintained, how should issues of cooperation and decision-making be dealt with, are there policies to save water, who is judging and who is allocating Give. Accordingly, the current water crisis is the result of inadequate governance, not water scarcity itself. The focus should be on who decides, who consumes them, and how they are protected.

In governance, the government is not the only actor. In its 2009 World Water Development Report No. 3, the United Nations Water Coordination Organization called the global water crisis a governance crisis. The United Nations Development Program summarizes the difficulties of water governance as follows:

"Every organization thinks that water belongs to it, but no one has the will to protect it." Some argue that the only reason for the shortcomings is the lack of water governance, and that what we need is good governance, not a perfect, flawless government. Effective water governance is considered necessary for continuous water management. In this regard, the following five characteristics can be considered for good governance:

Leadership bound to ethics (with a focus on the vulnerable and vulnerable),

- Timely and reliable information for decision making and accountability,
- Mechanisms for discussing, discussing and resolving disputes in the presence of civil society,
- Decentralized discontinuity in public policy coverage,
- Fight against corruption.

Another important point is that it is important to review the structure of water management from the perspective of water governance. Unstable exploitation of groundwater and surface resources in most of Iran's watersheds is a "managerial" trend and, to put it better, a "rule".

Although efforts have been made to provide practical solutions to this situation and to achieve effective governance of water resources in the country, but in practice, the direction of movement during the last half century has always led to a worsening crisis. Has been mentioned. Cases such as comprehensive and integrated management of water resources, public participation, demand management, intrinsic value of water and finally compliance with economic criteria to increase water efficiency in the national economy, are common speeches and hearings that plan more than two decades. The water has filled the country, but little practical result has been achieved. As a result, the country is now and will continue to be increasingly affected by the unpleasant consequences of this trend. It seems that the quality of legislative, executive and private institutions and how they relate to each other and the lack of will and serious determination

TABLE 7.1 The main aspects of water management structure reform according to the effective and appropriate water governance approach.

Efficient and decent governance	Current situation	The main aspects
Bilateral flow at different levels	Command, from top to bottom	How to decide and execute
Responding to stakeholders, stakeholders and their demand, controlling and managing demand	Supplier-oriented	Relationships governing water management
Participating and seeking justice through transparent, transparent and accountable organizations and open processes.	Closed specialized fan base organization	Orientation of institutions
Connectivity	Partiality	The prevailing attitude towards the organization

to implement the policies contained in the programs, has had a decisive effect on the failure to achieve the goals of the programs. From the point of view of the components of governance, this quality can be better analyzed and explained.

Table 7.1 summarizes the main aspects of water governance in the current situation and the desired conditions.

According to the United Nations Development Program, good water governance is more than just legislation, regulation and water-related institutions, although they are all important. Good water governance also oversees processes that promote people's partnership in design and decision-making on water systems and heritage.

The unique value and position of water resources in ensuring the welfare and well-being of society, the need for continued access to sufficient and safe water for all at all times and strong links between economic development, development goals of the third millennium and sustainable development with water resources management and development. The need to respond appropriately to the growing needs of structural and fundamental water management has made the country's water management inevitable.

Institutional developments such as the establishment of the Ministry of Water and Electricity in 1963 and the Water Law and its nationalization in 1963 and the policies adopted and pursued in the planning system such as the establishment of agro-industries, agricultural joint stock companies and cooperatives The production at the bottom of the dams clearly shows the movement towards concentration and the governmentalization of water affairs from the 1940s onwards.

The most important solutions and proposed measures in the field of governance in the water sector.

The idea of forming a "local water council" in order to achieve successful governance in the water part is one of the proposed solutions. Based on its policy text, the formation, challenges and principles of the "Local Water Assembly" for the country's catchment areas have been clearly explained. The powers of this House of Water go beyond the High Council of Water or the Provincial Council for the Protection of Water Resources, which will have more members than the people's representatives, especially the exploiters. The main benefits of this assembly will be networking among stakeholders, a common perception of the problem and a clear algorithm for collective decision-making in water management.

In the policy text "Evaluation of the country's water governance system and proposal of institutional reforms, (implementation of the WGG model or improved IDA model)", after reviewing some of the world's leading research in the field of assessment and pathology of governance in the water sector In particular, Dinar and Salt's research in 2004 and 2005 respectively, which surveyed 39 countries, explored how governance in the country's water sector was explored using the aforementioned patterns and presented the areas that needed to be improved and the policy tools needed by each sector. The results of the implementation of this model have shown that the combined government-market-oriented policy is more suitable for organizing the country's water sector. Also, a package of improvement and improvement tools in the framework of final reforms has been proposed in such a way that the role of each of the governing, market-oriented and community-oriented water institutions in the decision-making or decision-making space of the country's water sector is 40, 33 and 27 percent, respectively.

The following 5 general policy packages can be proposed:

- Creating a national discourse with determination and practical will for the country's water crisis
- Replacing a water-based development approach with a water-based development
- Decreased domestic production of cereals, especially wheat
- Water Economics and Tariff System Reform
- Improving the country's water management structure

Water management and governance should start in the short term from local units (development of agricultural water allocation program taking into account the constraints such as the environment) and then in the medium term to catchment areas and developed hydrological areas (localization of water cultivation and valuation model) and in the long run to end the reform of the country's water management structure (formation of a smart system and real-time water management).

In order to arrive more sustainable management of water resources in the country, the following two proposals in the field of governance are presented:

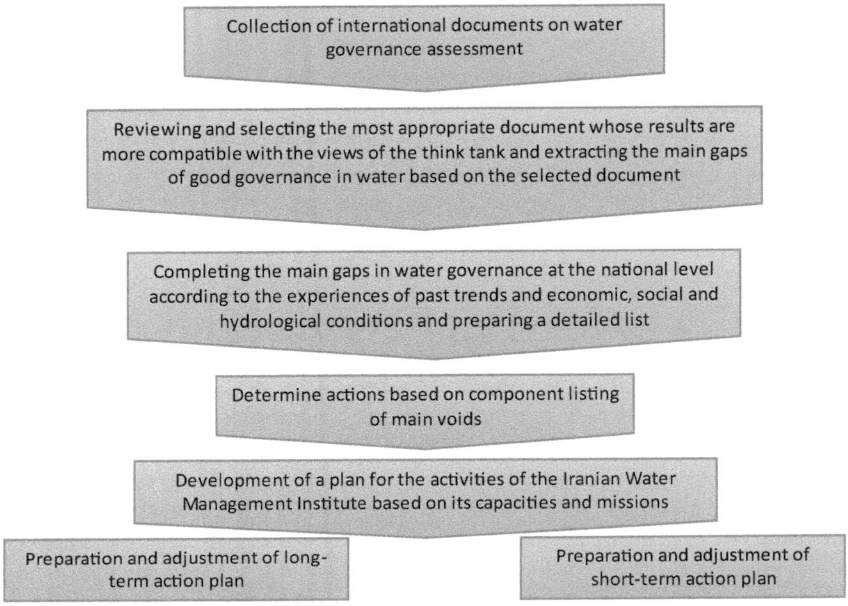

FIGURE 7.1 The process of brief evaluation of the country's water governance and formulation of the action plan of the Iranian Water Management Institute.

Acceptance of legal validity for local water governance in parallel with and during national water governance.

A specific judicial system should be established on the issue of water at the macro level and national water governance to follow the laws approved by the Islamic Consultative Assembly and the approvals of the institutions covered by the executive branch with special supervision and prosecute.

Development of agricultural lands and construction of new agro-industrial units (instead of increasing water efficiency and increasing production per unit area.) Any budget allocation, especially with financing, which means indebtedness of the country, for the above measures.

7.7 Working method and flowchart

According to the Iranian Water Management Institute, in order to assess the state of water governance in the country, we follow the steps shown in Fig. 7.1.

As can be seen in Fig. 7.2, there are eight main gaps for the country's water governance, which are identified as the main factors in the weakness of the country's water governance system Fig. 7.3.

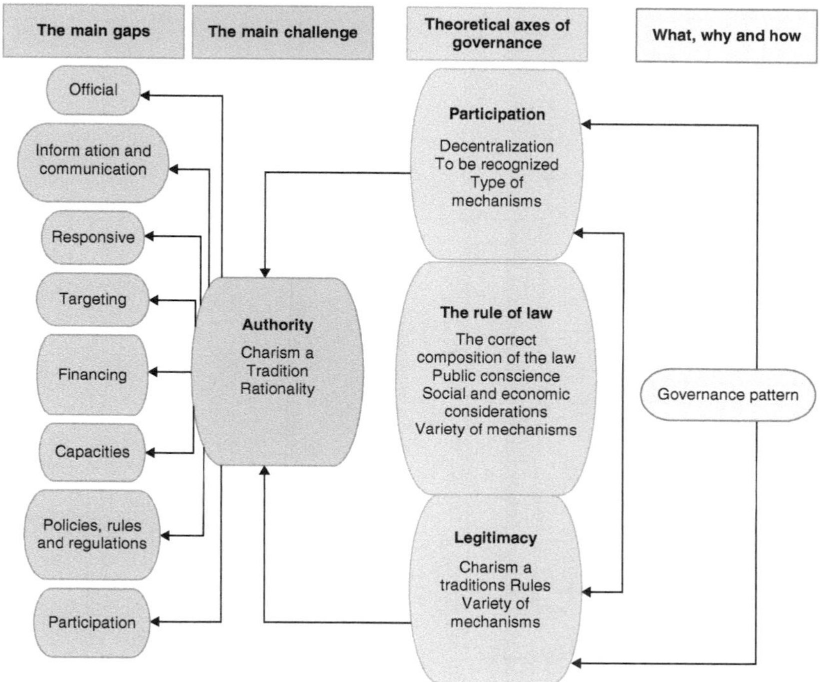

FIGURE 7.2 Completion of the main gaps based on the conceptual model of the Iranian Water Management Institute.

7.8 Explain some practical examples

Some groundwater management studies in South Asia pay attention to self-regulation as a model of governance. For example, in Gujarat, India, Saurashstra when government efforts failed to adequately address groundwater abstraction, declining groundwater levels, and salinity of coastal groundwater, some members of the local society was beginning to addition rain water harvesting. During the Manson period, they were fed water. Inspired by local Hindu spiritual leaders, the People's Movement grew rapidly under official laws. The first acceptors acted like a thinker's brain, inspiring the farmers of the surrounding villages and acting like a pervasive network. Although the initial focus was on growing water resources, some demand-side methods were accepted.

Social cohesion in the countryside and religion helped ensure compliance. The general movement, identified by the Gujarat government, led to a pluralistic coexistence between the official rules of groundwater governance and the rules and norms of the informal movement. Technical support from the Gujarat government and local NGOs also helped improve the movement's effectiveness.

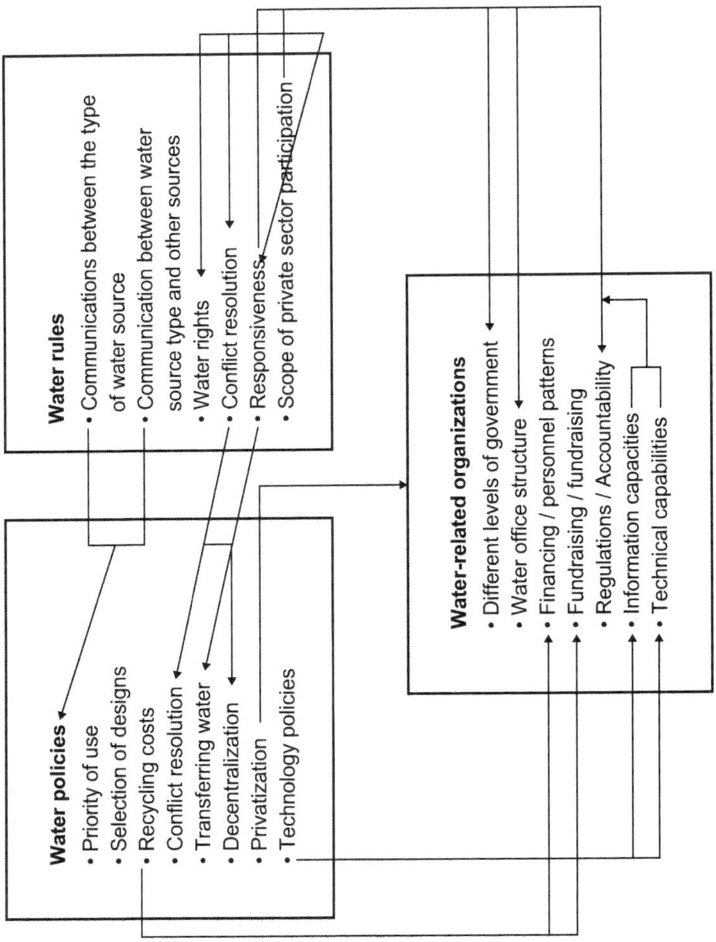

FIGURE 7.3 Internal links of water policy and governance institution.

Financial support from diamond traders from these rural areas was another help. This can be seen as a public-private participation with government agencies, citizens, communities, religious communities, non-governmental organizations and the private part.

Groundwater management strategies in India can be compared to the strategies used in China. Since the 1950s, both countries have faced the problem of high population density, as well as the issue of food self-sufficiency and agricultural-based economies, and have relied on groundwater to irrigate their agricultural products.

This has led to over-harvesting of groundwater resources, resulting in declining water levels in many areas under its control. In some cases, groundwater markets have emerged to allocate groundwater to people without active wells.

In general, today, despite the existence of IWRM programs and groundwater laws in India, there is poor governance. A recent World Bank study found that groundwater and urban water pollution is an emergency, and the weakest element in India's groundwater management chain is institutional capacity at the central and state levels.

In Spain, it is possible to understand how past methods and policies have affected the implementation of current groundwater governance. Spain, with its strong governance of groundwater infrastructure, has one of the highest per capita levels of surface reservoirs and transit volumes between river basins. This is due to past failures related to groundwater dependence in order to supply water for public needs (for example, the failure of the subterranean aqueduct's water supply system to supply Madrid's water in the 19th century). As a result, groundwater has become part of an old-fashioned stereotype that the use of groundwater has always led to depletion of resources, making it unreliable. Groundwater was largely used in rural areas where there were insufficient surface water systems.

Considering private ownership as well as land ownership, property owners were able to pump as much water as they could unless a third party was affected. However, excessive use of groundwater (which was partially promoted by the government in the 1950s and 1960s) has led to a dangerous situation. Under the 1985 law, groundwater ownership returned to the public sector. Therefore, those who had the right to harvest at that time were granted the right to harvest based on the rate at which they had previously harvested, provided that their license was registered and that each new operator required a permit from the relevant water authority. But many exploiters in Spain are still unregistered. Part of the problem is the lack of capacity and resources to do so. But part of it is also due to the concept of private ownership of groundwater that has existed for a long time, and this prevents the desire to register and obtain a license. The National Water Project in 2001 raised the challenges of the old model of surface water infrastructure development.

For the first time, the project strongly supported groundwater management. The project states that before extensive research on groundwater and the evaluation of various options, the Ebro River water cannot be transferred to any of the overflowing aquifers. The project also promotes the organization of groundwater exploitation societies and educating the public about groundwater.

A case study of climate change adaptation and groundwater governance in East Anglia, UK, shows the implementation of the IWRM the basics and the relevance of the EU Water Guidelines. In this region, which is one of the fertile agricultural areas of England, agriculture and food processing industries, and of course the supply of drinking water, depend on groundwater. Groundwater resources are provided by the Environmental Protection Agency (EA, a non-governmental organization responsible for the environment, food and rural affairs, which is affiliated with the State Office of the Prime Minister), through the licensing process. They are managed with 12 years of credit and with priority given to the allocation of water for drinking, environmental and irrigation purposes. Although it is difficult to understand why EA has started climate change adaptation in the green (and often humid) region of the UK, the bitter truth is that groundwater is entirely dependent on the agricultural or food processing sector. Undergrounds that have political power are allocated.

Future droughts and groundwater shortages mean huge economic losses in the region. As a result, EA has partnered with local operators in policy and policy-making with water supply organizations. In the participatory method, the following adaptation measures were developed:

Reduce license time to 6 years.

- Changes in land reclamation methods to decrease groundwater pollution
- More efficient irrigation techniques and winter harvesting on the farm.

This illustrates how the IWRM-based approach meets environmental needs and how regulatory powers use general participation as a tool for improvement.

Another study is the state of Arizona in the United States, which is an arid and highly dependent area of groundwater. In addition to the standard standards for drinking water and draining at the federal level, each state has its own guidelines for determining groundwater rights and its own management systems. With the passage of the GMA in 1980, Arizona became the national leader in groundwater management among selected areas for active management (AMAs) (Calby and Jacobs, 2007). The characteristic of these areas was the excessive discharge of groundwater.

Groundwater rights and licensing regulations were assigned to major sectors of groundwater exploitation, namely drinking, industry and agriculture, and each sector was required to comply with Arizona Department of Water Resources' ten-year plans.

The key to GMA was to establish the legal objectives of water management and water security, which required the active management of modern urban development to demonstrate the existing legal, physical, and permanent resources for water supply for the next 100 years. Groundwater use should be in line with management objectives, which in most areas of five-year active management, these efforts are aimed at achieving reliable drainage (meaning long-term harvests that do not exceed the amount of feed). The natural, economic, and water resources of these areas are different, and there are some capabilities and capabilities to set rules for recognizing different conditions.

A legal framework for nutrition and recovery makes it possible to store water for the future, and groundwater supply as a treatment mechanism for the Colorado River is costly rather than therapeutic.

Arizona Underwater Management Law has been hailed as an advanced approach to groundwater regulation and provides remarkable novation and variety in water use approaches (Calby and Jacobs, 2007). However, after 30 years of experience, the challenges of water management remain remarkable.

Even in large areas of active management where secure drainage is a management goal, there are areas where local drop in water level is a concern. The reliable water supply framework allows the pumping of groundwater even from a depth of 1000 feet (303 m), so that most of the harvested values are recovered by surface water sources. However, the Arizona Central Water Protection Authority, which is responsible for nutrition and recovery, does not have the long-term renewable resources to commit to groundwater supply. Among these concerns are the issue of the cost of feeding and recovering these resources in the future. There are no penalties for failing to achieve management targets, and a small number of wells are exempt from regulation.

Although about 80 percent of Arizona is home to about 6.5 million people living in active management areas, some other states in the state face disbalance in supply and demand. Problems and gaps in information related to determining the water needs of natural systems and the lack of knowledge of the environmental needs of water (Magdalen et al., 2011) have led to a lack of policy to assess nature's needs for water. Although the Water Management Law provided a strong base for groundwater management in 1980, very remains to be done.

Arizona's experience, in fact, refers to the complexities of groundwater management in growing and semi-arid regions, where water supply is an important issue.

7.9 Summary

In the absence of effective policymaking, the design and implementation of water reform faces many obstacles, so the acute conditions of water resources in the country require a new approach and a smart approach. Due to the water crisis in Iran, we need centralized and professional research resources in this field. Collecting and integrating information and creating a centralized database, provides the possibility of specialized review and accurate and realistic analysis for specialized management. Experts believe that water policy in the country needs multidimensional and comprehensive management. An attitude that should include technical, engineering, economic, social, ecological, historical, political, cultural, etc. technical issues.

In the field of water, futurism is a systematic effort for the future in the field of policy, so water policy requires knowledge, monitoring, careful analysis, evaluation of water resources. Efficient water policy is the only way to survive and make an impact, which is achieved through a revision of the country's water policy. Experts consider the passage of the water crisis to be a "sustainable water supply" and believe that all the country's policies should be based on this. It is necessary to change the attitude towards "water" especially water policy makers.

Despite some positive orientations over the past two to three decades towards decentralization and acceptance of the private and cooperative sectors, and voluntary and participatory motivations, these orientations have not yet been properly addressed. In response to the question that if this hypothesis is accepted, to what extent can the reconstruction of the water governance system, especially with the aim of creating more diversity in the country's water governance mechanisms, be effective in the fundamental reform of affairs, it should be said that to be more active in the private sector and civic institutions, it is very important to strengthen the country's water governance. Because a review of general policies and quality objectives (20 to 25 years ago) and available evidence clearly shows that the main challenge of the country's water governance system is the authority and implementation of decisions to overcome the problems and shortcomings. It should not be forgotten that the main factor in the implementation of programs is human beings, who have a complex individual and collective motivational structure. But the decades-long experience of the country's centralized water management clearly shows that even to implement the best and most progressive policies, there is no central and main emphasis on coercion (also through the imported import administrative system transformed into indigenous organizational culture). It can be a solution for making decisions. If, instead of coercion, the people are accompanied and the orientations adopted in the upstream documents of the country's water sector through the creation of trust, learning, persuasion and collective agreement become social demands and are pursued by the public, failures and shortcomings. Existing water resources are unlikely to be replicated, and there will be much hope, even in improving effective oversight and increasing efficiency in the public sector. If economic incentives are properly incorporated into program implementation mechanisms, the satisfaction of these incentives with hidden methods will be reduced by corruption and bribery and the level of resistance and non-cooperation.

Evaluation results show that in order to improve water policy and governance, first of all, orientations should be changed and in the second step, assistance should be used in planning to change and restore water governance order. It is far more difficult to transform the existing system of governance than to upgrade and improve it. Therefore, in view of similar international experiences, it is obvious that in the reconstruction of water governance, a long path must be followed, in which different stages are: convergence and collective agreement in recognizing the desired framework of water governance in society, assessing the situation. Available based on the imaginary pattern of optimal governance, setting priorities, selecting the right starting point, and gradually advancing and transforming and gradually improving things.

Necessary measures at each stage should be determined based on the requirements of a multifaceted framework (economic, social and political). Prioritizing priorities and prioritizing simpler goals and scheduling reforms and prioritizing measures that facilitate achieving simpler goals are important steps in this style, are considered planning and management. Using the results of water governance evaluation and governance gaps and combining these results by predicting the stages of change in water governance system in the country can pave the way for proper orientation of activities in the short, medium and long term.

Moving to such a participatory level requires a change in the structure of the country's water management and the need to review and redefine the roles and powers and responsibilities (from the perspective of good water governance theory). Although some elements of this purpose appear in the official documents and programs of the country and apparently do not seem to be much of a dispute, taking practical steps to achieve it requires equipping the national will to achieve effective and decent water governance. Achieving the desired level of national will and will is essentially and essentially dependent on public and political awareness and awareness and the creation of social capital.

References

Alcañiz, Isabella, 2010. Bureaucratic networks and government spending: a network analysis of nuclear cooperation in Latin America. Lat Am Res Rev 45 (1), 148–172.

Andishkadeh Tadbir Water Institute (2014)

Azimi, Dezfuli, Seyed Ali, Akbar, Abdolreza Rokanuddin, Eftekhari, 2014. Integrated futurism, an approach to overcoming the challenge of water and food crisis. In: National Conference on Water Crisis Prerequisites in Iran and the Middle East.

Colby, B.G., Jacobs, K.L., 2007. Arizona Water Policy: Management Innovations in an Urbanizing, Arid Region. RFF Press.

Froon, Mohammad Taghi, 2016. Environmental Management. KWN Monthly Ninth Year No. 97.

Ho, C.C., 2005. The application of system dynamics modeling to study impact of water resources planning and management in Taiwan. In: Proc., 23rd International Conference of the System Dynamics Society July 17–21.

Huerta, J.M., Serra, M., Sandoval, R., 2001. ProEstado-MAUA, A system dynamics approach for the design of sustaianble water use strategy in the Middle Lerma-Chapala Basin. In the lerma–Chapala watershed, evaluation and management, 201. Edited by A.M. Hansen and M.V. Afferden. Dordrecht, the Netherlands: Kluwer Academic/Plenum Publishers.

Karbasian, A.H., 2014. Public Policy. Baqerul Uloom Research Institute, available at.

Kickert, W. J., Klijn, E., & Koppenjan, J. F. (Eds.) (1997). Managing complex networks: Strategies for the public sector. SAGE Publications Ltd, https://www.doi.org/10.4135/9781446217658.

Lindblom, Charles E., 1968. The Policy-making Process. Englewood Cliffs, New Jerdey: Prentice-Hall,Inc.

Megdal, S., Nadeau, J., Tom, T., 2011. The forgotten sector: arizona Water Law and the Environment. ArizonaJournal of Environmental Law and Policy 1 (2), 243–293. http://www.ajelp.com/.
Rhodes, RAW, 1996. The New Governance: Governing without Government. Political Studies. 44 (4) 652–667. doi:10.1111/j.1467-9248.1996.tb01747.x
Ruth, M., Pieper, F., 1994. Modeling spatial dynamics of sea level rise in a coastal area. Syst. Dyn. Rev. 10 (4), 375–389.
Simonovic, P.S., Fahmy, H., Ei-Shorbaghy, A., 1997. The use of object-oriented modeling for water resources planing in Egypt. Journal of Water Resources Manage. 11, 243–261.

Chapter 8

Shared water resources management

Fahimeh Mirzaei-Nodoushan[a], Omid Bozorg-Haddad[a] and Assefa M. Melesse[b]

[a] *Department of Irrigation and Reclamation Engineering, Faculty of Agricultural Engineering and Technology, College of Agriculture and Natural Resources, University of Tehran, Karaj, Iran.*

[b] *Department of Earth and Environment, Florida International University, Modesto A. Maidique Campus, Miami, FL, United States*

8.1 Introduction

Water resources including river and lake basins and aquifers may cross international borders by which shared (transboundary) water resources emerge. These resources are shared with two or more countries having different interests in their water use and development. Shared water resources may lead countries to cooperate or dispute and management of these resources are one of the most complex social, economical, and political issues in recent years for a variety of reasons like the absence of regional treaties for water allocation in some countries, the lack of global participation to enforce a convention on shared waters, and static water treaties which are contrary to the variability of water resources conditions. Powerful and active institutions are needed both for competing interests over water allocation and for managing scarce waters. Lee (1995) focused on shared waters in Latin America and reviewed the management of the shared waters mentioning economic factors. It was explained that there is little interest in international water management unless there would be economic benefits for the countries. Emphasizing successful transboundary water management, Blomquist and Ingram (2003) attempted to recognize and define the distinctions between international and national settings of groundwater resources. They concluded that the capacities developing by people to resolve water resource problems are essential to improving water management and achieving peace and justice in communities. Parhi and Sankhua (2013) presented issues of riparian responsibilities over shared rivers. They mentioned that sustainable transboundary water resources management needs sharing water-related benefits, recognizing the main water needs of ecosystems, setting competing uses of water resources,

Economical, Political, and Social Issues in Water Resources. DOI: 10.1016/B978-0-323-90567-1.00012-7

developing capacity for integrated water resources management (IWRM), protecting water resources during wars and improving knowledge of political approaches to prevent conflicts. Sullivan (2014) examined the challenges of managing transboundary basins and studied the Murray-Darling Basin plan located in Australia. As Sullivan (2014) expressed, the plan provided the water needs of water holders, recognized ecosystem water demand and considered the importance of socio-cultural features, which is still work in progress and such a plan is essential for all transboundary basins.

It is obvious that if shared water resources are managed cooperatively, besides sustainable water management it would make a joint effort to global unity and peace. Successful cooperation in shared water management leads to many benefits for all riparians participants. Cooperation can reduce the potential of conflict, improve the quality of the shared waters, generate secure livelihoods, strengthen economic ties and also develop trust among basin countries. Mylopoulos and Kolokytha (2008) implemented the European Union (EU) Water Framework Directive in Greece's shared water resources. Their study expressed that the new directive integrated the total of parameters that is related to the shared water resources in river basin level and proceeding to agreements for Greece in shared waters would bring benefits for all riparian sides. Biswas (2011) focused on the benefits of cooperation on transboundary rivers and the cost of non-cooperation between countries in South Asia. The experience from South Asian countries indicated that the countries have no other alternatives but to cooperate in managing their shared waters. Seeing the Nepal-India-Bangladesh experience, the cost of non-cooperation in shared water resources would be paid not by the politicians, but by hundreds of millions of poor people. Leb (2015) analyzed the dynamics of international cooperation to manage and develop the shared water resources. The study represented that international treaties are tools to provide a stable process of cooperation because a treaty binds a country, not just the individual government that agreed to it. Leb (2015) concluded that cooperation between countries leads to greater benefits for all basin states. He (2015) explored China's state practice in shared water cooperation and water diplomacy. The study concluded that China and its neighbors would need the political attempt of policy-makers, the modification of economic development strategies, the development of shared water resources management strategies, and the support of riparian sides. Petersen-Perlman et al. (2017) detailed the nature of water conflict and water cooperation. To be prepared for new challenges in managing shared water resources like climate change and growing economy, they suggested that parties should engage in conflict resolution mechanisms and invest in institutional capacity with their neighbors. Skoulikaris and Zafirakou (2019) applied a river basin management framework in four transboundary river basins where Greece is the downstream country and the upstream countries are not all member states of the European Union. They demonstrated that significant conflict situations in the past including water quality and water quantity issues could be resolved using continuous and reliable datasets for sustainable water management. The management of shared water resources considers not only the scientific and technical aspects of water resources management, but also the specific legal, economical, cultural, social, and historical aspects of shared water resources management. Shared water management would bring benefits for everyone to share such as food security, economic growth, climate change adaptation, international trade, improved governance, and regional integration.

8.2 Definitions and terms

Words and idioms which are used in this chapter related to share water resources management are explained as follow:

Convention "begins as an international meeting of representatives from many nations that results in general agreement about procedures or actions they will take on specific topics, e.g. wetlands, endangered species, etc. " (Guertin, 2009).

Hydro-hegemony is "hegemony at the river basin level, achieved through water resource control strategies such as resource capture, integration, and containment, whereas hegemony is considered as leadership buttressed by authority" (Zeitoun and Warner, 2006).

Hydropolitics is "a systematic analysis of interstate conflict and cooperation regarding international water resources" (Elhance, 1997).

An international drainage basin is "a geographical area extending over two or more states determined by the watershed limits of the system of waters, including surface and underground waters, flowing into a common terminus"[1] (ILA, 1966).

A protocol is "an agreement that diplomatic negotiators formulate and sign as the basis for a final convention or treaty. The treaty itself may not be completed for many years" (Guertin, 2009).

Riparian states mean "the states bordering the same transboundary waters" (UN, 1999).

Transboundary waters are "any surface or ground waters which mark, cross or are located on boundaries between two or more states; wherever transboundary waters flow directly into the sea, these transboundary waters end at a straight line across their respective mouths between points on the low-water line of their banks" (UN, 1999).

Treaty is "an agreement where the parties negotiate on it to reach common ground and avoid further conflict or disagreement. Usually, it is ratified by the lawmaking authority of the government whose representative has signed it. In the United States, the Senate must ratify all treaties" (Guertin, 2009).

[1] International Law Association

Water conflict is "an all-embracing notion covering the entire spectrum of possible situations where the interests of states may collide: from minor differences in opinion to the other extreme of situations of tension and hostility that may threaten international peace and security" (Vinogradov et al., 2003).

Watercourse means "a system of surface waters and groundwaters constituting under their physical relationship a unitary whole and normally flowing into a common terminus", therefore International watercourse means "a watercourse, parts of which are situated in different states" (UN, 1997).

Water diplomacy is "an approach that enables a variety of stakeholders to assess ways to contribute to finding solutions for joint management of shared freshwater resources. It is a dynamic process that seeks to develop reasonable, sustainable and peaceful solutions to water management while promoting or informing cooperation and collaboration among riparian stakeholders" (Klimes et al., 2019).

Water disputes limited to "those conflicts involving the use of transboundary water resources, both surface and ground waters" (Vinogradov et al., 2003).

8.3 Theories about the use of shared water resources

There are four main theories which describe the use of water from shared basins by riparians. These theories have been used only for surface waters and are as follow (Aguilar and Iza, 2011):

1) Absolute Territorial Sovereignty: This theory express that the countries which are part of a shared basin have absolute power over the waters located within their territory. Thus, upstream countries in the shared basin may change the quality and quantity of water and do not respect the effects on downstream states.[2] As is clear, this hypothesis has no practical application.
2) Absolute Territorial Integrity: Unlike the theory of absolute sovereignty which favors upstream countries, this theory gives downstream countries a right to natural water flow. Any changes in the natural flow of shared waters which would adversely affect the territorial integrity of the downstream countries are illegal.[3]
3) Equitable and Reasonable Use: This theory that has been the most widespread in practice is based on the idea which shared basins are shared natural resources that are subject to equitable use. According to this theory, countries have equal rights and shared sovereignty over the water stream and there is a need to find a balance of interests by taking into account the demands and uses of the water from all riparian countries.

[2] This theory was prepared by the Attorney General of the United States of America during the conflict over the diversion of Rio Grande water and it is now referred to as the Harmon Doctrine.

[3] This theory draws on the Anglo-Saxon principle of riparian rights.

4) Joint Management: This theory emphasizes integrated development and joint regulation of the river which goes further than the theory of equitable and reasonable use. The theory offers a community of interests between all of the countries located in a shared water basin and considers the establishment of international institutions with different powers and responsibilities.

8.4 Importance of shared water resources

Freshwater is one of the most important resources in human life. Scarcity, non-uniform distribution and contaminant of freshwater resources made many regions around the world facing a water crisis. Because of rising water demand, extending the use of water and climate change, more and more countries would experience water stress. Therefore, improved management and redistribution of existing freshwater resources is essential.

Shared river basins with the number of 263 host about 40 percent of the world's population and nearly 50 percent of global land area and 40 percent of freshwater resources are located in shared water basins (Giordano and Wolf, 2003). Some of the largest rivers around the world are split between two or more countries. Moreover, shared fresh groundwater resources are crucially important, sometimes more than international rivers and lakes, especially in water-scarce regions. Although the certain number of shared aquifers is still unknown, 273 shared aquifers have been identified including about 155 aquifers located in the European continent, 68 in the Americas, 38 in Africa and 12 in Asia by the International Shared Aquifer Resources Management (ISARM) project, a joint project of United Nations Educational, Scientific and Cultural Organization-International Hydrological Programme (UNESCO-IHP) and Organization of American States (OAS) (Stephan, 2009).

Where water resources are shared between countries, shared water management remains a major challenge to supply water needs. Lack of cooperation and unilateralism between riparian states certainly would affect water availability, producing destructive consequences to ecosystems, and vital human needs. Non-cooperation in using shared water resources would also harm upstream countries, not only downstream countries. For instance, reductions in river flow due to diversions upstream make downstream desertification, by which moving dust would hurt upstream countries, furthermore, even local climate may be changed resulting in vast changes in fauna and flora of the sea.

Allocating water among users of different countries is a politically challenging project. Theoretically, the optimal approach is to manage water resources according to the objectives of water users such as agricultural and hydropower objectives. In practice most of the shared waters have no institutions to resolve differences and share benefits, and concepts like trust and strategic decisions are scarce in government policy.

Countries vary in their hydrological interdependence on shared water which is related to geographic and hydrologic terms in a country. For example, only 6 percent of the Ganges-Brahmaputra-Meghna Basin lies in Bangladesh but the basin occupies three-quarters of the country. Also, one-fifth of the Mekong Basin located in China while the basin occupies less than 2 percent of China's area (UNDP, 2006a).

Political power is a key constraint in using shared waters in which a particular country may have a dominating position in receiving more than an equitable share of the available water. Interestingly geographical position is not the only term for having more power on shared waters. As an illustration of the power of upstream positions, Turkey and China have dominated power in Tigris and Euphrates Rivers and Mekong River, respectively. As mid-stream countries, Israel and Egypt have dominated position in Jordan River and Nile River, respectively (UNDP, 2006b).

Increasing water demands have led various countries to maximize utilization of the available water in their territory so that sometimes the common water resources have been over-exploited without noticing the negative effects on downstream countries. Such actions have created tensions between several countries over the past years, as well as the increasing water crisis. Hence, these water resources are of great importance, requiring intelligent policies, management, and maintenance. It is necessary to manage shared water resources among the countries taking into account the different conditions of all consumers.

8.5 International water agreements

The growth in water withdrawals brings the need for international water resources management and cooperation among countries. Management of shared water resources is a complex problem due to its close relationship with political and judicial approaches which sometimes heterogeneous and conflicting national laws might make management more difficult. Meanwhile, shared water agreements typically take two forms of global and regional agreements.

Global agreements are general principles of international behavior and law. One of the first principles that would prohibit riparian states from causing harm to other states was the Helsinki Rules on the field of the international rivers adopted by the International Law Association (ILA) during 1966. The principle of "reasonable and equitable utilization" of the waters of an international drainage basin among riparian states was the basic principle of international water law established by Helsinki Rules (Salman, 2007). The Helsinki Rules were followed by the convention on the Law of The Non-Navigational Uses of International Watercourses adopted by the UN General Assembly during May 1997. So far, this convention is the strongest international legal instrument regarding shared water management but it has not yet obtained the necessary number of signature to enable it to enter into force (Salman, 2007).

Almost no upstream country has signed on and only downstream countries have signed the convention. Eventually, the most recent international rules for shared water management were the Berlin Rules adopted during 2004. The Berlin Rules were an attempt to collect all relevant international law together.

Regional agreements are laws and specific bilateral or multilateral treaties that are negotiated for particular international resources. Many bilateral and multilateral water resources treaties have been signed by countries to arrange power, allocate water, control water quality, and manage other aspects of shared waters. Because of the lack of enforceable international frameworks for shared water management, currently, regional agreements are the strongest mechanism for promising cooperation in the management and conservation of shared water resources. As an example of the regional agreement, Boundary Waters Treaty (BWT), the treaty relating to boundary waters and questions arising between the United States and Canada, was signed during 1909 between the United States and Great Britain to prevent disputes related to the use of shared waters. Another important regional agreement is the agreement on cooperation for the sustainable development of the Mekong River Basin in 1995 which was signed by Thailand, Cambodia, Vietnam, and Laos (the four countries in the Lower Mekong Basin) to achieve optimum use and to prevent waste of the waters in the basin. The convention on the protection of the Rhine (CPR) during 1998 for reducing river contamination is another example for the regional agreement which was signed by France, Germany, Luxembourg, Netherlands, and Switzerland governments and by European Community.

Despite groundwater's importance, shared groundwater resources have received less attention than shared rivers and lakes and there are few international laws and agreements that especially focused on these vital resources. A review of international water laws demonstrates that groundwater resources have been usually considered as subordinate to surface water. UN Watercourse Convention 1997 defines watercourse as a system of surface waters and groundwaters which are interconnected. This definition of the convention not only does not apply to all groundwater resources but also the rules of the convention are fit with surface waters. There are some agreements regarding shared aquifers such as the agreement on the Genevese Aquifer signed by France and Switzerland during 1977 and re-negotiated during 2007 which is covering controlled groundwater extractions, controlled artificial aquifer recharge operations, pollution control, and the apportionment of all relevant costs. As another example, the agreement of the Iullemeden Aquifer System (IAS) made by Mali, Niger, and Nigeria during 2009 which was replaced by a later agreement during 2014 by contribution of Algeria, Benin, Burkina Faso, Mali, Mauritania, Niger and Nigeria for the Iullemeden and Taoudeni/Tanezrouft Aquifer Systems (ITAS) and focused on promoting the integrated management of groundwater resources and emphasizing the environment-support function of groundwater (Burchi, 2018).

8.6 Management of shared waters

Water resources management includes a set of legislation, policies, regulations, engineering practices, and traditions (Jain and Singh, 2003). Managing water resources need to solve various water-related problems such as fluctuation of water quantity and quality, flooding, drought, etc. The complexity of water resources management would increase when water resources are shared by two or more countries.

Management of shared waters has different aspects to study, such as hydrological sciences, hydraulic engineering, environmental sciences, public health, economic sciences, international law, and social sciences. Both surface and groundwater are important resources in international issues to have sustainable shared water management. From the scientific, legal, social and economical points of view, management of the shared water resources is a complicated process and it becomes more complex because of different needs, uses, cultures, even approaches and interpretations of countries located in shared water resources.

A step by step guidance is recommended to effectively identify, solve and manage shared water resources problems. Five steps of shared water resources management are explained as following which may be adapted to any particular case study in the world:

8.6.1 Shared water problem analysis

The first step in the shared water management process is to get a clear view of the actual situation of water resources. To solve the shared water problems, it is needed to identify, discuss and understand problems of a shared water resource. The objective is to gain an overview of the water and land resource issues and ongoing activities in a basin or aquifer. In problem analysis, some questions may be asked about a shared water resource such as: Where the resource is located? Which countries are located in the shared basin or aquifer? What is the hydrological and environmental situation of the water resource (flood, drought, water pollutants, etc.) and how critical is it? How dependent are the riparians on the water resources in terms of food security and livelihood? Which conflicts are between the countries over the shared water resources? Which cooperation and benefit-sharing issues are between the countries? Which water projects are being done, ongoing and will be done in the future by the countries of a shared water resource?

8.6.2 Data and information collection

Data and information on shared water resources are basic prerequisites for transboundary water management. Having sufficient data and information about shared waters informs countries sharing common water on general conditions of the basin or aquifer. Also, real data sharing and monitoring can support objective water resources assessment, effective operation of flood and pollution warnings, proper practices in water scarcity and drought conditions, etc. Despite countries' awareness of the benefits of sharing data and information for water resources development and management, actual cooperation in data sharing and joint monitoring is inadequate in many shared basins and aquifers. Generally, this inadequate data and information are due to incorrect or defective data and a lack of willingness among the riparian countries to share basin or aquifer data. The reluctance of national authorities to share information with neighboring countries is mostly because of the strategic status of water resources in an area with scarce resources or economic values of water, regarding irrigation, hydropower, fishing, and navigation.

For shared waters, information and data are usually collected from the national monitoring systems instead of gathering from joint monitoring systems. Some data and information needed from the shared basin or aquifer depending on water issues (e.g. drought, flooding, pollution, sedimentation, etc.) are as follow:

- Physical characteristics (topography, soils, etc.);
- Land use;
- Hydrology and hydrogeology data;
- Water quantity and quality data;
- Water use, withdrawals, and discharges (e.g. irrigation, hydropower, and drinking water);
- Environmental characteristics;
- Point and non-point sources of pollution;
- Population;
- Social and economic characteristics;
- Hydropolitics information.

Remote Sensing (RS) as a possible method of obtaining water resources data could be used successfully to monitor quantity and quality characterizations of rivers, lakes, aquifers. RS could be practical for issues such as meteorological data, land-use issues, and water-extent changes. These types of data have different spatial and temporal resolutions (e.g. 0.5° and months, respectively) and need particular verification and correction. On the other hand, various global water-related models have been developed (e.g. WaterGAP by Frankfurt University) which are easily and freely accessible and their results could be derived from the internet.

Overall, there are some world databases and datasets available that are gathered by different world organizations. Available data and information systems are provided for coordinating datasets and using common data to be able to coordinate results. There is a long list of data holders presented by the United Nations Environment Programme (UNEP) made up of a mixture of different sources and undoubtedly is not comprehensive. The source data and partners are represented by organizations, such as UNEP, Food and Agriculture Organization

(FAO), UN Children's Fund (UNICEF), World Meteorological Organization (WMO) and International Groundwaters Assessment Centre (IGRAC) which some of them have some gaps or are not updated. These data are collected by assuming through country partners and/or RS from different satellites.

8.6.3 Engineering assessment and modeling

Modeling shared water resources has to face two types of difficulties. One is of national water resources modeling such as correctly evaluating the resource, its variations with time, weather and climate, its quality and pollution and its economic and financial dimensions such as costs of infrastructures, risks of investments and many other parameters. Other difficulties are different issues shared among countries such as different uses, different legal and also different scientific approaches for a similar problem.

Many scientific disciplines have developed various methods and tools to assess shared basins and aquifers. Different technical models have been recommended to simulate and evaluate water resource status, screen alternative policies and strategies, optimize network design, determine the risks to human health and ecosystems, etc. Models should be carefully calibrated and validated to have reliable results on the behavior of the basin or aquifer.

The following issues can be solved by modeling shared water resources:

- Assessing the actual situation of water resources and water-related processes;
- Assessing the long-term trends in quantity and quality water variables;
- Forecasting water variables in short and long periods;
- Detecting pollution from point and non-point sources;
- Examining the effects of policy practices;
- Providing early warning in an event of flooding or accidental pollution to protect riparian states and water users;
- Assessing short and long-time droughts;
- Investigating climate change impacts on watercourses.

Many technical water-related models have been applied which would satisfy some of the above objectives and are addressed in various studies around the world. This chapter presents models that are applied for specific case studies.

8.6.4 Scenario analysis, conflict resolution, and decision making

Shared waters as scarce resources in many arid and semiarid countries need to be allocated among competing countries and therefore a decision-maker has to ensure that all the demands and concerns of different users are taken into account when making decisions. It is important to consider the conflicting needs of consumers who are affected by the decisions and conflict resolution is a part of the decision making process.

To understand the water-related conflicts and to find the optimal or acceptable solution for all consumers, various approaches have been developed by the engineering, law, economic, political and social point of view. There are models formulated as optimization or based on the team and game theories which are developed by engineers and management experts. Most of these models are based on the basic economic concept of Pareto optimality which suggests optimal solutions and tries to find a way to eliminate conflict by an equitable resolution between beneficiary users (Ganoulis et al., 1996).

Optimization models such as linear and nonlinear programming find optimal solutions for decision making which should be formulated correctly to provide the right decision over local, regional and political aspects. Game theory is a framework that investigates the strategic practices of decision-makers to explore some acceptable solutions for water allocation in shared water resources. Game theory reflects the behaviors of beneficiary actors and users and provides a certain solution for a decision-maker, while optimization techniques consider several objectives solved multi decision maker problems and provide a Pareto front (Madani, 2010).

It should be mentioned that the success in the practice of this type of modeling is generally dependent on the acceptance of involved parties of the model assumptions. Therefore, a set of objectives and the relative weights or preferences related to the goals of beneficiaries play an important role to make a decision.

8.6.5 Cooperation vision, plans, and investment development

Cooperation on a shared water resource can bring many benefits for all involved parties. Many types of benefits such as social, economic, environmental and political benefits can be achieved through the cooperative use of shared rivers and aquifers with different objectives and tendencies. The optimal manner of cooperation for a shared water resource depends on a set of elements consisting of hydrologic characteristics, economic and investment facets including benefit and cost of cooperation and relationship between riparian sides. The uniqueness of each shared water resource will offer different cooperative practices and benefit-sharing.

Reviewing the historical records, conflicts over water and its benefits do arise and lead to political tensions but most conflicts are peacefully and cooperatively resolved, even if the negotiation process took a long time (Wolf et al., 2005). Three types of cooperation stretch from coordination (e.g. data-sharing) to collaboration (e.g. developing adaptable national plans to decrease regional costs and capture regional gains) to joint action (e.g. joint project assessment and design and joint investment) (Wolf, 2010).

Beyond bargaining over volumetric allocations, countries on shared water could exchange a set of benefits that lead to more welfare of all sides. The set of benefits could be investing in financial projects, trading in energy resources,

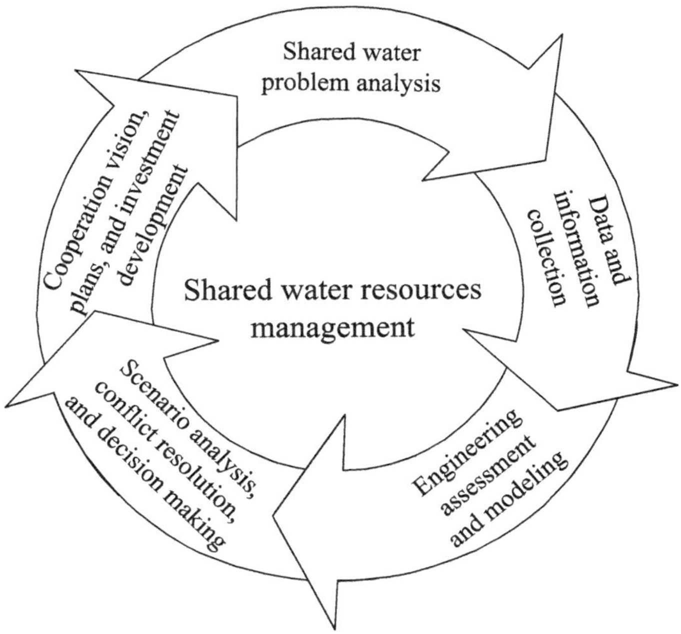

FIGURE 8.1 Steps of shared water resources management process.

sharing data and linking countries and would help countries look beyond focused goals of self-reliance. Of course, strong institutions are needed to manage conflicts and enforce provisions (Fig. 8.1).

Besides, the management process must be iterative and adaptive to have sustainable management. In each shared basin or aquifer after problem analysis, data and information collection, basin or aquifer assessment and modeling, and conflict resolution and decision making, a process going on until approaching cooperation and joint management. Continue the process; it is necessary to constantly analyze the transboundary water situation to have dynamic management. To illustrate the recommended management process, the management of two well-known case studies of the shared river basins and a great shared aquifer are presented below.

8.7 Case study

8.7.1 The Nile river basin

8.7.1.1 Shared water problem analysis

The Nile River is world's longest river flowing in northeastern Africa which runs through eleven countries (Burundi, Rwanda, Tanzania, Kenya, Uganda,

Democratic Republic of Congo, Egypt, Ethiopia, Eritrea, South Sudan, and the Republic of Sudan) and its basin covers about 10 percent of the African continent area. The Nile River which is about 6,800 km long has two main river systems, the White Nile and the Blue Nile. The White Nile originates from Great Lakes (Victoria, Kyoga, and Mobutu) and the Blue Nile originates from Lake Tana located in Ethiopian highlands. The two rivers coming from regions with abundant moisture are connected at the north of the Sudanese capital of Khartoum which has arid conditions. The river flows north onto Egypt and finally into the Mediterranean Sea (NBI, 2012) (Fig. 8.2).

The Nile has 84 billion cubic meters (BCM) of average annual flow but it unevenly varies by spatial and temporal situation. The Nile's discharge is largely provided by its Eastern Nile Basin headwaters which 59 percent of it comes from the Blue Nile (Waterbury, 2002). The Blue Nile flow is extremely seasonal and most of its flow occurs between July and September, while the White Nile is approximately stable over the year due to regulating the effect of its upstream lakes. A significant portion of the basin (downstream) is located in the arid and semi-arid region that leads to more water dependency of involved countries on Nile River flows. Each year the average amount of 12–14 BCM flow reaches the Mediterranean (NBI, 2016).

The Nile River was home to one of the world's earliest civilizations and today the basin supports nearly 160 million people depending on its waters for irrigation, hydropower navigation, drinking water, fishing, and tourism. All the countries located in the Nile Basin have mainly agricultural economies and water demand would increase by growing population estimated to nearly double by 2030. Almost 90 percent of the Nile Basin's renewable water resources are used by the riparian states (NBI, 2016). Egypt and Sudan have the most dependency on the Nile River flow because they are located in the arid region and there are no other reliable flows in those countries. The water withdrawals of Egypt and Sudan are 57 and 31 percent of the total renewable water withdrawals, respectively. Most other involved countries are close to the water crisis due to rapid population growth and urbanization. Generally, nearly 5×10^6 ha of a potential total area in the Nile Basin (10.2×10^6 ha) is irrigated and 98.7 percent of the area is located in dry Egypt and northern Sudan. Three of the eleven countries, Ethiopia as the primary supplier and Egypt and Sudan as the dominant consumers, are the most important partners to cooperate meaningfully in sharing water (Swain, 2011).

Numerous treaties have been done over the Nile River which many of them were from the colonial era. During 1959, a bilateral agreement between Egypt and Sudan was signed to govern Nile water by the Nile Waters Treaty. Egypt and Sudan shared water between themselves with the amount of 55.5 km^3 and 18.5 km^3 per year under this agreement, respectively. The cost of water-related projects would be split equally between Egypt and Sudan. Prior treaties were not inclusive and left Ethiopia, the major source of Nile water, in the 1929 and 1959 agreements. To resolve disputes and review claims made by other countries, the

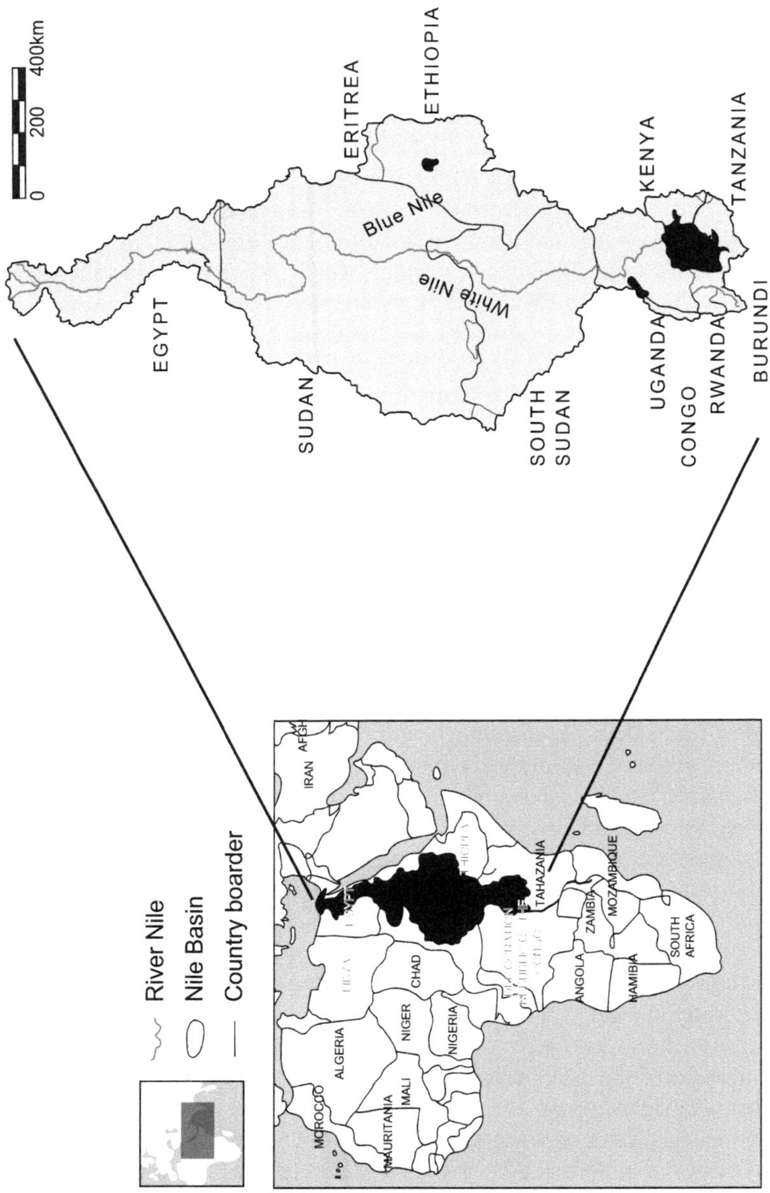

FIGURE 8.2 Nile River Basin Location (Ding et al., 2016).

Nile Waters Treaty established a Permanent Joint Technical Commission. The other task of this commission was devising a fair water allocation scheme for persistent low-flow periods (Cooley et al., 2009).

The Nile Basin Countries have designed numerous activities and initiatives to promote cooperation among them. The Hydromet Survey Project established during the 1960s and designed to collect hydrometeorological information to have more effective management of the Nile water. During 1999, the Nile Basin Initiative (NBI) was established by the water ministers of 9 countries sharing Nile water – Ethiopia, Egypt, Sudan, Uganda, Tanzania, Kenya, Rwanda, Democratic Republic of Congo, and Burundi – to develop the river cooperatively, sharing remarkable socio-economic benefits, and raising peace and security in the region. The Initiative has financed projects to make regional trust and cooperation among riparian countries and share technical data and information. NBI attempts to share many benefits such as energy generation, industrial use, and navigation (Cooley et al., 2009).

The Nile Basin water resources management has been studied for the existing water resources system and water resources development planning. Different water resources issues have been investigated, such as the impact of upstream development on downstream water availability, the economic impact of water developments, the impact of alternative dam filling and dam operation policies on downstream and the impact of climate change on water availability which could be studied deeper and more innovative (Melesse et al., 2009a; 2014; Melesse, 2011).

8.7.1.2 Data and information collection

In Nile River Basin, there are numerous data deficiencies and there is a reluctance to share data by countries placed in the basin. As an illustration of collecting data, Belete et al. (2018) used remote sensing data to model the water balance property of the whole Nile due to data-scarce. They applied openly monthly accessible precipitation data from WorldClim with a spatial resolution of about 1 km^2. Also, the annual ETo data was provided from The Consultative Group on International Agricultural Research, Consortium on Spatial Information (CGIAR-CSI) dataset with a 1 km^2 spatial resolution. For estimating root restricting layer depth, the Harmonized World Soil Database was used. Land Use and Land Cover (LULC) datasets were provided by the University of Maryland with a 1 km^2 spatial resolution.

Ahmed et al. (2019) investigated and assessed the effects of upstream human activities coupled with drought events on the flow regime of downstream countries of the transboundary Nile River Basin. The data collection of the study is as follows. Time series of discharge data (1900–2012) in the northern area of Sudan, downstream at the connection of Blue and White Nile, and the beginning of the main Nile were obtained from the Egyptian Ministry of Water Resources and Irrigation. Hydrologic data for the Nile Basin are incomplete and non-uniform distributed. There are 98 stations in Egypt that only 49 are working, 131 stations in Sudan that 31 are operational, and one station in Ethiopia. The NBI has presented the incomplete annual sub-basin Nile data. The FAO has provided monthly data of the Nile Basin for cross-checking purposes. Ahmed et al. (2019) collected, compared, and verified between the collected land-based data, the Global Precipitation Climatology Centre (GPCC), and the Climate Forecasting System Reanalysis (CFSR).

8.7.1.3 Engineering assessment and modeling

Various modeling approaches have been used to technically assess the Nile River system. Different physical and environmental aspects could be assessed for surface water and groundwater issues including streamflow modeling, sediment dynamics, teleconnections and river flow, land use dynamics, climate change impact, groundwater flow modeling, and demand analysis. Many technical models have been developed for the whole Nile Basin and a large number of models applied for component sub-basins especially the Blue Nile Basin (Melesse et al., 2009b; 2011; Abtew et al., 2009a; 2009b; Abtew and Melesse, 2014a; 2014b; 2014c; Yitayew and Melesse, 2011; Dessu et al. 2014). Aiming to show the variety of models used to assess Nile Basin, some of the studies over the basin assessment are presented as follow:

Soil and Water Assessment Tool (SWAT) is a physically-based model that simulates the water balance of a basin. The input data of the model are geographic information on the basin including topography, soil and land use and weather data. Dessu and Melesse (2012) assessed the applicability of the SWAT for simulating rainfall-runoff processes in the Mara River Basin located in the Nile River Basin which is shared with Tanzania and Kenya. Satellite rainfall data were combined by in-situ rainfall data for simulation. The model was calibrated and validated using 10 years of discharge data from three stations. The results showed that the model performance in the Mara River Basin relied on the accuracy and completeness of discharge data.

Dessu and Melesse (2013) studied climate change on the hydrology of the transboundary Mara River Basin using General Circulation Models (GCMs). A GCM is a type of climate model that employs a mathematical model of the general circulation of a planetary atmosphere or ocean and forecasts climate change. Various scenarios could be considered to project/predict climate changes which include greenhouse gas forcing in the long term. In the study, daily simulation outputs of rainfall and maximum and minimum surface temperature of five selected GCMs were combined with observed historical data to assess the impact of the climate change on the hydrology of the basin. Using three scenarios of climate change (A1B, A2, and B1 SRES), the results showed that the Mara Basin would be considerably affected by climate change by the non-uniform change of rainfall and temperature pattern. Also, Dessu and Melesse (2013) investigated the response of the basin hydrologic system to future changes in rainfall and temperature. Therefore, the SWAT model was used to analyze the hydrologic changes in the basin. The results demonstrated that the increased rainfall and warmer temperature would lead to a comparatively high range of peak and low river flow over the coming decades and it would put pressure on the basin. It is expressed that the existing infrastructure in the basin may need to be improved and strengthened.

Belachew et al. (2015) simulated the Eastern Nile Basin system employing the Hec-ResSim model. Hec-ResSim model is one of the water system simulation models that simulate reservoir operations and is a decision support tool for reservoir project studies and reservoir regulators during real-time events. The model was developed for Abbay –Blue Nile, Tekeze-Setit-Atbara, Baro-Akobo-Sobat, and Main Nile Sub-basins. The model was calibrated by gauge stations selected from each of the sub-basins. Different development scenarios including proposed dams, irrigation schemes and cascade options for each sub basin was considered and simulated from 1960 to 2002. The results indicated that there is an investment opportunity to optimum benefit-sharing among riparian countries of the Eastern Nile Basin.

Kahsay et al. (2015) estimate the direct and indirect economic impacts of operating the Grand Ethiopian Renaissance Dam (GERD) on downstream countries' economies using a Computable General Equilibrium (CGE) modeling framework. CGE models are economic simulation models that analyze the impacts of large-scale developments on interrelated economic systems. Three climatic and hydrologic scenarios were considered in two project phases of the impounding phase and the long-term operation phase of the dam. The results showed that there are benefits for Ethiopia and partly Sudan and negative effects on Egypt during the impounding phase. While the operation phase of the dam has significant economic benefits and improving welfare for all the Eastern Nile countries.

8.7.1.4 Scenario analysis, conflict resolution, and decision making

The Nile River system, similar to many other international rivers, has insufficient water to satisfy the water demands of all planned development projects. Each of the countries sharing the river is unilaterally managing and utilizing the water in its territory, while unilateral management of water resources would decline the potential benefits of cooperation for joint water management. Water resources system analysis in the Nile Basin would be helpful in conflict resolution and understanding conflicts and cooperation options which leads to reasonable and reliable decision making. Various techniques are applied for conflict resolution and water allocation in the Nile Basin in studies which some of them are illustrated as follow:

Whittington et al. (2005) optimized the economic benefits of the entire Nile Basin. The optimization model considered the potential gross economic benefits of Nile water utilization in irrigation and hydropower generation. Results showed that direct economic value of cooperation would be increasingly significant and it would accelerate many development gains through the region by having a cooperative perspective.

Madani et al. (2011) used non-cooperative game theory solution concepts to understand the conflict in the Nile River Basin. The Graph Model for Conflict Resolution (GMCR) (Kilgour et al., 1987; Fang et al., 1993) modeled the conflict as a game, found its equilibrium, provided strategic insights and identified the stable outcomes for players of the game. The GCMR model is a game theory technique that represents a conflict as moving from state to state via transitions based on decision-makers' priorities. The results demonstrated that Egypt's preferences are very determinative for stable status and the existing status is not stable.

Construction of GERD on the Blue Nile is a project which is planned by the Ethiopian government during 2011. Arjoon et al. (2014) analyzed the hydrologic and economic risks faced by downstream countries (Sudan and Egypt) when the GERD will be online. They used a stochastic hydro-economic model of the entire Eastern Nile River Basin to analyze different development and management scenarios. The model is formulated to maximize economic benefits (irrigation and hydropower benefits) related to the allocation decisions over a given planning period which took account of the hydraulic infrastructure and water demands. The optimization problem is solved by stochastic dual dynamic programming (SDDP) algorithm, an extension of stochastic dynamic programming (SDP). Results showed that the benefits of GERD operation for related countries especially Ethiopia and Sudan are significant and there would be positive externalities in Sudan and Egypt during dry years.

Degefu et al. (2016) solved the water problem in the Nile River Basin using a water allocation framework which consisted of asymmetric Nash bargaining solution concepts and the bankruptcy theory. The Nash bargaining theory (Nash, 1950; 1953) is a game-theoretic model that was used to divide common water resources among claiming partners. Also, bankruptcy theory (O' Neill, 1982) is an economic game theory technique which would be applied when the amount of shared resource available is less than the resource demand. The results obtained water allocation outcomes under water-scarce conditions.

8.7.1.5 Cooperation vision, plans, and investment development

Certainly, due to growing water demands and developing more and more hydraulic infrastructure in the Nile Basin, the region would meet undergo substantial changes. The total storage capacity of dams in the basin would be doubled by 2040–50, the total area under irrigation would grow to about 60 percent of the current area of irrigation, and the installed capacity of hydropower plans would grow from a current value of 5600 MW to over 25,000 MW (NBI, 2016). Thus the basin will be in a critical situation and this situation needs a very high degree of trust, cooperation and sharing water and benefits between involved countries.

Batisha (2011) investigated possibilities for cooperative management of shared water resources options in the Nile Basin. It is mentioned that there are many common benefits in ecological, economic and political features when there is cooperation vision in the region. Also, momentous time, effort and resources were invested by countries through NBI plans. The paper resulted that NBI would be an effective institute for cooperation, peace, stability, and joint investment. Hence, cooperative management would bring greater ecological, economical, social and political benefits for all countries besides water-related benefits.

Degefu and He (2016) explained about water bankruptcy in the Nile River Basin. The current and predicted water demands of the consumers were considered to emphasize the critical situation of basin water and the importance of the management of the basin. The authors believed that water bankruptcy is inevitable and recommended to apply the scenario of water bankruptcy for water allocation and water rights determination. Therefore, it would be caused by avoiding water conflicts in the future and ensuring the fairness and sustainability of allocation rules.

Al-Saidi and Hefny (2018) focused on beneficial regional cooperation on water, energy, and food priority issues in the Eastern Nile Basin despite the current focus on the transboundary water cooperation issues related to equitable consumption and hydropower generation. They emphasized that issues such as trade, certain services of ecosystems, and agricultural technologies could help enhance cooperation and are more feasible politically and economically which also mitigate distrust and disputes in the region.

Onencan and de Walle (2018) tried to reconstruct the Nile Basin water allocation dialogue. As the first step to address the deadlock of water allocation in the region, new agendas were defined and nine possible water allocation methods were developed. Then, the factors of equitable and reasonable utilization were applied to compare the methods which results showed that none of the nine methods satisfied all of the factors. Therefore, a new agenda was recommended and a method which is a combination of three water allocation methods was identified to solve the problem. The agenda included a change in the recipient of the water resources, the allocated item, and the process of decision-making.

8.7.2 Euphrates-Tigris river basin

8.7.2.1 Shared water problem analysis

The Euphrates-Tigris River Basin is shared among Iraq, Turkey, the Syrian Arab Republic, the Islamic Republic of Iran, Saudi Arabia and Jordan which is located in West Asia. The Islamic Republic of Iran is only in Tigris sub-basin and Jordan and Saudi Arabia are only in Euphrates sub-basin. The Euphrates and Tigris

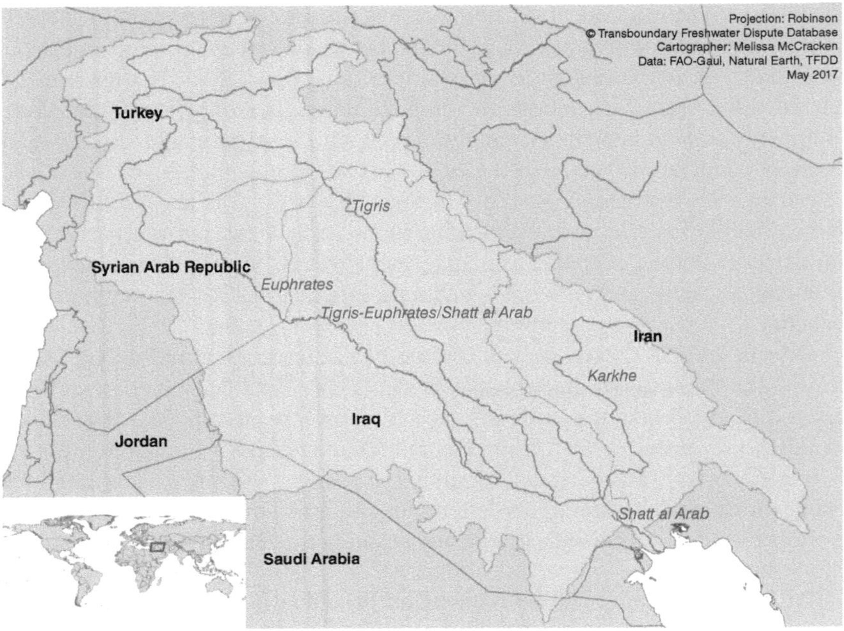

FIGURE 8.3 Euphrates-Tigris River Basin Location (Transboundary Freshwater Dispute Database (TFDD) 2017).

rivers constitute a single shared water system and are connected by their natural course near Qurna in Iraq and make Shatt-al-Arab flow which flows into the Persian Gulf (FAO, 2009) (Fig. 8.3).

The mean annual flow of the Euphrates is 32 billion cubic meters which about 90 percent of it emanates in Turkey and the rest from Syria. The Tigris has 52 billion cubic meters of average annual flow. Nearly 40 percent of the Tigris flow originates in Turkey, about 51 percent from Iraq and 9 percent from Iran. Although the 40 percentages of Euphrates Basin and 54 percentages of the Tigris Basin lies in Iraq, Iraq is a downstream country and its water supply from these rivers strongly depends on the upstream countries. The geographical situation gives Turkey certain advantages in regional water disputes and more power in hydropolitics is of Turkey (FAO, 2009).

Riparian countries of the basin use water of the two rivers for irrigation and increasingly for hydropower generation. Irrigation water withdrawal is about 68 km^3 and the total area of irrigated lands in the basin is about 6.5–7 million ha which approximately the percentage of each country is as follows: 53 percent Iraq, 18 percent the Islamic Republic of Iran, 15 percent Turkey and 14 percent the Syrian Arab Republic. Countries located in the basin constructed a large number of dams to produce hydropower and store water for large-scale irrigation (FAO, 2009).

There are some international legal frameworks to manage international water conflicts which were explained before. These frameworks cannot be significant while transboundary water problems are dynamic and uncertain. When Convention 1997 was approved, Iraq and Syria ratified it. Turkey has not ratified the Convention and also voted against approval in the diplomatic conference that opened it for signature (Cahan, 2017).

Two water-related developments play important roles in the Euphrates and the Tigris Basin. From the mid-1980s, Turkey engaged the Southeastern Anatolia Development Project (GAP in the Turkish acronym) in its territory which is consisted of multiple dams on Tigris and Euphrates Rivers to develop irrigation and energy productions. GAP will include the construction of 21 dams and 19 hydroelectric projects and 1.65 million ha of land will be irrigated and 26 billion kWh will be generated yearly. When all GAPs dams are built, it would negatively impact Syria and Iraq's water resources in quantity and quality terms. Moreover, another broad development that is very effective on water problems is the deterioration of stability in Iraq and Syria. Instability in Iraq and Syria such as changing the Iraqi regime, the Syrian civil war, and the movement calling itself the Iraq and Sham Islamic State (ISIS) made a reasonable policy impossible over issues in the Euphrates and the Tigris Basin (Wolf and Newton, 2007).

Turkey, Syria, and Iraq negotiated over water problems from a long time ago and reached some diplomatic agreements. Between Turkey and Iraq, a protocol of the Joint Economic Committee was made during 1980 which allowed for joint technical committee meetings related to water issues. During 1983, Syria Joined the meetings too. An agreement was signed between Syria and Turkey which Turks guaranteed a minimum flow of 500 m^3/s across the border with Syria. Some trilateral meeting between Syria, Turkey, and Iraq was held but there were a few beneficial results and no agreement was signed between the three governments (Wolf and Newton, 2007).

8.7.2.2 Data and information collection

Data and information of the Euphrates-Tigris Basin collected by the governments are protected and hard to access because water data considered as a bargaining chip in regional negotiations. It is difficult to obtain up-to-date information on the status of the water resources in the region due to the lack of studies carried out or published in international journals and the confidentiality of information. This part mentioned two data and information collection which have been gathered by researches to study and assess the Euphrates-Tigris Basin.

Kucukmehmetoglu (2012) presented a linear programming model to allocate water resources in the Euphrates and Tigris River Basin. There was needed extensive data to solve the problem which is explained as follows. Data on the water contribution of each riparian country was available in other academic studies. The return flow rate was assumed 35 percent for agriculture and 80 percent for urban water use. Evaporation rates of reservoirs were computed based on observed annual evaporation figures given the reservoir surface areas for the major reservoirs in the Euphrates Basin (extract from another study). Total planned irrigation land areas for all countries were available in another study but irrigation land areas were available only for Syria and Iraq which made the author use existing irrigation maps. Agriculture productivity was assumed to be the same throughout the region due to a lack of information. To determine urban demands, historical population data of cities have been used to estimate population by 2040. Water values for agriculture and urban use was derived from other studies. Also, maximum and minimum water withdrawal and average electric generation rates were estimated similar to another study.

Hasan et al. (2018) monitored changes in the extent of surface reservoirs along the Tigris and Euphrates Rivers when conflicts, drought or dam mismanagement occurred. The data used to analyze the basin consisted of information on the amount of water in the reservoirs, discharge information along the rivers, precipitation data and information on when conflicts and droughts occurred. Landsat ortho-rectified top-of-the-atmosphere (TOA) was used to measure the water surface area of rivers and reservoirs which its resolution is $30 \times 30\ m^2$. Altimetry data was collected from Ocean Surface Topography Mission, TOPEX/Poseidon, and ENVISAT satellites. Rainfall satellite data was extracted from the PERSIANN Climate Data Record. Moreover, archived monthly discharge records of several Iraqi river stations accessible from a study, information about cross-sections of the rivers were not available and depth-height-surface area data from Mosul reservoir was available from another study.

8.7.2.3 Engineering assessment and modeling

It is necessary to have an assessment of the water balance and water availability in the Euphrates-Tigris Basin under water deficit and flood conditions to develop a framework for decisions on water allocation among countries located in the basin.

Jones et al. (2008) studied the hydrologic impacts of engineering projects on the Euphrates-Tigris Basin and its marshlands. The SWAT model was used to understand the dynamics of the natural flow system and to demonstrate the impacts of decreasing flow in the marshes. The results of the model were used to compute the temporal variations of the average monthly flow rate, the average monthly peak flow rate, and the annual flow volume of the rivers. Variation of marsh size was extracted from temporal satellite data around the same period that major dam construction projects were preceded and the results showed the sustainable reduction of marshes in the basin.

Kavvas et al. (2011) investigated water balances over the Euphrates-Tigris Basin. First, a regional hydroclimate model (RegHCM) was developed to perform land hydrologic water balance computations for infiltration, soil water storage, direct runoff, and evapotranspiration and to estimate irrigation water demands. RegHCM has five major modules: an atmospheric module, a snow module, a soil/vegetation module that needs necessary atmospheric inputs such as precipitation, the temperature in time and space, and a river module. Then a hydrologic model was implemented to route streamflows within the river system. Finally, the basin was assessed over water characteristics.

Bozkurt and Sen (2013) studied climate change impacts in the Euphrates-Tigris Basin. Three different models of GCM (ECHAM5, CCSM3, and HadCM3) and emissions scenario (A1F1, A2, and B1) simulations were dynamically downscaled. All scenario simulations showed an increase in winter surface temperature in the entire basin while there was a decrease in winter precipitation in the highlands and an increase in the southern parts of the basin. Moreover, in all simulations projected annual surface runoff changes presented that the territories of Turkey and Syria have the most vulnerability to climate change by decreasing surface runoff of the river. However, the downstream country (Iraq) may suffer more based on the water use of upstream countries.

Chao et al. (2018) estimated the groundwater depletion and droughts in the Euphrates-Tigris Basin to be useful for sustainable water management. Monthly changes in the total water storage of the basin were collected from the Gravity Recovery and Climate Experiment (GRACE) satellite mission. Then altimetry and parameters of surface water from land surface models were used to remove from total water storage and determine the total groundwater depletion, human-driven groundwater depletion, and climate-driven groundwater depletion. The results showed that the total groundwater depletion trend from 2007 to 2015 was 3.6 times that from 2003 to 2006 and human-driven depletion made 98 percent of total depletion. Also to estimate the drought, they used total storage deficit index (TSDI) and the results presented that the basin had been undergoing a severe drought since 2007.

8.7.2.4 Scenario analysis, conflict resolution, and decision making

The Euphrates-Tigris Basin has potentially high risks for continued disputes in the future. Water conflict could be occurred by the effect of hydrological factors, national and international political factors, or a combination of these factors. Equitable and reasonable allocation of water resources in this basin among riparian countries is a complex and important process in conflict resolution.

Hipel et al. (2014) gained insight into the causes and resolutions of disputes that occurred along the Euphrates River. Three water conflicts occurred during 1975, 1990 and 1998 involved Turkey as the upstream country which was analyzed using Graph Model to demonstrate the importance of conflict resolution of coalitions and third-party interventions. The results depicted the conflict along the Euphrates River as a complex one and represented that bilateral and tripartite negotiations would continue with mixed success which was determined in the study.

Zagonari and Rossi (2014) described a prototype Negotiation Support System (NSS) (Nandalal and Simonovic, 2003) based on the negotiation setting in the Euphrates-Tigris Basin. Bilateral negotiation between Turkey and Iraq, annual analysis, the assumption that Iraq and Syria have no inherent water rights, differences in water entitlements or needs, analysis related to the basin current irrigation technologies in Turkey and Iraq, and negotiations on water quantity and quality that account for national dam construction plans were proposed by Turkey and were analyzed based on all principles of the 1997 UN convention. The authors tried to achieve a balance of interests among riparian countries and provide insights into water allocation rules, side payments, water requirements of the parties and cooperation. The results indicated that there were two exploring potential negotiation paths: allocation should meet Iraqi equitable use constraints and remaining water should be allocated to Turkey for agriculture utilization or allocations should meet Turkish maximum agricultural water needs and then remaining water could be allocated to Iraqi's agricultural use.

Mianabadi et al. (2015) applied a weighted bankruptcy solution method to solve the problem of demands exceeding resources in the Tigris River shared among Turkey, Syria, and Iraq. Three alternative bankruptcy rules including Proportional Rule (PRO), Constrained Equal Awards (CEA) and Constrained Equal Losses (CEL) were compared to the weighted method for water allocation and also Shapley Value (Shapley, 1953) was used as a cooperative game theory method to reallocate the water resources to riparian countries. The results indicated that the proposed method allowed considering the contribution rate of states and their relative weights as important factors. The authors claimed that the method could help policymakers to facilitate negotiation in conflict management over water allocation problems.

8.7.2.5 Cooperation vision, plans, and investment development

Regional drought has affected all riparian countries in the Euphrates-Tigris Basin and the current situation indicates that water withdrawals will continue while water quality will further decline. Yet, cooperative mechanisms are not well established and the progress soon is going slow, unless more collaborative efforts are made towards water resources management of the basin.

Kibaroglu (2017) reviewed transboundary water governance in the Euphrates-Tigris Basin. It is noted that the present challenge in the basin is to coordinate water resources management and establish suitable transboundary water governance in the current critical state of the basin. A significant development in the basin is the Euphrates-Tigris Initiative for Cooperation (ETIC) which was established during 2005 by a group of professionals from the three main riparian countries. The ETIC implemented joint training and programs during 2006–11 for professionals and government officials from Iraq, Turkey, Syria, and Iran on dam safety, river hydrology, conflict management, and geographic information systems. The ETIC showed that it is possible to design and implement technical and political activities across national borders.

Al-Muqdadi (2019) evaluated the impact of the political dimensions of Euphrates–Tigris system such as hydropolitical cycle and negotiation concepts to reach an agreement and developed a strategy to support the transformation from conflict to cooperation. The results demonstrated the challenges in conflict management and a sustain toolkit were proposed to reach a collaboration situation in the basin. The author believed that water conflict between the riparian countries of the basin might be exaggerated while the countries have more shared benefits to cooperate especially noting their national security. Currently, there is a willingness to collaborate in terms of economic and security issues from all countries which may enter into promising agreements.

8.7.3 Guaraní aquifer system

Groundwater management is a global necessity while these resources and their exploitation are being severely neglected because of a lack of knowledge about their functioning and dynamics, groundwater unknown vulnerability, and careless human agricultural activities which decline quantity and quality of groundwater. Thus, a groundwater case study is explained in this section to realize types of studies have been done over transboundary groundwater management.

8.7.3.1 Shared water problem analysis

The Guaraní Aquifer System (GAS) is one of the largest groundwater reservoirs in the world. The aquifer includes areas of Brazil, Uruguay, Paraguay and Argentina in the eastern and south-central zone of South America with an estimated total surface area of about 1.2 million km^2 (71 percent in Brazil, 19 percent in Argentina, 6 percent in Paraguay, and 4 percent in Uruguay). It includes parts of eight Brazilian States and approximately fifteen million people live within the aquifer's area of influence (UNDP, 2011) (Fig. 8.4).

The aquifer has about 40,000 km^3 storage volume and 40–60 km^3 annual recharge rate. Nearly 80 percent of the groundwater extraction in the region is used for public water supply (home supply and tourism) for a total population of 5,500 million people for 200 years at a rate for 100 liters per day per person. The aquifer is located in the Paraná and Chaco-Paraná Basins and its thickness ranges from a few meters to 800 m and its transmissivity coefficient varies from 5 m^2/day to 500 m^2/day (Puri et al., 2001).

The importance and complexity of the Guaraní Aquifer System are because of its high dimensions, water quality, temperature, and renewal rate which arise from hydrogeological aspects of the environment and legal and institutional aspects of this transboundary groundwater. The major future threats for this resource are irrational abstraction and uncontrolled point and non-point source pollution in using groundwater.

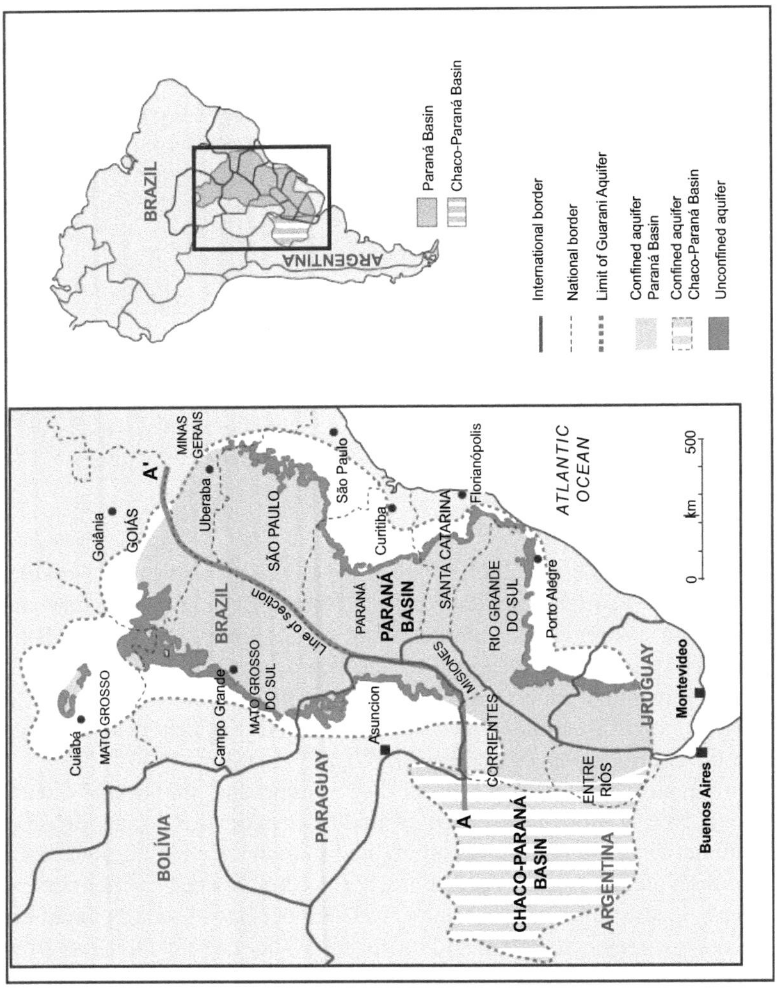

FIGURE 8.4 Guaraní Aquifer System location (Puri et al., 2001).

To groundwater management in the Guaraní Aquifer System, there are few procedures in the aspects of allocation and use licensing or decision-making. There is no binding legal agreement to govern the use of the Guaraní Aquifer System. Of the four countries, only Paraguay has signed the 1997 UN Convention. In Argentina and Brazil, groundwater management is limited to individual states or provinces. Uruguay is the only country that has determined a specific decree regulating the use of the aquifer. The riparian countries adopted the La Plata River Basin treaty during 1969 where the basin overlies much of the Guaraní Aquifer. The initial goals of the treaty were applied over surface waters and did not pay attention to groundwater. The four countries initiated a project called the Environmental Protection and Sustainable Development of the Guaraní Aquifer System during 2003 and the aim of the project was the creation of a common institutional, legal and technical framework that was capable to manage and protect the aquifer for present and future generations. The project was launched under the supervision of the World Bank, coordination of the OAS and support of the International Atomic Energy Authority (IAEA). The Guaraní Aquifer Steering Council was established with national officials from each country who are responsible for groundwater resources. During 2008, a Strategic Action Plan was approved by the council that established actions for each country to take to maintain a GIS database of the aquifer and a monitoring network. The Strategic Action Plan was aimed to evolve into a permanent structure that will co-evaluate and negotiate main aquifer development under potential transboundary effects (Cooley et al., 2009). During 2010, the Guaraní Aquifer Agreement was signed by four riparian countries to build dialogue and cooperation between countries. This agreement aimed to reduce asymmetric information on the GAS, to establish a regional commission, and to provide the responsibility to set an additional protocol defining mechanisms for dispute resolution which is not in force but has been ratified by Argentina and Uruguay (Silva and Hussein, 2019).

8.7.3.2 Data and information collection

The sharing data and improving knowledge is an essential step to assess and characterize the Guaraní Aquifer as a shared resource and it would be available when there are international agreements between riparian countries. However, there are many attempts to study this aquifer besides the lack of data and information.

Gómez et al. (2010) developed a conceptual model to estimate groundwater recharge in the cities of Rivera (Uruguay) and Santana do Livramento (Brazil) located on the Guaraní Aquifer. The elevation dataset was used to generate the Digital Terrain Model (DTM) which was provided by Shuttle Radar Topography Mission of National Aeronautics and Space Administration (NASA) and the National Geospatial-Intelligence Agency from the USA. Drainage network was identified from DTM. Only one weather station located in Rivera was available within the case study. There were no gauging stations for streamflow within the area, so the Manuel Diaz gauging station was used as the creek outside of the study area. Stratigraphic profiles from 141 boreholes, geologic maps, hydrogeochemical data, and water levels measured at 69 wells were integrated into the model.

Lucas et al. (2015) assessed the suitability of using remote sensing data in the water budget equation to estimate recharge inter-annual patterns in the Guaraní Aquifer area. Precipitation and evapotranspiration data was extracted from Tropical Rainfall Measuring Mission (TRMM) and Moderate Resolution Imaging Spectroradiometer (MODIS) satellite observations, respectively. Observed climatological data including rainfall, solar radiation, wind speed, sunshine duration, air temperature, and water level depth were provided by the Center for Water Resources and Applied Ecology of the University of São Paulo (CRHEA/USP). Results showed significant agreement between estimation data based on remote sensing and regional observed data which could be used where data are scarce or nonexistent.

8.7.3.3 Engineering assessment and modeling

In large aquifers, regional groundwater modeling is challenging due to geologic complexities and spatial and temporal data scarcity. The Guaraní Aquifer System contains an enormous volume of water but it is not well known. Therefore, it is difficult to assess the impact of exploitation currently used to supply over 25 million riparians and there are few studies for assessment and modeling of the aquifer.

Gómez et al. (2010) developed a numerical model using MODFLOW (McDonald and Harbaugh, 1988) to estimate recharge rates and other parameters of the aquifer system located in the southern Brazil/northern Uruguay border. MODFLOW is a finite-difference flow model and simulate and predict groundwater conditions and river-aquifer interactions. Also, the water balance EASY-BAL (Vazquez-Suñé and Castro, 2002) was employed for estimating recharge rates to use multiple methodologies. EASY-BAL evaluates water balance per unit of soil area using precipitation, potential evapotranspiration, temperature, and irrigation. The software gives the deficit and the recharge of the aquifer. The simulating results showed that estimated recharge rates were comparable to previous researches.

To analyze the flow system and water balance components in the full extent of the Guaraní Aquifer, Rodríguez et al. (2013) used a transient regional scale model based on five conceptual models represented by different hydraulic conductivity parameterizations. The code TRANSIN (Medina et al., 1996) was used to simulate groundwater flow and solute transport. Thus, five hydraulic conductivity zonations were explored to improve model calibration along 39 years based on an annual time step. The results showed that recharge and stream/aquifer fluxes were the dominant components representing about 84.2

percent of total inflows and 61.4 percent of total outflows. The water balance of the model demonstrated that the current rate of groundwater withdrawals did not exceed the rate of recharge on a regional scale of the aquifer.

8.7.3.4 Scenario analysis, conflict resolution, and decision making

In the case of shared groundwater resources, the hidden nature of the resource makes the monitoring and managing the aquifer more difficult. Whereas the management should be integrated due to transcending the groundwater to other national borders, in practice, there is a tendency to neglect or have a narrow management scope over aquifers. There are few studies about conflict resolution or decision-making over groundwater resources especially Guaraní Aquifer.

Katic (2014) developed a theoretical model to estimate welfare gains from optimal groundwater management in the Guaraní Aquifer. The author examined well location decisions and extraction paths in a spatially dynamic model and then quantified the importance of spatial effects in groundwater management. A three-dimensional package MODFLOW was used to extract hydrologic components of the aquifer and then the aquifer response became a constraint in a hydro-economic optimization model which was solved by a Monte Carlo technique. The results showed that locating new wells away from interference areas increased the economic benefits of groundwater exploitation. Also, it is demonstrated that unregulated well location may lead to underestimation of the welfare and hydrological costs.

Gorgoglione et al. (2019) developed a new scenario-based framework to improve the sustainability and equity of water allocation among riparian countries which was following the UN convention. The shared watershed of Cuareim/Quaraí River shared by Uruguay and Brazil were considered as a case study which is located in the area of the Guaraní Aquifer and the exchange between surface water and groundwater is negligible. The large-scale MGB-IPH hydraulic/hydrologic model (Pontes et al., 2017) was applied to simulate the streamflow. Then the river basin management system-irrigation module (SiGBaH-Irriga) (Kayser et al., 2017) was implemented to analyze the feasibility of new water licenses for direct water intakes and reservoirs at any point and to improve water management in situations of water deficit. Groundwater demands and groundwater extraction were included in the implementation of the water-management model. In the next step n possible optimal water allocations and m economic, social and environmental factors were selected to be the input of the multi-criteria decision analysis (MCDA). The results obtained from MCDA were used to consider different scenarios with the aid of What-If Scenario Analysis (WISA) to solve a conflict in the basin. The final results introduced various policy options that were tested with different scenarios and reached an objective decision in a water-sharing issue.

8.7.3.5 Cooperation vision, plans, and investment development

In South America, there are no conflicts over shared aquifers due to a lack of or little information about the aquifer's situation. International cooperation for joint groundwater management could be a positive force to promote knowledge of aquifers, to monitor aquifer by well networks, to improve groundwater management, and to allocate and protect groundwaters. The Guaraní Aquifer case showed that international cooperation could be promoted, even though there is neither water scarcity nor water-related conflicts.

Castro (2011) described the shared management of the Guaraní Aquifer and analyzed the Agreement on the Guaraní Aquifer signed by riparian countries during 2010 and its compliance with the terms of international water law. The 2010 GAS agreement arrived at a time of great withdrawal of groundwater in the region to protect this delicate and it represented unequivocal recognition by the riparian countries to water utilization and conservation. The author believed that the agreement would impose a need for more active participation of governments and civil society in managing water resources of the region which is inseparable from political stability, environmental protection, and trade.

Silva and Hussein (2019) applied a hydro-hegemony framework to demonstrate the impact of hydropolitical regional actions on shared water cooperation in the La Plata River Basin and the Guaraní Aquifer System. It was expressed that the Guaraní Aquifer System was believed to be not all connected in a united system before. When riparian countries accepted the aquifer as a system, the local management of projects turned to the regional scale of management and led to the 2010 GAS agreement. Also, it was explained that Brazilian hegemony was an important term in regional negotiation with neighbors especially Argentina. The analyses showed that new hydropolitics vision must be provided with the overlapping of surface and groundwater users and new models of governance are needed to deal with increasing pressures on both systems.

8.8 Summary

This chapter presented a practical way to manage shared water resources which answers to different questions of shared surface water and groundwater resources management. To get acquainted with shared water resources management, first, we presented some studies over this type of management, brought definitions and terms which are highly used in this issue, expressed importance of shared waters and their management, and described international agreement over these shared resources. Then we provide a guideline to technically address, assess, understand, and resolve problems of shared water resources. Eventually, characteristic case studies from around the world including river basins and an aquifer were illustrated to show the technical and dynamic process of shared water resources management. This guideline may be applied to any particular case study of shared waters.

References

Abtew, W., Melesse, A.M., Desalegn, T., 2009a. Spatial, inter and intra-annual variability of the Blue Nile River Basin rainfall. Hydrol. Process 23 (21), 3075–3082. https://doi.org/10.1002/hyp.7419.

Abtew, W., Melesse, A.M., Desalegn, T., 2009b. El Niño southern oscillation link to the Blue Nile River Basin hydrology. Hydrological Processes, Special Issue: Nile Hydrology 23 (26), 3653–3660. https://doi.org/10.1002/hyp.7367.

Abtew, W., Melesse, A.M., 2014a. The Nile River Basin. In: Nile River Basin: Ecohydrological challenges, climate change and hydropolitics. (Editors: Melesse, A.M., Abtew, W., and Setegn, S.), *Springer International Publishing*, New York.

Abtew, W., Melesse, A.M., 2014b. Climate teleconnections and water management. In: Nile River Basin. *Springer International Publishing*, New York, pp. 685–705.

Abtew, W., Melesse, A.M., 2014c. Transboundary rivers and the Nile. In: Nile River Basin. *Springer International Publishing*, New York, pp. 565–579.

Aguilar, G., Iza, A., 2011. Governance of shared waters: Legal and institutional issues. *Switzerland* (IUCN), 230p.

Ahmed, Y., Al-Faraj, F., Scholz, M., Soliman, A., 2019. Assessment of upstream human intervention coupled with climate change impact for a transboundary river flow regime: Nile River Basin. Water Resour. Manage. 33, 2485–2500. https://doi.org/10.1007/s11269-019-02256-1.

Al-Muqdadi, S.W.H, 2019. Developing Strategy for Water Conflict Management and Transformation at Euphrates–Tigris Basin, 11. *Water*, p. 2037. doi:10.3390/w11102037.

Al-Saidi, M., Hefny, A., 2018. Institutional arrangements for beneficial regional cooperation on water, energy and food priority issues in the Eastern Nile Basin. J. Hydrol. (Amst.) 562, 821–831. https://doi.org/10.1016/j.jhydrol.2018.05.009.

Arjoon, D., Mohamed, Y., Goor, Q., Tilmant, A., 2014. Hydro-economic risk assessment in the eastern Nile River Basin. Water Res. Econ. 8, 16–31. http://dx.doi.org/10.1016/j.wre.2014.10.004.

Batisha, A.F., 2011. Cooperative shared water resources management in the Nile River Basin: the engine for growth. *Fifteenth International Water Technology Conference*, IWTC-15 2011, Alexandria, Egypt.

Belachew, A., Mekonen, Z., Ibrahim, Y, 2015. Eastern Nile Basin water system simulation using Hec-ResSim model. *11th International Conference on Hydroinformatics*, HIC 2014, New York City, USA.

Belete, M., Deng, J., Zhou, M., Wang, K., You, S., Hong, Y., Weston, M., 2018. A new approach to modeling water balance in Nile River Basin, Africa. Sustainability 10 (810). doi:10.3390/su10030810.

Biswas, A.K., 2011. Cooperation or conflict in transboundary water management: case study of South Asia. Hydrol. Sci. J. 56 (4), 662–670. doi:10.1080/02626667.2011.572886.

Blomquist, W., Ingram, H.M., 2003. Boundaries seen and unseen. Water Int. 28 (2), 162–169. doi:10.1080/02508060308691681.

Bozkurt, D., Sen, O.L., 2013. Climate change impacts in the Euphrates–Tigris Basin based on different model and scenario simulations. J. Hydrol. (Amst.) 480, 149–161.

Burchi, S., 2018. Legal frameworks for the governance of international transboundary aquifers: pre- and post-ISARM experience. J. Hydrol. Reg. Stud. 20, 15–20. https://doi.org/10.1016/j.ejrh.2018.04.007.

Cahan, J.A., 2017. Water Security in the Middle East: Essays in Scientific and Social Cooperation. *Anthem Press*, UK.

Chao, N., Luo, Z., Wang, Z., Jin, T, 2018. Retrieving groundwater depletion and drought in the Tigris-Euphrates Basin between 2003 and 2015. Groundwater 56 (5), 770–782. doi:10.1111/gwat.12611.

Castro, D., 2011. The shared management of the Guarani Aquifer: the South American example in global governance over water resources. Y. B. Int. Environ. Law 22 (1), 140–157. doi:10.1093/yiel/yvs067.

Cooley, H., Christian-Smith, J., Gleick, P.H., Allen, L., Cohen, M., 2009. Understanding and Reducing the Risks of Climate Change for Transboundary Waters. Pacific Institute for Studies in Development, Environment and Security, California, 65p.

Degefu, D.M., He, W., 2016. Water bankruptcy in the mighty Nile River Basin. Sustain. Water Resour. Manag. 2 (1), 29–37. doi:10.1007/s40899-015-0035-2.

Degefu, D.M., He, W., Yuan, L., Zhao, J.H., 2016. Water allocation in transboundary river basins under water scarcity: a Cooperative bargaining approach. Water Resour. Manage. 30 (2), 4451–4466. doi:10.1007/s11269-016-1431-6.

Dessu S.B. and Melesse, A.M. (2012). Modeling the rainfall-runoff process of the Mara River Basin using SWAT. Hydrol. Process, 26(26), 4038–4049, https://doi.org/10.1002/hyp.9205.

Dessu, S.B., Melesse, A.M., 2013. Impact and uncertainties of climate change on the hydrology of the Mara River Basin. Hydrol. Process 27 (20), 2973–2986. doi:10.1002/hyp.9434.

Dessu, S.B., Melesse, A.M., Bhat, M., McClain, M., 2014. Assessment of water resources availability and demand in the Mara River Basin. Catena 115, 104–114. https://doi.org/10.1016/j.catena.2013.11.017.

Ding, N., Erfani, R., Mokhtar, H., Erfani, T., 2016. Agent based modelling for water resource allocation in the transboundary Nile River. Water (Basel) 8 (4), 139. https://doi.org/10.3390/w8040139.

Elhance, A.P., 1997. Conflict and co-operation over water in the Aral Sea Basin. Stud. Confl. Terror. 20 (2), 207–218.

Fang, L., Hipel, K.W., Kilgour, D, 1993. Interactive Decision Making: The Graph Model for Conflict Resolution. *Wiley*, New York.

Food and Agriculture Organization (FAO), 2009. AQUASTAT: Transboundary river basins – Euphrates-Tigris River Basin. Food and Agriculture Organization of the United Nations (FAO), Rome, Italy.

Ganoulis, J., Duckstein, L., Literathy, P., Bogardi, I., 1996. Transboundary water resources management: institutional and engineering approaches. Part of the NATO ASI Series, Environment 7, 481p.

Giordano, M.A., Wolf, A.T., 2003. Sharing waters: post-Rio international water management. Nat. Resour. Forum 27 (2), 163–171. https://doi.org/10.1111/1477-8947.00051.

Gómez, A.A., Rodríguez, L.B., Vives, L.S., 2010. The guarani aquifer system: estimation of recharge along the Uruguay–Brazil border. Hydrogeol. J. 18 (7), 1667–1684. doi:10.1007/s10040-010-0630-0.

Gorgoglione, A., Crisci, M., Kayser, R.H., Chreties, C., Collischonn, W., 2019. A new scenario-based framework for conflict resolution in water allocation in transboundary watersheds. Water (Basel) 11 (6), 1174. doi:10.3390/w11061174.

Guertin, S., 2009. International Protocols, Treaties, and Conventions. International activities: Fish and wildlife service, USA.

Hasan, M., Moody, A., Benninger, L., Hedlund, H., 2018. How war, drought, and dam management impact water supply in the Tigris and Euphrates Rivers. Ambio 48 (3), 264–279. https://doi.org/10.1007/s13280-018-1073-4.

He, Y., 2015. China's practice on the non-navigational uses of transboundary waters: transforming diplomacy through rules of international law. Water Int. 40 (2). doi:10.1080/02508060.2015.1011455.

Hipel, K.W., Kilgour, D.M., Kinsara, R.A., 2014. Strategic investigations of water conflicts in the Middle East. Group Decis. Negot. 23 (3), 355–376. doi:10.1007/s10726-012-9325-3.

International Law Association (ILA) (1966). Helsinki rules on the uses of the waters of international rivers. Retrieved on December 29, 2009 from http://webworld.unesco.org/water/wwap/pccp/cd/pdf/educational_tools/course_modules/reference_documents/internationalregionconventions/helsinkirules.pdf.

Jain, S.K., Singh, V.P., 2003. Water Resources Systems Planning and Management. Developments in Water Science, 51. *Elsevier*, 882p.

Jones, C., Sultan, M., Yan, E., Milewski, A., Hussein, M., Al-Dousari, A., Al-Kaisy, S., Becker, R., 2008. Hydrologic impacts of engineering projects on the Tigris–Euphrates system and its marshlands. J. Hydrol. (Amst.) 353, 59–75. doi:10.1016/j.jhydrol.2008.01.029.

Kahsay, T.N., Kuik, O., Brouwer, R., van der Zaag, P., 2015. Estimation of the transboundary economic impacts of the Grand Ethiopia Renaissance Dam: a computable general equilibrium analysis. Water Res. Econ. 10, 14–30. doi:10.1016/j.wre.2015.02.003.

Katic, P.G., 2014. Groundwater spatial dynamics and endogenous well location. Water Resour. Manage. 29 (1), 181–196. doi:10.1007/s11269-014-0834-5.

Kavvas, M.L., Chen, Z.Q., Anderson, M.L., Ohara, N., Yoon, J.Y., Xiang, F., 2011. A study of water balances over the Tigris–Euphrates watershed. Phys. Chem. Earth 36, 197–203. doi:10.1016/j.pce.2010.02.005.

Kayser, R.H.B., Chreties, C., Crisci, M., Collischonn, W., 2017. Estimation of the interannual production of irrigated crop in a watershed from hydrological modeling. In: Proceedings of the XXII Simpósio Brasileiro de Recursos Hídricos. 26 November–1 December 2017. (In Portuguese).

Kibaroglu, A., 2017. State-of-the-art review of transboundary water governance in the Euphrates–Tigris River Basin. Int. J. Water Resour. Dev. 35 (1). https://doi.org/10.1080/07900627.2017.1408458.

Kilgour, D.M., Hipel, K.W., Fang, L., 1987. The graph model for conflicts. Automatica 23 (1), 41–55. doi:10.1016/0005-1098(87)90117-8.

Klimes, M., Michel, D., Yaari, E., Restiani, P., 2019. Water diplomacy: the intersect of science, policy and practice. J. Hydrol. (Amst.). https://doi.org/10.1016/j.jhydrol.2019.02.049.

Kucukmehmetoglu, M., 2012. An integrative case study approach between game theory and Pareto frontier concepts for the transboundary water resources allocations. J. Hydrol. (Amst.) 450–451, 308–319. http://dx.doi.org/10.1016/j.jhydrol.2012.04.036.

Leb, C., 2015. One step at a time: international law and the duty to cooperate in the management of shared water resources. Water Int. 40 (1), 21–32. doi:10.1080/02508060.2014.978972.

Lee, T.R., 1995. The management of shared water resources in Latin America. Nat. Resour. J. 35, 541–553.

Lucas, M., Oliveira, P.T.S., Melo, D.C.D., Wendland, E, 2015. Evaluation of remotely sensed data for estimating recharge to an outcrop zone of the Guarani Aquifer System (South America). Hydrogeol. J. 23 (5), 961–969. doi:10.1007/s10040-015-1246-1.

Madani, K., 2010. Game theory and water resources. J. Hydrol. (Amst.) 381 (3–4), 225–238. doi:10.1016/j.jhydrol.2009.11.045.

Madani, K., Rheinheimer, D., Elimam, L., Connell-Buck, C, 2011. A game theory approach to understanding the Nile River Basin conflict. In: "A Water Resource" Festschrift in honor of Professor Lars Bengtsson. (Edited by: Person K. M). Division of Water Resources Engineering, No. 3253, Lund University, Lund, Sweden. ISBN 978-91-7473-221-4.

McDonald, M.G., Harbaugh, A.W., 1988. A modular three-dimensional finite-difference groundwater flow model. US Geological Survey (part 2) doi:10.3133/ofr83875.

Medina, A., Galarza, G., Carrera, J., 1996. TRANSIN-II: FORTRAN code for solving the coupled flow and transport inverse problem in saturated conditions. El Berrocal Project: characterization and validation of natural radionuclide migration processes under real conditions on the fissured granitic environment, European Commission Contract no. FI2W/CT91/0080, Technical Report 16, ENRESA, Spain.

Melesse, A.M., Athanasios, G.L., Senay, G., Yitayew, M., 2009a. Climate change, land-cover dynamics and ecohydrology of the Nile River Basin. *Hydrological Processes, Special Issue: Nile Hydrology, 23*(26), 3651–3652, doi:10.1002/hyp.7522.

Melesse, A.M., Abtew, W., Desalegne, T., Wang, X., 2009b. Low and high flow analysis and wavelet application for characterization of the Blue Nile River system. Hydrol. Process, 24(3), 241–252, doi:10.1002/hyp.7312.

Melesse, A.M., Abtew, W., Setegn, S., Dessalegne, T., 2011. Hydrological variability and climate of the upper Blue Nile River Basin. In: Nile River Basin: Hydrology, Climate and Water Use, Melesse, A.M. (Ed.). *Springer Science Publisher*, pp. 3–37, doi:10.1007/978-94-007-0689-7_1.

Melesse, A.M., 2011. Nile River Basin: Hydrology, Climate and Water Use. *Springer Science and Business Media.*

Melesse, A., Abtew, W., Setegn, S.G., 2014. Nile River Basin: Ecohydrological Challenges, Climate Change and Hydropolitics. *Springer Science and Business Media.*

Mianabadi, H., Mostert, E., Pande, S., van de Giesen, N., 2015. Weighted bankruptcy rules and transboundary water resources allocation. Water Resour. Manage. 29 (7), 2303–2321. doi:10.1007/s11269-015-0942-x.

Mylopoulos, Y.A., Kolokytha, E.G., 2008. Integrated water management in shared water resources: the EU water framework directive implementation in Greece. Phys. Chem. Earth 33 (5), 347–353. doi:10.1016/j.pce.2008.02.031.

Nandalal, K.D.W., Simonovic, S.P, 2003. State-of-the-Art Report on Systems Analysis Methods for Resolution of Conflicts in Water Resources Management. *UNESCO*, Paris.

Nash, Z., 1950. The bargaining problem. Econometrica 18 (2), 155–162.

Nash, Z., 1953. Two-person cooperative games. Econometrica 21 (1), 128–140.

Nile Basin Initiative (NBI), 2012. State of the River Nile Basin report. Nile Basin Initiative: Entebbe, Uganda.

Nile Basin Initiative (NBI), 2016. Chapter 3: the socio-economic profiles of the countries. In the Nile Basin Resources Atlas. Editors: Akol, P.J., Galla, R.P.Z., and Wanyony, S.; Nile Basin Initiative, Entebbe, Uganda.

O' Neill, B., 1982. A problem of rights arbitration from the Talmud. Math. Soc. Sci. 2 (4), 345–371. https://doi.org/10.1016/0165-4896(82)90029-4.

Onencan, A.M., de Walle, B.V., 2018. Equitable and reasonable utilization: reconstructing the Nile Basin water allocation dialogue. Water. (Basel) 10, 707. doi:10.3390/w10060707.

Parhi, P.K., Sankhua, R.N., 2013. Beyond the transboundary river: issues of riparian responsibilities. J. Inst. Eng. (India): Series A 94 (4), 257–261. doi:10.1007/s40030-014-0067-x.

Petersen-Perlman, J.D., Veilleux, J.C., Wolf, A.T., 2017. International water conflict and cooperation: challenges and opportunities. Water Int. 42 (2), 105–120. doi:10.1080/02508060.2017.1276041.

Pontes, P.R.M., Fan, F.A., Fleischmann, A.S., Dias de Paiva, R.C., Buarque, D.C., Siqueira, V.A., Jardim, P.F., Collischonn, W., 2017. MGB-IPH model for hydrological and hydraulic simulation of large floodplain river systems coupled with open source GIS. Environ. Modell. Softw. 94, 1–20. https://doi.org/10.1016/j.envsoft.2017.03.029.

Puri, S., Appelgren, B., Arnold, G., Aureli, A., Burchi, S., Burke, J., Margat, J., Pallas, P., 2001. Internationally shared (transboundary) aquifer resources management: Their significance and sustainable management. *UNESCO*, Paris.

Rodríguez, L., Vives, L., Gomez, A., 2013. Conceptual and numerical modeling approach of the Guarani Aquifer System. Hydrol. Earth Syst. Sci. 17, 295–314. doi:10.5194/hess-17-295-2013.

Salman, S., 2007. The Helsinki rules, the UN Watercourses convention and the Berlin rules: perspectives on international water law. Water Res. Dev. 23 (4), 625–640.

Shapley, L.S., 1953. A value for n-person games. In: Kuhn, HW, Tucker, AW (Eds.), Contributions to the Theory of Games II, Annals of Mathematics Studies. Princeton University Press, Princeton, pp. 307–317.

Silva, L.P.B., Hussein, H, 2019. Production of scale in regional hydropolitics: an analysis of La Plata River Basin and the guarani aquifer system in South America. Geoforum 99, 42–53. https://doi.org/10.1016/j.geoforum.2018.11.019.

Skoulikaris, C., Zafirakou, A., 2019. River basin management plans as a tool for sustainable transboundary river basins' management. Environ. Sci. Pollut. Res. 26, 14835–14848. https://doi.org/10.1007/s11356-019-04122-4.

Stephan, R.M., 2009. Transboundary Aquifers: Managing a Vital Resource. The UNILC Draft Articles on the Law of Transboundary Aquifers. UNESCO, Paris.

Sullivan, C.A., 2014. Planning for the Murray-Darling Basin: lessons from transboundary basins around the world. Stoch. Environ. Res. Risk Assess. 28, 123–136. doi:10.1007/s00477-013-0789-8.

Swain, A., 2011. Challenges for water sharing in the Nile Basin: changing geo-politics and changing climate. Hydrol. Sci. J. 56 (4), 687–702.

Transboundary Freshwater Dispute Database (TFDD) 2017. *Oregon State University*, Available online at: http://www.transboundarywaters.orst.edu/

United Nations (UN), 1997. Convention on the law of the non-navigational uses of international watercourses. General Assembly Resolution. New York: United Nations 51/229, UNGA, 51st session, UN Doc A/RES/51/229.

United Nations (UN), 1999. UNECE Draft Protocol on Water and Health to the 1992 Convention on the Protection and Use of Transboundary Watercourses and International Lakes. MP.WAT/AC.1/1999/1. New York: United Nations final draft adopted 17 June 1999.

United Nations Development Programme (UNDP), 2006a. Human Development Report. Beyond scarcity: Power, poverty and the global water crisis, New York, USA.

United Nations Development Programme (UNDP), 2006b. Human Development Report. Managing trans-boundary waters for human development, New York, USA.

United Nations Development Programme (UNDP), 2011. International Waters: Review of Legal and Institutional Frameworks. UNDP-GEF International Waters Project, UBC, White and Case, 309p.

Vázquez-Suñé, E., Castro, A., 2002. EASY-BAL (Spreadsheet). Hydrogeology Group, UPC-CSIC, Barcelona.

Vinogradov, S., Wouters, P., Jones, P., 2003. Transforming potential conflict into cooperation potential: the role of international water law. Technical documents in hydrology (UNESCO) PC-CP, series 2, 106p.

Waterbury, J., 2002. The Nile: National Determinants of Collective Action. *New Haven: Yale University Press*.

Whittington, D., Wu, X., Sadoff, C., 2005. Water resources management in the Nile Basin: the economic value of cooperation. Water Policy 7 (3), 227–252.

Wolf, A.T., 2010. Sharing Water, Sharing Benefits: Working Towards Effective Transboundary Water Resources Management. *Oregon State University (UNESCO)*, 278p.

Wolf, A.T., Newton, J.T., 2007. Case Study of Transboundary Dispute Resolution: the Tigris-Euphrates Basin. Appendix: C of the book on Transboundary Dispute Resolution by the same authors. Oregon State University; Institute of water and watersheds (2008).

Wolf, A.T., Kramer, A., Carius, A., Dabelko, G.D., 2005. Managing water conflict and cooperation. State of the World: Redefining Global Security, Chapter 5. 80–99.

Yitayew, M., Melesse, A.M., 2011. Critical water resources management issues in Nile River Basin. In: Nile River Basin: Hydrology, Climate and Water Use (Editor Melesse, A.M.), *Springer Science Publisher*, Chapter 20, pp. 401–416, doi:10.1007/978-94-007-0689-7_20.

Zagonari, F., Rossi, C., 2014. A negotiation support system for disputes between Iraq and Turkey over the Tigris-Euphrates Basin. J. Hydrol. (Amst.) 514, 65–84. http://dx.doi.org/10.1016/j.jhydrol.2014.04.003.

Zeitoun, M., Warner, J., 2006. Hydro-hegemony – a framework for analysis of trans-boundary water conflicts. Water Policy 8 (5), 435–460. doi:10.2166/wp.2006.054.

Chapter 9

Water, culture, civilization, and history

Seyedeh Habibbeh Hosseiny[a], Omid Bozorg-Haddad[b] and Daniele Bocchiola[c]

[a] *Department of Range Management, Gorgan University of Agricultural Science and Natural Resources, Gorgan, Iran.* [b] *Department of Irrigation & Reclamation Engineering, Faculty of Agricultural Engineering & Technology, College of Agriculture & Natural Resources, University of Tehran, Karaj, Tehran, Iran.* [c] *Department ICA, Politecnico di Milano, Milan, Itely & EVK2CNR Association, Bergamo, Italy*

9.1 Introduction

Water is one of the four main elements, and certainly the most important one. The human body is composed of water for two-thirds, and three-fourths of the earth globe, are covered in water. Waer is therefore deeply connected to human life, and have had a profound and far-reaching impact on all its aspects.

In the following, it is attempted not only to examine the impact of water on civilizations, beliefs, religions, and cultures, but also to examine their interactions with water resources exploration, operation, and conservation.

Considering the need for water to survive, human kind has realized that conservation and proper allocation are very important for welfare, and access to safe water. Accordingly, by planning and promoting beliefs, they tried to build a collective thinking of the society so as to use the empathy and companionship of the people to conserve resources (Pourkhaleghi, 2009; Musharraf Ghafrokhi et al., 2011).

The key to any sustainable society is access to safe water. Almost all of the great civilizations wordwide have grown up around water, necessary for supply, hygiene, agriculture, commerce, transportation, and defense.

9.2 Water in the ancient civilizations

Civilizations such as the Roman Empire, the Egyptians, the Venetian Empire, Mesopotamian Civilization, and the Awadhi Dynasty all based upon easily accessible water, which made it possible for their populations to survive and expand (Sher Afroos, 2011).

Economical, Political, and Social Issues in Water Resources. DOI: 10.1016/B978-0-323-90567-1.00010-3

About 10,000 years ago, when men adopted a farming lifestyle, humans created permanent settlements. This new habits spread everywhere, and the population began to expand more rapidly than ever before.

Such sedentary agricultural life made villages, towns and eventually nations, all dependent upon water. This created a whole new relationship between humans and water.

The pathogens transmitted by contaminated water became a serious threat to the health of sedentary farmers. Jericho is located in a Valley with the Jordan River to the east and Jerusalem to the west and was the first known permanent settlement (8000–7000 BC) near natural springs.

Traces of wells in Egypt, of canals in the Mesopotamian rocks have been reported since 3000 BC. Archaeologists have found hundreds of ancient wells, water pipes, and toilets from the early Bronze Age in Mohenjo Daro, Pakistan. The earliest evidence of purposeful construction of water supplies, baths, toilets, and drainage facilities in Europe dates back to the Bronze Age of Minoan (and Mycenaean), in Crete (the oldest Minoan civilization in Europe) in the second millennium BC.

Ever since, ensuring freshwater for people has become a prerequisite for successful urbanization and government formation (Juuti et al., 2007; Vuorinen et al., 2007).

Great civilizations have historically developed along rivers, with water was used for drinking, washing, cooling, cooking, and traveling.

The first farmers needed water to irrigate their land more than 8,000 years ago in Africa and Asia. Large rivers such as the Nile, Tigris, and Euphrates flooded the land several times a year, but crops needed water almost daily during the crop season. So people started digging canals from the rivers to get constant water. The men dug these canals using the simplest tools, such as shovels, dibber, wooden utensils and earthenware. Drinking water was obtained directly from the rivers. In areas with frequent droughts, prehistoric men soon learned to collect rainwater for storage purposes, and dug ponds, known as reservoirs in the soil, and in stone.

The underside of reservoirs, made of water-resistant soil, was covered with sand that was used to purify water and eliminate residues of vegetation and other impurities. In fact, by filtering the mud, the water was potable. But when the farms needed irrigation, the flow had to be increased so the canals had to be built with some slope.

However, in times of flooding, water destroys everything, and even leeves, and boundaries. For this reason, the ancient Egyptians, Sumerians, and Babylonians invented and perfected the art of geometry, so that they could redraw borders, and boundaries of their land at any time, if necessary. As a precautionary measure, prehistoric men built riverbanks along the rivers, and as well built dwellings on high ground. Others used the riverbanks on the margins of their villages (Glick, 1970).

Humans learnt how to dig deep wells, to build underground passages, canals, and water reservoirs, bridges and aqueducts. Those living nearby ice caps, learnt how to gather water from melting ice tongues.

Humans also started recycling wastewater, extracting drinking water from seawater, and even from air. In ancient civilizations, circular and elliptical staircases, resembling small amphitheaters were built to extract water from air. Such methods were also used in medieval castles where it was often necessary to withstand prolonged siege with few resources.

Throughout history all civilizations used to dig wells. Nowadays, in the industrial age, modern water sources are increasingly being provided from deep underground, and because of such excessive use water resources are rapidly drained, and polluted by human activities (such as industry, waste disposal, mines, pesticides, acidic rainfalls).

In some desert areas, "ancient" or "fossil" groundwater persists, stoted hundreds or thousands of years ago when the climate was wetter. Today, water is extracted from deep wells in many countries such as Libya, Saudi Arabia, Israel, and the United States. However, fossil water may eventually dry up, because of low rainfall in these arid areas, and it would be hardly possible to gather water therefater (Laureano, 2001).

Wells (mainly deep holes in the ground through which water can be harvested) were the most traditional, and simplest way to supply water since prehistoric times. A series of handmade water wells in Kyssonerga, Cyprus, near Paphos (an area without springs or rivers) dates back to the late 10th and ninth millennia, and seems to be the oldest known water wells (Peltenburg et al., 2000, 2014). Most of the population of Iran, and other arid countries were largely dependent on freshwater harvesting using wells (Voudouris et al., 2013).

In Table 9.1, we report a number of relevant studies focusing upon the role of water in past, and present civilizations.

9.3 The indian valleys

History of water in Indian well valleys follows the history of human habitation, and how they used water. This also shows the strong correlation between the emergence of civilization and water throughout history.

The natives were the first consumers of water collected in these valleys, dating back ca. 10,000 years. India has long been plagued by water shortages (Dugger, 2006). Water that is rapidly pumped from the aquifers is one of the important factors in the water supply. In the first century AD, the banks of the main rivers became steep due to formation of long and shallow complexes of wells. The same approach was used in the construction of a new type of wells. The earliest stairs are likely dating to AD 550, but the most famous were built in the middle Ages. It is estimated that more than 3,000 stairwells were built in the two northern states of India, and are now common in West India. In New Delhi alone, there are more

TABLE 9.1 Recent studies focusing on the role of water in past and present civilizations.

Reference	Topic	Main output
Delli Priscoli (1998)	Using history to build a new ecological realism.	The paper illustrates how water requires us to reconsider the concepts of security and interdependence.
Laureano (2001)	Traditional knowledge to combat desertification.	This book examines different methods used by early civilizartions for the purposes of storing rainwater in areas with frequent droughts.
Vuorinen et al. (2007)	Water and health from ancient to modern civilizations.	This paper summarizes the general outline and some of the main results of two academic, multi- and cross-disciplinary projects. Special emphasis is given to the first urbanization of ancient civilizations, focusing on ancient Greece and Rome.
Angelakis and Spyridakis (2010)	Water and wastewater management in ancient Hellas.	The paper discusses the most characteristic forms of ancient hydraulic works, and related technologies and their uses in past Greek civilizations.
Shir-Afroos and Mohammadpour (2011)	Role of Water in Civilization and Human Culture.	The paper examines the role of water in the emergence and development of civilizations, such as the Roman Empire, Egyptian civilization, and the Venetian Empire.
Mushref Ghahfarkhi and Jabbari (2011)	Survey of Water and Culture Interaction in History.	The paper examines the impact of water on the culture of societies, and explains how beliefs are formed in societies.
Krause and Strang (2016)	Thinking Relationships Through Water.	The paper discusses how social and hydrological relationships are interconnected, and mutually constitutive. Also, it gives a deeper understanding of the role of water in human social life.
Voudouris et al. (2017)	Water Wells in ancient Civilizations.	The paper reviews the evolution of wells through the centuries. It also discusses how the extensive social relations in the Balkan, Indian and Chinese civilizations developed around wells.

than 30 cases. In these wells, water can be reached by moving down a staircase. They may be covered, and often display large architectural importance.

Many stairs were also used to relieve the heat of the day for leisure. Ancient Indian scriptures refer to the construction of wells, canals, reservoirs, and dams and their efficient maintenance. Irrigation as a tool for food production has been in use therein for more than 5,000 years (Voudouris et al., 2017).

Temples and rest areas with beautiful carvings were also built in many wells. At the beginning, many of them were painted in light, and lime-based colors, and no longer traces reamians of the ancient colors (Centre Water Commission, 2006).

The staircase conditions began to deteriorate during the years of presence in Rajkot of the English, that frightened with unsanitary conditions for bathing and drinking water, began installing pumps and pipes, and eventually using stairs in some places was inhibited. Chand Baori in Abhaneri, near Jaipur, is a largest and perhaps most spectacular stairwell. It has a deep tetrahedral structure, an unparalleled 10th-century temple built by Hirsh Mata, the goddess of happiness.

The city of Monjujo-Daru (India) was built in a semi-arid environment around 2450 BC, and it was supplied by at least 700 wells which were one well for every three houses (De Feo, 2011; Voudouris et al., 2017).

9.4 Water in the ancient china

Wells have been discovered in many areas, indicating that they were popular during 3000–1000 BC. In general, early Chinese civilization emerged within the Yellow River, and Yangtze River basins, and wells therein date back to that time.

Ancient China had a long history of making and using wells. After the Yuan Dynasty, the use of wells for irrigation and crop production, as well as for daily collection of water from rural to urban areas was very popular, and many of the traditional foods were dependent on wells. Even, foods were dependent on the well water to maintain its taste and quality. For example, bean curd (i.e. *tofu*) production in Jinshui, Greece, has been dependent on a special well for hundreds of years to maintain its taste and quality.

Distribution of water to households began with the construction of water supply networks built in the late 19th century in some major cities. Until then the wells, and rivers near the wells were supplied with water. The wells were mostly for everyday use, because they provided stable water in amount, and quality as compared to river water, also easier to access. People believed that some wells had good aromatic water, and used river water mainly for irrigation. Besides China, as an independent country in agriculture, used these wells to meet irrigation needs in rural areas.

According to discoveries, the use of wells in China for irrigation purposes goes back to the second century BC. The people not only regarded wells as water facilities, but even as special and sacred places, where the dragon was the

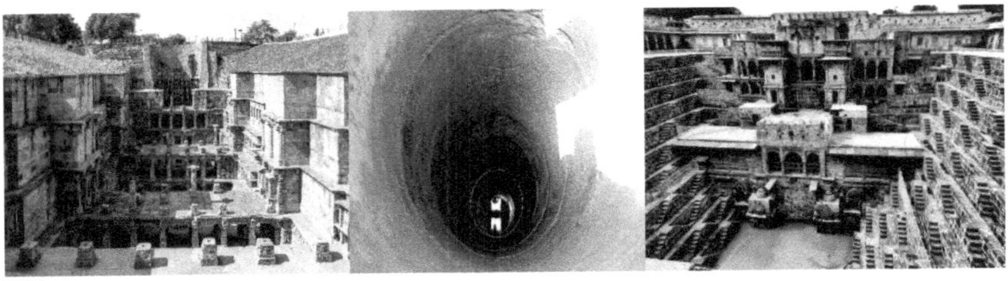

(a) (b) (c)

FIGURE 9.1 View into a hand-dug well cased with concrete rings: (a) multistory step well (b) Dug well, and (c) Chand Baori stepwell in Rajasthan.

water god. On that basis, people made extraordinarily beautiful wells, in many fashions.

Widespread social relationships around wells were being met on a daily basis, and forced to manage the economic benefits of well water, and worship in the community. However, wells were cultural, religious, social centers, and well making became an art. Well construction was the first step of building a village. Even today, tap water is common in public areas, and still many villages use this tradition (Voudouris et al., 2017).

9.5 Water in the ancient greece, and balkans

Wells were very well developed in ancient Hellas (Byzantine province of southern Greece) since prehistoric times, and were always in accordance with local hydrogeological conditions. Thus, in the southeastern settlements of Crete (eg Zakros, Palaikastro, and Comos) water supply was dependent on groundwater. The ancient site of the Conus has also had wells since the earliest times of Minoan civilization (Mays et al., 2007; Angelakis and Spyridakis, 2010). During the first kings' reign (ca. 1900 to 1700 BC), several wells were used to supply drinking water, and at least six were reported (Evans, 1964). Wells' depth did not exceed 20 m, and their diameter was less than 5 m (Buffet and Evrard, 1950). Water supply of dry, water limited areas was a most prominent feature of Greek civilization, as e.g. in the town of Policastro in eastern Crete (Angelakis et al., 2012).

Private or public wells with an array of different traits (such as size, cross-sectional shape, and structural features) have existed repeatedly in ancient times in every Greek city until recently. They were located along the main streets, in houses, baths, palaces, shelters, even in cemeteries for burial.

Wells had also great value to people, and were always equipped with holes for cleaning (Kaiafa, 2008). In the Croatian Zadobravlje region, a Neolithic (i.e. from the Stone Age) water well was found, with 1.5 m in diameter, and at least 4.9 m depth. From C14 analysis of the well's coal, it was estimated that construction may date to 6610–6340 BC (Fig. 9.1).

A water well has recently been discovered in the Chalkiti town of Solenitsa, Bulgaria (Nikolov, 2012). On the eve of the Balkans' northwest, a constant tradition of well construction can be found in present-day Slovenia, where several wells were drilled in prehistoric times, with some of them dating back to about 4,000 BC in Lipovuci, and other coming from the Bronze and Iron Age. The wells are of two types, a first type simply dug in the ground, and a second type almost quadrilateral.

Various traditions and features of culture can be found in the Balkans in the Roman period. Elsewhere in the Balkans, culture reflects the evolution of local Protetonic traditions. The Venice Plain and the Western Balkans, in Protonist and Roman times, are among the richest lands for wells, with more than two hundred specimens (Vigoni, 2011).

9.6 Aqueducts, arteries of drinking water

Aqueducts are widespread throughout the world from Japan to Chile. The largest aqueducts however, serve as a main source of water supplies in the deserts in Iran. Most of Iran's population is dependent upon aqueducts (De Feo, 2011; Voudouris et al., 2017). In this area, the aqueduct technology has evolved in every possible way. In this region, the weakness of the factors of production has caused the weakness of cultural, social and economic structures. Therefore, thanes, feudal landowners or great herdsmen, were never present in this area.

Towns and villages in the region before modern technology were dependent on the strength of aqueducts, and they were not able to create national or transnational governments, either of thane or feudal type, due to their low production capacity. People around the desert of Iran turned to handicrafts, and commerce due to the weakness of natural potential therein.

The industrial, commercial and retailer spirit has led people towards peace, friendship, conversation, and negotiation. Yazdi et al. (2000) call the area around the Iranian deserts an aqueduct civilization. First, the area has a distinctive culture with its surroundings. Second, with the drying up of the aqueducts, the characteristics of the aqueducts culture and civilization have not disappeared.

The Sumerians, Hittites, Cretans, Greeks, and Romans all built aqueducts that were beneficial to humans, animals, and plants. In ancient cities, countless springs provided freshwater for everyone. However, not all houses had water, and usually only the rich could afford it. However, in Greece and Rome there was scarcely any water shortage, and everyone had access to public fountains. The aqueducts were several kilometers long, with gently sloping channels where water moved by gravitational force. The Romans, for example, built magnificent aqueducts with bridges and arches, examples of which are still admired today, one of them built over the French Guard River, more than two thousand years ago. After providing water to cities and homes, it became clear that underground chambers known as sewers used to remove wastewaters were important for health (Glick, 1970).

Besides, a well or aqueduct was even a cultural symbol in history, related to local ritual and custom. Study of the ancient wells can provide important information from the experience and knowledge of ancient civilizations.

In different ages, beliefs such as water worship, celebrations such as water weddings and Abangan (Thanksgiving to God for the water), and believing that water has gender have grown. These beliefs have contributed to the formation of water culture. Whenever tenets are presented in a way that is rooted in the culture of nations and societies, they make the collective thinking of the society in one form, and cause empathy and companionship of the people to build a coherent and efficient system. In other words, man does what he believes, and belief does not arise except in knowledge. So it is better to get help with old, or even today's deformed beliefs to promote water protection. In many cultures, water was so important that water suppliers and protectors were subject to tax exemptions (Ghafrokhi et al., 2011).

Jung (one of the mythical gods of China) sees water in a dream as a symbol of the unconscious, which forces us to rely on ourselves, and our behavior and internalize these thoughts. It can be said that the continued quest for integrated and multipurpose rivers, and watersheds management, is an external social manifestation of long-term growth processes (Delli Priscoli, 1998; Haught, 1996).

9.7 Vocabulary definitions

9.7.1 Culture

The concept of culture is very complex and has a great deal of meaning (Edward, Tylor, 1958). Edward Taylor defines culture as a complex set of sciences, beliefs, arts, laws, morals, habits, and everything that a person derives from his or her community as a member of society. Culture is transmitted to the next generation through education (Kroeber and Kluckhohn, 1952).

9.7.2 Civilization

Civilization is derived from the Latin word *civis* (citizen), and its adjective *civilis* (attribute of a citizen), that is to say the lifestyle of a group of people, who lived in a historical region in a geographic area and had common goals and achievements. The sentence *Civis Romanus sum* (I am a roman citizen) implied full possession of the rights, and duties of the people belonging to the Roman Empire. The word civilization basically indicates the dwelling of people within a town, or an area (state, nation) governed by recognized authorities, and the legal status of that community. Historically, civilization has often been identified with larger and more advanced cultures (Wright, 2004 Ronald).

9.7.3 Aqueduct (*Qanat*)

The ditch, or channel is a waterway, or canal, which is dug underground, to intercept a deep aquifer, and allow water to flow to the surface. Such ditch or channel is deep in the ground enough to connect a number of wells, originating from the submerged aquifer. This aqueduct can run for several kilometers under the surface, and the water outlet is called the *qanat* outlet (World Heritage Committee, 2016; English, 1968).

9.7.4 Water governance

Water governance is the controlling behavior that is applied through management measures or by regulating water (across a range of political, social, environmental, economic and administrative systems) and ultimately it leads to allocation and improvement of water use conditions (operation, Global Water Partnership, 2002). Water governance determines who, and under what conditions, has access to water, what is the quality and quantity of water protection, what is the rule of water management. Also water governace includes assessment of the role of the stakeholders in making decisions, the policies of productivity and savings in water consumption, and the choice of a reference authority (the judge) of disputes and conflicts (Omranian-Khorasani, 2015).

9.8 Basics and logics

Water has been together an element of life, a means of transport, a life-threatening force, and often even a means of social and cultural activity throughout history (Krause and Strang, 2013). Today, water has a prominent position in research, because several places worldwide will suffer from shortage, or lack of water, and pollution and excessive use have endangered water resources. Therefore, researchers are discovering the profound consequences of water scarcity for human societies and cultures. Just as biophysical life without water is unthinkable, so is social and cultural life (Krause and Strang, 2016).

Much of the current literature implies that water is "… one of the most important environmental concerns, natural resources" (Tempelhoff et al., 2009), and water is known as a political challenge because of "… the plurality of world-views, ideologies, interests, and water-related discourses" (Molle et al., 2008). Wiley Interdisciplinary Reviews of water discusses "… *the interpretations we, as a society, have brought to water through art, religion, history and which in turn shapes how we come to understand it.*" (Ingold, 2000; Kohn, 2013). Water can be viewed as an integral part of social, cultural and political relations.

Recent research emphasizes the deep influence of water on social and cultural life (Fontein 2008; Linton, 2010; Chen et al., 2013; Strang, 2014). For example, the analysis of simultaneous social and material processes are the bases of watershed management (Carse, 2012), and of the reasonings that providing drinking water is connected with "pressures" that are simultaneously physical and political (Anand, 2011). Bijker (2012) recommends that human societies may be called "water cultures" (Bakker, 2012), and others even describe our environments as "water worlds" (Hastrup 2009; Orlove and Caton, 2010; Barnes and Alatout, 2012).

9.9 A short history of the relation of water to culture and civilization

Although water is a controversial issue, and there have been a lot of water-related conflicts around the world, water has created the conditions for humanity to learn to build a community. The water distribution program, and the creation of an infrastructure of water networks have increased public confidence and created a sense of the need for teamwork. Climate events, such as droughts and floods throughout history, have demonstrated the need for social organization in river basins and watershed management, and have led to extensive collaboration between communities.

Water is still teaching to humans. The population growth, the high water demand for wealth creation and development, and the need for more water infrastructure such as water treatment plants, along with environmental concerns, climate change, drying up of aquifers and lakes, and growing water threats (such as water-related terrorism and conflicts) have made it impossible for humans to deal with water problems using traditional systems. Accordingly, mankind has to tackle such issues with innovative approaches, based upon international co-operation, and more comprehensive strategies (Wittfogel, 1955; Priscoli, 1998).

According to the old motto "those who ignore history are doomed to repeat it'. If humans want to actively create the future of water, they should be looking to the past. Given that the destiny of humanity is inextricably linked with water, just as water creates great civilizations, neglecting, and forgetting the history of water will destroy such civilizations.

History provides large evidence of the impact of water on the growth of civilization, and forecasts of the future indicate that water will be crucial for human development in the XXI century, and further. In the past and present, human development depends on advances in water sciences, and water engineering and technology were of use to human societies. It can be said that scientific advances in water have always led great changes in society. Steam power, water filtering and sanitation, wastewater treatment, water pumping, hydropower discovery. Every step that is taken forward in water studies, subsequently promotes progress in other areas such as the economy, society, law and environment.

The existence of water resources has led to the formation of civilizations and cultures in different regions. Undoubtedly, throughout history, the development and expansion of societies and the evolution and Promoting of culture and civilization, has had an unbroken relationship with water resources.

Along with the growth of civilizations and cultures, the creation and management of water systems has been a foundation of modern societies, and there is a lot of evidence of such in the specific history (i.e. about ancient water structures and water management in ancient times) across the globe (Ghafrokhi et al., 2011).

As widely known, culture and civilization have no final definition, and all aspects of human life, such a religion, politics, beliefs, norms and anomalies, philosophies, health, water use, water governance, and all scientific, literary, epistemic, epistemological, education topics have shaped societies' culture and civilization throughout history. It could be claimed that civilization is built based on the use of water. In the following, some key findings and relationships of water with other foundative topics of human societes are discussed.

9.10 Water and hygiene

Long before antiquity, human beings have recognized the importance of drinking water quality, and of sanitary hazards caused by contaminated water. Both such items have historically depended on, and have defined the scientific and technological level of a society. Since the 19th century onwards, new methods for studying water quality were introduced, including chemical and microbiological studies. Ever since the beginning, various methods such as screening, filtering, and boiling have been used to improve the quality of drinking water. However, it was as late as in the 19th century that drinking water treatment became common practice in urban centers. Disinfection of drinking water was pursued in various ways in the early 20th century. With the proper use of water treatment technologies, it was possible to increase the quality of drinking water. The first "modern" urban sewers' system was built in England, and such use quickly spread around the world. It should also be considered that water systems need proper maintainance for long term operation (Vuorinen et al., 2007; Grmek, 1989).

In different historical periods, the community health of societies depends on the level of access to safe water. As an instance, the quality and quantity of bathing may differ in different civilizations. Regardless of whether drinking water is supplied from plumbing systems or wells, water must be healthy and safe for human health and environmental issues. The same is true for wastewater.

In Rome, since the sixth century BC, wastewater flowed into a number of sewer branches, eventually delivering into the *Cloaca Maxima* (the largest sewer) which was one of the earliest sewage systems worldwide.

In ancient Rome, bathing was very important. So much so, that at the end of the republican age (first century BC), providing freshwater and bathing places was the biggest challenge of urban life. The presence of public baths, and drinking water was a way for the hereditary empires of Rome to show off their power and ability to their citizens, and their enemies. The most famous Roman baths were built during the reign of the Diocletian and Caracalla. Water has been widely used as an element of architectural building designs, such as in public baths in the passages, gardens, shop centers, and libraries.

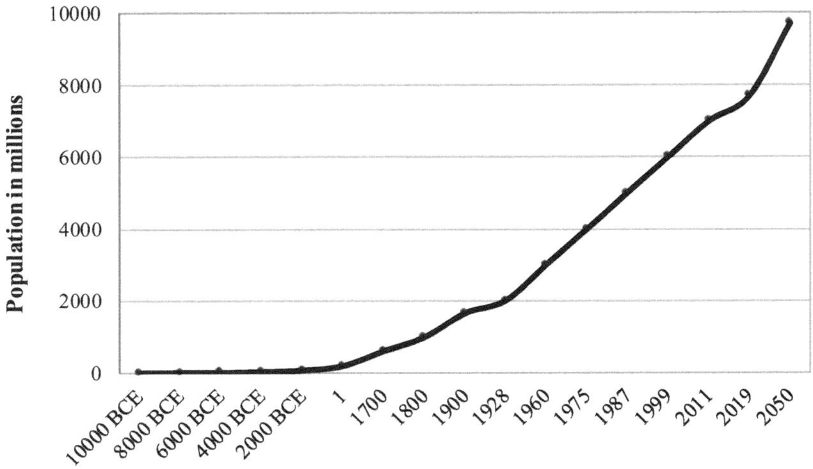

FIGURE 9.2 Estimated human population growth from 10,000 BC until the year 2050 (Source: Adapted from Juuti et al., 2007).

The importance of clean water is evident in ancient cultures. Due to lack of proper wastewater disposal, and of public toilets or baths, various types of intestinal diseases occurred, as reported by ancient physicians (Greek or Roman, Nutton, 2005). Two major diseases, malaria and schistosomiasis were closely linked to water hygiene, and water management in ancient times. Mosquito breeding depended on water, which was a serious and widespread problem throughout the Mediterranean in ancient times, and there is strong paleontological evidence of schistosomiasis in human remains from ancient Egypt.

Maritime trade was high, especially in the Mediterranean Sea during 200 BC to AD 200, and the Mediterranean became a pool of infectious diseases (McNeill, 1979; Vuorinen et al., 2007).

In Europe, after the fall of the Roman Empire, water and wastewater systems experienced major changes. The agricultural world began to industrialize since the 18th century onwards. As a result, world population growth onset (Fig. 9.2), with profound effects on water supply. The onset of industrialization, and the related growth of population within cities, created conditions that overwhelmed the local governments far more than ever before, and modern technology was often seen as the solution. In the 19th century, Great Britain was considered to be a trailblazer of modern water and sewage systems, but these innovations soon spread to Germany, and other parts of Europe, the United States and later elsewhere(Vuorinen, 2007).

The 20th century was a period of massive population growth. The global population has almost quadrupled, while the urban population increased by 13 times (Fig. 9.2). By 2000, in almost every country, more than half of the population moved to urban areas. Over the centuries, industrial production has increased 40 times, and energy consumption has increased dozens of times. Water and sewage services played a key role in this rapid economic, and social change across the globe.

Anxiety about chemical, and radioactive contamination hazards and their impact on human health that arose in the 1960s. In the early 20th century, health problems associated with water pollution were resolved by the development of chlorination, and other water treatment methods in industrialized countries. Water microbiological contamination was (and is) considered a large problem in the developing world. However, in the late 20th century the risk of the water biological contaminations reappeared in the post-modern Western world (Vuorinen et al., 2007).

In today's world, about 10,000 people die every day due to diseases, such as bloody diarrhea, cholera and others, due to the lack of safe water and poor hygiene. Many of those who die are the children, and the elderly, so their deaths are considered almost "natural", the same happening with people more or less marginalized in their societies (such as refugees and poor people).

9.11 Water and religion

Water is one of the most complex symbols in ancient symbology. In most civilizations water is considered the basis of creation, and sometimes the creator of the universe. In many cases water is a symbol of life or fertility or birth.

The beliefs and opinions of different nations in the ancient world are related to water, because of its importance and its place in human life. In many areas, such as West Asia and the eastern Mediterranean, people praised a goddess of water in ancient times. She was pictured as a fat woman, a symbol of fertility and birth, and the goddess of rain, rivers, and seas.

Also, the growth of plants, the fruiting of trees, and the growth of children in the wombs were all blessings and mercy of the god of water. A statue of this goddess was found in the hills of Sarab city of Kermanshah province in Iran, dating back to 9000 years ago.

This goddess had various names, for example, *Ardwisarā Anāhitā* in ancient Iran, *Anāhit* in Avestan, *Aphrodite* in Greece, and *Inanna* in Sumeric lands. This goddess has been very important and worshipped among the people.

The kings of ancient Iran considered themselves the guardians of the temples of *Anāhitā* (Seyedi, 2009; Mushref and Jabbari, 2011). In ancient Iran, even before Zarathustra, water had a special place in people's beliefs, and keeping water clean was one of their beliefs. They considered water along with fire, wind, and soil (called the four *Akhshijs*) to be elements of life and the creator of the universe. The origin of this theory is conceived alongside Aryan civilization, and this is the belief that water is the basis of existence and the origin of life (Mushref and Jabbari, 2011).

Smith and Ali (2006) noted that water use patterns were highly characterized by religious practices, particularly for Jewish and Muslim groups. Also, they claimed that religion has a fundamental bearing on how people use water, and even on how they think about water.

In Animism, the perception of water is that it is an entity that connects others and creates relationships. In Hinduism, Water is a sacred place, because it is believed to hold purifying and cleansing powers. Buddhism embodies the calmness and serenity of water by practicing water offerings at Buddhist shrines. In Christianity, water is primarily linked to the ritual of Baptism, where a follower professes his/her faith by bathing in 'holy water'. This symbolizes rebirth and purity. In Islam and the holy book of the Quran, water symbolizes wisdom. It recognizes that water is a part of life; that we are made up of, live with, breathe and consume water (Zipwater, 2017).

9.12 Water and poverty

Historically, it seems that major problems affect more poor people than other, richer ones. Poor people suffered most from the lack of sanitation, of drinking water, with health problems, poor education, and often lack of good governance and democracy.

In most less developed countries, due to the lack of water supply systems, often women, children, and especially the youngest girls carry water from distant sources (such as rivers, wells, and springs). This routine task often prevents them from going to school, learning, reading, and generally from fully living their young lives. They may grow up being unable to act as active citizens, and this will seriously affect their lives and cultures.

Also, it is not possible to have standard and proper sewage systems everywhere, for economic and other reasons, while obviously it is necessary in largely populated areas. In rural areas, at least alternative systems may be used, especially under local conditions. All of these reasons caused that poor areas with poor facilities have more health problems that hinder their cultural development.

The effects of climate change, which have been highlighted in many studies (Akbari-Alashti et al., 2018), will increase environmental problems. This is a serious and relatively new danger that threatens the whole world, especially the poor and less developed societies.

9.13 Water and culture of water use

Water used in large quantities has been regarded as a vital part of civilized human life at different times: Roman baths needed as much water as the current lifestyle (like shower, and Jacuzzi). The waste and consumption of water is very large, especially in places where there is no proper maintenance. Evidences show that as soon as the drinking water system, and even the sewage system, is maintained at real costs, the water consumption will be significantly reduced. Although most of the water is globally used for irrigation, the highest priority is for the purposes of drinking water, and consumption for society water supply, requiring a highest standard of quality. Therefore, throughout history, the cultural strategies of each civilization for the optimal use of water have been important and practical (Vuorinen et al., 2007).

Throughout history, water has been regarded as a goal. In fact, when water is abundant, it is more likely to be regarded as a functional factor of production and application. In arid areas it is less likely, and water probably becomes a principle of organizing society. But in the end, often beyond the economic debate, water has had great respect and value throughout history. There have been various solutions to ensure adequate water supply for human settlements throughout history. The natives have been very innovative in drawing their water. They have considered water very important and often a sacred element. In the long run, the availability of a sufficient amount of water has been an important factor in the development of a society and cities.

9.14 Water and sanctity

Water's sanctity refers to those aspects of water that are full of mystery and unknown, and the elements appear in our epistemology. But, understanding the intrinsic and extensive value of water does not take place in a traditional way of using it. Water shouldn't be religious. We must realize that water is not a means for other purposes and it is an important goal by itself.

The sanctity of water isn't a new thing. From ancient people's respect for water to Thales and Hippocratic interpretations of water as the source of life, Christian fathers' conceptions of water as a vitality, the innovation of Herodotus to predict the status of gifts in the Nile, to the Fontina Renaissance (men who combined the knowledge of hydraulics, physics, and hydro-mythology), until the nineteenth century, all pointed to "conquer", and democratization technologies of water. Humans were constantly worshiping and using water (Priscoli, 1998; Haught, 1996).

Water plays an unmatched role in all great mythical systems, and its symbology in all aspects of human life, from the emergence of the life, to life after death is quite obvious. We now know that water is the source of life on earth, however, it seems that the earlier civilizations had already believed this fact.

Many created myths suggest that the ancients have considered the first eclipse as the origin of the world. The value of water in human society is linked to its culture, whether a part of religious ceremony or art, it has continued to inspire art artistically and spiritually. Water permanent movement, shape, color, and changing amount have always played an essential role in myths and religious rituals around the world. From the Scandinavian Vikings religion to the Middle East monks, the value of water in different cultures is obvious. Ganga's mother's role in Hinduism, and the importance of the great ceremony of burning the dead in water in the holy city of Varanasi are all examples of water sanctification among human cultures (Eulisse, 2010).

9.15 Water and beliefs

Humans have been dealing with water from their birth until the last moment of their life. In all parts of the world and in all religions, water is used to wash and purify babies after birth, just like washing human bodies after death. And the reason is the belief in the purity of human nature that must step to this world, and go to their God cleanly. In Iran, even today, people use to pour water container behind travelers, which means the clear direction of travel, the health of the travel and the early return of the travelers. This is a long-standing tradition of water history and culture.

In eras when water Gods were not worshiped, myths helped to draw people's attention to their importance, by creating or exaggerating real stories, or trying to create beliefs that influence public opinion toward them. They managed to obtain people's empathy and cooperation in accessing, managing, and proper use of resources.

One of the stories about the creation of the aqueduct in Iran originated from the myths, according to which aqueducts were built with the help of demons. There are aqueducts all over Iran, and the indigenous people of those areas considered them to be made by myths. In Oman, people believed that the aqueducts were built by Prophet Solomon. Other manifestations of water culture are the belief in male and female water. In the past, Iranian peasants and farmers in low-water areas of Iran, had come to the conclusion that water is female if it moves smoothly and quietly, and crops can be watered and more fertilized by this water, and that water is male instead if it roars. Male water causes damage to irrigation (Mushref and Jabbari, 2011).

Wedding for the aqueduct was another belief of villagers, especially Iranian farmers, who believed that the aqueduct would be stimulated and this excitement would increase its water supply. In some parts of Iran, the custom has been that a widow would be taken to the aqueduct, and one night she would have to sleep naked in the aqueduct, and in return people would support her for the rest of her life. She was not allowed to marry anymore, because people believed that she was the wife of the aqueduct and it would never divorce her wife. The origins of this ceremony seemed to come from the shrines of the Mitraavism temples (Fazeli, 1998; Sedaghat-Kish, 2003).

In many cultures of the world there have been water commemorations, some of which were held annually, or revived in the new era, such as celebrations in Thailand, India, etc. There have been water celebrations in Iran too. Iranians have had sprinkler ceremonies in most of the celebrations, but about two celebrations have been reported in the writings: one is a Tyrgan celebration on the first day of July, and the other is the Abangan celebration on the first day of November. Remarkably, the date of these two celebrations exactly coincides with the lowest rainfall months in the Iranian plateau. For each of these celebrations, mythical and fascinating stories are quoted, and as well very beautiful ceremonies are held (Aurang, 1956).

9.16 Water and art

Among the ancient buildings all over the world, the fountain is one of the most conspicuous human artifacts. It attracts any visitor, from the large pools at the foot of the mansions to the small underground pools, and even fancy fountains in today's reception halls, they are all the spirit of a building. Water has flowed in poems and literary texts, in travelogues and paintings by various nations, like a clear river, and has taken them to the most prestigious places. Water also has a special place in poetry, literary texts, travelogues, and paintings in different nations (Papli-Yazdi, 2000)

9.17 Water and technical choices

Humans have also been fighting against water since the beginning of life, and the excess of water must be monitored and counteracted. This was certainly hard and time-consuming, so people learned to build ditches, and banks around their villages. They could withstand floods for weeks, and even months and protect themselves, cattles and food. Such structures were mostly made without the help of machines. Lakeside settlements, built from the early times around the world, if high enough give humans the advantages of living on water (communication, protecting enemies and wild animals, for washing, cooling, etc.) without turning them into floods victims.

It was difficult to preserve some fresh foods near the water, so they had to be dried in fresh salt. However, sometimes it was really useful. Water in contact with the grain planted inside the soil made it grow. Therefore, human beings were awake and aware of all water faces, good and bad, so a "hydraulic order" was created, a principle leading to civilization. Hydraulic order is a balance between land and water, which is achieved by regulating water through dams and drains, and avoiding harm to human activities.

Human's intelligence has been able to use water to spin mill wheels, or turbines to generate hydroelectric energy.

The ancient Romans used watermelon weights to crush their grains and olives. During the Middle Ages, monks had made many mills, which facilitated the hardships of human labor. From the ninth to the 16th centuries, Islamic societies from Spain to Oman experienced the golden age of science and technology.

A most important technology is hydrology, i.e. the assessment of water quantity and timing. The advent of cities such as Cordoba, Damascus, Baghdad,

Fez, and Morocco, all required complex water supply methods (Eulisse, 2010; Wolf, 1992).

A debatable issue, that will be very important in the long run, is how to use groundwater, or surface water for social purposes, or what resources to use in general.

Groundwater is generally sufficient for the use of less populated areas, but for larger populations surface water needs to be controlled, and utilized. Growth of such systems throughout history was economically significant and increased the wealth of population.

Muslim water engineers began an agricultural revolution in the Arab countries in the early seventh century, and in Spain around the tenth century.

As recent discoveries at Pompei showed, Roman aqueducts used taps that were similar to those used today. Unfortunately, the lead used in water pipes polluted the water and poisoned people over time (saturnism, or lead poisoning disease).

Later it was found that pipes needed to be made of steel, stone or plastic, so that the water would not be polluted. Nowadays water in aqueducts is carefully controlled and often has a very high quality (El-Faïz, 2005).

Arabs, Chinese, Indians, Khmer, Venetians and Dutch were very vigilant in terms of the hydraulic order of their territories. They managed to change the course of rivers, they created high dams, artificial lakes, and pools that even became places for people to relax and have fun. To defend their cities and villages, they built ditches under the rivers. From the earliest times, humans have built dams that contain water and can also be used to generate energy if needed.

More than one thousand years ago in Cambodia, the Khmer people built a huge city called Angkor, the remains of which are still seen today. They cultivated and irrigated vast fields of rice, and fed more than one million people. Since rice only grows in marshy fields, they used ponds to store excess water, which could be used as storage when rainfall was low.

Hindus people, who lived along the Great Indian River, used the natural banks of the river to protect against floods.

About 500 years ago, a fully developed secret ditch flowing under the Arno River was built by the people of Florence, Italy.

About forty years ago, a huge dam which was built on the Nile River created an artificial lake in Aswan, Egypt, that put a number of ancient monuments at risk of being submerged (Eulisse, 2010).

In some cases, the choices may be incorrect or less successful. An example is the lead-water pipe used in the past and recently in home fittings.

Water and wastewater systems need constant maintenance and adequate repair. This was already evident in Roman aqueducts: the accumulation of calcium carbonate in the ducts requires constant removal, otherwise it would stop water flow.

The same is true for modern systems. They need to be taken care of, and in fact a major problem with these systems is the need for continuous replacement and repair, so human and technical forces were trained in different civilizations. This led to the promotion of aquaculture and increase public awareness.

More than two thousand years ago, the Romans dug canals to drain the stagnant and unsanitary water of the Velino River to the downstream of the Nara River through the Marmore Falls in Umbria region, now the highest waterfall in Europe (165 m).

The water of the ancient city of Petra (Jordan) was collected and stored in special underground ponds, by collecting moisture from the night air, previously gathered on the rock walls. It was a precious traditional method that unfortunately is not used anymore.

The amount of water distribution in ancient Rome was approximately one million cubic meters (more than 25 million gallons) per day, which was transported by canals to the city center and flowed through a large number of reservoirs and supplied the entire city's water. The first canals therein date back to 312 BC. The Venetians built many luxury villas between the 15th and 18th centuries. For example, the Villa Contarini di Piazzola, on the Brenta River (Padua, Italy). Therein, fishing ponds, and large water ponds were used to regulate many surrounding irrigation canals, and sometimes for fishing and simulated wars, as well as for noblemen's entertainment. Marmore Falls and Villa Contarini are examples of how people can build structures that also protect the environment and landscape (Laureano, 2001; Eulisse, 2010).

9.18 Water and governance

The failure of past water management policies, under pressures from development has led to a reform movement, for revision and pathology of past behavior, and to find ways to increase countrywise water management efficiency.

It is widely believed that the current water (scarcity) crisis is the result of improper governance. Among others, the UN World Water Development Report 2003 pointed out how the water crisis is largely due to lack of proper management, and thereby claimed need for good governance. People should be encouraged to use their experiences and abilities to solve their problems. Involving all stakeholders in decision-making in their appropriate roles is likely to have more sustainable results than any top-down approach.

There are many development directions and solutions in urban water supply and sewage. While future solutions are considered, conditions, traditions and local people must be at the core of decision making.

Since water resources of each place, city, and town have their unique location and quality, and each place has its own unique social and administrative location, solutions for one place may not work for another, and one needs to identify the most practical and applicable principles and procedures (Vuorinen et al., 2007; Grmek, 1989).

Venice's history has always been intertwined with the sea, due to the city's special conditions, where approximately 120 small islands are formed in the middle of the lagoon and located along the Po River. From the beginning, water played a crucial role in the city life, and later it was considered as the source of economic growth and wealth for the cities. After the ninth century, Venice expanded to a port, and became the center of the Levantine goods to enter Europe. A full-fledged lifestyle was set up in Venice's wetland. Although the city has always had to contend with rising water and mud, and although water is Venice's foundation and the reason for attracting tourists today, the city is still slowly sinking in the lagoon.

The city's advantage that grew gradually in time was its port location, at the crossroads of Eastern Europe, the Middle East, and Central Asia. Throughout the Middle Ages, native merchants were able to monopolize wealth from the Middle East and Asia, and the city could expand itself at an imperial level.

Without the Nile, the land between the Red Sea and the Atlantic Ocean would have been a vast desert. If there were no Nile, there would be no Egyptians. The river is the reason for the high population density and cultural integration in the region, which have always been the main characteristics therein. Traces of human habitation in this area go back to 5,000 years ago. People have migrated from arid lands over the centuries and eventually found a safe place on the beach (Goddess River). The Nile River water enabled them to live, communicate and irrigate their crops and every year, and the river's muddy floods fertilized and revived the land. During the Pharaohs' age, Nile was worshiped as a manifestation of divine. It was even an area for measuring flood levels, so the tax on workers was based on this. The more floods, the higher expectation for crop production.

The Nile River during its material and spiritual life has been called by numerous names according to the seasons and water levels. The ancient Egyptians' calendar was based on the floods that they believed to be from the gods.

Pharaohs had to honor and respect these gods and have a good relationship with them, so that floods could occur every year. The Nile was both a valuable source of agriculture and the only means of communication and travel in Egypt, which may explain e.g. why camels were not used in this part of the Middle East for a long time.

Today, the Nile's mud no longer determines the life rhythms of Egyptians. Aswan Dam, founded in 1960, made Egypt a modern country in this sense. Controlling the water in this river could eliminate the destructive effects of occasional floods, but also prevent famine caused by drought.

The reclamation of desert lands significantly increased the amount of arable lands. Today, Egypt benefits greatly from unrivaled resources of hydroelectric powers, and the river can be navigated year-round.

For the population, grown to 6 million in just one century, this was a tremendous blessing. But it cost them much. All of Nubia's low lands were submerged by the formation of Nasser Lake and About 60,000 people were stranded.

The Nile delta is eroding (no longer compensated by settling silt), and delta faces is increasing in salinity, with proliferation of water hyacinths, home to a snail causing cochlear parasitic fever. Valuable mud and sludge had to be replaced with industrial chemical fertilizers., which also caused the salinity of water.

For visitors of arid and uncultivated fields of Syria and Iraq, it may be hard to believe that during the Umayyad Empire (661–720) this region was a rich agricultural land, extending from Spain to the edge of China. As the water supply problems were skillfully resolved, these lands became fertile areas, and their populations were able to reside in the land, due to the presence of water that sometimes comes from distances and is sufficiently stored (Shir-Afroos, 2011).

9.19 Water and politics

Since social and hydrological relationships come together, an environmentally and politically sensitive approach to water resources management has long been developed (Chen et al., 2013). The fact that we share water or ban it from other humans as well as animals, plants and soils has made water a strategic element during history, and the social and environmental domains are intertwined.

Gomez Tanzio in Senegal describes the efforts of the locals, to obtain and maintain drinking water resources. In this context, drilling and managing wells is not the only issue for water supply, but even social relations with employees, and the bureaucracy must be considered.

Bonelike Robinson writes that in the Caribbean (Senegal), drinking water traditionally came from shallow wells, and digging and using wells meant proving the ownership of the lands around the wells, so people built houses there to use high-quality water of the wells.

But in the late 1970s the arrival of plumbing water resources disrupted this relationship, and water was pumped directly into people's homes. Government officials, staffs and engineers believe that the former system was the most sustainable strategy for the permanent use of water resources, and local people have expressed dissatisfaction, even through frequent sabotage against the infrastructure of the public water system.

It is therefore necessary to study the relationships that water creates. Bonelike Robinson said that wells, pumps, and pipes are not just for water, but also create complex and different relationships between people, water, and the earth.

Krause (2013) studied the concerns of residents in southwestern England, and said that floods have historically affected people's social and political relationships, especially in terms of structural changes in the landscape, including properties on flood path and road embankments. Krause concludes that more effective flood risk management is the result of an understanding of the intertwined relationship between humans and water (Krause, 2016).

Many claim that (the fight for) water may cause war or probably will cause conflicts, of even war in the next century (Bocchiola et al., 2017). It is thus necessary to focus upon areas of shared waters, and upon use of water as a weapon. The root of the word "competitor" is from the Latin *competitores*, which means joint use of a river, or stream of water (Priscoli, 1998; Biswas, 1993). The struggle for access to water within and among countries, may lead to violence.

Also, examples exist of how vulnerable are seaside civilizations to the invasion from foreign armies.

Water played a vital role in the survival of Rome. In the fifth century, when the barbarians invaded Italy, they first attempted to break the Roman fortifications, and to destroy their water resources. This scarcity of water prevented the retrieval of the glorious Roman period until the Renaissance, when Roman architects began to rebuild Roman water resources and made Rome a densely populated city again (Shir-Afroos, 2011).

Polypus, the Greek historian in the third century BC, noted that in the Achaemenid era, the five generations of people who excavated or cultivated land, or rebuilt an arid land were exempted from taxes. In the Parthian and Sassanid times, which paid special attention to agriculture, the construction of dams, creeks, and aqueducts was their first priority.

The Achaemenid kings used taxes to maintain and repair aqueducts, also done nowadays, i.e. a part of the income from the sale of water, or products irrigated with water is used for this purpose.

In the book of Genesis, struggles in the Negev (semi-Arab region of southern Israel) over water wells with the Palestinians are described.

Herodotus describes how the Persian cities succumbed by filling wells and water tunnels. In 1187, Saladin could deprive the Crusaders of water access, and defeat them in Hatin Horn.

We have recently witnessed the bombing of irrigation systems, and hydroelectric facilities in modern warfare. During the Gulf War, desalination plants and water distribution systems were targeted (Wittfogel, 1955; Hillel, 1994; Priscoli, 1998). On the other hand, water has helped to build early societies, and to bring these societies together for larger goals. Such social networks were the main impetus for the growth of civilization. Dams were not targeted during the Vietnam War.

Water was actually one of the great learning fields of human beings to create society. Hillel (1994) proposed a quotation from Arabic: "*Someone with a hand in the water is not like one with a hand on fire*." In his view, this shows that people managing the water may not be involved in the war, and that the thirst for water may be more convincing than the motivation to seize it.

Between 805 and 1984, more than 3,600 FAO non-navigable water treaties were identified. Since 1945, about 300 treaties related to water management, or allocation in international waters have been negotiated.

Even in the most productive Middle East, perhaps the most prominent meeting place in the world for high water policies and tensions, only one event can be mentioned that water has caused conflict. In fact, water agreements have prevented major subcontinental conflicts between Pakistan and India (Wittfogel, 1955; Priscoli, 1998).

According to historical experiences the threats to water systems are real. Terrorists target people, water supply systems, and other important facilities with different motives.

So, it is necessary to consider safety and protection actions (Gleick, 2019). Water forces us to reconsider the concepts of security and mutual interdependence. Increasing mutual interdependence through water supply programs, infrastructure, and water systems may be seen as a way to increase vulnerability, dependence, and thus to reduce security.

However, they can also be seen as networks that enhance the flexibility and ability to deal with natural emergencies, and reduce our vulnerability to events such as droughts and floods, and thus increase security.

In fact, this point of view has always been fundamental to the evolution of civilization. We gain power by sharing our vulnerabilities. This fear and insecurity have promoted humans to become builders and engineers. Somehow, water forces us to act deeper than with familiar hostile positions, and to consider what we really have in common. In the next step, water constantly leads us to higher concepts of integration and social connection (Lees, 1973; Wolf, 1992; Priscoli, 1998).

Water is not just a matter of social relations, or just a natural resource, but political conflicts over it have been formed throughout history. A deeper understanding of the role of water in human social life is obtained by studying social and hydrological relationships.

Most of the previous researches on water has raised the need to reconsider the relationship between society and natural resources (Linton, 2010). Here, hydrology and society are considered together. Historically, water has been together an element, a stream, a means of transportation, a living substance and a life-threatening force, and often a means of social and cultural activity (Hahn et al., 2012; Krause and Strang, 2013).

9.20 Summary

Throughout history, water has been a great impact on human to build civilization. In ancient times, humans built watermills to grind wheat, reservoirs, canals, wells, aqueducts (*qanats*), drains, water pipes, and levees to protect against floods. Humans also created regular systems for water distribution and protection of water structures. Humans realized that all these actions required a proper and regular society, so governments and laws are created.

To increase welfare, humans constantly developed technology and tools, and on the other hand they realized the importance of social life. Over time, they have developed new sciences, such as civil engineering, hydrology, and water resources management with a focus on development, operation, control, and conservation, and this trend continues.

Water is one of the most vital natural resources that has had a profound effect on all aspects of society, such as agricultural and industrial development and cultural and religious values. Throughout history, water has been a prerequisite for health, economic development, cultural goals and growth, and development of human civilization.

There is a mutual relationship between water and culture. Cultural and religious differences make a difference in the perception, value, and managing of water in different societies. For example, water consumption patterns are directly related to the religious and cultural values, and features of a society.

Also, today, to find solutions to water problems, there is a need for continuous research on the complexities of the relationship between water and human history and civilization. Intercultural dialogue, to become familiar with the capacity of different cultures, is an important principle for raising awareness, developing educational tools. Cultural diversity is also an opportunity to solve regional and global water problems.

References

Akbari-Alashti, H., Soncini, A., Dinpashoh, Y., Fakheri-Fard, A., Talatahari, S., Bocchiola, D., 2018. Operation of two major reservoirs of Iran under IPCC scenarios during the XXI century. Hydrol. Process 32 (21), 3254–3271.

Anand, N., 2011. Pressure: the politechnics of water supply in Mumbai. Cultural Anthropology 26 (4), 542–564. doi:10.1111/j.1548-1360.2011.01111.x.

Angelakis, A.N., Spyridakis, D.S., 2010. Water supply and wastewater management aspects in ancient Hellas. Water Science and Technology, Water Supply 10 (4), 618–628.

Angelakis, A.N., Dialynas, M.G., Despotakis, V., 2012, Evolution of water supply technologies in Crete, Hellas through the centuries. Evolution of water supply throughout millennia. In: Angelakis, A.N. (Ed.). IWA Publishing, London, UK, Chapter 9, pp. 227–258.

Aurang, S.M., 1956. Celebrations of Ancient Iran, Second Edition. Rasti Publications, Tehran, Iran. (In Persian).

Bakker, K., 2012. Water: political, biopolitical, material. Soc. Stud. Sci 42 (4), 616–623. doi:10.1177/0306312712441396.

Barnes, J., Alatout, S, 2012. Water worlds: introduction to the special issue of social studies of science. Soc. Stud. Sci 42 (4), 483–488. doi:10.1177/0306312712448524.

Bijker, W.E., 2012. Do we live in water cultures? A methodological commentary. Soc. Stud. Sci 42 (4), 624–627. doi:10.1177/0306312712441690.

Biswas, A.K., 1993. Management of international water: problems and perspective. UNESCO, 41–55.

Bocchiola, D., Soncini, A., Pelosi, M.G., 2017. Effects of hydrological changes upon cooperation in transnational catchments: the case of Syr Darya. Water Int. 42, 1–22.

Carse, A., 2012. Nature as infrastructure: making and managing the Panama Canal watershed. Soc. Stud. Sci 42 (4), 539–563. doi:10.1177/0306312712440166.

Centre Water Commission (CWC), 2006. Government of India. Water Resources of India, New Delhi, India, 34–48.

Chen, C., McLeod, J., Neimanis, A., 2013. Thinking with water. McGill–Queen's University Press, Montreal, Canada.

De Feo, G., Mays, L.W., Angelakis, A.N., 2011, Water and wastewater management technologies in ancient Greek and Roman civilizations. Treatise on water science. In: Wilderer, P. (Ed.). Academic, Oxford, pp. 3–22.

Delli Priscoli, J., 1998. Water and civilization: using history to reframe water policy debates and to build a new ecological realism. Water Policy 1, 623–636.

Dugger, C., 2006. Need for water could double in 50 years, U.N. Study Finds. New York Times, August 22, 2006, NY, USA.

El-Faïz, M., 2005. Les maîtres de l'eau. Histoire de l'hydraulique arabe. Actes Sud.

English, P.W., 1968. The Origin and Spread of Qanats in the Old World. American Philosophical Society, 112 (3), pp. 170–181.

Eulisse, E., 2010. Water, Culture and Civilization in the Mediterranean Italy, in Losing Paradise. The Water Crisis in the Mediterranean (G. Holst-Warhaft and T. Steenhuis Eds.), Ashgate.

Fazeli, A., 1998. A Baggage of the History of Golpayegan and its People. Abjad Publications, Tehran, Iran (In Persian).

Fontein, J., 2008. The power of water: landscape, water and the state in Southern and Eastern Africa: an introduction. J. South. Afr. Stud 34 (4), 737–756. doi:10.1080/03057070802456730.

Gleick, P.H., 2019. Water as a Weapon and Casualty of Armed Conflict: A Review of Recent Water-Related Violence in Iraq, Syria, and Yemen. Wiley Interdisciplinary Reviews: Water, 6(4): e1351, doi:10.1002/wat2.1351.

Glick, T.F., 1970. Irrigation and Society in Medieval Valencia. Harvard University Press, Cambridge.

Grmek, M.D., 1989. Diseases in the Ancient Greek world. The Johns Hopkins University Press, Baltimore Translated by Muellner, M. and L.

Hahn, H.P., Cless, K., Soentgen, J., 2012. People at the Well: Kinds, Usages and Meanings of Water in a Global Perspective. Frankfurt-on-Main, Germany: Campus Verlag.

Hastrup, K, 2009. Waterworlds: framing the question of social resilience. In: The Question of Resilience: Social Responses to Climate Change. (K. Hastrup Ed.), Royal Danish Academy of Science and Letters, Copenhagen, Denmark.

Haught, J.F., 1996. Christianity and ecology. In: This Sacred Earth: Religion, Nature, Environment. Gottlieb, R.S. (Ed.), Routledge, New York.

Hillel, D., 1994. The Rivers of Eden. Oxford University Press, New York.

Ingold, T., 2000. The Perception of the Environment: Essays in Livelihood, Dwelling and Skill. Routledge, London, UK.

Juuti, P.S., Katko, T.S., Vuorinen, H.S., 2007. Environmental History of Water – Global View of Community Water Supply and Sanitation. IWA Publishing, London.

Kaiafa, A., 2008. Urban Water Supply and Drainage Systems in Hellenisic and Roman Macedonia. Doctoral dissertation, Dept. of Architecture. Aristotle University, Thessaloniki, Hellas (in Greek).

Kohn, E., 2013. How Forests Think: Toward an Anthropology Beyond the Human. University of California Press, Berkeley, CA.

Krause, F., Strang, V., 2013. Living water: the powers and politics of a vital substance. Special issue. Worldviews 17 (2), 95–185.

Krause, F., Strang, V., 2016. Thinking relationships through water. Society and Natural Resources 29 (6), 633–638. doi:10.1080/08941920.2016.1151714.

Kroeber, A.L., Kluckhohn, C., 1952. Culture: A Critical Review of Concepts and Definitions.

Laureano, P., 2001. The Water Atlas. Traditional Knowledge to Combat Desertification. IPOGEA-LAIA-UNESCO.

Lees, S.H., 1973. Hydraulic development as a process of response. Hum. Ecol. 2, 159–175.
Linton, J., 2010. What is water? The History of a Modern Abstraction. UBC Press, Vancouver, BC, Canada.
Mays, L.W., D Koutsoyiannis, D., Angelakis, A.N, 2007. A brief History of water in antiquity. Water Science and Technology, Water Supply 7 (1), 1–12.
McNeill, W.H., 1979. Plagues and Peoples. Penguin Books, Harmondsworth.
Molle, F., Mollinga, P.P., Meinzen-Dick, R., 2008. Water, politics and development: introducing water alternatives. Water Alternatives 1 (1), 1–6.
Mushref Ghahfarkhi, P., Jabbari, M.M., 2011. Survey of Water and Culture Interaction in History, International Conference on Traditional Knowledge of Water Resources Management, International Center of Aqueduct and Historical Structures, Yazd, Iran. 21 February 2011. https://www.civilica.com/Paper-TKWRM01-TKWRM01_163.html. (In Persian).
Nutton, V., 2005. Ancient Medicine. Routledge, London and New York.
Omranian-Khorasani, H., 2015. Good Governance and water management. Water and Sustainable Development 1 (3), 94–95 (In Persian).
Orlove, B., Caton, S.C., 2010. Water sustainability: anthropological approaches and prospects. Annu. Rev. Anthropol 39, 401–415. doi:10.1146/annurev.anthro.012809.105045.
Papli-Yazdi, M.H., Labaf Khaniki, M., 2000. The role of aqueducts in the formation of civilizations, the theory of sustainability of culture and Karizi civilization. In: International Conference on Cultural Approach to Geography, Mashhad, Iran, 14 October 2011, 1, 1–3. (In Persian).
Peltenburg, E., Colledge, S., Croft, P., Jackson, A., McCartney, C., Murray, M.A., 2000. Agro-pastoralist colonization of Cyprus in the 10th millennium BP: initial assessments. Antiquity 74, 844–853.
Peltenburg, E., Croft, P., Cucchi, T., 2014. Excavations of Neolithic Kissonerga-Mylouthkia, 2000-2006. Report of the Department of Antiquities Cyprus, Nicosia, Cyprus.
Pourkhaleghi chatrodi, M., 2009. Proceedings of the National and Interdisciplinary Conference on Water and its Cultural Values. University of Ferdowsi, Mashhad, Iran 1-3 November 2009. (In Persian).
Sedaghat-Kish, J., 2003. The Holy Aqueducts of Iran, the. Month of Art Journal 57 (9), 34–42 (In Persian).
Seyedi, V., 2009. The Goddess of Water under Greek and Roman Mythology. In: National and Interdisciplinary Conference on Water and Its Cultural Values. University of Ferdowsi, Mashhad, Iran, 1-3 November 2009. (In Persian).
Shir-Afroos, A., Mohammadpour, M., 2011. The role of water in civilization and human culture. In: Proceedings of the National Conference on Archeology and Architecture of Water Structures. Dezful, Iran, 30 January 2011. https://www.civilica.com/Paper-NCAAD01-NCAAD01_040.html. (In Persian).
Smith, A., Ali, M, 2006. Understanding the impact of cultural and religious water use. Water and Environment Journal 20 (4), 203–209.
Strang, V., 2014. Fluid consistencies: meaning and materiality in human engagements with water. Archaeological Dialogues 21 (2), 133–150. doi:10.1017/s1380203814000130.
The World Heritage Committee, 2016. The 40th session, 10–20 July, Istanbul, Turkey.
Tylor, E.B., 1958. Primitive culture. Published by Harper and Row; Part 1 edition, 1 (2), 5–9.
Vigoni, A., 2011. Pozzi antichi nel Veneto: tipologia e diffusione. In: Archeologia e tecnica dei pozzi per acqua dalla pre-protostoria alleta moderna (S. Cipriano, and E. Petteno, Eds.), Antichita altoadriatiche 70. Editreg, Trieste, Italy, pp. 19-52. (In Italian).
Voudouris, K., Kaiafa, A., Xiao Yun, Z., Kumar, R., Zanier, K., Kolokytha, E., Angelakis, A, 2017. A brief history of water wells focusing on balkan, Indian and Chinese Civilizations. In: IWA 2nd

Regional Symposium on Water, Wastewater and Environment, pp. 22–24 March, Cesme-Izmir, Turkey.

Voudouris, K., Christodoulakos, Y., Stiakakis, M., Angelakis, A.N., 2013. Hydrogeological characteristics of Hellenic Aqueducts-like Qanats. Water (Basel) 5 (2), 1326–1345.

Vuorinen, H., Juuti, P.S., Katko, T.S., 2007. History of water and health from ancient civilizations to modern times. Water Science and Technology: Water Supply 7 (15), 49–57.

Wittfogel, K.A., 1955. The hydraulic civilizations. In: Man's role in changing the face of the earth. (W.L. Thomas, Jr, Ed.), University of Chicago Press, Chicago, pp. 152–164.

Wolf, A., 1992. Water wars and water reality: conflict and cooperation along international waterways. NATO workshop on environmental change, Budapest 9–12 October.

Wright, R., 2004. A short history of progress. house of Anansi. Global Water Partnership, Towards Water Security: A Framework for Action, GWP, March 2000. 115–212. ISBN 9780887847066

Zipwater, 2017. The importance of water in different cultures. https://www.zipwater.com/ zip-effect/the-importance-of-water-in-different-cultures.

Chapter 10

Water quality, hygiene, and health

Omid Bozorg-Haddad[a], Mohammad Delpasand[a] and Hugo A. Loáiciga[b]

[a] *Department of Irrigation & Reclamation Engineering, Faculty of Agricultural Engineering & Technology, College of Agriculture & Natural Resources, University of Tehran, Karaj, Tehran, Iran.* [b] *Department of Geography, University of California, Santa Barbara, California*

10.1 Introduction

The existence of sufficient water with desirable quantity for the survival of human life is essential. Primitive humans discovered the importance of water and created civilizations around water resources, which, in addition to meeting vital needs, also met the needs of agriculture and transportation. But the recognition of the importance of water quality was made much later and the urbanization and industrialization of societies has stressed freshwater source and degraded water quality. Significant amounts of waste and pollutants enter water bodies that exceed their self-healing capacity. The availability of healthy water resources is an essential prerequisite for environmental conservation, and for economic, political, social and cultural development. On the other hand, water resources are susceptible to pollution and excessive use. Water quality assessment has many dimensions, and a key one concern the transmission of water diseases. More than 250 million people become infected with water-based diseases annually worldwide, resulting in 5–10 million deaths. This highlights the importance of preserving water quality and imposing high sanitation standards. We cite below several studies that have been conducted to estimate the quality of water resources worldwide' Fan et al. (2010) examined the water quality of the Pearl River Delta in China using multivariate statistical techniques. Monitoring station in the northern, eastern, and western regions were divided into clusters that were characterized by high pollution, moderate pollution, low pollution, and good water quality, indicating the quality of river water. The former authors concluded that Principle Control Analysis (PCA) and cluster analysis techniques were the appropriate methods for assessing water quality.

Economical, Political, and Social Issues in Water Resources. DOI: 10.1016/B978-0-323-90567-1.00008-5

Fiorentino et al. (2012) applied the Lagrangian adaptive structure to assess the quality of coastal waters in Hobie Beach in Miami, USA, and found that its water quality has high levels of microbial contamination from human bathing, runoff, and sand outflow.

Peiyue et al. (2012) evaluated the water quality of the Yinchuan Plain in China using three methods: multi-criteria decision-making, open accreditation, and prioritization techniques based on similarity. They found that groundwater in the Yinchuan plain was contaminated with F^-, TDS, SO_4^{2-} and NO_2 and must be treated before use for domestically purposes.

Ding et al. (2014) used methods such as hybrid neural networks and genetic algorithms to predict the water quality of the rivers flowing into the Taihu Lake in China. They combined the methods of analysis of the basic components, the neural network of reproduction and the genetic algorithm. They applied the genetic algorithm to optimize neural network parameters and reported a model that had good training speed and predictive power and was able to predict the water quality with good accuracy. The model also be used for other classification issues.

Azhar et al. (2015) examined the water quality of the Muda river in Malaysia with multivariate analysis. The applied three multivariate statistical techniques to evaluate nine water monitoring stations in the river and classified water quality into two clusters.

Jung et al. (2016) evaluated the water quality of the Nakdong River in South Korea with multivariate analysis. They evaluated 20 river stations and discovered that the points in the middle and lower reaches of the Nakdong River are the most polluted, receiving high levels of pollution from the city and from residential sewage. They also determined priorities for improving the water quality needed in future watershed management.

Barakat et al. (2016) assessed seasonal and spatial variations in the water quality of the Oum Er-Rbia River in Morocco using multivariate statistical techniques, and found that some stations water quality parameters did not meet Moroccan water quality standards. Changes in the concentration of water combinations are mainly due to point source pollution (domestic and industrial waste water), non-point source pollution (agricultural activities), and natural processes (weathering of soil and rock). River water was potentially dangerous for the health of consumers, and recommended sustainable agriculture and treatment of urban and industrial wastewater in order to prevent harmful effects on health.

Sener et al. (2017) evaluated the water quality of the Aksu River in Turkey using the Water Quality Index (WQI) and the Geographic Information Systems (GIS) in Turkey. Their results showed that the water in the river has overall a suitable quality when compared with criteria set by the World Health Organization and Turkey's domestic standards. However, the water quality in the northern and southern parts of the river basin is poor, and water in these regions is affected by point sources of pollutants.

Li et al. (2016) investigated the Qu River water quality in China on the basis of the fuzzy index of water pollution. This approach provides a comprehensive ranking of water quality, and describes the status of water quality quantitatively. The authors collected data from four stations from 2006 through 2010, and the results of the assessment showed that the water quality of the river declined and was in its worst quality in 2010.

Rostom et al. (2017) examined the water quality of Mariot Lake in Egypt by means of remote sensing and laboratory analyzes. Using laboratory analysis and a prediction model for water pollution with heavy metals and based on spectral specification they concluded that nickel, lead, and chromium concentrations in this area exceed maximum contaminant levels, and the location of the highest manganese and iron accumulation is in the upstream stations due to human and industrial activities. Also, the eastern part of the area is contaminated with heavy metals due to discharge drainage water. The authors recommend that their model be used for other Egyptian lakes to be calibrated under changing new conditions.

D'Ugo et al. (2018) examined the microbiological quality of the Italian medical device industry. Due to the manufacture of medical devices water directly contacts the product, and microbiological quality control of water is essential. Their results showed a permanent presence of *Staphylococcus aureus*, an irregular presence of fecal *E. coli* and the absence of *Salmonella*. The authors recommended regular inspections of the sewage drainage system to prevent the risks associated with microbial contamination.

Long et al. (2018) examined the effects of wet and dry seasons, as well as land use, on drinking water quality in the state of Yucatan in Mexico. For this purpose, 21 wells were sampled in the studied area. The studied parameters included pH, temperature, EC, Ca^{2+}, Mg^{2+}, Na^{+}, K^{+}, $HCO3^{-}$, $CO3^{2-}$, Cl^{-}, $SO4^{2-}$, alkalinity and fecal coliforms. The data were analyzed using various geochemical, statistical, and spatial methods, and it was concluded that 1- some of the parameters are diluted in the rainy seasons, but this does not apply to all of them, 2- the geochemical structure of water in the wet seasons is complex, 3- fecal coliform concentrations is higher in the wet seasons, 4- water quality in the dry season is at risk of nitrate pollution, and in the wet season is at risk of fecal coliform. They ultimately recommended that more studies be conducted employing different methods to verify the relation between land use and water quality.

Zhang et al. (2018) applied Laser Decomposition Spectroscopy and the Machine Learning Algorithm to estimate the Fe and Cu values at the Ebinur Lake in China, and found that the Cu content in this basin is generally higher than Fe. The highest Fe and Cu concentrations are found in the Ebinur Lake Basin and the lowest in the Jing River. The authors Estimated Retreat Model exhibited good accuracy and a strong estimation capability in the Cu estimation of the target area. In general, the laser-based decomposition spectrometric estimation improves the accuracy and efficiency of estimating Fe and Cu.

Xiao et al. (2019) estimated the resources and characteristics of water quality and health risk in river and well water wells in the Plateau Loess area of China. They analyzed well and river samples using several methods and indicators, including analysis of main components, correlation analysis, sodium and WQI adsorption ratio, and the risk ratio, to assess the water quality risk index and health risks. Their study has shown that the water quality of the river was poor and was accompanied by high levels of alkalinity and salinity. Contamination from arsenic in wells has contaminated river water, which poses health hazards to the community, especially for children.

Bouaroudj et al. (2019) evaluated the quality of the water of the Beni Haroun Dam in Algeria and its effect on irrigation and soil characteristics of agricultural lands. Their analyses established that water had low levels of cadmium and lead, and also contained large amounts of chloride and biofilms, the pH varied between 6.5 to 8.4, and the electric conductivity level in 39 percent of the samples exceeded the standard for irrigation water. The Wilcox chart established that 58 percent of samples were ofgood quality and 42 percent were unsuitable for irrigation. The authors concluded that efficient measures of protection, along with appropriate water and water parameters, are needed to reduce the harmful effects of irrigation water in agricultural lands.

Tian et al. (2019) employed the WQI to assess water quality in the upstream and midwest reaches of the Luanhe River in China, and investigated twelve water quality parameters in the summer and spring. The results revealed that the WQI varied from bad to excellent values. Agricultural and urban activities were found to be important factors in water quality decline in the study area, and in order to improve water quality urban wastewater as well as fertilizer consumption, especially in agricultural lands along the river, must be controlled.

10.2 Definitions and terms

10.2.1 Physical parameters of water quality

Physical parameters are those properties of water that are detectable by visual inspection, touching, tasting or smelling. Physical parameters important properties of water.

10.2.1.1 Suspended solids

Solids can be in suspended and soluble forms in water. These include organic and inorganic substances such as plankton, soil, and silt. Suspended solids concentration in surface waters depends on the seasons and runoff fluctuations. Some soluble solids are detectable by physical means.

Suspended solids sources

Suspended solids may be composed of organic and inorganic particles. Mineral solids such as clay, sludge and other soil constituents are found in surface waters. Organic matter such as plant roots and biological solids (algae cells, bacteria, etc.) are also constituents of surface water. Due their size suspended solids are rarely found in groundwater. Municipal sewage usually contains large amounts of suspended solids, mostly organic. Industrial uses of water may also result in impurities of an organic or mineral nature in the water.

Effects of suspended solids

The presence of suspended solids in water may be esthetically damaging to water and may also be adsorbed to chemical and biological surfaces. Suspended organic solids may be biodegradable and eventually produce undesirable products. Biologically active suspended solids may include pathogenic organisms such as organisms that are toxin-producing algae chains.

Application of suspended solids

The suspended solids parameter is used to measure the quality of the effluent inlet to control a number of treatment processes and to measure the effluent quality. The US Environmental Protection Agency (EPA) has set maximum suspended solids standard equal to 30 mg/l for most of the effluents.

10.2.1.2 Turbidity

Turbidity is a measure of the transparency of water and is a measure of the amount of light absorbed or the scattering of light by suspended matter. Light absorption and dispersion are affected by the size and surface properties of suspended matter; therefore, turbidity is not a direct quantitative measurement of solid suspended particles. For example, a small particle inside a glass of water does not actually create any turbidity. If the particle is broken into many smaller particles of colloidal size the turbidity may reach a level that is measurable even though the mass of solids remains unchanged.

Turbidity sources

Much of the turbidity in surface waters is due to the erosion of colloidal materials such as clay, sludge, crumbs, and metal oxides from the soil. Plant residues and microorganisms also play an important role in increasing turbidity. In addition to industrial and domestic wastewater, it may also contain turbidity materials, soaps, detergents and emulsifiers that produce durable colloids that ultimately lead to turbidity.

Effects of turbidity

Turbidity colloidal materials provide surfaces for the absorption of biological organisms or harmful chemicals or an unpleasant taste and odor agent. Disinfection of opaque waters is a difficult task due to the absorption properties of some colloids and also because solids may protect microorganisms against disinfectants. In natural water sources, the opacity may cause the water to turn gray or brown. The color of water depends on the properties of light absorption by solids and can be altered by light penetration or photosynthetic reactions in rivers and lakes.

Application of turbidity

Measurements of turbidity are usually performed in clear waters. Natural waters may have opacities varying from a few Furmazine Turbidity Units (FTUs) to several hundred FTUs. USEPAs standards for drinking water require a maximum of one FTU.

10.2.1.3 Color

Potable water, that is, suitable for human consumption, is colorless, but the water found in nature is usually colored by external materials. The blue that results from the effect of suspended matter is called the clear color, and the color that results from soluble solids and remains in the water after when suspensions were separated is called true color.

Color sources

Water is found to be brownish upon contact with fine components of organic matter such as leaves of trees, weeds, or wood by absorbing substances such as oak apple ink and humic acid. Iron oxides give a pale red color and manganese oxides cause brown or dark water. Industrial wastewater from textile and dyeing industries, pulp and paper production, food processing, chemical production, ore extraction, refining and slaughterhouse operations may also add color to the streams and rivers.

Color effects

Colored water is not aesthetically acceptable to the general public. In fact, if consumers have a choice, they prefer to use clean, colorless water. Colored water is not suitable for applications such as washing clothes, dyeing, paper making, dairy production with other foods, textile industries and plastic production. Organic compounds that give rise to true color may increase the need for chlorine in water and ultimately reduce the effect of chlorine on water as a disinfectant.

Color applications

Color is not usually a factor in wastewater analysis. In the analysis of drinking water only the true color created by organic acids caused by the decomposition of plants in water is usually measured. The resulting color can be regarded as the amount of humic substances in water.

10.2.1.4 Taste and odor

Potable water is odorless, colorless, and tasteless. Taste and Odor are often related to each other and are often confused. The substances that smell in the water almost always produce taste too, but the opposite is not true because there are many minerals that taste but do not smell at all, like non-volatile agents such as sodium chloride that have no effect on the taste of the water.

Taste and odor sources

Many substances that contact with water in nature or may be added by human uses give water a distinctive taste and smell. These substances, include minerals, metals, soil-derived dyes, end-products of biological reactions, and wastewater components. Inorganic compounds in most cases produce odorless flavors. Alkalis give a bitter taste to water, while metal salts can give a salty or bitter taste. Organic matter, on the other hand, tends to produce both taste and odor properties. These organic substances that taste and smell in water are products which produced in the petroleum industry. Biodegradation of organic matters can also result in the creation of liquids and gases that add taste and odor to the water. In all of the above, the most important products that produce a taste and odor in the water are sulfuric compounds that produce odor and taste like rotten eggs. In addition, some species of algae release an oily substance that has a smell and taste. Sometimes a combination of two or more substances, neither of which can create a taste or smell alone, can cause taste and odor problems.

Taste and odor effects

For consumers taste and smell are unpleasant for obvious reasons. Taste and odor are associated with pollution, and, thus, water consumers prefer tasteless and odorless water. Some organic substances that produce taste and odor may also be carcinogenic.

Taste and odor applications

Taste and odor characteristics are regulated in drinking water. The USEPA has not set a maximum standard for taste and odor threshold [Threshold Odor Number (TON)], but a maximum level of an odor threshold equal to three TON has been recommended by the World Health Organization.

10.2.1.5 Temperature

Temperature is not used in the direct assessment of drinking water or sewage; it is still one of the most important characteristics of water in nature. Surface water temperature largely control the biological species present in the water and the intensity of their activity, and affects many of the chemical reactions that occur in natural water systems. Temperature also has a significant effect on the solubility of gases in water.

Temperature sources

The temperature of natural water systems is affected by many factors including atmospheric temperature. In general, surface water reserves are more affected by temperatures than deep groundwater. The use of water as a coolant in the industry and the discharge of hot water can lead to drastic changes. Harvesting the bottom leaves of forest trees and returning irrigation streams in turn can increase the temperature of natural water streams.

Temperature effects

Relatively cool waters usually support a wide variety of biological species. At lower temperatures, biological activity, such as the utilization of food resources, growth, reproduction, etc., slows down. If water temperature rises the biological activity increases, and if adequate nutrients are available, a 5° C increase is usually sufficient to double the biological activity. The rapid growth of algae is usually seen in warm waters. The natural secretion of oils by algae and products from the decomposition of dead algae cells can also cause problems with taste and odor, and deplete dissolved oxygen causing a condition called eutrophication. Larger species of aquatic organisms such as fish are significantly affected by water temperature and the amount of dissolved oxygen in the water, while the amount of dissolved oxygen in the water is a function of temperature and its chemical composition. Most chemical reactions, such as dissolving of solids, accelerate as temperatures rise. On the other hand, the solubility of gases decreases with increasing temperature. Sufficient dissolved oxygen is necessary to maintain aerobic aquatic organisms. Temperature also affects other physical properties of water. For example, with decreasing temperature, the water viscosity increases, and water density varies with its temperature, being largest at about 4° C in freshwater. This phenomenon is unique among all liquids. Temperature and density affect planktonic microorganisms in natural water systems.

10.2.2 Chemical parameters of water quality

Water is recognized as a general solvent and chemical agents are related to its water solubility capabilities. The total amount of soluble solids, alkalinity, hardness, fluoride, metals, organic matter and nutrients are key chemical water quality characteristics.

10.2.2.1 Total dissolved solids (TDS)

Substances that remain in the water after filtration are considered soluble materials.

Dissolved solids sources

Dissolved solids arise from the solubility of water on solids, liquids and gases. As suspensions soluble materials may be in forms of an organic, inorganic or mineral material. Dissolved inorganic materials in water, include minerals, metals and gases. Also, the decomposition of plant material forms dissolved organic chemicals and gases.

Dissolved solids effects

Many Dissolved Solids in water are undesirable. Water-soluble minerals, gases, and organic matter may be aesthetically non-pleasing in color, taste, and odor of water. Some chemicals may be toxic and some soluble organic components are carcinogenic. Sometimes two or more two soluble substances (especially organic matter and halogen members) may combine to form substances whose properties are more dangerous than any of the raw materials. However, not all water-soluble materials are undesirable. For example, distilled water, which is basically pure, is tasteless. Acidic water is corrosive and rapidly dissolves the substances it contacts.

Dissolved solids applications

Since there is no distinction between its constituent components the TDS gives a generic measurement of solute content. Most often, however, knowing the composition of the water-soluble constituents intended for drinking, agricultural, and industrial use may be necessary.

10.2.2.2 Alkalinity

The concentration of ions in the water that neutralize the hydrogen ion is known as alkalinity. The most well-known alkalinity components are bicarbonate, carbonate, and hydroxide, respectively. These compounds are obtained from the decomposition of minerals in the soil or the atmosphere. In addition to the mineral origin of the said components these materials can be obtained from the dissolution of carbon dioxide and from the microbial decomposition of organic matter.

Alkalinity effects

High alkalinity gives a bitter taste to water. At the same time, however, the major concern about water alkalinity is related to the reactions that may occur between alkalinity and certain cations in water. The resulting sediment can lead to clogging of pipes and other water supply network accessories.

Alkalinity applications

Alkalinity measurements are usually carried out in the analysis of natural waters to determine their buffering capacity. Such measurements are also frequently used as a process control variable in water and sewage treatment. The maximum alkalinity values for drinking water or sewage effluents have not yet been formulated by the USEPA.

10.2.2.3 Hardness

Hardness is defined as the concentration of polyvalent metal cations in solution. In the supersaturated state, the hard cations react with the anions in the water to form a solid precipitate. Hardness is categorized as carbonate hardness and non-carbonate hardness (depending on the anion that it contains). A hardness equivalent to alkalinity is known as carbon hardness, and in this case any remaining hardness will be called non-carbon hardness. Carbonate hardness is sensitive to heat and precipitates rapidly at high temperatures.

Hardness sources

Multivalent metal ions that are more abundant than other ions found in natural waters, include calcium and magnesium. Other ions may include Fe and Mn in reduced forms (Mn^{2+}, Fe^{2+}), strontium (Sr^{2+}) and aluminum (Al^{3+}). Al^{3+} ions are usually much lower than calcium and magnesium, and in practice can be roughly equal to the sum of calcium and magnesium ions.

Hardness effects

Soap consumption in combination with hard water indicates an economic disadvantage for the water user. Sodium soaps react with polyvalent metal cations and give rise to a yellow precipitate. The foaming process does not occur until all the hard ions have precipitated. In this case, the so-called water is said to be softened by soap. Precipitates formed by the hardness and function of the soap adhere to the surfaces of toilets, milk, and dishwashers and may cause stains on clothing, dishes and other objects. Sediment residues resulting from the hardness of soap and water may remain in the openings on the skin and cause itching and tenderness. In recent years, these problems have been largely solved by the development of soaps and detergents that do not react with hard water. A layer of hardness in boilers, heaters, and hot water pipes which results from the hardening of carbonate can cause mass clogging. Changes in pH in water distribution systems can also lead to sedimentation in such systems. Bicarbonates begin to convert to less soluble carbonates at pH values above 9. Magnesium hardness, especially with sulfate ions, can cause diarrhea in people who are not used to it, as well as causing kidney stones in susceptible people. Concentrations of magnesium less than 50 mg/l are favorable for less potable water. In fact, hard water appears to be beneficial to the human circulatory system, where areas with light drinking water have a significantly higher number of patients with cardiovascular disease.

Hardness applications

Hardness decomposition is usually performed on natural waters and waters intended for drinking and special industrial processes. In practice, hardness can vary from zero to hundreds or even thousand parts per million. Although, the acceptable level of hardness, varies according to consumer statements, but a general classification which is acceptable in accordance to Public health standards, limit the maximum hardness content to 500 mg/l in drinking water, but the maximum value is still not stated by USEPA standards.

Water softening

The process of separating solutes (especially calcium and magnesium) from water, is called lightening or softening. The main and most common ways to soften water are:

1. with lime water and sodium carbonate ash.
2. with ion exchangers

Other ways to soften water include electrodialysis, distillation, freezing, and reverse osmosis.

10.2.2.4 Fluoride

Fluoride is generally found in nature and in some types of sedimentary or igneous rocks. It may be naturally in water or may be added to the water during a controlled treatment. Large concentration of fluoride is detrimental to human health and animals, while low concentrations can be beneficial, and of about 1 mg/l in drinking water help prevents tooth decay in humans. Fluoride is usually added to drinking water to compensate for a deficiency if it is not sufficient to aid in dental health. High fluoride discolors teeth. Large concentration of fluoride can cause bone fluorescence and other skeletal abnormalities, but concentrations less than 5 mg/l in drinking water usually do not lead to or cause bone fluorescence.

10.2.2.5 Metals

All metals are partially soluble in water. However, large amounts of each metal can endanger human health, and some are considered toxic and others are considered non-toxic. Metals are produced in natural waters by mineral dissolution; they may pollute water by their presence in municipal, industrial, and agricultural wastewaters. The amount of metals in the water is usually measured by atomic photometry.

Non-toxic metals

In addition to the ions which cause hardness in water (calcium and magnesium), there are naturally other non-toxic metals, namely sodium, iron, manganese, aluminum, copper and zinc. Sodium is a metal that is more abundant than other metals in natural waters. It is abundant in the earth's crust and has a strong tendency to combine with other elements. Sodium rocks are soluble in water. High concentrations of sodium salts in water can cause a bitter taste in water, in addition to being harmful to heart and kidney patients. Sodium, in addition to the above, corrodes metal surfaces at high concentrations and is toxic to plants at high concentrations. Iron and manganese are in many cases interconnected and do not pose known threats to human health. Of course, this is true for ordinary concentrations of these two metals in water. Iron and manganese can also cause water color problems to a very small extent. However, iron concentrations of about 0.3 mg/l and manganese concentrations of 0.05 mg/l can cause water color problems. In addition, some bacteria use iron and manganese as an energy source, and sludge produced by the growth of such bacteria can cause unsuitable taste and odor. Significant amounts of iron are only found in water that contain little oxygen, such as groundwater or the lower surface of lakes with thermal layers. Other non-toxic metals are usually found in very small amounts in natural water systems. Many of these metals cause an undesirable taste in water before their concentration reaches a toxic level. However, copper and zinc exacerbate each other, and can be toxic to many organisms even at low concentration.

Toxic metals

Toxic metals are harmful to humans and other organisms even at low concentration. Water-soluble toxic metals include arsenic, cadmium, lead, mercury, barium, chromium and silver. Some, such as arsenic, cadmium, lead and mercury, are particularly hazardous. These metals are condensed by the food chain, thereby posing the greatest danger to organisms near the top of the food chain.

Fortunately, toxic metals are present only in very small amounts in most natural water. Although there are natural sources in which toxic metals are found, significant concentrations in water are often caused by the discharge of mineral, industrial, and agricultural effluents.

10.2.2.6 Organic matter

Many organic compounds are soluble in water. Such compounds in natural water systems may come from natural sources or from human activities. Most natural organic compounds are composed of the decomposition of organic solids, while synthetic organic materials are usually the result of sewage disposal or agricultural operations or industrial waste. Water-soluble organic matter is usually divided into two groups: Biodegradable and Non-Biodegradable.

Non-biodegradable organic matter

Some organic compounds are resistant to biodegradation. Tannic and clinic acids, cellulose and phenols are commonly found in natural water systems. These constituents of the woody plants decompose so slowly that they are usually called non-decomposable. Molecules that have extremely strong bonds (such as some polysaccharides) and molecules with a ring structure (such as benzene) are essentially non-decomposable. Many of the organic compounds present in petroleum that are also found in refining processes contain benzene and are essentially non-biodegradable. Some organic compounds are insoluble due to their toxic properties to microorganisms. These include organic insecticides, some types of industrial chemicals and hydrocarbon compounds that are combined with chlorine. Pesticides, which include insecticides and herbicides, are widely used in today's advanced society for urban living. Improper and subsequent use and heavy rains and floods can cause contamination of natural water streams. Many insecticides are toxic accumulating substances and because of this property cause acute problems at the bottom of the food chain. For example, it is a species of extinct brown pelican, which feeds on fish and other large aquatic species contaminated with insecticides.

Biodegradable organic matter

The degradable material consists of those organic substances that can be used as a food source by natural microorganisms over an acceptable period of time. Soluble matters include starches, fats, proteins, alcohols, acids, aldehydes and esters. Such matters may be end products of primary microbial degradation of plant or animal tissue, and may also be caused by the discharge of municipal or industrial wastewater. Although some of these matters can cause problems with color, taste and odor, the main problem is the secondary effect of microorganisms on these matters. Consumption of soluble organic matter by microorganisms may be accompanied by oxidation (i.e., increased oxygen or removal of hydrogen from organic matter molecule) or reduction (increase of hydrogen or removal of oxygen from the structure of the organic molecule). Although it is possible for both processes to occur simultaneously the oxidation process is more effective and superior to the other processes if oxygen is present. In aerobic environments where oxygen is present the final products of microbial degradation of organic matter are stable and do not harm human health. The anaerobic decomposition that occurs in the absence of oxygen will lead to unstable and controversial products. If oxygen is replenished in water anaerobic end products will be oxidized to aerobic products. The need for oxygen to decompose organic compounds in natural water systems is of utmost importance. When the rate of oxygen consumption exceeds the rate of dissolution of oxygen in the atmosphere in the water anaerobic conditions mayseverely affect aquatic ecosystems.

10.2.2.7 Nutrients

Nutrients are the elements that are essential for the growth and reproduction of plants and animals. Aquatic species are also dependent on the surrounding water for food supply. A wide range of minerals and trace elements can be classified as nutrients, the substances most commonly consumed by aquatic species are carbon, nitrogen, and phosphorus. Carbon dioxide comes from the atmosphere, alkalis, and products from the decomposition of organic matter. In many cases nitrogen and phosphorus are nutrients which are a limiting factor in the growth of aquatic plants.

Nitrogen

Nitrogen gas (N_2) is a predominant component of Earth's atmosphere and is an extremely stable gas. Under extremely high energy conditions this gas reacts with oxygen (electric discharge or flame burn) and as a result of this reaction nitrogen oxides are produced. Other sources of nitrogen in aquatic systems include fecal matters, chemicals (especially chemicals used as fertilizers), and organic effluents. Nitrogen from these sources may penetrate directly into natural streams of water or waterways. Nitrogen compounds can be oxidized and converted to nitrate by the bacteria in the soil. These compounds can also penetrate underground aquifers. Inside the aquifers, nitrates may move rapidly. Groundwater contamination by nitrogen from animal waste has been reported in many cases. Nitrate poisoning in young mammalian species can also cause serious problems and even death. Nitrite, on the other hand, tends to bond with hemoglobin more than oxygen, so it easily replaces oxygen in the blood complex, thereby depriving the body of essential oxygen for life and, in acute cases, the victim suffocates. A severe oxygen deficiency causes the body to stain blue, therefore, nitrate poisoning is termed the so-called "blue baby syndrome".

Fortunately, however, natural nitrite oxidation to nitrate occurs so rapidly that no significant amounts of nitrite are found in natural water sources.

Phosphorus

Phosphorus is found exclusively in phosphate (PO_4^{2-}) in aquatic environments. There are several forms of phosphate, including orthophosphate, dense phosphate (pyrrole, meta, polyphosphate) and organic phosphate compounds. These compounds may be soluble or solid. They can even be a component of plant or animal tissue. Phosphate is a component of soil types and is widely used in fertilizers to replace or supplement natural values in arable lands. Phosphate as part of animal waste may also be found in the soil where the waste is added to or burned. Phosphate also enters the surface waters as a result of washing crops. The tendency of the soil to absorb phosphate will block its movement in soil moisture and groundwater, but it will enter surface water through soil erosion. Urban wastewater is another major source of phosphate entry into surface waters. Polymeric phosphates are widely used in detergent construction, and organic phosphates are components of body waste and food residues. Other sources include industrial effluents, in which phosphate compounds are used for purposes such as boiler water treatment. Although phosphates are not toxic and do not pose any direct threat to human health or other organisms, they are a serious and indirect threat to water quality. As mentioned, phosphate is usually a limiting nutrient in surface water and phosphate can also interfere with water treatment processes. For example, phosphate at concentrations of about 0.2 mg/L is inhibits turbidity coagulation.

10.2.3 Biological parameters of water quality

Water is an environment in which numerous biological species coexist. Aquatic organisms vary in size and complexity from the smallest single-celled microorganisms to the largest fish. All members of the biological community are partially influenced by water quality because their presence or absence reflects the quality of water source characteristics. For example, the general quality of water in the streams in which the salmon live must be excellent: well oxygenated, of cool temperature, and free of organic and inorganic pollution. Biologists typically use an index of species richness (which is related to the number of species and the relative abundance of organisms in each species) as a qualitative factor for streams and lakes. A water stream that encompasses a large number of species with a balanced number of species is considered a healthy ecosystem. Some organisms can be used as indicators of the presence of pollutants.

Several parameters related to the chemical properties of water, namely, Dissolved Oxygen (DO), Biological Oxygen Demand (BOD), and Chemical

Oxygen Demand (COD) are reviewed prior to presenting a summary of the biologic properties of water.

10.2.3.1 Dissolved oxygen (DO)

DO is the amount of oxygen in aquatic environments that is accessible to fish, invertebrates, and all organism in the water. Most aquatic plants and animals require oxygen to survive; fish, for instance, cannot survive for long in water with dissolved oxygen less than 5 mg/L. The low level of dissolved oxygen in water is a sign of contamination and is an important factor in determining water quality, pollution control and treatment process. The DO in a saturated solution varies with the water temperature and elevation. For example, cold water has a higher DO than warm water. At sea level and at 20° C the DO value is 9.1 mg/L in freshwater. At a constant temperature, the higher the elevation, the lower the DO. The introduction of organic waste, especially domestic and animal sewage, industrial waste from the activities of paper mills, leather manufacturing, slaughterhouse sewage and crop wastewater, dramatically reduce the DO in water. The wastes in these industries cause oxygen demand, and they are broken down and decomposed by bacteria into oxygen. Most oxygen-demanding waste is organic waste. The oxidation of 3 of mg/L of carbon requires 9 ppm of soluble Oxygen. Dissolved oxygen is measured with an oxygen measuring device (DO-meter).

10.2.3.2 Biological oxygen demand (BOD)

BOD is defined as the amount of molecular oxygen required for the biological oxidation of organic matter in water and, on average, the oxygen demand for organic waste material which are aerobically degradable. Biological oxygen demand testing is very useful in managing pollution control of streams and assessing their self-purification capacity. BOD is the amount of oxygen that aerobic bacteria need to break down biodegradable organic matter into end products such as water and CO_2. If the BOD is low, the water is has low content of organic matter and has low counts of microbial organisms. The size of biological treatment facilities, especially the amount of aeration in aeration ponds, can be calculated based on BOD. BOD tests provide a good estimate of the n quality of water from the perspective of its oxygen content.

10.2.3.3 Chemical oxygen demand (COD)

BOD_5 testing takes five days to complete and organisms may be chemically rather than biologically oxidized, the COD measurement is commonly used. It consists in the oxygen demanded exerted chemically by organic substances in water. For example, cellulose derived from the effluent of paper mills which is biologically oxidized at slow rates, is chemically oxidized at high rates.

10.2.3.4 Pathogenic microorganisms

In terms of human use and consumption pathogenic organisms in the water are pf major concern because of their disease-causing capacity. These microorganisms are either capable of transmitting diseases to humans or causing them to become infected. They are found in the aquatic system, and usually require a host animal for growth and reproduction. These microorganisms can, however, be transmitted by the natural water system and are therefore a members of the aquatic community. Many pathogenic species are able to survive in the water and maintain their pathogenic properties for a considerable period of time. These pathogenic aquatic organisms include bacteria, viruses, protozoa and, parasitic worms.

Bacteria

The word bacterium is derived from the Greek word bar and solid, a form that is characteristic of most bacteria. Bacteria, on the other hand, are single-celled microorganisms that are usually colorless and the lowest form of life capable of synthesizing protoplasm from the environment. In addition to the rod-shaped (bacilli) mentioned above, the bacteria may be spherical (cox) or spiral-shaped. Stomach and intestinal diseases are clear examples of most diseases transmitted by pathogenic aquatic bacteria. Cholera, for example, is a disease that has killed many people the world over since immemorial times. This disease is transmitted by a bacterium called *Vibrio cholera*. Cholera is one of the most serious diseases caused by aquatic bacteria that causes nausea and diarrhea and, if left untreated, the body becomes dehydrated and eventually dies. Typhoid, another disease caused by a pathogenic aquatic organism called *Salmonella*, includes diseases of the stomach and intestines, severe fever, intestinal bleeding, and possibly neurological disorders. Although immunization and disinfection of water sources have eradicated cholera and typhoid in many parts of the world, there are still areas in some developing and overpopulated countries with poor health and sanitation conditions which continue to be at high risk for these two diseases. Temporary negligence in health considerations can in some cases lead to intestinal swelling caused by the activity of pathogenic bacteria. There are other diseases that have bacterial causes.

Fecal coliform It is a form of bacteria found in human and animal excretory waste, which is the first source of water-borne diseases.

Viruses

Viruses are the smallest known biological constructs that have all the genetic information needed to reproduce themselves. The viruses are so small that they can only be seen with the help of electron microscopes. Viruses are parasites that rely on hosts to survive. Symptoms that are associated with viral infections of aquatic origin usually include nervous system disorders rather than gastrointestinal disorders. Pathogenic viruses of aquatic origin are believed to be the cause of polio and infectious liver infections. Immunization in advanced countries has greatly reduced the number of polio patients. The prevalence of hepatitis is more prevalent every year, with about 60,000 cases reported in the United States, compared to polio. Viral disinfection procedures are available; yet, it is difficult to determine their efficacy because due to the difficulty in their detection. The uncertainty of viral disinfection is a major problem for the direct reuse of treated sewage, which is one of the reasons for the increase in land disposal methods for sewage treatment.

Protozoa

The smallest and lowest forms of animal life can be seen in protozoa. Protozoa are single-celled organisms that are more complex in activity than bacteria and viruses. Protozoa are complete and self-sufficient organisms that can live free or parasitic lives and exhibit pathogenic or non-pathogenic properties at microscopic and macroscopic sizes. Only a small number of aquatic protozoa are pathogenic; they are highly adaptable to the environment and are widely distributed in natural waters. Infections caused by protozoa are usually slightly milder in the gastrointestinal disorders than the bacterial infections herein discussed. However, infections caused by protozoa can be serious, as the prevalence of such infection in Chicago in year 1993 proved. That year, more than 1,400 people became infected with protozoa, and 98 deaths were reported from drinking contaminated water. Many cases of giardiasis have been reported in recent years among people who have drank stream water. Under acute environmental conditions aquatic protozoa form resistant networks that are difficult to disable by disinfection. In such cases purification including filtration is necessary to eliminate these networks. Protozoan-caused diseases include Amoebiasis, Toxoplasmosis, and Giardiasis.

Parasitic worms

The life cycles of parasitic worms usually involve two or more humans or animals as hosts, and water contamination may result from humans or animal feces containing parasitic worms. Contamination can also be spread through aquatic species to other hosts such as snails and insects. While aquatic systems can be a means of transmitting diseases caused by parasitic worms' new water treatment methods are very effective in destroying these organisms. It is thus seen that parasitic worms primarily threaten people who are in direct contact with untreated water. Operators of sewage treatment plants, swimmers in lakes with polluted water, contact with stormwater and livestock sewage, and workers hired for irrigation operations on arable land are at risks of parasitic worms. Diseases caused by parasitic worms include ascariasis, hookworms, and guinea fever.

10.2.4 Water-borne diseases

These diseases are caused by pathogenic bacteria, viruses, protozoa, and worms in humans. Resources and reservoirs such as rivers, water canals, lakes and dams, especially in the tropics, are suitable sites for the spread of these diseases. These include cholera, typhoid, diarrhea, amoebic diarrhea, giardiasis, hepatitis, and ascariasis.

10.2.5 Pathogenicity index

Water analysis to measure the pathogenic hazard of water can be very time consuming and costly. Testing for certain species of pathogens, called indicator organisms, is usually performed to assess the pathogenic hazard in water. An indicator organism is such that its presence indicates the occurrence of contamination, and in addition, it makes clear the prediction of the nature and extent of other pathogens. The ideal pathogenicity index would have the following properties:

1. Can be used for all types of waters.
2. Always present in places where pathogens accumulate.
3. Are absent in places where pathogens are not present there.
4. It is possible to perform quantitative tests on it without interfering with other organisms.
5. Do not pose a health risk to laboratory workers who take necessary precautions.

Most waterborne pathogens get into the water through fecal matter. Thus, any organism that resides in the human intestinal tract and has the above-mentioned characteristics would be a good indicator organism. Organisms that possess many of these properties belong to the fecal coliform group.

10.2.6 Water quality index (WQI)

The water quality index was developed in the 1970s, and is used to monitor changes in water quality in specific water resources over a period of time to detect changes in aquatic ecosystems. This index measures the health of watershed resources at various points, and can be used to track water-quality changes over time. It is also be used to compare the quality of waters across regions. The WQI is a quantitative measure that converts the results of water tests into water quality factors, and is reported in numeric form. The WQI index value decreases with increasing water pollution.

10.2.7 Human health risk

It measures the harmful and potentially harmful effects on human health for those exposed to pollutants in the environment, including water.

10.2.8 Health

Health measures the biologic and mental functioning of a person.

10.2.9 Water pollution

The introduction of any foreign matter into the water that may disrupt the specific use of water or alter its physical, chemical, and biological properties, which may have adverse effects on human and aquatic life. Some of the water pollutants are oxygen-demanding wastes, pathogens, plant nutrients, synthetic organic compounds, petroleum, chemicals, minerals, minerals, radioactive compounds, and heat. Sometimes contaminated water contains several sources of contaminants at the same time.

10.2.10 Pesticides

Pesticides encompass insecticides, herbicides, fungicides, herbicides, rodenticides. Pesticides are introduced into natural waters in a variety of ways, including their direct use to control plants and insects in water due to water passing through agricultural lands and industrial wastewater. Pesticides accumulate in the body and may cause and chronic health problems.

10.2.11 Toxic elements

Elements such as arsenic, cyanide, and pesticides are toxic substances in water. The toxicity of an element in water depends on various factors. Most toxic substances enter the drinking water sources due to poor water treatment or poor control over the disposal of industrial wastewater. The various stages of water treatment may have minor effect reducing the concentration of these substances; therefore, the maximum permissible concentrations for these substances are usually considered in treated and untreated water.

10.2.12 Acidity

Acidic water is a type of water that has a pH of less than seven. Acidity in unpolluted water is usually caused by dissolved carbon dioxide in the water. Humic and fulvic acids and other organic acids resulting from the decomposition of plants are also a cause of water acidity.

10.2.13 Electrical conductivity

Electrical conductivity is a characteristic that measures the capacity of a water solution to conduct electricity; this capacity is a function of the presence of ions in a solution, which is easily measured with an electric conductivity meter, a reported in micro siemens per centimeter.

10.2.14 Corrosion

Corrosive water damages metals and concrete. Corrosive water occurs when one or more of the following occur:

1. Low pH or acidic water
2. High dissolved CO_2
3. Low hardness and alkalinity

Corrosive waters include marshy waters, low pH shallow well water with low hardness, saline waters, high CO, gypsum and limestone waters, green sand and coal waters, waters with high content of chlorine or residual free chlorine.

10.2.15 Water treatment

Water treatment is the process by which suspended solids, harmful chemicals, biological pollutants, and gases are removed from contaminated water to make it usable for humans, animals, agricultural use, and for return to the environment.

10.2.16 Urban pollution sources

Wastewater produced from household, commercial, office and industrial sources within an urban area are the main sources of urban pollutants; oxygen demand, fats, oils, and suspended solids are removed in primary and secondary sewage treatment processes.

10.2.17 Agricultural pollution sources

Food is an important limiting factor in plant growth. The rate and frequency of plant growth, along with other factors, also depend on the amount of nutrients available. The nutrients required by plants include carbon, nitrogen, phosphorus, potassium, and a number of trace metals that reach the plant through natural processes. The direct pollution caused by agricultural activities depends largely on the amount of agricultural activity. As such, intensive agriculture of non-organic type produces environmental pollution.

10.2.18 Industrial pollution sources

Industrial pollutants have different physical and chemical properties due to the variety of raw materials consumed and the variety of production processes. For example, the food industry's wastewater contains significant amounts of nitrogen and phosphorus; biological methods are effective in treating these wastewaters. On the other hand, industrial wastewater has high levels of oil and grease, especially in the petrochemical industry, which need to be pre-purified to reduce these pollutants.

10.2.19 Groundwater contamination

Pollution can enter aquifers in many ways. Contamination may occur by accident, such as oil or chemical spills due to road accidents; on the other hand, contamination may be intentional, such as burying harmful substances in the ground or injecting liquid waste into aqueducts or wells. Contamination can issue from the infiltration of fertilizers and other chemicals.

10.2.20 Water purification in natural systems

Self-purification mechanisms in natural water systems include physical, chemical, and biological processes. The speed and completeness of these processes depend on the many variables that are specific to each aquatic system. Hydraulic properties such as volume, velocity, and turbulence, physical properties of substrates, fluctuation of sunlight and temperature as well as chemical properties of water are all variables that affect the water self-purification process. In natural waters such processes are spontaneously occurring and are rarely modifiable.

10.2.21 Direct transmission of diseases by water

The water is contaminated by feces and other foreign bodies and is transmitted to the human body by ingesting pathogens. The most important diseases that are transmitted by ingestion include typhoid, cholera, infectious diarrhea, bacilli and amoeba, infectious jaundice and polio, and other viral diseases transmitted through the gastrointestinal tract such as hepatitis.

10.2.22 Indirect transmission of diseases by water

In this instance water creates a favorable environment for the growth and propagation of the vectors of disease, and does not directly play a role in the transmission of the disease. Vector-transmitted diseases are malaria, dengue, filariasis, yellow fever, and a variety of encephalitis.

10.2.23 Primary pollutants

These are pollutants of concern to public health and consumer health.

10.2.24 Secondary pollutants

Contaminants that have adverse effects on the appearance and appeal of water.

10.3 Basics and logic

Water is one of the most abundant substances found in nature and covers almost three-quarters of the Earth's surface. In spite of this abundance various factors such as lack of proper management, population growth limits the amount of water available for human consumption. Water is always moving in the hydrological cycle, atmospheric water returns to earth after condensation as rain, snow, and ice. Water flows into rivers, lakes, and eventually into the oceans, or joins underground water bodies as a result of infiltration into the soil, eventually entering surface water and evaporating from surface water sources or evaporation from plants. Water molecules return to the atmosphere and the hydrological cycle repeats itself endlessly. In nature, water in the vapor state reaches its highest purity, and since the condensation phenomenon usually requires a surface or core, it is likely that water will capture airborne matter at the moment of condensation. Further impurities due to the movement of water in other parts of the hydrological cycle may be added to liquid water, especially when water is in contact with airborne substances or when water flows on or below the ground. Water is a natural solvent that can dissolve substance that comes into contact with it. Human activities, including industrial production, generation of household waste, release of industrial chemicals, agriculture, contribute to water pollution. Eventually, these impure waters evaporate and return to the atmosphere as relatively pure water molecules. However, water quality is of the utmost importance as it moves through lakes, rivers, and aquifers, since it is at this stage that human use of water occurs. Early humans measured water quality only through physical senses such as vision, taste, and smell, and therefore, until biological, chemical, and physical measurement devices were developed no methods of determining water quality and of knowing its effects on human health were available. In the mid-19th century the connection between human sewage, drinking water, and disease was discovered. In 1856 Dr John Snow, a public health expert, discovered the link between cholera and water use from wells. According to the researcher's findings cholera infestation occurred in the vicinity of the well, and in other nearby areas. At the end of the 19th century and at the beginning of the 20th century advances in the understanding of pathogens were made by Pasteur and other researchers, and the field of microbiology became accepted at about the same time that water chemistry emerged. Many chemicals used in industrial and agricultural processes were identified in water. However, there were many chemical compounds that were previously found in very small quantities in many water sources, and whose effect on human health has become understood in recent years.

10.4 Importance and necessity

Water in nature is never pure and always contains soluble substances. Although more than three-quarters of the planet's water is contaminated, a small proportion of the available water can be used for health and agricultural purposes. With the advancement of science and technology it is no longer concealed that one of the most important causes of the transmission of diseases from the environment to humans is water, which, if contaminated, can pose a serious threat to human health. Growing consumption in all areas of consumption, including drinking, industry, services, and agriculture, is having the consequences of quality change and degradation, with many annually dying from unhealthy drinking water. With the growth of population, the provision of safe and healthy water is one of the basic needs of mankind. But the various pollutants from urban, agricultural, and industrial activities enter the environment and areas near water supplies, indicating the necessity to prevent contamination of water resources and identify pollutant sources. Due to population growth, consumption growth, spreading of pollutants and irregular uptake of water resources, drinking water quality is declining with simultaneous increase of deleterious compounds in drinking water. Improvement of public sanitation is one of the fundamental concerns of the international community. Water quality in agriculture has also been taken into consideration by planners and water management officials in order to prevent the contamination of agricultural products and their transmission to humans through the food chain. It should be noted that in arid and semi-arid regions it is important to evaluate the quality of water for drinking and agriculture. Nowadays, access to water of suitable quality and in sufficient quantity is an obstacle to sustainable development in many regions. Large investment in water resources development and related projects such as dam construction, irrigation and drainage, watershed management, flood control, and groundwater recharge are devoted to mainly to quantify water quantity, while the study of water quality is often neglected. The main purpose of qualitative analyzes of drinking water is to preserve public health and to protect consumer health. Lack of water resources of suitable quality is a key concern in arid and semi-arid regions. Safe guarding of human health is a central pillar of sustainable development, and water resources must always be evaluated in terms of quality.

10.5 Materials and methods

10.5.1 Suspended solids measurement

There are several experiments to measure solids content. Most of these tests are based on gravimetry and involve measurement of sediment mass. The total solids test determines the amount of all solids suspended or dissolved in water, organic or inorganic. This parameter is determined by sample evaporation to the drying stage and the residual mass measurement. The total amount of residual mass is expressed in milligrams per liter (mg/l) based on the solids mass. In this process a drying temperature that is slightly above the boiling point (104° C) is required to remove the liquid as well as the water absorbed at the surface of the solid particles. However, for evaporation of water trapped between particles, a temperature of about 180° C is required. Most solids can be separated from the water by filtration. Therefore, the suspended solids in a water sample can be estimated by filtering water, drying the residual mass, and filtering to a constant weight at 104° C and determining the residual mass on the filter. The results of this test are expressed in terms of dry mass in volume (mg/l). The number of dissolved solids passing through the filter is also expressed in milligrams per liter, which is the difference between the total solids in the water and the solids suspended in a water sample. After drying and measuring the samples the organic matter content of the suspended solids and total solids is obtained by burning the residual mass at 600° C for one hour. The remaining organic matter is converted to carbon dioxide, water vapor, and other gases and released to the environment. The rest of the resulting matter is composed of minerals.

10.5.2 Turbidity measurement

Turbidity is calculated photochemically by measuring the scattered or absorbed light at a given intensity, which is the basis of the standard method of determining the water turbidity using the Jackson turbidimeters. The basis of this device is based on absorbing light using a standard tube and standard candle. The candle is placed under a glass tube and then placed in a black metal layer so that the light from the candle can only be seen from the top of the device. The water sample is then slowly poured into the tube and this is continued until the candle is no longer visible, so that complete absorption is achieved. The glass tube is then calibrated using data obtained for the turbidity of the silicon oxide (SiO_2) particles. One Jackson Turbidity Unit (JTU) is equivalent to the turbidity created by one milligram of SiO_2 in one liter of distilled water. This measurement method is suitable for opacities above 25 units and is currently historically important. Purified water turbidity should be less than one; therefore, other methods are required to measure low turbidity, using nephelometry and absorptiometry methods. The nephelometer measures the intensity of light scattering in a particular circle and at approximately the right-hand angle of the light's surface. This allows the turbidity of less than 0.05 units to be measured and the color of the water to be unaffected. But in the absorptiometry method, the amount of light absorbed is measured by the particles in the water. The sensitivity of the adsorption method to the very low turbidity is not as great as the nephelometric method and the color of the water is effective. The primary standard used to regulate turbidity is Formazin, which can match Jackson units to those of other methods used to measure low turbidity. The Formazin turbidity unit (FTU) and the nephelometer turbidity unit (NTU) can be converted to each other When Formazin is used to calibrate the nephelometer device.

10.5.3 Color measurement

There are several methods for measuring color; methods of comparison with standard color materials are used to the maximum extent possible. Color comparison tubes contain a set of standards that can be used to directly compare water samples that have previously been filtered to lose their apparent color. The results are expressed in True Color Unit (TCU), in which, one unit is equivalent to the color produced by 1 mg/l of platinum in the form of chloroplatinate ions. Specific spectrophotometric techniques are usually used for colors other than brownish yellow, and especially for colored waters resulting from industrial effluent outflows. In work environments, devices equipped with color glass disks that are calibrated to color standards are used. Specimens are tested at most 72 h after collection, because of the biological and physical changes which may affect the color storage during storage.

10.5.4 Taste and odor measurement

Direct measurement of substances that give rise to taste and odor is possible only if their constituents are known. By doing some analysis, sufficient flavoring agents can be measured. Measurement of taste and odor of organic matter as well as oil contamination can be done by gas or liquid chromatography. Chromatographic decomposition takes a long time and requires expensive equipment to complete, and therefore such tests are not typically performed on water samples. However, in the case of suspended organic matter, such tests are necessary. Quantitative experiments using human taste and smell properties can be used for this purpose. The Threshold Odor Number (TON) is calculated by pouring variable amounts of odorant water into containers and diluted to make a mixture by volume of 200 mg with sufficient distilled water. An installed device that has 5 to 10 noses is used to determine the mixture whose odor is detected, and the TON of the test sample is calculated using Eq. (10.1):

$$TON = \frac{A + B}{A} \tag{10.1}$$

where A = the volume of saline water in milliliters and B = the volume of odorless water needed to prepare a mixture of 200 ml.

10.5.5 Dissolved solids measurement

Direct measurement of the total solid dissolved solids by evaporation until drying is possible after the water is filtered to separate suspended solids. The residual sediment is weighted and represents the total amount of dissolved solids. TDS is expressed in milligrams of dry mass per liter of water. Organic and inorganic components can also be determined by burning the sediment at 600° C. Approximate TDS can also be obtained by determining the electrical conductivity of water, because the capacity of water to conduct electricity, which is called specific conductivity, is a function of the ionic strength of water. Specific conductivity is measured by the conductivity-meter device. The standard method to determine the electrical conductivity is to measure the conductivity in a 1cm field at 25° C and to express the results in milli-siemens (mS/m). The specific conductivity and TDS are not in general related to each other, however, because only ionized materials create special conductivity. Organic molecules and compounds that dissolve without ionization are TDS that cannot be measured with electrical conductivity measurements. In addition, the amount of specific conductivity varies with the ions present in the solution, their mobility, and their relative number. Temperature also has an effect on specific conductivity, thereby increasing the specific conductivity as the temperature increases. mS/cm are converted to mg/L by multiplying the former by a coefficient that varies between 0.055 and 0.09. To use specific conductivity as a quantitative test adequate analysis of the filtrable sediment must be performed to determine the conversion factor. Specific conductivity is therefore often used in quality assurance to control TDS changes that occur in natural streams or during water treatment processes.

10.5.6 Alkalinity measurement

Alkalinity is measured by titrating water with an acid and determining the hydrogen equivalent. Alkalinity is expressed in milligrams per liter of $CaCO_3$. If H_2SO_4 acid is slowly added to water and the pH is measured and reported for each increase of the acid a titration curve is obtained. The turning points of this curve, occurring at approximately pH values of 4.5 and 8.3, have special importance as the conversion of carbonate to bicarbonate is essentially completed at pH 8.3. However, since bicarbonate is an alkaline species, acid should also be added to it to achieve complete neutralization. Therefore, carbonate neutralization at pH = 8.3 is only partially accomplished. The conversion of hydroxide to water is accomplished at pH = 8.3. All of the hydroxide and half of the carbonate are measured at pH = 8.3. At pH = 4.5, all bicarbonate is converted to carbonic acid, including bicarbonate obtained from the reaction of acid and carbonate. Thus, it is found that the amount of acid needed to titrate a sample to pH = 4.5 equals the total alkalinity of water.

10.5.7 Hardness measurement

The hardness can be measured by spectrophotometric or chemical titration techniques used to determine the amounts of calcium and magnesium ions in a given sample. The hardness can be measured directly by titration with Ethylene Diamine Tetra Acetic Acid (EDTA) with the black reagent Eriochrome Black T (EBT). EBT reacts with divalent metal cations to form a red complex. EDTA substitutes EBT in the complex and when the substitution is complete, the soluble color changes from red to blue. Each milliliter of titrant solution measures 1 mg hardness as $CaCO_3$ if 0.01 molar EDTA is used.

10.5.8 Metals measurement

The atom is made up of protons, neutrons, and electrons orbiting it. Electrons move in specific energy circuits. Electrons in metal atoms, due to the lower nucleus charge, can temporarily move to higher energy levels with little energy. Of course, this state is unstable, and the atom tends to return to its initial state soon after losing extra energy. This property is used in an atomic absorption device. The larger the amount of metal atoms absorbed, the greater the amount of radiation absorbed. By measuring the amount of absorbed radiation the concentration of the metal ion in the solution is determined. In other words, the electrons in the capacitance layer of the atoms with the electron mutations they make absorb or emit electromagnetic radiation in the ultraviolet and visible range, and, therefore, when the energy is absorbed or emitted by the atoms separate spectral lines are observed that are the basis of the spectroscopic methods. Atomic. Employing the atomic spectrum (emission, absorption, or fluorescence) the sample is placed in a cell and its absorption, emission, or fluorescence are measured at a specific wavelength and measured according to the Beer Lambert law. In this method the samples must be in solution. In the first step of the experiment the solution containing the element is vaporized at 2° Celsius by a flame which burns with air and acetone. As the steam evaporates most of the elements in the solution become neutral. After steaming the neutral atoms are absorbed by the cathode lamp (the lamp for each element). Finally, the concentration of the unknown element in the solution is determined by calculating the amount of radiation absorbed by the detector and by calibration curves.

10.5.9 Non-biodegradable organic matter measurement

Determination of non-biodegradable organic matter is usually accomplished by Chemical Oxygen Demand (COD) testing. Non-degradable organic matter is estimated by analyzing Total Organic Carbon (TOC). The COD and TOC parameters measure the degradable component of organic matter. The specific organic compounds are quantitatively and qualitatively identifiable by quantitative decomposition and gas chromatography techniques.

10.5.10 Nitrogen measurement

Tests for various types of nitrogen in water repeatedly include ammonia analysis (both ammonia and ammonium), nitrate, and organic nitrogen. The results of the analysis are usually expressed as mg/L of any specific nitrogen content. Experiments on ammonia and organic nitrogen are usually common in wastewater and other contaminated water, but in clean water samples and treated wastewater nitrate testing is common.

10.5.11 Phosphates measurement

Phosphates are usually measured by colorimetry. Orthophosphates can be measured directly, whereas polymeric phosphates must be converted to orthophosphate by acid hydrolysis, and organic phosphates must be converted to orthophosphate by acid digestion. It should be noted that the results are reported in milligrams per liter of phosphate in terms of phosphorus. The important point here is that careful preservation of the specimens is essential before decomposition. For example, acid-washed glass bottles should be used for sampling since the washed bottles can contaminate the samples with phosphate-containing detergents.

10.5.12 Dissolved oxygen (DO) measurement

Dissolved oxygen of water is measured by an oxygen measuring device (DO-Meter).

$$C + O_2 \rightarrow CO_2$$

In the above reaction for every 12 gs of carbon, 32 gs of oxygen are consumed.

10.5.13 BOD measurement

BOD measures the amount of oxygen consumed by in a sample enclosed in a sealed container kept for a specified period of time in a controlled environment.

In a standard test a 300 ml of microbially cultured water at 20° C for 5 days is used. Algal growth, which results in oxygen in the bottle, is prevented by not exposing the test water to light. The oxygen saturation in water at 20° C is approximately 9 mg/L, thus, the test water is diluted with BOD-free water and saturated water to measure BOD values that are only slightly above a few milligrams per liter. The BOD value of a sample diluted is calculated with Eq. (10.2):

$$BOD = \frac{DO_i - DO_f}{P} \tag{10.2}$$

where, DO_i and DO_f are the initial and final concentrations of dissolved oxygen in milligrams per liter, respectively, and P = percent of the original sample in 300 ml bottles.

10.5.14 COD measurement

Potassium dichromate is used as a oxidizing agent to measure COD. The test lasts for three hours but cannot measure the amount of organic matter that is inherently or biologically oxidized. It should be noted that inorganic substances such as Cl can affect the results. If such substances are in the samples they must be removed prior to COD testing.

10.5.15 Coliform bacteria measurement

There are relatively simple experiments to determine the presence of coliform bacteria in water and to count them. Experiments used to determine whole coliform organisms have different culture media and temperatures lower than those used to identify common fecal coliform. The membrane filter method, which is a very favorable method for environmental engineers, enables direct counting of coliform bacteria. According to the latter method a small sample of the sample is passed through the membrane. The diameter of the membrane apertures is less than 0.45 μm. Bacteria remain on the filter and are then placed in specific environments for the growth and formation of bacterial clusters along with the inhibition of growth of other species. Membranes that are cultured for 6 h at the appropriate temperature allow the coliform bacteria to become visible colonies. After that, the bacteria are counted. Results are reported in terms of number of organisms in 1 ml of water, or as the most probable number per 100 mL (MPN/100 mL). Another method commonly used by microbiologists is the multi-tube fermentation test. Coliforms are known as lactose fermentation. As a result, one of the final products is gas. A culture medium containing lactose and other substances that inhibit non-coliform organisms is placed in a number of test tubes. These test tubes are then cultured with decimal components of 100 ml (100, 10, 0.1, 0.01, etc.) of water sample. These tubes are cultured at the appropriate temperature and examined for gas. The first step of this procedure is called probability testing. This test assumes that the gas-producing pipes have coliform organisms. A similar test known as a confirmed test is then performed to confirm the presence of coliforms. The presence of coliform organisms in water indicates that some water components have recently been linked to soil or plant debris or that have entered the water from the intestines of warm animals, thus assuming that pathogens are associated with coliform bacteria.

10.5.16 WQI calculation

Nine parameters are needed to determine the WQI index: BOD, DO, fecal coliform, nitrate, pH, temperature, TDS, phosphate and turbidity. The value of each of the nine parameters is calculated by measurement; each of the nine parameters has its own weight factor depending on its importance in water quality. The WQI is calculated by Eqs. (10.3) to (10.6):

$$WQI = \sum Q_i W_i / \sum W_i \tag{10.3}$$

$$Q_i = 100\left[\frac{V_i - V_o}{S_i - V_o}\right] \tag{10.4}$$

$$W_i = K/S_i \tag{10.5}$$

$$K = 1/\sum (1/S_i) \tag{10.6}$$

where, Q_i = degree of each parameter's quality, V_i = Estimated value of each parameter, V_o = Ideal value of each parameter in pure water taken for pH 7, for DO equal to 14.6, and for all other parameters equal to zero, S_i = Standard value, is the recommended value for each parameter, and K = constant is calculated according to Eq. (10.6).

10.5.17 pH measurement

A pH-meter is used to measure this parameter. The pH is measured at standard temperature of 25° C. For this purpose, temperature is measured in each experiment and temperature correction is performed according to the temperature table. The pH range must first be known to the meter. Buffers of known pH are used. If the sample is alkaline, buffers of 7 and 9 (for very basic samples up to 10) are used. Also, two buffers with pH, 4 and 7 are used to calibrate the device if the sample is acidic.

10.5.18 EC measurement

An EC-meter is used for this purpose. The EC measurement is carried out at a standard temperature of 25° C, and since the ambient temperature varies, it must be corrected using the EC temperature correction table after reading the relevant number. EC meters have temperature-corrections capacity. An electrode connected to the device is used to measure the electrical conductivity. The electrode is first thoroughly washed and dried with deionized water. The sample is well mixed and the electrode is inserted into the sample. It should be noted that the electrode must be in the center of the container and not adhere to the wall or bottom of the container. The number displayed on the device's display is equivalent to the sample's electrical conductivity.

10.5.19 Corrosion determination

Marble or gypsum powder is immersed in water for a period. The initial and final pH are then measured and the pH change indicates corrosive properties. If calcium is dissolved in water this means that the water is likely to have corrosive effects on iron and steel as well as cement.

10.5.20 Atomic absorption

Atomic absorption is a quantitative decomposition method that is not widely used in qualitative analysis. This type of spectroscopy is widely used in the quantitative analysis of metals, minerals, non-metallic and metallic mineral products, ceramics, cements, minerals in living systems such as serum and food and various components of plants, surface and ground water and industrial effluents. Therefore, this device method is used to measure about 30 elements with an appropriate detection limit below 1 ppm.

10.5.21 Titration

Titration is one of the methods of volume decomposition. In volumetric analysis, first, a certain volume of solution (sample) with unknown concentration is compared with another solution whose concentration and volume are known (standard solution). The standard solution in titration is slowly added from a borate to a solution containing a specified volume or weight of unknown concentration. Increasing the standard solution continues until the standard solution is titrated equivalently to the concentration of the solution which is called the equivalence point. In theory, the equivalence point is the endpoint of the titration process. The equivalence point in practice is characterized by physical change (e.g. color change). The point at which this color change occurs is the endpoint of the titration. Finally, the volume and concentration of the standard solution and the volume of the titrant solution are calculated as the concentration of the titrant solution.

10.5.22 Chromatography

Chromatography is a method of separating materials from a mixture. Chromatography allows the constituents to be separated, purified, and identified close to the complex mixtures, many of these separations being impossible by other methods. Chromatography is based on the selective diffusion of different components between the mobile and stationary phases. The uneven movement of the detachable objects due to the inequality of the two forces of motion and the deterrent on the part of the separating agent is the basis of these methods. The static phase can be either liquid or solid. The moving phase can be either gas or liquid.

10.5.23 Spectrophotometry

Spectrophotometry is the capture of luminous photons by molecules of a solution. A spectrophotometer is a device designed to pass a beam of light at a specified wavelength inside a sample and to measure the amount of light passing through it. The amount of incoming light is absorbed by the sample and changes from the base surface to the excited surface. Part of the light absorbs matter. It should be noted that the amount of light absorbed by the material is related to its concentration.

10.5.24 Flame photometry

Flame photometry is used to quantitatively and qualitatively quantify several types of cations, particularly those that are readily excited and move to relatively high energy levels at relatively low flame temperatures such as Cu, Cs, Rb and Ba. The technique uses a flame that vaporizes the solution as well as the metal and then transfers the electron to a higher energy level. As the electron returns to its original state light is produced at a specific wavelength for each metal.

10.5.25 Cl measurement

The titration method can be used to measure chlorine concentration. To standardize chlorine a 0.01 normal (colorless) silver nitrate solution is used. The volume of 25 cc of the sample is taken by pipette and poured into the beaker to prepare the titrated solution (sample of the desired water. Subsequently, 5 cc of potassium chromate is added as a reagent to change the color of the solution from colorless to yellow. After preparing the sample solution titrate it with silver nitrate. To do this the beaker is placed under the borate device, the valve is opened, and silver nitrate is slowly added to the solution. The endpoint of the titration process is when the yellow color of the sample solution changes to brick color.

10.5.26 Calcium measurement

Calcium concentration is measured by the titration method. To determine the amount of calcium ions 5 cc of sample are added to the sample, and then 0.1 times of the sample volume of NaOH buffer (0.5 cc) is added to the sample. The 3 gs of Murexide powder is then added to the sample solution as a reagent. Pink appears due to the combination of murexide with calcium present in the sample. The addition of EDTA as the titrating agent to the borate in the sample solution removes the calcium from the medium and the color of the sample solution changes from pink to purple.

10.5.27 Magnesium measurement

It is difficult to measure magnesium; the concentration of magnesium ions is determined by measuring the sum of Ca^{2+} + Mg^{2+}. To measure Ca^{2+} + Mg^{2+} 5 cc of ammonia buffer is added to 5 cc of the sample. The reagent EBT is added to the solution and the sample color changes from colorless to cherry red. For beaker titration the titration solution is placed under the burette. With the burst valve open 0.01 normal EDTA solution is added to the sample solution to turn the color from cherry red to blue.

10.5.28 Carbonate (CO_3^{2-}) and bicarbonate (HCO_3^{-}) measurement

To measure these two parameters, it is best to first determine the pH of the water because at a pH below 8.2 no carbonate is present and no measurement is needed, or if the phenolphthalein solution is added to the sample the sample will change color to pink indicate carbonate, otherwise there is no carbonate. If the pH is higher than 8.2, the determination of carbonate content proceeds by adding 5 cc of the sample in the beaker. A few drops of phenolphthalein reagent in the beaker are added to the sample to change its color to pink. The solution containing the sample is titrated with 0.01 normal sulfuric acid and its color changes from pink to colorless. After carbonate measurement, the same beaker solution is used to measure bicarbonate. Methyl orange is added to the beaker solution as a reagent to change from colorless to yellow.

10.5.29 Nitrate measurement

A spectrophotometer is used to perform this test. The specified standards must first be established, where the solution contains 0.5 liters of sodium nitrate to a volume of 1 liter in a volumetric flask, containing 100 mg nitrogen. Apply 2, 4 and 5 cc of the standard solution of nitrate to distilled water to 100 cc and the prepared standards are 0.02, 0.04 and 0.05 mg/L respectively. Then, 10 mL of distilled water are poured into the container as a control. 10 cc of the sample that have already been passed through the filter are poured into the container, the standards, controls and water samples are placed into the water container and the following solutions are added. First, add 2 cc of the sodium salt solution and stir to mix. Then add 10 cc of the sulfuric acid solution and shake slowly and cool down after a few pauses. Add 0.5 ml of brucine sulfanilic acid solution and mix well. The containers are placed in a Ben Marie for 20 min at a temperature not lower than 95° C, the dishes are removed and placed in cold water until the temperature reaches room temperature. The spectrophotometer is set at 410 nm and adjusted with the control, read the standards and then the concentration of the samples.

10.5.30 Phosphate measurement

Spectrophotometry is also used to perform this experiment. To 100 cc of the sample add 4 cc of ammonium molybdate solution and shake well, tadd 0.5 cc or 10 drops of Tin chloride and shake well, after 10 min and before 15 min the absorption or passage percentage of the sample is obtained by spectrophotometer at 690 nm. Prepare standard solutions of 0.1, 0.3, 0.5, 0.7, 0.9 and 1 mg/l of phosphorus for this experiment. The standard is plotted and calculated using the standard curve of the concentration of sample phosphorus in milligrams per liter of phosphate ion.

10.5.31 Sodium and potassium measurement

A flame photometer is used to measure sodium and potassium. After setting up the device the sodium or potassium filter is placed in its proper place. The device with distilled water is set to zero by the dedicated button. Using a pre-prepared sensitivity button and standard solution the galvanometer is set to a number, standard solutions are poured one at a time from bottom to top one by one. The galvanometer numbers are read. The galvanometer reading is determined from the device table in milligrams or milli equivalent.

10.6 Case study

10.6.1 Comparison of seven water quality assessment methods for the characterization and management of highly impaired river systems

Ji et al. (2016) evaluated seven different variables in order to evaluate the quality of the Wen Rui Tang river water in China. This river is the main source of water for industry, agriculture, and municipalities in this region. Due to the economic development of the last 20 years sewage discharges and industrial wastewater have caused severe pollution and color changing in this river. The most important parameters which considered in this study are as follows. DO, BOD_5, COD, ammonium nitrogen, total phosphorus, total nitrogen, and oil. The evaluation methods include Single-factor assessment method, Water quality grading method, comprehensive pollution index, Nemerow pollution index, Principle Component Analysis (PCA), Fuzzy comprehensive evaluation method, and Comprehensive water quality identification index. The results showed that the time variation of water quality was due to rainfall, and the spatial change was affected by human, industrial and agricultural activities. Also, upstream flow of the river was affected by its upstream cumulative pollution. The authors concluded that Single-factor assessment method and Water quality grading method had a constraint that would reduce the impact of the highest and lowest pollution parameters. The Nemerow pollution index does not accurately describes water quality because it focuses on one pollutant factor. Principle Component Analysis is a suitable tool for complex data analysis, but it has no good accuracy and reliability in water quality analysis. The fuzzy comprehensive evaluation method with objective observations of water quality matches well, however, it is unable to classify the quality of water in more than five classes. The comprehensive water quality identification index is able to classify water quality in all classes as well as the quality and quantity of water. Therefore, comprehensive water quality identification index was introduced as the best method to evaluate the quality of water of this river. The authors stated that their study has indicated valuable information to control water pollution and manage water resources in surface water contaminated with severe pollution problems especially in regions where urbanization and industrialization are increasing.

10.6.2 Tailor-made protocol for assessing water quality of irrigation canals: Case study of El-Nubaria canal, Egypt

Habash et al. (2018), examined the El-Nubaria canal in Egypt with the aim of introducing the custom protocol made for estimating the water quality of irrigation canals in accordance with the regulatory goals of the accumulation of pollution. The developed protocol consists of statistical series tools to facilitate the data management and then transforming them into useful information. These tools include descriptive statistics and data compliance with standards. Cluster analysis was used to classify regions with similar water quality status. The clusters were displayed on the map by GIS. For this purpose, quantitative and qualitative data of water samples were collected monthly from 2008 through 2016. 12 samples were collected and analyzed in areas with local pollution, and in areas that were mixed with canal water. 34 water quality parameters were investigated with special attention to biological parameters, nutrients, cations, metals, and trace elements along with conventional water quality parameters on the samples. The TMP development employed the Phyton programming language for high speed, providing spatial data for water quality status in the channels, rapid application of statistical analysis for water quality, and for providing the main results to the form that is easily saved. In five phases, the investigation has been carried out that the first phase is to apply descriptive statistics for the measured parameters and compare them with standard values. The second phase was to examine the load of pollution containing TDS, DO, BOD5 and COD loading. The third phase, the percentage analysis, estimated the percentage of measured parameters that range from 25 percent to 100 percent with an average of 71.85 percent. The fourth phase, cluster analysis, divided the quality of channel water into five clusters, the fifth phase mapped the water quality by GIS. The map showed the BOD and COD exceeded water quality criteria, and the status of DO and TDS was normal in the canal. The authors concluded that TPM is a practical tool in reducing human error, high speed, capability to support GIS, and low cost and provides a dynamic system to display results. Water quality in this river has high salinity and is under the influence of biological load of drainage basins. Compared with the Egyptian water quality standards the condition of water quality was moderate. It was recommended that the national networks of water quality monitoring in Egypt be generalized to irrigation and drainage basins.

10.7 Summary

Water is essential for life, and its physical and chemical properties must be optimal drinking. Water is one of the main constraints to achieving sustainable development, especially when it is in addition to quantitative constraints, and water resources are threatened. Quantitative constraints are naturally created by population growth, while qualitative constraints arise due to inefficient use of water resources and inadequate management of production effluents. Pollution is often caused due to the non-sanitary disposal of waste and sewage from urban and industrial uses, including the kinds of pollutants, as well as pesticides and fertilizers used in agriculture. These pollutants include different kinds of pathogens, minerals and organic substances that may be expensive and difficult to treat. Therefore, there is a need to make the best use of this resource with regard to the extent and the nature of the water, with a view to preventing water contamination. The unsanitary disposal of human waste and industrial waste are important factors in the degradation of water quality in developing countries; therefore, with the introduction of technologies available, including appropriate technologies and to build facilities for wastewater treatment improvement can improve significantly. The present study first deals with general descriptions of water pollution in recent years and its impact on human health, as well as the availability of safe water resources, which are essential prerequisites for environmental protection and economic, political, social and cultural development of countries. A number of recent studies on water quality estimation in the world have also been briefly explained. All the physical, chemical, and biological parameters of water and common terms related to this topic are described. The tables also provide permitted values of water quality parameters declared by the WHO, permissible values of parameters for agricultural and irrigation purposes declared by the FAO, as well as the means by which the nature and amount of pollutants in water are measured and expressed. In the following, water circulation in the hydrologic cycle and how water is contaminated and the relationship between human wastewater, drinking water and diseases are discussed. Considering that one of the most important causes of transmission of diseases from the environment to human beings is water and in the event of contamination and intolerance can pose a serious threat to human health, the reasons for the importance of water quality and the necessity of evaluating the qualitative parameters of water resources are described. The method of measuring all physical, chemical and biological parameters as well as how to calculate water quality index is fully described and three cases of water quality related studies in India, Egypt and China were selected. They are summarized and briefly explained.

To reduce the amount of wastewater production and, in other words, to reduce water pollution, the following strategies are adopted.

1. Optimal consumption of raw materials in the industry that results in contaminated water.
2. Proper construction and operation of water and wastewater treatment systems.
3. Control and treatment of agricultural drains.
4. Developing a suitable cultivation pattern with each climatology for optimal consumption of fertilizer and toxin in agriculture and preventing unnecessary and unnecessary irrigation.
5. Replacing fertilizers and biological pesticides instead of using pesticides that are highly persistent in nature and have long-term consequences for human health.
6. Prevention of unnecessary water harvesting by users, which increases the concentration of pollution in water resources and its associated environment.

References

Azhar, S.C., Zaharin Arisa, A., Yusoff, M.K., Firuz Ramlia, M., Juahir, H., 2015. Classification of river water quality using multivariate analysis. Procedia. Environ. Sci. 30, 79–84.

Barakat, A., El Baghdadi, M., Rias, J., Aghezzaf, B., Slassi, M., 2016. Assessment of spatial and seasonal water quality variation of Oum Er Rbia River (Morocco) using multivariate statistical techniques. Int. Soil and Water Conservation Res. 4 (4), 284–292.

Bouaroudj, S., Menad, A., Bounamous, A., Ali-Khodja, H., Gherib, A., Weigel, D., Chenchouni, H., 2019. Assessment of water quality at the largest dam in Algeria (Beni Haroun Dam) and effects of irrigation on soil characteristics of agricultural lands. Chemosphere 219, 76–88.

Ding, Y.R., Cai, Y.J., Sun, P.D., Chen, B., 2014. The use of combined neural networks and genetic algorithms for prediction of river water quality. J. Appl. Res. Technol. 493–499.

D'Ugo, E., Marcheggiani, S., D'Angelo, M., Caciolli, S., Puccinelli, C., Giuseppetti, R.., Marcoaldi, R., Romanelli, C., Mancini, L., 2018. Microbiological water quality in the medical device industry in Italy. Microbiological J. 136, 293–299.

Fan, X., Cui, B., Zhao, H., Zhang, Z., Zhang, H., 2010. Assessment of river water quality in Pearl River Delta using multivariate statistical techniques. Procedia. Environ. Sci 2, 1220–1234.

Fiorentino, L.A., Olascoaga, M.J., Reniers, A., Feng, Z., Beron-Vera, F.J., MacMahan, J.H., 2012. Using Lagrangian Coherent Structures to understand coastal water quality. Cont. Shelf. Res. 47, 145–149.

Habash, A., El-Molla, A., Shaban, A., Abdelall, M., 2018. Tailor-made protocol for assessing water quality of irrigation canals: case study of El-Nubaria canal, Egypt. Water Science 32 (2), 380–399.

Ji, X., Dahlgren, R.A., Zhang, M., 2016. Comparison of seven water quality assessment methods for the characterization and management of highly impaired river systems. Environ. Monit. Assess. 188, 15.

Jung, K.Y., Lee, K.L., Im, T.H., Lee, I.J., Kim, S., Han, K.Y.., Ahn, J.M., 2016. Evaluation of water quality for the Nakdong River watershed using multivariate analysis. Environ. Technol. Innov. 5, 67–82.

Li, R., Zou, Z., An, Y, 2016. Water quality assessment in Qu River based on fuzzy water pollution index method. J. Environ. Sci. 50, 87–92.

Long, D.T., Pearson, A.L., Voice, T.C., Polanco-Rodriguez, A.G., Sanchez-Rodriguez, E.C., Xagoraraki, I., Concha-Valdez, F.G., Puc-Franco, M., Lopez-Cetz, R., Rzotkiewicz, A.T., 2018. Influence of rainy season and land use on drinking water quality in a karst landscape, State of Yucatan, Mexico. Appl. Geochem. 98, 265–277.

Peiyue, L., Jianhua, W.U., Hui, Q., 2012. Groundwater quality assessment with two multi-criteria decision-making methods. Int. J. Geomatics and Geosci. 2, 868–877.

Rostom, N.G., Shalaby, A.A., Issa, Y.M., Afifi, A.A., 2017. Evaluation of Mariut Lake water quality using Hyperspectral Remote Sensing and laboratory works. The Egyptian J. Remote Sensing and Space Sci. 20 (1), 539–548.

Sener, S., Sener, E., Davraz, A., 2017. Evaluation of water quality using water quality index (WQI) method and GIS in Aksu River (SW-Turkey). Sci. Total Environ. 584-585, 131–144.

Tian, Y., Jiang, Y., Liu, Q., Dong, M., Xu, D., Liu, Y., Xu, X., 2019. Using a water quality index to assess the water quality of the upper and middle streams of the Luanhe River, Northen China. Sci. Total Environ. 667, 142–151.

Xiao, J., Wang, L., Deng, L., Jin, Z., 2019. Characteristics, sources, warer quality and health risk assessment of trace elements in river water and well water in the Chinese Loess Plateau. Sci. Total Environ. 650 (2), 2004–2012.

Zhang, X., Zhang, F., Kung, H., Shi, P., Yushanjiang, A., Zhu, S., 2018. Estimation of the Fe and Cu Contents of the surface water in the Ebinur lake basin based on LIBS and a machine learning algorithm. Environ. Res. Public Health 15 (11), 2390.

Chapter 11

Water and society

Alireza Rezaee[a], **Omid Bozorg-Haddad**[a] **and Vijay P. Singh**[b]
[a] *Department of Irrigation & Reclamation Engineering, Faculty of Agriculture Engineering & Technology, College of Agriculture & Natural Resources, University of Tehran, Karaj, Alborz, Iran.*
[b] *Department of Biological and Agricultural Engineering & Zachry Department of Civil Engineering, Texas A&M University, College Station, TX, United States*

11.1 Introduction

"Water is the source of life" is a simple yet basic phrase that illustrates the relationship between water cycle and living beings.

The availability and supply of quality water for living organisms is one of the most important prerequisites for life and social and community development. In human history, water sources have undoubtedly been the founders of ancient civilizations, such as the civilizations of Mesopotamia, Egypt, India, and the Inca, which all developed around large rivers, such as the Tigris and Euphrates, the Nile, the Indus, and the Amazon. Just as the earliest human civilizations, the Sumerians founded one of the oldest human civilizations in the Tigris and Euphrates basin. After the Sumerians, Assyrian, Babylonian, and Egyptian civilizations were established alongside the rivers. With increasing population and high density of people along the rivers, cities were gradually established and human thought took root, such as the concept of surplus production that in turn led to the emergence of cities.

The water cycle is open and observable in societies that are always associated with access to and return to nature. Today, for many reasons, such as climate change, population growth, human activities, pollution, etc., the water cycle has undergone changes (Asia, 2001; Ratna Reddy and Syme, 2014). These changes need to be managed, given the increasing need for food supply and abatement of contaminants. These changes are causing depletion of water resources and environmental pollution that are related to social and natural indicators (Walker et al., 2015).

At present, the interactive relationship between water resources and community is at an early stage of research and lacks a comprehensive, unified, and scientific foundation. The only known relationship between water resources and community is "extraction and collection, consumption and discharge and return

Economical, Political, and Social Issues in Water Resources. DOI: 10.1016/B978-0-323-90567-1.00004-8

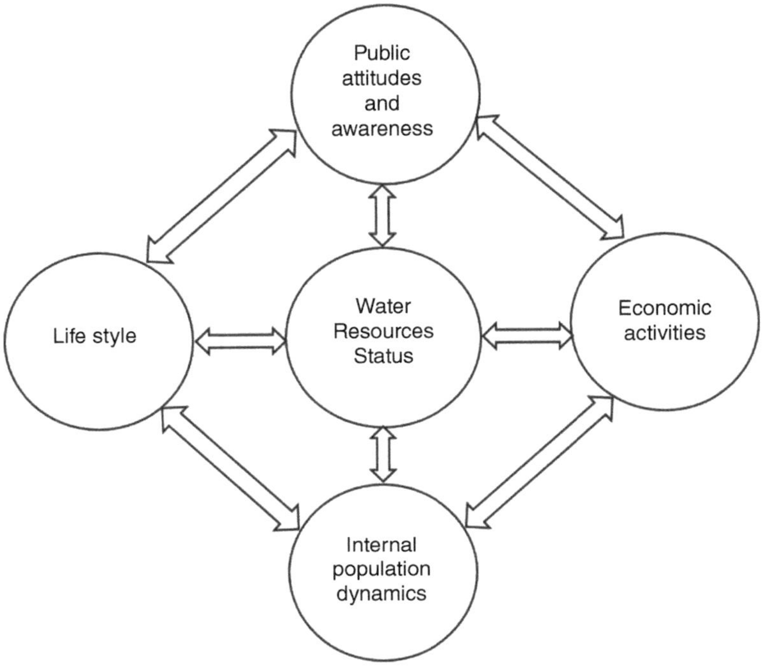

FIGURE 11.1 General model of interconnection of water resources and society.

of water," but this concept has not been fully deduced and requires further study, including initial concept, structure and processes.

The impact of socio-economic parameters on water resources has been addressed to some extent, but time management is effective when interactions are examined so that the influence of changes in water resources on social and economic parameters is also evaluated. Researchers have considered the relationship between water resources and community as an unknown black box (Lu et al., 2016), due to the complexity of relationship between them and the influence of various parameters, which necessitates further study to evaluate, analyze processes, and clarify mechanisms and their relationships. A general model for the interactions is illustrated in Fig. 11.1.

This chapter discusses some of the key factors affecting the status of society, such as population, employment, GDP and income index, and mortality index. Identification and trend of changes in water resources in different regions against influencing parameters have been studied and analyzed and why this two-way communication is made is discussed.

From Fig. 11.1, one can fully understand the concept of two-way relationship between water resources and society. How societies are affected by changes in internal population dynamics, such as deaths, births, wars, economic activities, attitudes, value systems, and lifestyles, and are mutually influenced by changes in the status of water resources. For example, demographic changes in a catchment under the influence of internal components or influenced by external factors, such as war, violence or natural disasters, cultural changes, lifestyle changes, and economic changes, can be through changes in housing patterns, changes in employment structure, and land use. Consequently, changes in drinking water demand, agriculture, industry, and changes in effluent and wastewater production, affect the quantity and quality of water resources and the environmental conditions of the basin, and thus affect the basin water management. Basin water management also responds to changes in the community by responding to changes in the social landscape, as appropriate to the status of water resources.

The objective of this study is to examine the impact of water resources on factors that lead to the emergence, development, maintenance, and enhancement of satisfaction, such as population, employment, education and health, and the relationship between water resources and socioeconomic status.

11.2 Definitions

11.2.1 Society

Cowan (2005) points to five major interpretations of society (Cowan, 2005):

- A group of people with common interests
- A place and its people
- A place where people live (especially in the US). It does not mean the closeness or interaction between inhabitants
- A group of people living under the control of a government
- Contrary to the term institution or institution

It is also possible to introduce a community by combining two elements:

A network of interpersonal relationships; relationships that are often mutually reinforcing

Index of commitment to a set of shared values, states, meanings, a shared history, and identity in a particular culture

11.2.2 Water resources management and social issues

All communities are dependent on water; water is needed for agricultural production, energy generation, health service, and industrial manufacture. Sustainability of communities depends directly and indirectly on quantity, quality, reliability, and affordability. On one hand, water systems decisions can create social challenges, but on the other hand, social behaviors can change the status of water systems (Sivapalan et al., 2012).

11.2.3 Water consumption approaches

Components of total water per capita are: domestic, public, commercial, and industrial consumption, landscape, and actual water loss. The most important type of consumption per capita is related to household consumption. Water use approaches can be divided into four groups: socio-cultural, technical-engineering, economic-tariff, and legal-regulatory. The socio-cultural approach encompasses all the methods, techniques, and tools that are needed to raise public awareness and change consumer behavior in order to optimally use and conserve water and support the plans in question. These techniques are used to motivate and create positive attitudes and appropriate norms which are a prerequisite for other water management activities. The following points are important in adopting this approach:

Reducing water consumption is linked to education and culture.
Reducing water consumption is a collaborative effort.
In reducing water consumption, it is important to acknowledge the differences of community groups.
The paradigm that governs people's current attitudes needs to change.

People's attitudes and intentions towards saving water depend on their awareness of water issues. In order to be successful in water management, it is, therefore, necessary to increase people's awareness and attitude. Since consumption is a behavioral phenomenon and since no behavioral phenomenon is independent of the culture of a society appropriate cultural behaviors must be modified in order to prevent unnecessary consumption of water.

Public education entails a wide range of cultural and educational programs that should lead to behavior modification and improvement of consumption patterns as well as to the pattern of water consumption (Aerts et al., 2018). These methods can include content in textbooks, the use of audiovisual, propaganda and public media, and the use of specific and popular groups.

The technical approach includes all technical and engineering measures to improve the pattern of drinking water consumption. The use of reducing equipment and methods for improving home appliances are emphasized in this approach.

In the economic approach one of the effective parameters in water consumption pattern and management is water price. Determining the right price for water will encourage consumers have both its optimum use in commercial and industrial uses and the revenue it generates from selling water to water and wastewater companies which can be used to cover the bulk of water and wastewater treatment costs.

11.3 Importance and necessity of the subject

Water systems management is difficult because no academic discipline covers all aspects, so an interdisciplinary approach is needed. Despite the uncertainty about the status of future water resources and the community's response to it, it will be difficult to accomplish it (Walker et al., 2015). A wide range of human sciences is needed to solve water issues, including economics, behavioral and cognitive sciences, decision making, social values, community psychology, and politics.

The need to examine the interaction of water and community systems is not only one of cultural status, that is, learning more about social systems and their effects on water systems, but also learning how decision making in water systems drives the society toward social responses and movements (Mosavi, 2015). Developing an analytical model can help predict how society responds to water system decisions.

Considering all social issues in water resources management is simply not possible, because some social criteria are difficult to quantify. It may be possible to simulate the effect of selective solutions and decisions on aquatic systems using simulation models and sociological theories, but it is difficult to understand community activities in future scenarios with thousands of socio-economic and environmental impacts. It should also be noted that any event in the world can change values and issues that are important to a community over time, so it is important to consider cultures and values in decision making.

11.4 Basics and logic

Understanding the very close and complex relationship between water and the lives of individuals in the community has become increasingly important. Social markers can act as influencing forces on other water-related indicators and facilitate or accelerate environmental or economic processes in the water domain. For example, population changes affect the amount of wastewater produced, thereby producing pollution and affecting the natural environment. On the other hand, unequal economic development of regions can cause population changes in the form of migration and consequently changes in water demand in a region.

In this section, considering the indices and parameters set by the World Bank and considering the importance of water resources changes in indices and basic parameters of people's lives in societies, we analyze some of the parameters and socio-economic indices of the community based on studies. Statistical observations correlate with per capita amount of water resources, as shown by data from 14 countries (Albania, Belarus, Bosnia and Herzegovina, Bulgaria, Croatia, Estonia, Georgia, Hungary, Latvia, Lithuania, Poland, Romania, Serbia and Ukraine) available for over a 20-years period (1998–2017) from

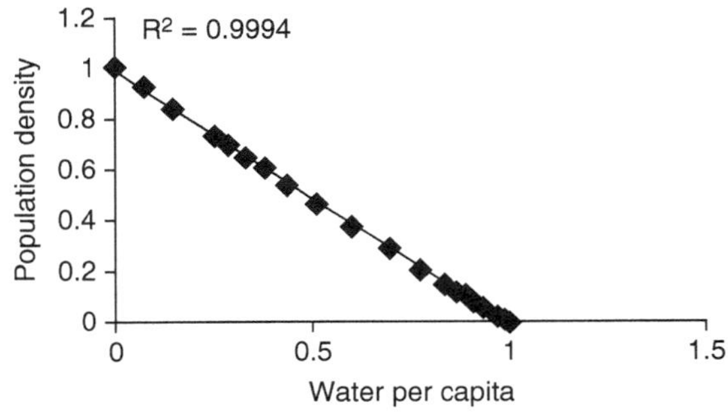

FIGURE 11.2 Correlation between water per capita and population density.

Knoema (2021). The relevant data were normalized after extraction as

$$X_N = \frac{(X_i - X_{min})}{(X_{max} - X_{min})} \tag{11.1}$$

where X_N is the normalized value; X_i is the real value; X_{min} is the minimum value; and X_{max} is the maximum value.

All relevant parameters are then analyzed and analyzed against changes in water resources. These parameters have been studied and analyzed in accordance with consumption management and water consumption approaches. These approaches are presented in the fields of culture, education, engineering, economics, and in accordance with the laws of the region. It is also worth noting that this information is from 14 countries but other countries may not necessarily follow these trends.

11.4.1 Population density

The distribution of population in each country varies according to its natural parameters and characteristics. One of the main factors in the distribution and density of the population is climate. Therefore, rainfall and in particular access to water resources play an important role in population density and the emergence of rural and urban centers.

As mentioned, the annual per capita water use per person in a given region depends on the number of people and density of population of that area and with the increase in population, the amount of water per person per unit area decreases. Per capita consumption of renewable water using Albanian country data is shown in (Fig. 11.2).

According to Fig. 11.2, the population density is a measure of population per unit area or unit volume of living organisms, especially humans. As can be seen from the figure that, with the increase in per capita water resources in Albania, the population density is declining and there will be no need to focus on the huge

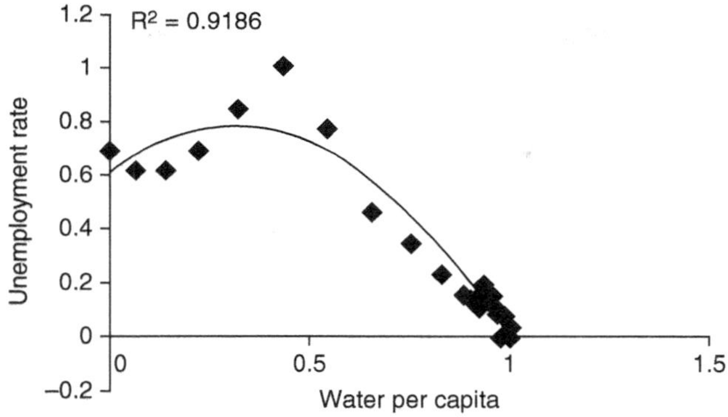

FIGURE 11.3 Correlation between water per capita and unemployment rate.

flood population in a particular area to use water resources. The population will live on a gentle slope in different parts of the country.

11.4.2 Employment and unemployment rates

An unemployed person in the economy is a person who is of working age (15 to 65 years old) and is looking for a job but does not find a job or a source of income. Children and the elderly are not considered active because they are unable to work. Housewives and students are not considered active if they are not job seekers and even those who are not working despite an insufficient source of income. Among the causes of unemployment are population growth and the consequent lack of job opportunities for employment.

An important factor in unemployment is the increase in international competition in the industries on which the growing progress of the West was based. Another factor in unemployment is the lack of proper government and family planning, which is the main reason for unemployment in some countries.

Research shows that the number of unemployed people is increasing. One of the reasons for this could be the declining trend of water resources in the region and the elimination of job opportunities and job security that require water resources. Also, according to a study by Scanlon et al. (2016) in Africa, who studied the effect of water availability on the potential for livelihoods and employment, and concluded that people's employment and quality of life depended heavily on water. This trend is shown in Fig. 11.3 using Belarusian data.

Studies demonstrated that the availability of water has an impact on increasing livelihoods, employment, and quality of life. Reducing water resources and making them unavailable impede the industry, agriculture, and services, resulting in rising unemployment. As a result, employment in that region is rising, and the

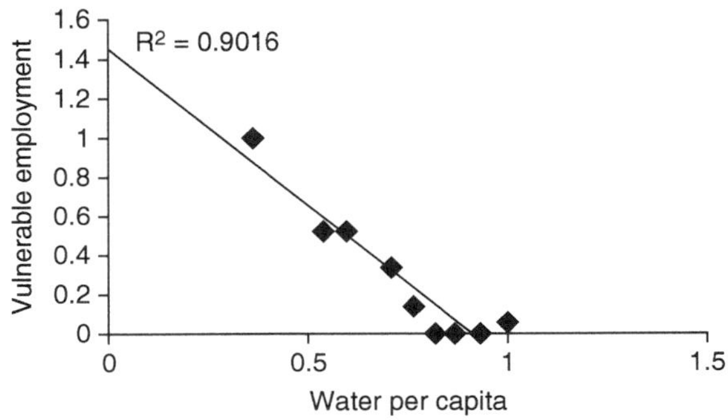

FIGURE 11.4 Correlation between water per capita and vulnerable employment.

unemployment rate is declining, according to Fig. 11.3, which looks at countries with a per capita increase in water resources.

11.4.3 Vulnerable employment

In the definition of pathology, it can be said that pathology is the study of the basics of dysfunctional functions. In fact, organizational damage is the cause of fundamental problems that put the organization in crisis. In general, there are two main problems in employment and working life: job opportunities and job security. According to the Hungarian data, the trend of changes in the vulnerability of jobs and per capita water resources is shown in (Fig. 11.4).

As mentioned in the section on employment and unemployment, with the increase of water resources and the creation of more job opportunities, job security will be much more provided. As a result, job vulnerabilities are declining and people working in these areas will have higher job security.

11.4.4 GDP and income index

GDP represents the total value of final goods and services produced in a country over a period of time. GDP is one of the social and economic criteria and is the best way to compare countries economically. Per capita income is the number obtained by dividing the percentage of GDP by the region's population, and it simply means how much per capita income there is in a country. However, it should be noted that the per capita income cannot provide an accurate picture of the economic situation of one country compared to another country, because the economic situation does not only depend on the income but also on the costs and

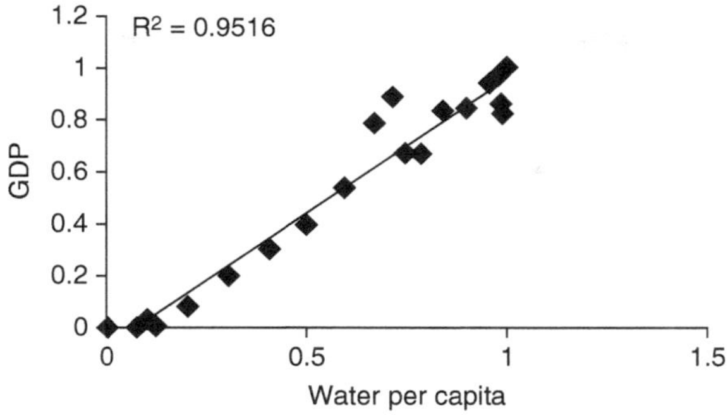

FIGURE 11.5 Correlation between water per capita and GDP.

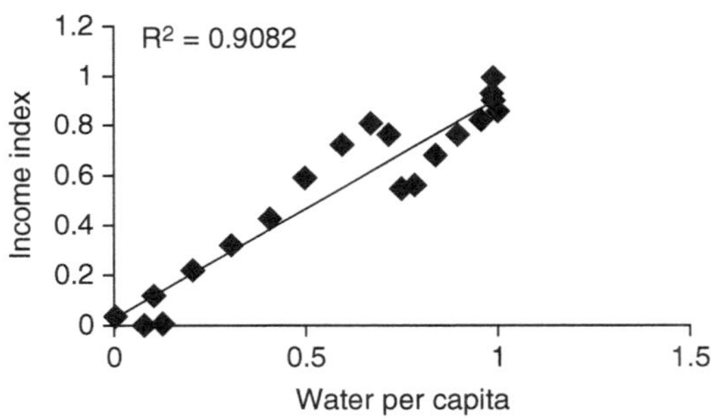

FIGURE 11.6 Correlation between water per capita and income index.

expenses of that country. The increase in the per capita income index is related to the employment and economic situation of that country. Changes in GDP and income index per capita water resources in Estonia are shown in Figs. 11.5 and 11.6, respectively.

As explained in the analysis of the employment sector, despite the increasing trend of water resources and employment, the income of people working in agriculture, industry, and services naturally increases and leads to more development and investment. As a result, according to Fig. 11.5, with the per capita increasing trend of water resources, the income index of the people of that region also increases.

Subsequently, according to the existing definitions and studies conducted in Estonia, in accordance with Fig. 11.6, the increase in per capita income is accompanied by an increase in GDP. As a result, with increasing per capita water

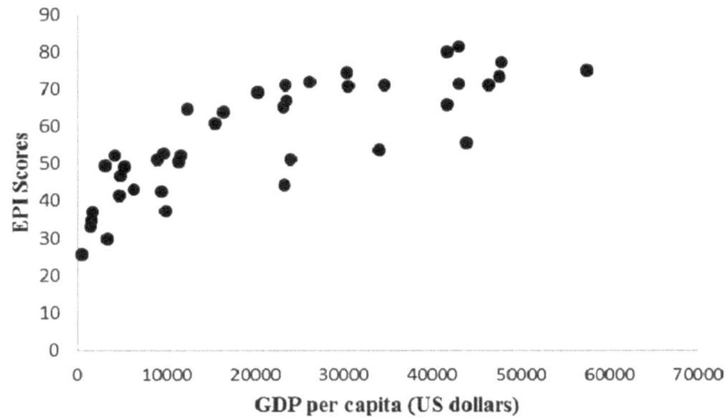

FIGURE 11.7 Correlation between GDP and EPI.

resources, the region will progress economically and will generate more revenue and production.

Also, according to a study conducted by the Environmental Assessment Team conducted in 180 countries in 2018 (Wendling et al., 2018), there is a positive correlation between GDP and Environmental Performance Index (EPI), and the reason for this is that by increasing the amount of GDP in a region, it increases the region's potential to create the necessary facilities and equipment to supply drinking water and wastewater management, thus improving the amount of water resources. The bio-performance index will be environmental. The graph of GDP and EPI indicators for 40 countries from different continents is shown in Fig. 11.7, which shows a positive correlation between the two indicators.

11.4.5 Human development index (HDI)

The Human Development Index (HDI) is an indicator of social standards that consists of three areas: life expectancy, access to education and literacy, and living conditions and income. The index depends on a number of factors based on development, including the ability to perform tasks, such as access to proper nutrition, health, occupation, education, and community participation. The relationship between the HDI and the per capita water resources of Ukraine is shown in Fig. 11.8.

According to research conducted in different countries, the results obtained in accordance with the data of Ukraine have an upward trend with per capita water resources. As a result, the availability of water increases the quality of life, improves the economic situation, and develops relevant indicators.

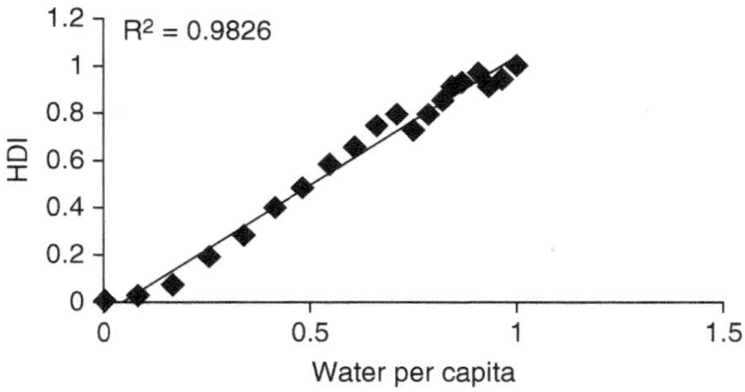

FIGURE 11.8 Correlation between water per capita and HDI.

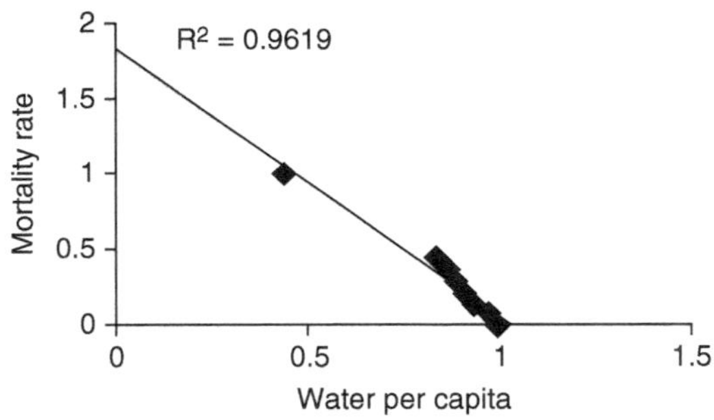

FIGURE 11.9 Correlation between water per capita and mortality rate.

11.4.6 Mortality rate index

In general, health indicators are divided into three categories: birth or fertility rate, mortality rate, and infection rate. This unit index is used to measure the number of deaths (in general or for a specific reason) divided by the total population per unit time. Mortality rates are typically expressed as the number of deaths per 1,000 people per year. The leading cause of death in the world is cardiovascular disease, which kills nearly 18 million people a year. One of the causes of aggravation of this disease is the contamination of water resources and, of course, the lack of production of high quality products which has a great impact on the health of people of an area. According to the study of Keshavarz et al. (2013) on people living in two villages in Fars province in Iran, a large and important impact of water shortage on the health of people in the region resulted. According to the data available in Albania, Fig. 11.9 shows the per capita impact of water resources on the mortality rate.

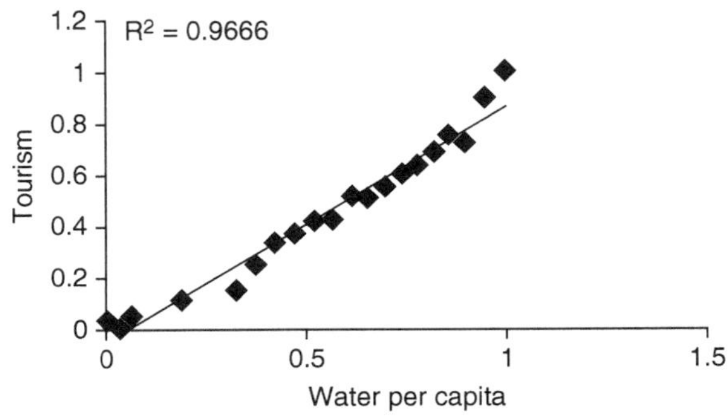

FIGURE 11.10 Correlation between water per capita and tourism.

According to the figure, using healthy nutrition and lifestyle modification, having access to adequate and adequate water resources, and strengthening the agricultural sector can reduce the majority of this number of diseases. As a result, increasing the quality and quantity of available water resources improves health and subsequently increases life expectancy and reduces the number of deaths and injuries to people with water-related diseases.

11.4.7 Tourism

When tourism enters a country, it must pay for accommodation, entertainment, and nutrition in that country. As a result, the currency that enters the host country leads to economic prosperity, which is why tourism is also called an invisible export. The development of tourism provides the basis for job creation, and this is very useful for countries facing young people and applicants. One in 10 tourists entering the host country will have a job opportunity.

The impact of tourism on per capita water resources is shown in Fig. 11.10 for data from Bulgaria.

From Fig. 11.10, it can be concluded that increasing the amount of water resources has attracted more tourists in the region, and on the other hand, with the inference made above, the tourism can be considered one of the important factors of economic development. That's why the government has the ability to invest and buy more advanced equipment to reduce water loss and better manage water resources systems. On the other hand, economic development increases employment and welfare in communities.

11.5 Conclusion

Water resources are important resources in terms of economic, social, cultural, and environmental values for human beings. Due to the declining trend of water resources in recent decades, socioeconomic considerations have played a crucial role in the use of these resources.

In this chapter, the impact of some parameters on the per capita water resources of the community was studied. In general, the function of the introduced parameters is placed in the cycle between the three cultural-social, structural-engineering and economic groups (Mylopoulos and Kolokytha, 1997). Of course, it should be noted that due to the complexity of water relations and issues, no specific relationship and model has been proposed between water and society, and the reason is the impact on various issues in this area. In general, given the vital value of water, more and more measures must be taken to manage its consumption which depends on the technology and behavior of people in the community. Education and culture can be used to create a platform for reducing waste and saving and increasing productivity in agriculture, industry, home and green spaces (Aerts et al., 2018). With proper policy and the application of capital to purchase more advanced equipment, it can increase the quality and efficiency of operation. In all of the above parameters, the relationship between these three groups is effective and the result of increasing water resources can be visualized objectively.

References

Aerts, J.C.J.H., Botzen, W.J., Clarke, K.C., Cutter, S.L., Hall, J.W., Merz, B., Michel-Kerjan, E., Mysiak, J., Surminski, S., Kunreuther, H., 2018. Integrating human behavior dynamics into flood disaster risk assessment. Nat. Clim. Chang. 8, 193–199.

Asia, E., 2001. Towards a sustainableAsia and the Pacific. Report of ECO ASIA Long-term Perspective Project Phase II. Published by the Institute for Global Environmental Strategies (IGES), Japan, p. 166.

https://knoema.com/atlas/topics/Economy/National-Accounts-Gross-Domestic-Product/GDP-per-capita (April, 10, 2021).

Cowan, Robert, 2005. *The dictionary of urbanism*. Streetwise Press.

Keshavarz, M., Karami, E., Vanclay, F., 2013. The social experience of drought in rural Iran. Land Use Policy 30 (1), 120–129.

Lu, S., Zhang, X., Bao, H., Skitmore, M., 2016. Review of social water cycle research in a changing environment. Renew. Sust. Energ. Rev. 63, 132–140.

Mosavi, J., 2015. Water Resource Management: Systematic Approach. Amirkabir University of Technology Publications, No. 1, Tehran, Iran.

Mylopoulos, Y.A., Kolokytha, E.G., 1997. Social and economic aspects of sustainable water supply policy: the city of Thessaloniki. Oper. Water Manage 41–45.

Ratna Reddy, V., Syme, G.J., 2014. Social sciences and hydrology: an introduction. J. Hydrol. 518, 1–4.

Scanlon, T., Uguru, O.P., Jafry, T., Chinsinga, B., Mvula, P., Chunga, J., Chungu, K., 2016. The role of social actors in water access in Sub-Saharan Africa: evidence from Malawi and Zambia. Water Resour. Rural Dev. 8, 25–36.

Sivapalan, M., Savenije, H.H., Blöschl, G., 2012. Socio-hydrology: a new science of people and water. Hydrol. Process. 26 (8), 1270–1276.

Walker, W.E., Loucks, D.P., Carr, G., 2015. Social responses to water management decisions. Environ. Process. 2 (3), 485–509.

Wendling, Z.A., merson, J.W., Esty, D.C., Levy, M.A., de Sherbinin, A., et al., 2018. Environmental Performance Index. New Haven, CT, Yale Center for Environmental Law and Policy.

Index

Page numbers followed by "*f*" and "*t*" indicate, figures and tables respectively.

FSC
www.fsc.org
MIX
Papier | Fördert
gute Waldnutzung
FSC® C083411

Kolibri 360 GmbH
Ferdinand-Jühlke-Straße 7
99095 Erfurt, Deutschland
produktsicherheit@kolibri360.de